Crystal Design: Structure and Function

Editorial Board

Crystal Design: Structure and Function

Perspectives in Supramolecular Chemistry
Volume 7

EDITED BY GAUTAM R. DESIRAJU
University of Hyderabad, Hyderabad, India

Telephone (+44) 1243 779777

E-mail (for orders and customer service enquiries): cs-books@wiley.co.uk
Visit our Home Page on www.wileyeurope.com
or www.wiley.com

Other Wiley Editorial Offices

John Wiley & Sons Inc., 111 River Street, Hoboken, NJ 07030, USA

Jossey-Bass, 989 Market Street, San Francisco, CA 94103–1741, USA

Wiley-VCH Verlag GmbH, Boschstr. 12, D-69469 Weinheim, Germany

John Wiley & Sons Australia Ltd, 33 Park Road, Milton, Queensland 4064, Australia

John Wiley & Sons (Asia) Pte Ltd, 2 Clementi Loop #02–01, Jin Xing Distripark, Singapore 129809

John Wiley & Sons Canada Ltd, 22 Worcester Road, Etobicoke, Ontario, Canada M9W 1L1

Library of Congress Cataloging-in-Publication Data

Crystal design: structure and function / edited by Gautam R. Desiraju.
p. cm. – (Perspectives in supramolecular chemistry; v. 6)
Includes bibliographical references and indexes.
ISBN 0-470-84333-0 (alk. paper)
1. Molecular crystals. 2. Crystal growth. 3. Crystallography. I. Desiraju, G. R. (Gautam R.) II. Series.

QD921 .C787 2003
548′.5–dc21 2002193382

British Library Cataloguing in Publication Data

A catalogue record for this book is available from the British Library

ISBN 0 470 84333 0

Typeset in 10/12pt Times by Kolam Information Services Pvt. Ltd, Pondicherry, India.
Printed and bound in Great Britain by TJ International, Padstow, Cornwall.
This book is printed on acid-free paper responsibly manufactured from sustainable forestry in which at least two trees are planted for each one used for paper production.

Contents

Contributors

Jerry L. Atwood, Department of Chemistry, University of Missouri–Columbia, Columbia, MO 65211, USA

Leonard J. Barbour, Department of Chemistry, University of Missouri–Columbia, Columbia, MO 65211, USA

Kumar Biradha, Graduate School of Engineering, Nagoya University, Chikusaku, Nagoya 464–8603, Japan

Dario Braga, Dipartimento di Chimica "G. Ciamician", Via F. Selmi 2, I-40126, Bologna, Italy

Lee Brammer, Department of Chemistry, University of Sheffield, Sheffield S37HF, UK

Eric Burkholder, Department of Chemistry, Syracuse University, Syracuse, NY 13244, USA

André De Cian, Laboratoire de Chimie de Coordination Organique, Tectonique Moléculaire des Solides (CNRS FRE 2423), Université Louis Pasteur, Institut Le Bel, F-67070 Strasbourg, France

Silvio Decurtins, Department of Chemistry and Biochemistry, University of Berne, Freiestrasse 3, CH-3012 Berne, Switzerland

Robert C. Finn, Department of Chemistry, Syracuse University, Syracuse, NY 13244, USA

Makoto Fujita, Graduate School of Engineering, Nagoya University, Chikusaku, Nagoya 464–8603, Japan

Ernest Graf, Laboratoire de Chimie de Coordination Organique, Tectonique Moléculaire des Solides (CNRS FRE 2423), Université Louis Pasteur, Institut Le Bel, F-67070 Strasbourg, France

Fabrizia Grepioni, Dipartimento di Chimica, Via Vienna 2, I-07100, Sassari, Italy

Stephen Hanessian, Department of Chemistry, Université de Montréal, C.P. 6128, Succ. Centre-Ville, Montréal, QC, H3C 3J7, Canada

Mir Wais Hosseini, Laboratoire de Chimie de Coordination Organique, Tectonique Moléculaire des Solides (CNRS FRE 2423), Université Louis Pasteur, Institut Le Bel, F-67070 Strasbourg, France

Agoston Jerga, Department of Chemistry, University of Missouri–Columbia, Columbia, MO 65211, USA

Robertus J. M. Klein Gebbink, Department of Metal-Mediated Synthesis, Debye Institute, Utrecht University, Padualaan 8, 3584 CH Utrecht, The Netherlands

Julien Martz, Laboratoire de Chimie de Coordination Organique, Tectonique Moléculaire des Solides (CNRS FRE 2423), Université Louis Pasteur, Institut Le Bel, F-67070 Strasbourg, France

Michel D. Meijer, Department of Metal-Mediated Synthesis, Debye Institute, Utrecht University, Padualaan 8, 3584 CH Utrecht, The Netherlands

Melanie Pilkington, Department of Chemistry and Biochemistry, University of Berne, Freiestrasse 3, CH-3012 Berne, Switzerland

Raffaele Saladino, Dipartimento di Agribiologia e Agrochimica, Università degli Studi della Tuscia, Via S. Camillo de Lillis, s.n.c., 01100 Viterbo, Italy

Gerard van Koten, Department of Metal-Mediated Synthesis, Debye Institute, Utrecht University, Padualaan 8, 3584 CH Utrecht, The Netherlands

Jon A. Zubieta, Department of Chemistry, Syracuse University, Syracuse, NY 13244, USA

Preface

Supramolecular chemistry, or chemistry beyond the molecule, has provided a wide canvas for a variety of studies of molecular materials in the solid state. The most orderly manifestations of the solid state are single crystals, and the earlier volume in this series with the present Editor, *The Crystal as a Supramolecular Entity*, sought to establish that the crystal is the perfect example of a supramolecular assembly, justifying as it were the earlier statements of Dunitz and Lehn in this regard.

Six years down the line, the supramolecular paradigm has continued to supply a reliable rubric for establishing the grammar of a new and rapidly growing subject, crystal engineering. The present volume is about crystal engineering, or design, and tries to establish connections between the structures of molecular materials and their properties. Crystal engineering links the domains of intermolecular interactions, crystal structures and crystal properties. Without interactions there cannot be structures, and without worthwhile properties as a goal, there cannot be sufficient reason for designing structures. In the process, many advances have been made in fabricating the nuts and bolts of crystal engineering. This is what is summarised in the present volume. So, if the earlier volume was conceptual in its theme, the present one has more to do with methodology and practice.

A major conclusion that emerges from this work is the great utility of defining a crystal structure as a network. This is true for all varieties of molecular crystals ranging from simple organics to labyrinthine coordination polymers that incorporate both inorganic and organic components. I hesitate to use the term 'building block' here, although several of the authors have done so, if only because in the softest of molecular solids, namely the pure organics, the building blocks are themselves pliable. This pliability is chemical rather than mechanical – a given organic molecule presents many faces to its neighbours and the slightest of modifications may mean that its recognition profile changes drastically. Accordingly, the term 'building block' is inappropriate for pure neutral organics, but it may be employed with increasing degrees of confidence as the intermolecular interactions become stronger and more directional. This, then, is the winning advantage of coordination polymers. The well defined coordination environment around the metal atom and the strong, directional nature of its interactions with the organic ligands that surround it mean that not only may a coordination polymer be designed reliably but also that its topological depiction as a network is the most natural one.

If coordination bonds to metal centres are robust design elements, the hydrogen bond or hydrogen bridge does not lag far behind. This master key of molecular recognition combines strength with suppleness and can be employed in a great many chemical situations. These interactions have been treated in some detail in this volume, in their organic, inorganic and ionic variations. Every stage in the development of the chemical sciences has witnessed much progress with respect to understanding hydrogen bonding and the supramolecular era is no exception. In addition to its use as an exclusive design element, a hydrogen bond may be used along with coordination bonds and even more precise structural control is obtained. Such combinations of interactions are always more effective than single interactions, however strong the latter may be. In the most favourable synergies, supramolecular synthons are obtained that may used with the highest levels of confidence in crystal engineering.

What now of properties? Does crystal engineering lose its innate character when the designed materials do not have any obvious property? Surely not – for how does one write a poem if one does not know how to arrange words together? The grammar of crystal design is devilishly complex. Most crystal structures, even those of coordination polymers, are not modular. The building blocks continue to twist and turn and interaction interference is always a danger. This, then, is the real goal of the subject – to identify systems that *are* modular, wherein a family of related molecules will yield a family of related crystal structures. Hierarchy is still elusive in most cases because of the supramolecular nature of the systems employed, and with the further complication that crystallisation is a kinetically controlled process rather than a thermodynamic one, issues of modularity and hierarchy will be the most difficult challenges for the crystal engineer for some time to come. Despite these limitations, and they are formidable ones, considerable progress has been made with respect to property design. The present volume describes materials that act as sensors, catalysts, microporous substances and molecular magnets. Polymorphism is addressed in this volume, although it is still deemed by most to be too intractable an issue with respect to design of either form or function.

Crystal engineering, which has now grown comfortably out of its organic origins to include inorganic compounds within its ambit, will no doubt further extend its scope from single crystals to micro- and nanocrystalline materials and to crystals of lower dimensionalities, and with this the transition from structure to properties will only become more complete. In the meantime, and as we anticipate the fuller coming together of crystal engineering, supramolecular chemistry and materials science, the perspectives provided in the present volume are ample enough for analysis and assessment.

Gautam R. Desiraju
Hyderabad, June 2002

Chapter 1

Hydrogen Bonds in Inorganic Chemistry: Application to Crystal Design

Lee Brammer

University of Sheffield, UK

1 INTRODUCTION

The foremost goal of crystal engineering [1] is to tailor the chemical and/or physical properties of crystalline solids through crystal design at the molecular level. A detailed modular synthetic strategy is employed permitting control over fabrication of the solid at the level of the repeating molecular pattern of which the crystal is comprised. Thus, microscopic control in principle permits macroscopic tunability of properties. It should of course be acknowledged that crystalline products may not always represent the most desirable form of a material with respect to a given function. However, we will concern ourselves only with crystalline materials in the context of this volume. Indeed, the focus of this chapter will be on inorganic crystalline materials, that is, metal-containing systems. Crystalline materials, of course, can exhibit widespread applications in areas including electronics, optics and magnetism and have the potential to provide new materials for use in areas such as separation science, catalysis and chemical sensing [2]. Because of their regular periodic nature, crystalline solids are amenable to precise structural characterization by diffraction methods. While characterization by spectroscopic methods or the characterization of physical properties (e.g. electronic, optical, magnetic) is also essential, the importance of accurate and detailed structural characterization cannot be underestimated. This permits more accurate

Crystal Design: Structure and Function, edited by G. R. Desiraju

structure–function correlations to be established, a key to the design of functional or so-called 'smart' materials.

The difficulties associated with synthesis of inorganic materials are captured in a recent paper by Tulsky and Long [3], who contrast the lack of predictive capability available using current methodologies with the high degree of predictability and control over complex systems that has been developed for the synthesis of *organic molecules*. Moreover, they reinforce the importance of improving synthetic control given the intimate link between solid-state structure and properties, as noted above. Their paper sets forth a well thought out systematic approach, referred to as *dimensional reduction*, applicable to the synthesis of a broad class of inorganic materials. Thus, reaction of binary solids, MX_x, with alkali metal salts, A_aX, yields ternary solids $A_{na}MX_{x+n}$ with the same metal (M) coordination geometry but lower network dimensionality. Through examination of ca 3000 crystal structures, they have been able to suggest strategies for enforcing some degree of control of the product ternary structure. Similarly, crystal engineering seeks to permit controlled synthesis of the crystalline product, though using a different approach focusing on modular assembly from molecular level building blocks. Such a modular approach to crystal design requires the use of (neutral or ionic) building blocks that can be linked in a predictable manner. Thus, a detailed knowledge of preferred intermolecular interactions is essential, and the study of such interactions so as to establish geometric preferences and interaction strengths is a vital part of crystal engineering. Construction of the final crystalline material is effected by self-assembly of the building blocks through deliberate molecular recognition between the building blocks. In the conceptually simplest case a single, self-recognizing building block is used. Two-component systems are also in widespread use, but the complexity introduced by competing modes of self-assembly rapidly increases the difficulty presented in the use of multi-component systems [4].

Crystal engineering has its roots in organic solid-state photochemistry [5], and indeed in its contemporary guise, wherein it broadly encompasses all aspects of modular crystal design, the use of organic molecular building blocks linked via noncovalent interactions has provided the predominant approach [1a]. However, the past decade, and especially the last 5 years, has seen a tremendous increase in the number of publications focused on inorganic crystal engineering [6], broadly defined as including metal ions in the supramolecular design in a structural and/or (potentially) functional role. The introduction of metals, particularly transition metals, has much to offer to the field of crystal engineering. On the structural side they can provide options for connectivity in the networks that make up designed crystalline solids that are unavailable in purely organic systems, viz. square-planar and octahedral coordination geometries. On the functional side, transition metals in particular can impart desirable electronic, optical or magnetic properties upon the final crystalline material. They also have the potential to serve as controlled-access reaction sites for catalytic transformations in a designed porous solid. In a series of recent reviews focusing on crystal engineering involving metal-containing building blocks linked by intermolecular forces, Braga and Grepioni have highlighted the special roles that metal ions can play in influencing the supramolecular assembly of these building blocks [7]. Such roles include pre-organization of intermolecular interactions through use of specific metal coordination geometries, tuning the ligand polarity or acid–base behaviour and reinforcement of intermolecular interactions through 'charge assistance' arising from the frequently ionic nature of metal complexes. These aspects will be elaborated upon in subsequent sections in the context of a domain structure for metal complexes and through the comparison of inorganic and organic building blocks.

2 SCOPE AND ORGANIZATION

Despite its youth, inorganic crystal engineering has already yielded an extensive literature. Thus, it will be necessary to focus the discussion on particular research avenues. The emphasis in this chapter will be placed upon hydrogen bonds as a means of connecting the molecular building blocks. An important aspect to consider will be the role of metal atoms. This will require examination of the influence of metals on hydrogen bonding and the potential role of metals in designed crystalline materials. It should be emphasized at this point that while p-block (i.e. main group) and f-block (i.e. lanthanides and actinides) metals may be mentioned occasionally in this chapter, the primary emphasis will be on d-block metals (i.e. transition metals). Thus, use of the word 'metal' throughout this chapter should be assumed to mean transition metal, unless specified otherwise.

The use of coordination bonds to form networks, so-called coordination polymers, perhaps represents the most widely studied form of inorganic crystal engineering. This approach has also been examined in a number of reviews [8] and is discussed elsewhere in this volume (see Chapter 5). While coordination chemistry will have an important role to play in a number of the hydrogen-bonded systems presented in this chapter, coordination polymers will only be discussed in the context of their cross-linking or their perturbation using hydrogen bonds.

The importance of analysing and understanding predominant hydrogen bonding geometries and patterns will be addressed, and in particular the use of the Cambridge Structural Database (CSD) [9] in obtaining such information. The use of different hydrogen bond types, the reliability of different means of molecular recognition between building blocks, viz. supramolecular synthons, and the similarities and differences between organic and inorganic building blocks will be discussed. It is not the intent of this review to be comprehensive in terms of cataloguing all inorganic crystals synthesized by means of molecular building blocks propagated by hydrogen bonded linkages. However, an effort has been made to classify the systems currently in the literature and to select illustrative examples from among these. Given the recent development of this research area the examples are almost exclusively taken from the past 10 years of the literature, and predominantly from the past 5 years.

There are a number of alternative ways in which this chapter could logically be organized. A structure has been chosen that places the primary emphasis upon the strength (and directionality) of different types of hydrogen bonds along with consideration of their likely abundance, since this reflects the likely usefulness of such hydrogen bonds in crystal engineering. Further divisions have been made by classifying different types of inorganic building blocks, e.g. based upon coordination compounds or organometallic compounds. In the later sections an examination is undertaken of what we might learn from 'mistakes' and unpredicted behaviour in crystal packing. The issue of polymorphism is considered only briefly since it is discussed elsewhere in this volume (see Chapter 8) [10]. The chapter concludes with sections that examine the extent to which functional inorganic crystalline materials have been designed and considers the prospects for future work in this area.

It should be noted that a number of reviews pertinent to the coverage within this chapter can be found either considering hydrogen bonding in inorganic or organometallic crystal engineering [7,11] or focusing on other aspects of hydrogen bonding in inorganic chemistry such as the direct involvement of metals in hydrogen bonds [12] or the formation of 'dihydrogen' (proton–hydride) bonds [13].

3 HYDROGEN BONDS

3.1 Definitions

Given the focus on hydrogen bonding in inorganic crystal design, the question of what constitutes a hydrogen bond needs to be addressed, as does the question of how hydrogen bonding may differ in inorganic and organic systems. All texts on hydrogen bonds address the issue of how to define them [14], although definitions vary in their degree of inclusiveness. A broad and inclusive definition will be adopted here, wherein a hydrogen bond, D–H $\cdots$ A, requires a hydrogen bond donor (D) that forms a polar σ-bond with hydrogen (D–H) in which the hydrogen atom carries a partial positive charge. This group interacts via the hydrogen atom in an attractive manner with at least one acceptor atom or group (A) by virtue of a lone pair of electrons or other accumulation of electron density on the acceptor. Thus, a hydrogen bond is a Lewis acid–Lewis base interaction, wherein D–H serves as the Lewis acid and A as the Lewis base. Limitations will not be placed, *a priori*, on the identities of the donor and acceptor atoms (groups). Thus, all hydrogen bond types, i.e. donor and acceptor combinations, will be considered given the limitation (in the context of this chapter) that inorganic, i.e. metal-containing molecules, must be involved, and that the system being discussed is pertinent in the context of crystal design. The question of how hydrogen bonds may differ in the context of inorganic rather than organic systems requires consideration of the influence of metal atoms on hydrogen bonding and even requires the introduction of classes of hydrogen bonds absent in a purely organic environment. These issues are addressed in Sections 3.3 and 3.4.

Hydrogen bonds exhibit a well-documented energetic preference for a linear D–H ··· A geometry and are arguably the strongest and most directional of noncovalent interactions. Hydrogen bonds with strengths in the range ca 0.2–40 kcal/mol are known, although not all are of significant importance in the context of crystal design. Hydrogen bonds are also flexible, in terms of both hydrogen bond length and geometry. It is for these combined reasons of strength, directionality and flexibility that hydrogen bonds are important to inorganic crystal engineering just as they are in organic crystal engineering [1,15] and for that matter to other structural fields such as structural biology [14c,e].

Returning now to terminology in use specifically in the field of crystal engineering, an important conceptual advance was the definition of so-called *supramolecular synthons* by Desiraju [16]. These are structure-directing recognition motifs involving noncovalent interactions. The intent is that they can be identified and used in *supramolecular synthesis* in a conceptually analogous manner to the use of *synthons* in the (covalent) synthesis of *organic molecules* [17]. Some examples [18] are provided in Section 3.2

3.2 Strong vs Weak Hydrogen Bonds

In considering the use of hydrogen bonds in inorganic crystal engineering, it is important to establish the applicability of different classes of hydrogen bonds. This will depend upon hydrogen bond strength, the reliability of hydrogen-bonded recognition motifs and how abundant or attainable the particular hydrogen bonds may be. While many texts classify hydrogen bonds as 'strong' and 'weak', the borderline between these classes, usually delineated in terms of hydrogen bond energies, often varies depending on the context in which hydrogen bonding is being discussed. The classifications provided by Desiraju and Steiner [14e], which are assigned in the context of the utility of hydrogen bonds in supramolecular chemistry, will be adopted here. These are documented in Table 1. The terms 'very strong', 'strong' and 'weak' hydrogen bond will be used in this frame of reference throughout the chapter.

Hydrogen bond types that are widely used in organic crystal engineering, primarily D–H ··· A where D, A = O or N, will inevitably be important in inorganic systems since the same functional groups that form such hydrogen bonds, i.e. carboxyl, amide, oxime, alcohol, amine, etc., can be present as part of organic

Table 1 Classification and properties of hydrogen bonds, D–H···A. © G. R. Desiraju and T. Steiner, 1999. Adapted from Table 1.5 in *The Weak Hydrogen Bond in Structural Chemistry and Biology* by Gautam R. Desiraju and Thomas Steiner (1999) by permission of Oxford University Press.

	Very strong	Strong	Weak
Bond energy (kcal/mol)	15–40	4–15	< 4
Examples	$[F\cdots H\cdots F]^-$	O–H···O=C	C–H···O
	$[N\cdots H\cdots N]^+$	N–H···O=C	N–H···F–C
	P–OH···O=P	O–H···O–H	O–H···π
IR ν_s relative shift (%)	> 25	5–25	< 5
Bond lengths	D–H $\approx$ H···A	D–H $<$ H···A	D–H $\ll$ H···A
Lengthening of D–H (Å)	0.05–0.2	0.01–0.05	≤ 0.01
D···A range (Å)	2.2–2.5	2.5–3.0	3.0–4.5
H···A range (Å)	1.2–1.5	1.5–2.2	2.2–3.5
Bonds shorter than H···A vdW separation (%)	100	Almost 100	30–80
D–H···A angles range (°)	175–180	130–180	90–180
Effect on crystal packing	Strong	Distinctive	Variable
Utility in crystal engineering	Unknown	Useful	Partly useful
Covalency	Pronounced	Weak	Vanishing
Electrostatic contribution	Significant	Dominant	Moderate

ligands used in metal-containing building blocks. These are strong hydrogen bonds (ca 4–15 kcal/mol) when formed between neutral ligands but can be stronger still when involving ionic species due to the additional electrostatic attraction between the ions, often referred to as 'charge-assistance' [1b,7]. Strong hydrogen bonds can be effective at directing association of building blocks and are therefore very valuable in crystal engineering. This is particularly so when they are part of reliable supramolecular synthons, some examples of which are provided in Figure 1.

The importance of weak hydrogen bonds ($<$ 4 kcal/mol), particularly those involving C–H donor groups, has been established and is recognized to be of importance in crystal engineering. The sheer abundance of C–H donor groups in organic compounds and thus organic ligands necessitates that C–H···A hydrogen bonds (particularly A = O, N) must be considered. Such hydrogen bonds often provide support, i.e. play a secondary role, to stronger hydrogen bonds. In support of this notion, Aakeröy and Leinen note that 'C–H···X interactions can tilt the balance between several options of stronger bonded networks, thus acting as an important "steering force" in the solid-state assembly' [19]. In fact, in the absence of stronger intermolecular interactions weak hydrogen bonds can be used to direct crystal design [20]. Indeed, many supramolecular synthons based upon weak hydrogen bonds have been identified, as illustrated in Figure 2. In many cases these are topologically analogous to supramolecular synthons that use strong hydrogen bonds.

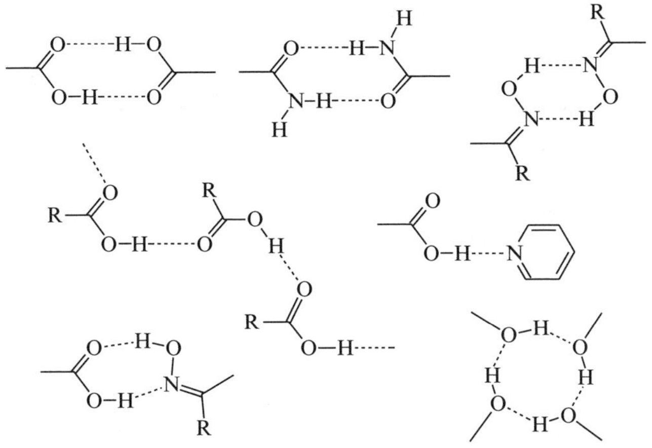

Figure 1 Examples of supramolecular synthons involving strong hydrogen bonds.

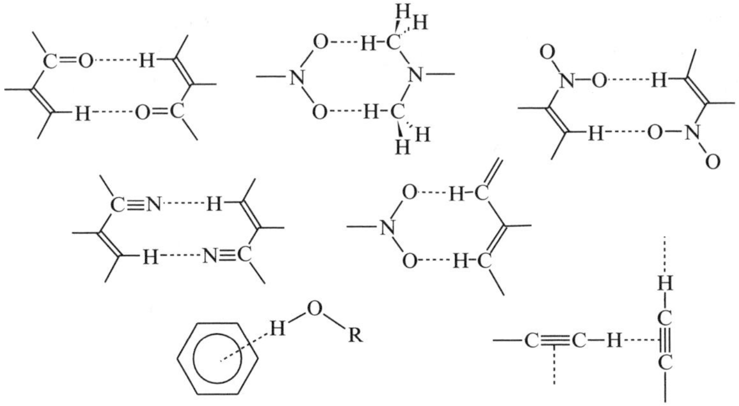

Figure 2 Examples of supramolecular synthons involving weak hydrogen bonds.

3.3 Hydrogen-bonding Domains in Metal Complexes and the Role of Metals in Hydrogen Bonding

In an earlier volume in this series, Dance provided an outstanding and extensive chapter on *Inorganic Supramolecular Chemistry*, in which he introduced a conceptual framework for considering supramolecular chemistry involving metals or metal complexes [21]. Dance suggested that metal complexes, that is, molecular entities comprised of metal atoms/ions coordinated by ligands, can be considered as consisting of a series of concentric, though not necessarily regularly shaped, domains. The central domain, known as the *Metal Domain*, consists of the metal atom itself for a mononuclear complex or a number of metal atoms if a metal cluster complex is being considered. Working outwards, next comes the *Ligand Domain*. This consists of the ligand atoms that surround the metal centre(s). The *Periphery Domain*, as its name suggests, is the outermost part of the complex, i.e. ligand atoms that are in a position to interact with the molecular surroundings, termed the *Environment Domain*. The *Environment Domain* consists of neighbouring molecules in the solid state, and would be comprised primarily of solvent molecules in solution phase supramolecular chemistry. Ligand atoms in the *Periphery Domain* are often, but not necessarily, remote from the metal centre(s). Herein a point of ambiguity arises in the definition of the *Ligand* and *Periphery Domains*. Thus, for very simple (e.g. monoatomic) ligands, the ligand atoms might be considered to simultaneously occupy both domains.

We will adopt a slightly modified version of this domain model to help focus attention on the role that metal atoms within inorganic building blocks play in the hydrogen bonds that link these building blocks together [22]. This modification affects the way in which ambiguities in the definition of the *Ligand* and *Periphery Domains* are resolved. Thus, monoatomic ligands such as halides or hydrides would be classified as belonging to the *Periphery Domain* in Dance's original model. While these atoms are clearly at the periphery of the metal complex, their behaviour in terms of hydrogen bonding interactions is dominated by the fact that they are directly bonded to the metal centre(s) (see Section 4.1). Therefore, we will consider such ligand atoms to belong to the *Ligand Domain*. For the same reason, other ligand atoms whose behaviour is strongly influenced by electronic interaction with the metal center, most often directly bonded to the metal, although not necessarily (e.g. the oxygen atom of carbonyl ligands), will be considered to be part of the *Ligand Domain* rather than the *Periphery Domain*. Thus, the *Periphery Domain* for our purposes in considering hydrogen bonding will consist of the parts of ligands whose properties that are not strongly influenced by an electronic interaction with the metal centre. This modified domain model applied to hydrogen bonding is represented in Figure 3.

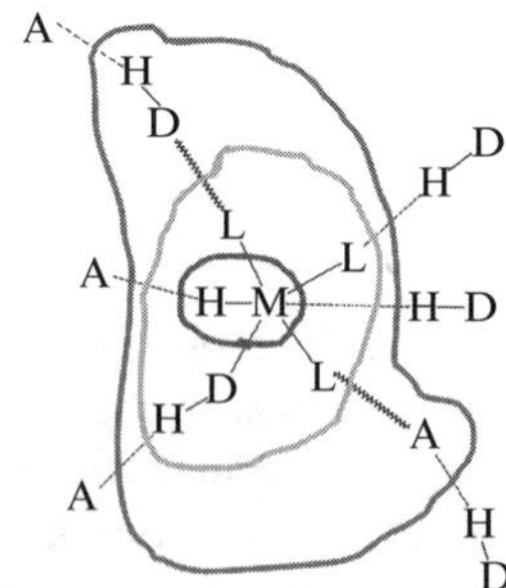

Figure 3 Domain model for hydrogen bonding involving metal complexes. *Metal Domain* (blue); *Ligand Domain* (green); *Periphery Domain* (red); *Environment* (cyan) (see also **Plate 1**).

Hydrogen bonding involving the *Metal Domain* requires that the metal (M) itself is part of the hydrogen bond, either as the hydrogen bond donor, M–H···A [23], or more commonly as the acceptor, i.e. D–H···M [24,25]. Such interactions are of necessity peculiar to inorganic systems, but while they may have some applications in crystal engineering these are likely to be very limited. This is not an issue of lack of strength of these interactions. Indeed, O–H···M hydrogen bonds between *neutral* species have been measured at up to 7 kcal/mol [25e]. Rather, the issue is that D–H···M hydrogen bonds are limited to specific types of metal complex in which sterically accessible filled metal-based orbitals are present. M–H···A hydrogen bonds are rarer still. Hydrogen bonds in the *Metal Domain* are discussed in more detail in Section 6.

Hydrogen bonding in the *Ligand Domain* infers that hydrogen bond donor (D) or acceptor atoms (A) are directly bonded to metal centres or have strong electronic interactions with metal centres, e.g. M–D–H···A or D–H···A–M. Metals can exert an electronic influence upon hydrogen bonds formed in this domain. Thus, the acidity of hydrogen bond donors and the basicity of hydrogen bond acceptors can be tuned via their coordination to metal centres. A good example, often taught in introductory undergraduate chemistry courses, is that water molecules become more acidic when coordinated to metals. Of course, such water molecules necessarily become stronger hydrogen bond donors. On the acceptor side, it has also been shown that halogens are excellent hydrogen bond acceptors when bound to transition metals (as metal halides) in contrast to their limited ability to serve as very weak hydrogen bond acceptors when bound to carbon (as halocarbons) [26]. Here it is the greater polarity of the M–X bond ($M^{\delta+}$–$X^{\delta-}$) relative to the C–X bond that is important. Coordination to a metal gives rise to a greater accumulation of negative charge on the halogen, thus enhancing its hydrogen bond acceptor capability. The importance of *Ligand Domain* hydrogen bonding in crystal engineering will be discussed in more detail in Section 4.1 and parts of Sections 4.3, 4.4 and 5.

In the *Periphery Domain*, hydrogen bonding involves organic functional groups associated with the ligands. The electronic influence of the metal centre is small and would typically depend upon the extent of through-ligand orbital overlap (conjugation) between the metal centre and the peripheral hydrogen bonding groups. However, the metal can still exert a spatial role in directing the hydrogen bonds. Thus, coordination of rigid ligands with hydrogen bonding groups at their periphery to a metal ion of well-defined coordination geometry can be used to direct hydrogen bond formation between neighbouring molecules [27]. One can consider the ligands as effectively amplifying the metal coordination geometry. The result in terms of network design is analogous to that of coordination polymers (networks), except that discrete molecular building blocks (coordination compounds) are linked via well-defined hydrogen bonds. In terms of inorganic crystal engineering, hydrogen bonds in the periphery domain account for the majority of systems studied to date. These systems are examined in Section 4.2 and parts of Sections 4.3, 4.4 and 5.

3.4 Similarities and Differences Between Inorganic and Organic Crystal Engineering

In a valuable series of studies, Braga, Grepioni, Desiraju and co-workers have examined patterns of hydrogen bonds in transition metal-containing crystal structures using the CSD [23e,25j,28]. These studies explore the similarities and differences between hydrogen bonds found in purely organic crystals and those found in inorganic systems. Thus, it is noted that hydrogen bonding patterns for carboxyl, alcohol and amide groups in crystals of metal complexes are similar to those in organic crystals [28a,c]. Importantly, this confirms that such functional groups, which when present typically will be in the *Periphery Domain* of metal complexes, can be used in inorganic crystal engineering in an analogous manner to their widespread use in the design and synthesis of organic crystals. Carbonyl ligands, which can serve as hydrogen bond acceptors, albeit weaker than their organic carbonyl counterparts, are abundant in organometallic compounds [11a]. However, while the carbonyl oxygen atoms can accept hydrogen bonds from strong hydrogen bond donors (O–H, N–H), it is the predominance of peripheral C–H groups that leads to the widespread importance of C–H $\cdots$ O$\equiv$C(M) hydrogen bonds in organometallic crystals [28b]. This topic will be taken up in Sections 4.1.4 and 5.

Clearly specific to inorganic systems are hydrogen bonds that directly involve metal atoms, M–H $\cdots$ A and D–H $\cdots$ M, i.e. those in the *Metal Domain*. A survey of crystal structures has illustrated that M–H $\cdots$ O hydrogen bonds appear to resemble C–H $\cdots$ O hydrogen bonds, although the former are of course far less abundant. The extent to which such hydrogen bonds may be useful in crystal engineering is addressed in Section 6.

Hydrogen bond acceptors found in the *Ligand Domain* are in a number of cases peculiar to inorganic systems in that it is the electronic influence of coordination to the metal centre that activates these ligands towards hydrogen bonding. Excellent examples are metal halides (M–X) and metal hydrides (M–H), both of which can serve as strong hydrogen bond *acceptors*, in contrast to their organic counterparts, C–X and C–H. The application of M–X and M–H acceptors in crystal engineering is discussed in Sections 4.1.2 and 4.1.3, respectively.

3.5 Abundant vs Rare Hydrogen Bonds – Their Importance in Crystal Engineering

The strength of hydrogen bonds and their ability to contribute to reliable supramolecular synthons are not the only criteria for judging the importance of different types of hydrogen bonds. Unless such hydrogen bonds are readily accessible, they will inevitably be of limited use, although perhaps of use in specialized cases. This will inevitably be the case with M–H $\cdots$ A and D–H $\cdots$ M hydrogen bonds, which are only accessible for certain classes of metal complex. At the other extreme are C–H···A hydrogen bonds, especially C–H···O. These are abundant in the crystal structures of many organometallic and coordination compounds and clearly play an important overall role in crystal cohesion and the overall optimization of the interactions between molecular units in a crystalline solid. Their weakness makes them more difficult to use in crystal design than stronger hydrogen bonds. However, their abundance ensures that they cannot be ignored. Indeed, not only do they guide the stronger intermolecular interactions in the crystal, as noted previously, but they can in some cases completely overwhelm stronger interactions as a result of their relative abundance [11d].

3.6 Analysis of Hydrogen Bonding Using the Cambridge Structural Database

The importance of identifying preferred hydrogen bond geometries, supramolecular synthons and packing arrangements in planning a crystal synthesis strategy based upon hydrogen-bonded building blocks is, of course, essential. The CSD [9] contains crystallographic data for all organic and organometallic crystal structures (245 392 crystal structures as of October 1, 2001). These data, combined with the search and data analysis tools that accompany the database, make it a veritable treasure trove of information that is invaluable to anyone considering research in the area of crystal engineering. It is unfortunate that the practice of 'data mining', the term sometimes applied to the derivation of trends from information stored in databases, is considered by some not to be original research. Such a short-sighted viewpoint fails to recognize that important *and unanticipated* trends inaccessible by examination of individual systems can only be identified by such means. Such analyses using crystallographic databases, particularly the CSD in the present context of hydrogen-bonded inorganic structures, are a vital part of crystal engineering and clearly help to lay the groundwork for crystal design strategies.

4 STRONG, STRUCTURE-DIRECTING HYDROGEN BONDS

Section 4 emphasizes the use in inorganic crystal engineering of strong donor groups, i.e. O–H or N–H, that can form hydrogen bonds with energies typically in the 4–15 kcal/mol range depending on the acceptor group employed. Such groups are common constituents of organic functional groups and thus can be incorporated into a wide range of inorganic building blocks based upon coordination compounds or organometallic compounds.

4.1 Ligand Domain

In the systems discussed in Section 4.1, the dominant role of the metal in terms of hydrogen bond formation arises form its strong electronic influence upon the donor or acceptor ability of the groups participating in hydrogen bonding.

4.1.1 Donors: (M)O–H *and* (M)N–H

Here we consider hydrogen bonds donated by coordinated ligands such as hydroxyl (OH), water (OH_2), alcohols (ROH) and amines (NH_3, NRH_2, NR_2H). The influence of coordination to a metal centre (M) on hydrogen donor OH_2 has already been commented upon (Section 3.3). Each of these ligands is a net electron donor to the metal centre (principally σ-donation). Thus, one should anticipate that coordination of the oxygen or nitrogen atom to a metal centre will result in increased polarity of the O–H or N–H bond, thus increasing the potency of the ligand as a hydrogen bond donor.

While this class of ligands participates extensively in hydrogen bonding, the application of such ligands in crystal synthesis has to date been somewhat limited. A common theme is one of using these ligands to form hydrogen bonds to a spacer molecule or anion that permits metal centres to be linked into networks. This approach presumably arises as a result of the small size of these ligands. In an early example, Zaworotko and co-workers used the tetrahedral metal cluster $[Mn(\mu_3\text{-OH})(CO)_3]_4$ to form a diamondoid network in which the face-capping hydroxyl ligands were linked to those on neighbouring clusters via O–H···N hydrogen bonds to a linear 4,4′-bipyridyl (4,4′-bipy) spacer (Figure 4) [29]. There are a number of examples of the analogous approach in which coordinated water molecules propagate a network via hydrogen bonds to spacer units [30], particularly leading to hydrogen-bonded cross-linking of coordination polymers (see Section 4.4.1).

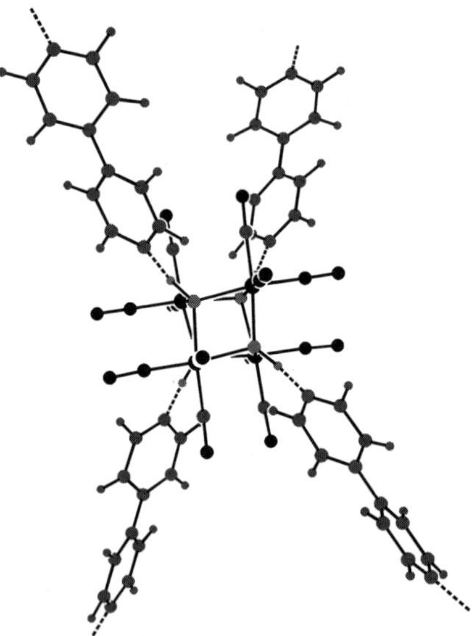

Figure 4 Diamondoid (M)O–H···N hydrogen-bonded network in crystalline $[Mn(\mu_3\text{-OH})(CO)_3]_4 \cdot 2(4,4'\text{-bipy}) \cdot 2CH_3CN$. O–H groups (red); 4,4′-bipy (blue) [29] (see also **Plate 2**).

Taking the idea one step further, in $[(4,4'\text{-bipy})H_2]_2[Ni(OH_2)_2(NCS)_4][NO_3]_2$, Chen and co-workers prepared a three-component hydrogen-bonded 'chicken wire' 2D grid network wherein coordinated water molecules form hydrogen bonds to nitrate anions, which in turn are hydrogen-bonded to the 4,4′-bipyridinium cations 30e (see Section 4.3.1). Since alcohols are weaker ligands than hydroxide or water, it is likely that they will be less effective for use in crystal design, although there are many examples in which networks are propagated due to M–O(R)–H···N or M–O(R)–H···O hydrogen bonds (Figure 5) [31].

Similarly, examination of the CSD reveals numerous examples of hydrogen-bonded networks arising from M–N(R_2)–H···A, M–N(R)(H)–H···A and M–N(H_2)–H···A hydrogen bonds (R = alkyl, A = N, O, Cl) associated with coordinated NR_2H [32], NRH_2 [33] or NH_3 [34] ligands. Hydrogen-bonded networks involving primary amines and ammine ligands are particularly abundant. However, little systematic effort seems to have been made to exploit such hydrogen bond donor ligands in crystal design. Many instances of these hydrogen-bonded networks arise in papers where the authors' interest in the crystal structure was in the molecular species rather than its intermolecular association. A distinct exception involves the study of second-sphere coordination involving the binding of M–NH_3 moieties to crown ethers via multiple N–H···O hydrogen bonds [35].

4.1.2 *Acceptors: halides*, M–X (X = F, Cl, Br, I)

Halide ions have long been considered good hydrogen bond acceptors, although in crystal engineering terms they are perhaps not so useful, or at least difficult to harness given their lack of directional interactions. Until recently, beyond halide ions halogens were frequently not considered in discussions of hydrogen bonding, as is apparent in Jeffrey's statement that 'while halide ions are strong hydrogen bond acceptors, there is no evidence from crystal structures supporting hydrogen bonds to halogens' [14d]. To be fair, the perspective of this statement is clearly an organic one, although it has been shown that even organic halides exhibit very

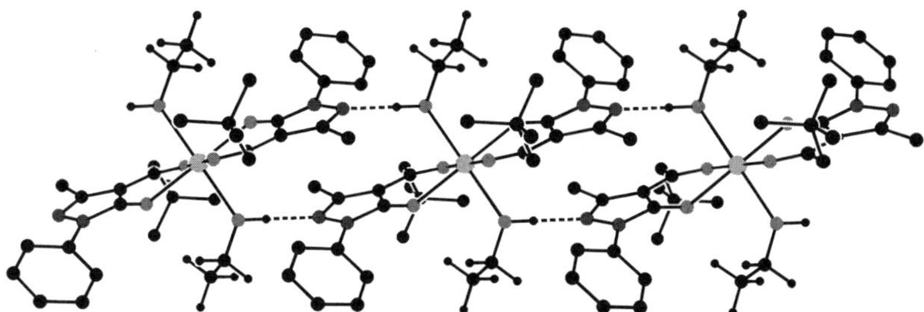

Figure 5 (M)O–H···N hydrogen-bonded network involving coordinated ethanol. O (red); N (blue); Cd (green) [31a] (see also **Plate 3**).

weak hydrogen bond acceptor behaviour [20a,26c,36]. In 1998, Brammer, Orpen and co-workers pointed out that contrary to the behaviour of carbon-bound chlorine, their metal-bound counterparts (i.e. inorganic chlorides) are good hydrogen bond acceptors [26a]. This arises from the greater polarity of M–Cl bonds relative to C–Cl bonds, leading to a much stronger electrostatic component for D–H $\cdots$ Cl–M hydrogen bonds than for the D–H $\cdots$ Cl–C case. These conclusions have since been generalized for all halogens through extensive studies using the CSD and based upon data from thousands of interactions involving halogens [26b,c]. Most pertinent to inorganic chemistry, geometric preferences and trends of D–H $\cdots$ X–M hydrogen bonds (D = O, N, C; X = F, Cl, Br, I) have been established. These are summarized in Table 2. The use of normalized H $\cdots$ X distances, R_{HX} [37], permits direct comparison of the hydrogen bond acceptor capabilities of the halogens and indicates that the trend in D–H $\cdots$ X–M bond strengths (for any given donor) is $F \gg Cl \geq Br > I$. This is in excellent agreement with the trend in intramolecular N–H $\cdots$ X–Ir bond strengths in $IrH_2X(pyNH_2)(PPh_3)_2$ determined though a combination of NMR spectroscopy and *ab initio* calculations by Peris *et al.* [38] (viz. X = F 5.2, Cl 2.1, Br 1.8 and I < 1.3 kcal/mol).

Equally, if not more, important from a crystal design perspective is the fact the terminal metal chlorides, bromides and iodides are distinctly directional acceptors of hydrogen bonds, with typical angles in the range H $\cdots$ X–M = 90–130°. Fluorides, although forming the strongest hydrogen bonds, are more isotropic in their acceptor behaviour, i.e. show less well-defined angular preference, although some preference for H $\cdots$ F–M angles in the range 120–160 ° is noted. These geometric preferences can be explained in terms of the M–X bonding interaction and its effect on the negative electrostatic potential around the halogen (Figure 6), which in turn is expected to guide the approach of the hydrogen atom in hydrogen bond formation [26b,c].

Examination of the electrostatic potential in the vicinity of *cis*-dichloride or *fac*-trichloride complexes illustrates that a cooperative effect between neighbouring halides arises because of overlap of the regions of negative electrostatic potential minimum from the individual halide ligands (Figure 7). This gives rise to a 'binding pocket' for the positively charged hydrogen atom between the set of two or three chloride ligands [39]. Similar observations are made for bromide and iodide

Table 2 Mean R_{HX} distances [37] for H $\cdots$ X contacts with $(R_{HX})^3 \leq 1.15$ (ca. $R_{HX} < 1.05$) in D–H $\cdots$ X–M hydrogen bonds (D = O, N, C; X = F, Cl, Br, I).

	Mean normalized distance, R_{HX} (No. of observations)		
X	O–H $\cdots$ X	N–H $\cdots$ X	C–H $\cdots$ X
F–M	0.703 (37)	0.776 (73)	0.943 (374)
Cl–M	0.799 (416)	0.853 (1341)	0.975 (7943)
Br–M	0.820 (30)	0.879 (205)	0.982 (3269)
I–M	0.868 (8)	0.923 (83)	0.997 (2429)

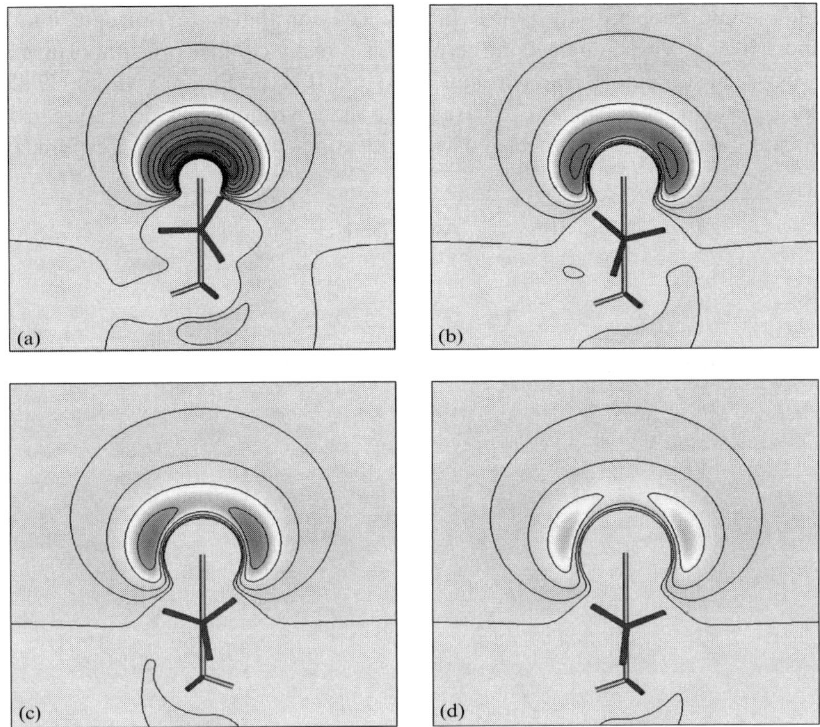

Figure 6 Calculated negative electrostatic potential for *trans*-[$PdX(CH_3)(PH_3)_2$] illustrating positions of potential minima; X = F (a); Cl (b); Br (c); I (d). Reprinted with permission from L. Brammer, E. A. Bruton and P. Sherwood, *Cryst. Growth Des.*, **1**, 277–90 (2001). Copyright 2001 American Chemical Society (see also **Plate 4**).

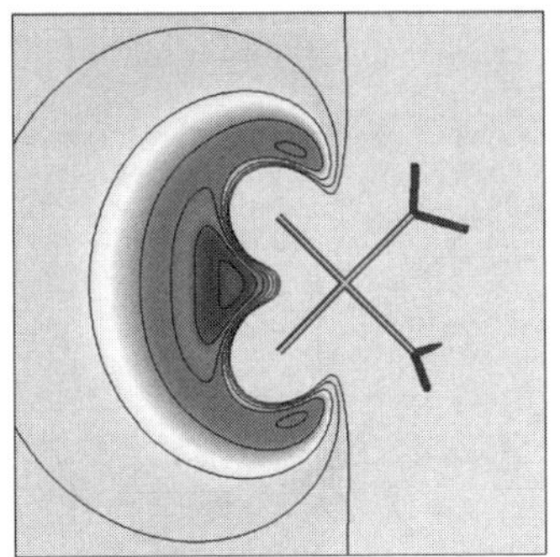

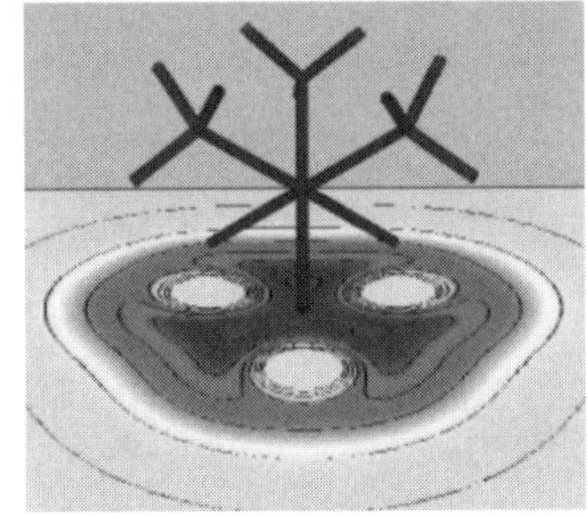

Figure 7 Calculated negative electrostatic potential for *cis*-[$PdCl_2(PH_3)_2$] (left) and *fac*-[$RhCl_3(PH_3)_3$] (right) identifying recognition sites (potential minima, deep blue) for hydrogen bond donors. Reproduced with permission from L. Brammer, J. K. Swearingen, E.A. Bruton and P. Sherwood, *Proc. Natl. Acad. Sci. USA*, **99**, 4956–61 (2002). Copyright 2002 National Academy of Sciences, USA (see also **Plate 5**).

ligands, while cooperativity is less pronounced for the more isotropic fluoride ligands [40]. In crystal engineering terms, this directly confirms the importance of the suggested supramolecular synthons **I** [41] and **II** [41a] (Figure 8). Figure 9 illustrates a crystal design strategy [39] based upon these synthons in which bifurcated or trifurcated acceptor sites associated with perhalometallate anions are populated

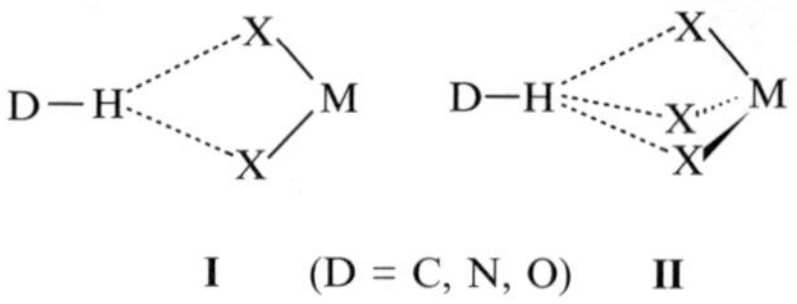

I (D = C, N, O) **II**

Figure 8 Supramolecular synthons D–H···X_2M and D–H···X_3M.

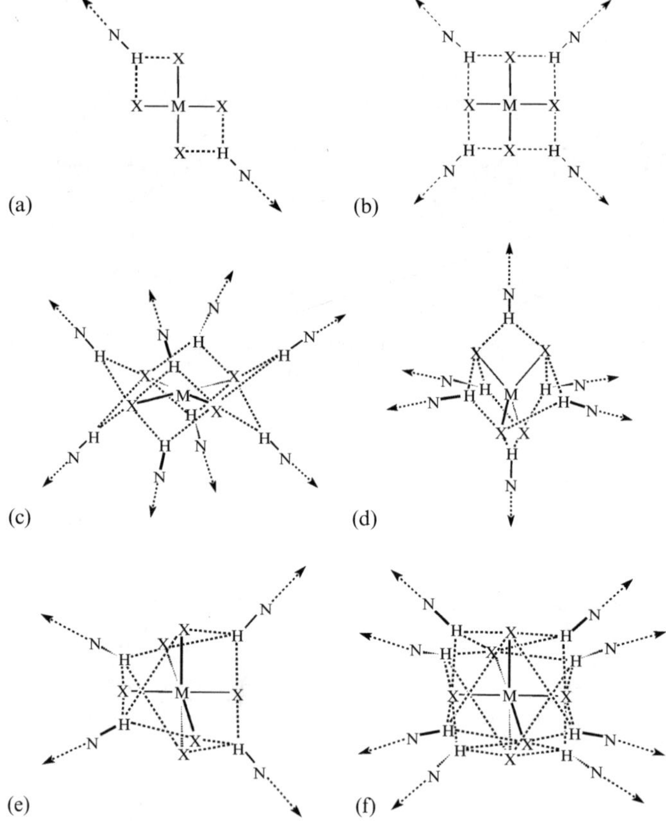

Figure 9 Perhalometallate ions as potential hydrogen-bonded network nodes. Reproduced with permission from L. Brammer, J. K. Swearingen, E. A. Bruton and P. Sherwood, *Proc. Natl. Acad. Sci. USA*, **99**, 4956–61 (2002). Copyright 2002 National Academy of Sciences, USA. (see also **Plate 6**).

by hydrogen bonds, permitting the anions to serve as nodes in a hydrogen-bonded network. Thus, in square-planar anions $[MX_4]^{n-}$ only (bifurcated) edge acceptor sites are available, whereas in octahedral $[MX_6]^{n-}$ anions, both edge sites and (trifurcated) face sites are accessible, the latter being slightly preferred. Less cooperativity between halogens arises in tetrahedral $[MX_4]^{n-}$ anions since the halide ligands are further apart, and hydrogen bonds are frequently asymmetrically bifurcated at one edge of the tetrahedron. This approach has been applied to the design of new inorganic materials constructed using charge-assisted hydrogen bonds between organic cations and perhalometallate anions [41a,b,42]. Examples of 1D and 2D hydrogen-bonded network structures so formed are shown in Figure 10.

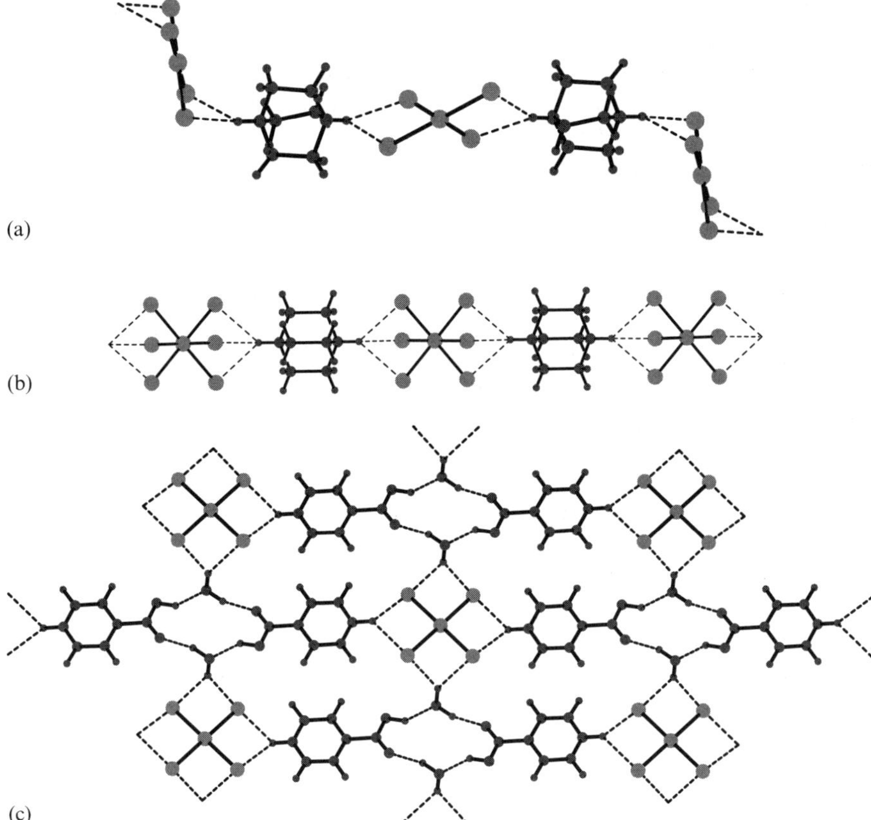

Figure 10 One-dimensional networks in $[(DABCO)H_2][PtCl_4]$ (a) and $[H_2(DABCO)][PtCl_6]$ (b) employing synthons **I** and **II**, respectively [41a]. Two-dimensional network in $[\{(\text{isonicotinic acid})H\}_2(OH_2)_2][PtCl_4]$ (c) employing synthon **I** [42b]. Reproduced with permission from L. Brammer, J. K. Swearingen, E. A. Bruton and P. Sherwood, *Proc. Natl. Acad. Sci. USA*, **99**, 4956–61 (2002). Copyright 2002 National Academy of Sciences, USA. (see also **Plate 7**).

In their work in this area, Orpen and co-workers have sought to prepare hydrogen-bonded halometallate salts using the planar 4,4′-bipyridinium dication [41b,42]. They note that while synthon **I** results from combination with the square-planar $[MCl_4]^{2-}$ anion, neither synthon **I** nor **II** arises when tetrahedral $[MCl_4]^{2-}$ or octahedral $[MCl_6]^{2-}$ anions are used, in contrast to the use of the $[H_2(DABCO)]^{2+}$ cation by Brammer *et al.* Importantly, however, their studies indicate that all of the 4,4′-bipyridinium structures can be rationalized as belonging to a larger homologous family of salts that include chloride and a variety of chlorometallates as counteranions [42c]. All form structures based upon ribbon motifs containing $NH \cdots (Cl)_2 \cdots HN$ interactions, into which synthon **I** can be accommodated, but is not required.

These hydrogen-bonded salts show distinct potential for the controlled design of new crystalline structures, and are applicable to incorporation of a wide variety of transition metal and main group metal ions. The work of Mitzi on perovskite structures in which perhalometallate layers are linked via organic alkyl ammonium cations that interact via N–H···X–M hydrogen bonds shows another area of potential application [43] (see Sections 4.4.2 and 7).

Finally, it should be noted that there are many examples, some designed deliberately [44], in which ligands bearing peripheral hydrogen bond donor groups have been coordinated to metals that bear a halide (often chloride) ligand, resulting in a 1D hydrogen-bonded tape [44]. In particular, an abundance of compounds of the type *cis*-MCl_2L_2(L = amine) have been crystallographically characterized as a result of research spawned by the discovery of anti-tumour agent cisplatin, *cis*-$[PtCl_2(NH_3)_2]$. These systems provide abundant information on *neutral* hydrogen-bonded networks propagated by N–H···Cl–M hydrogen bonds.

4.1.3 Acceptors: hydrides (D–H···H–M and D–H···H–E, *E* = *B, Al, Ga*)

The realization that hydridic hydrogen can serve as a hydrogen bond acceptor came about through work by the groups of Crabtree and Morris on transition metal hydrides in the mid-1990s [45]. It was subsequently established that main group hydrides, particularly from Group 3, formed analogous D–H···H–E (E = B, Al, Ga) hydrogen bonds [46]. This class of hydrogen bonds has been studied for its role in facilitating chemical reactions [47] and more recently with respect to applications in the area of crystal engineering. Thus, Morris and co-workers designed 1D hydrogen-bonded polymers propagated solely or at least in part by charge-assisted N–H···H–M hydrogen bonds [48]. These systems involve hydrogen bond donor cations, comprising K^+ ions encapsulated by azacrown ethers, combined with polyhydridometallate anions, as shown in Figure 11. There are clearly some analogies between these systems and the hydrogen-bonded halometallate salts (see above). Gladfelter and co-workers [46h,i) and Custelcean and Jackson [46c–e] have, respectively, prepared crystalline systems linked into networks via

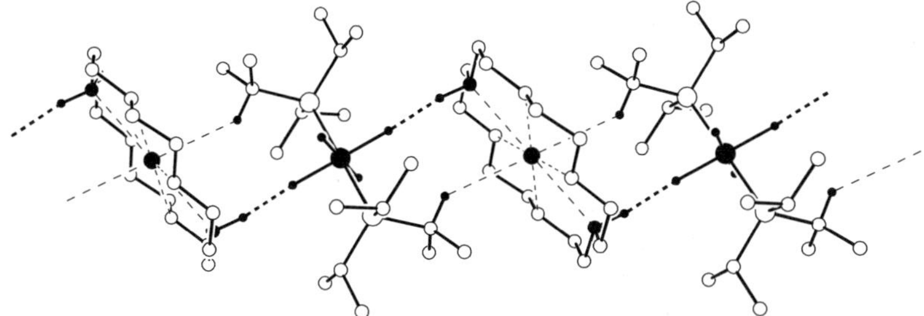

Figure 11 One-dimensional N–H···H–Ir network (H···H = 1.84 Å) supported by weak C–H···K interactions (H···K = 2.92 Å) in $[K(Q)][IrH_4(P^iPr_3)_2]$; Q = 1,10-diaza-18-crown-6. Iridium, potassium, nitrogen and key hydrogen atoms are shaded. Adapted with permission from K. Abdur-Rashid, D. G. Gusev, S. E. Landau, A. J. Lough and R. H. Morris, *J. Am. Chem. Soc.*, **120**, 11826–7 (1998). Copyright 1998 American Chemical Society.

N–H···H–Ga and N–H···H–B hydrogen bonds. These solids have then been shown to undergo topochemical formation of extended solids based upon Ga–N and B–N bonds, repectively, via solid-state reactions, harking back to the origins of crystal engineering [5].

4.1.4 *Acceptors: carbonyls* (D–H···OC–M)

The behaviour of carbonyl ligands as hydrogen bond acceptors in the solid state has been reviewed by Braga and Grepioni [11a], who have extensively examined the available crystallographic data by using the CSD. Their work shows that while the carbonyl oxygen does serve as an acceptor for strong hydrogen bond donors, i.e. O–H···OC–M and N–H···OC–M, such interactions are relatively uncommon. The principal reason is that the carbonyl oxygen is a reasonably weak acceptor and strong donors tend to form interactions with stronger acceptors where possible. However, the predominance in organometallic crystals of carbonyl ligands and of ligands with acidic C–H groups leads to an abundance of C–H···OC–M hydrogen bonds, which are of tremendous overall importance in organometallic crystal engineering (see Section 5). The reason for inclusion of carbonyl ligands in discussion of hydrogen bonds in the ligand domain is that their strong electronic interaction with the metal centre can lead to significant tuning of the basicity of the carbonyl oxygen and thus its capability as a hydrogen bond acceptor. This is most apparent when the acceptor behaviour of μ_3-CO and μ_2-CO bridging carbonyls is compared with that of terminal carbonyl ligands. Increased π-back-donation upon coordination of CO to additional metal centres is well established [49]. This leads to increased basicity of the carbonyl oxygen and in turn to shorter (stronger) hydrogen bonds. Average H···O distances for the most abundant C–H···OC–M hydrogen bonds are 2.44, 2.57 and 2.63 Å for triply bridging, doubly bridging and terminal CO ligands, respectively [28b].

4.2 Periphery Domain: Networks Formed by Direct Ligand–Ligand Hydrogen Bonds Between Building Blocks

In the *Periphery Domain*, the role of the metal in hydrogen bonding is less prominent than in the *Ligand Domain*. However, it can still exert an influence over directing hydrogen bond formation, in some cases exerting a weak electronic influence and in others serving more as a constituent of a pendant group in an organic hydrogen-bonded network.

This area of forming hydrogen-bonded networks using coordination compounds has recently been reviewed by Beatty [11f]. The review was organized primarily on the basis of the dimensionality of the network formed, i.e. 1D vs 2D vs 3D. In order to complement that review, the primary organization of this section instead focuses on the category of ligand used to provide the hydrogen-bonded links between building blocks.

4.2.1 *Monodentate ligands: directing of hydrogen bond formation by metal coordination geometry*

The use of monodentate ligands that are capable of binding to only a single coordination site at a metal centre, but also able to present an exterior functional group capable of forging a hydrogen-bonded link to a neighbouring molecule, permits metal-directed hydrogen-bonded networks to be prepared. This concept is illustrated by assembly **V** in Figure 12 for the case of a 1D network, and contrasted with the case of organometallic π-arene building blocks **VI** (see Section 4.5) in which the metal does not play such a structure-directing role.

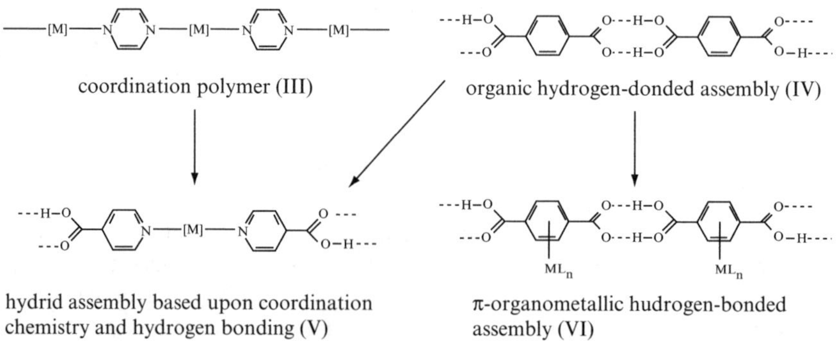

coordination polymer (III)

organic hydrogen-donded assembly (IV)

hydrid assembly based upon coordination chemistry and hydrogen bonding (V)

π-organometallic hudrogen-bonded assembly (VI)

Figure 12 Strategies for designing networks by combining hydrogen bonds with coordination chemistry or π-arene organometallic chemistry. Adapted from L. Brammer, J. C. Mareque Rivas, R. Atencio, S. Fang and F. C. Pigge, *J. Chem. Soc., Dalton Trans.*, 3855–67 (2000). Reproduced by permission of the Royal Society of Chemistry. (see also **Plate 8**).

The relationship between the 1D assembly **V** and either the coordination polymer **III** or the organic hydrogen-bonded assembly **IV** is readily apparent. Conceptually, the relationships involve replacement of linear N–M–N linkages by linear hydrogen-bonded linkages (carboxyl dimer) or vice versa. A linear assembly such as **V** results not only from the linear hydrogen-bonded link, but from the rigidity of the ligands and from the *trans* coordination of the ligands at the metal centre. Herein lies the metal's structure-directing role, namely orientation of the hydrogen bonding groups so as to direct the assembly of the building blocks. This contrasts with the situation suggested by **VI** in which the parent organic network remains essentially unchanged, and the metal-containing moiety (ML_n) is appended by coordination of the arene in a π manner. The concept embodied in **V** is in principle amenable to a variety of self-recognizing functional groups, such as carboxylic acids, amides and oximes, and is not limited to coordination of only two functional ligands at each metal centre (Figure 13).

4.2.1.1 ML_2 *building blocks*

A series of predominantly ID assemblies have been prepared by Aakeröy and co-workers using silver(I) ions [50], which have a tendency to adopt a linear two-coordinate geometry (i.e. AgL_2^+ building blocks, cf. **VII**, **XVII**, **XVIII**). These assemblies are obtained with ligands L = pyridine-4-carboxamide (isonicotinamide) [50a], pyridine-3-aldoxime [50b] and pyridine-3-acetoxime [50b], when using

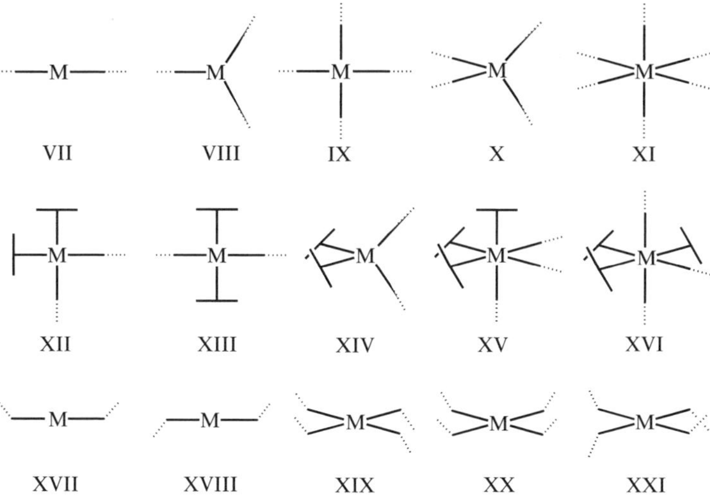

Figure 13 Schematic representation of coordination compounds with rigid monodentate ligands bearing hydrogen bonding groups (e.g. substituted pyridines). Arrangements **VII–XVI** are representative of 4-pyridine ligands and arrangements **XVII–XXI** of 3-pyridine ligands.

poorly coordinating counteranions such as ClO_4^-, BF_4^-, PF_6^- or $CF_3SO_3^-$ (Figure 14). Further assembly of the 1D tapes of [Ag(3-aldoximepyridine)$_2$]X (**XVII**) and [Ag(3-acetoximepyridine)$_2$]X (**XVIII**) (X = PF_6^-, ClO_4^-) into sheets is accomplished via additional C–H···O hydrogen bonds involving the oxime oxygen atoms and C–H···F/O hydrogen bonds involving the anions. The use of pyridine-3-carboxamide (nicotinamide) ligands leads to a 1D ladder structure in [Ag (nicotinamide)$_2$] CF_3SO_3, where a *cis* arrangement of amide groups arises (**XVII**), and a 2D sheet structure in [Ag(nicotinamide)$_2$]X (X = PF_6^-, BF_4^-), where the amide groups are mutually *trans* (**XVIII**). In each case the network is propagated by amide–amide hydrogen bonds [27b] (Figure 15). These examples illustrate the torsional flexibility of this class of metal–pyridine-based building blocks (Figure 16) and its consequences in terms of the final network formed.

Attempts to use nicotinic acid, isonicotinic acid and some derivatives in a manner similar to that described above for the corresponding amides and oximes

(a)

(b)

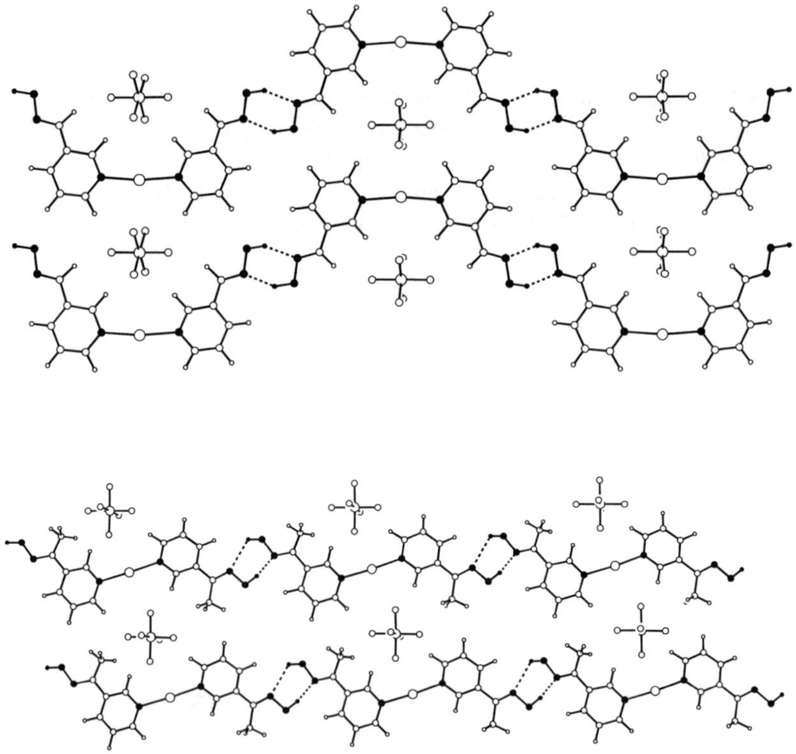

Figure 14 One-dimensional tapes linked via $R_2^2(6)$ O–H···N hydrogen-bonded oxime dimers in (a) [Ag(3-aldoximepyridine)$_2$]PF_6 (cf. **XVII**) and (b) [Ag(3-acetoximepyridine)$_2$]PF_6 (cf. **XVIII**) [50b]. Cross-linking of tapes via C–H···O and C–H···F hydrogen bonds not shown. Oxygen, nitrogen and key hydrogen atoms are shaded.

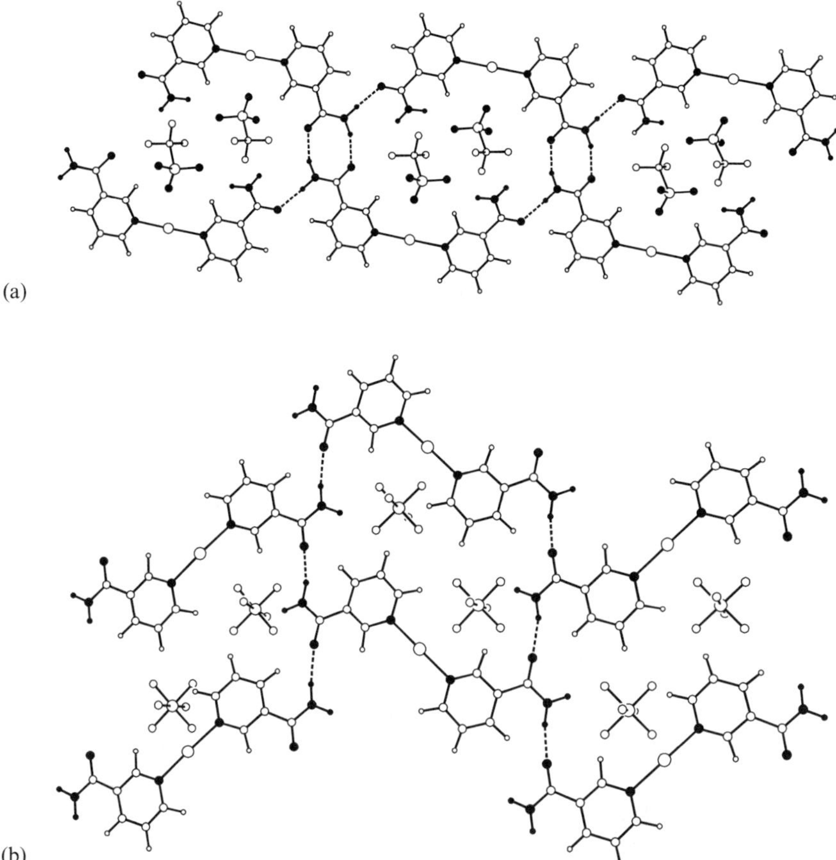

Figure 15 (a) Ladder structure in $[Ag(nicotinamide)_2]CF_3SO_3$ (cf. **XVII**) propagated via N–H···O hydrogen-bonded catemer with 'rungs' comprising $R_2^2(8)$ amide dimer interactions. Cationic ladders are interconnected via the oxygen atoms of the anions which form N–H···O–S–O···H–N hydrogen-bonded bridges and weak Ag···O···Ag bridges (not shown). (b) Hydrogen-bonded layer structure in $[Ag(nicotinamide)_2]PF_6$ (cf. **XVIII**). Cationic layers are cross-linked via N–H···F and C–H···F hydrogen bonds (not shown). Oxygen, nitrogen and key hydrogen atoms are shaded.

illustrates the fact that the metal can still exert some electronic influence on remote (peripheral) hydrogen bonding groups when linked via a conjugated ligand. Thus, the acid functions are frequently deprotonated upon coordination of the ligand to Ag(I) centres, although not reliably so, and Ag–O coordination is common [51]. Indeed, in an interesting recent example, the coexistence of an [Ag(L)] coordination network with an [Ag(L)(HL)] hydrogen-bonded network (L = isonicotinate) to yield a porous crystalline solid was reported [51b].

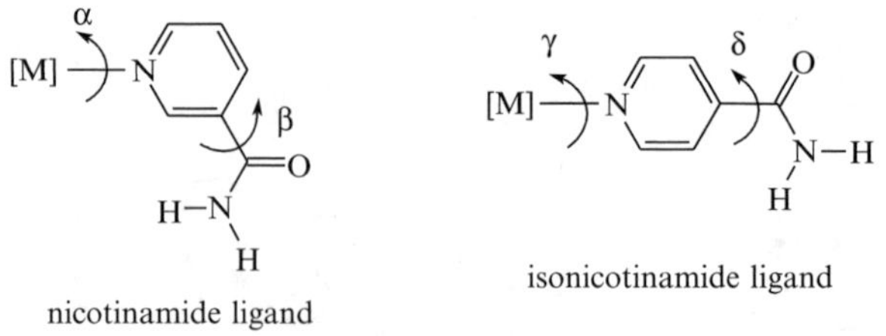

Figure 16 Nicotinamide (pyridine-3-carboxamide) and isonicotinamide (pyridine-4-carboxamide) ligands. Curved arrows indicated torsional flexibility at positions α–δ. Reprinted from L. Brammer, J. C. Mareque Rivas, R. Atencio, S. Fang and F. C. Pigge, *J. Chem. Soc., Dalton Trans.*, 3855–67 (2000). Reproduced by permission of the Royal Society of Chemistry.

Gold(I) is also well known for forming linear two-coordinate compounds as it is for forming attractive Au···Au interactions, so-called 'aurophilic' interactions, the strengths of which are comparable to those of many hydrogen bonds. Schmidbaur and co-workers have harnessed both aurophilic interactions and direct ligand–ligand hydrogen bonds in preparing examples of 1D and 2D networks using neutral Au(I) building blocks containing phosphine and thiolate ligands [52].

4.2.1.2 ML_3 *building blocks*

Aakeröy *et al.* were also able to prepare building blocks based upon Ag(I) centres coordinated in a trigonal planar manner by three isonicotinamide ligands. The isostructural perchlorate and tetrafluoroborate salts of these cationic complexes involve amide–amide hydrogen-bonded links from each cation to six others, resulting in 3D interpenetrated networks that resemble the network found in α-ThSi [53].

4.2.1.3 Square-planar ML_4 *building blocks*

Networks based upon homoleptic cationic building blocks consisting of square-planar Pt(II) or Ni(II) centres coordinated by the functionalized pyridine ligands discussed above have been prepared by the groups of Aakeröy and Brammer [27a,b,54]. Steric interactions between *ortho* hydrogen atoms require the pyridine rings to lie approximately orthogonal to the metal coordination plane (MN_4). Thus, whereas for 4-substituted pyridines the hydrogen bonding groups are projected in directions that effectively amplify the metal coordination geometry (cf. **IX**), for 3-substituted pyridines the hydrogen bonding groups can be oriented either above or below the coordination plane (**XIX**–**XXI**). In the structures of [Pt(nicotinamide)$_4$]Cl_2 and [Pt(nicotinamide)$_4$]$(PF_6)_2 \cdot H_2O$, the cations adopt a centrosymmetric arrangement with two amide groups 'up' and two 'down' (cf. **XIX**). Remarkably, the principal feature common to the two structures is the layer arrangement of the cations, which is sustained by interdigitation of the ligands involving by π–π and C–H $\cdots$ π interactions. In [Pt(nicotinamide)$_4$]Cl$_2$ [54a,b], all amide groups are linked to another in the next layer via the well-known $R_2^2(8)$ [55] hydrogen-bonded dimer supramolecular synthon leading to a 1D zig-zag hydrogen-bonded tape (Figure 17). This resembles the tape structures observed for [Ag(3-acetoximepyridine)$_2$]X (X = PF_6^-, ClO_4^-) [49b] and [Ag(L)(HL)] (L = nicotinate) [51a], except that for [Pt(nicotinamide)$_4$]Cl$_2$ two links occur between each metal centre along the chain. An analogous tape structure is also found in [Cu$_2$(μ_2-acetonato-O)-2-(6-cyano-3-methylpyridin-2(1*H*)-one](BF$_4$)$_2$ [56]. In [Pt (nicotinamide)$_4$](PF$_6$)$_2$ · H_2O [54a,c], the spacing between cation layers is increased to accommodate the larger anions and the water molecules. Amide–amide hydrogen bonds between the layers are no longer possible. However, rotation of the amide groups (angle β in Figure 16) permits weaker amide–amide hydrogen bonds *within* the cation layer.

The use of isonicotinic acid ligands leads to a threefold interpenetrated neutral square grid network in which [Pt(L)$_2$(HL)$_2$] building blocks (L = isonicotinate) (cf. **IX**), resulting from deprotonation of half of the acid groups, are linked by carboxyl–carboxylate hydrogen bonds (Figure 18) [27c,57]. A network of the same

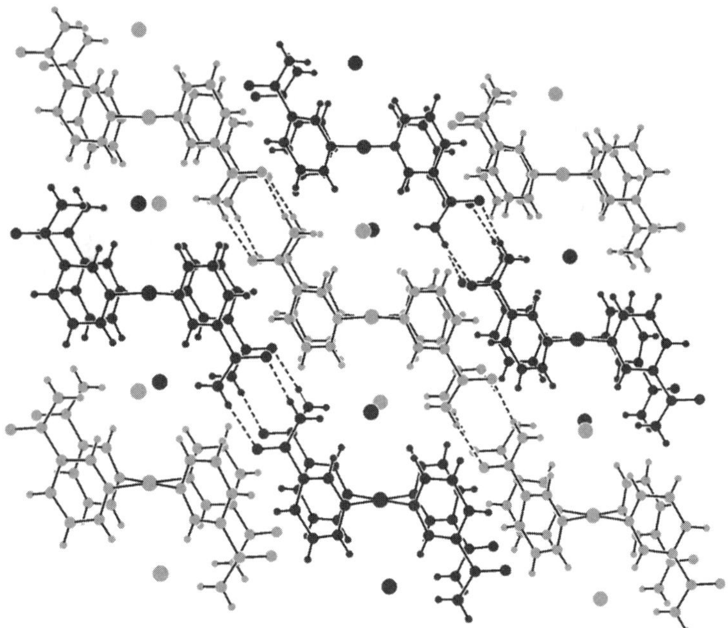

Figure 17 Hydrogen-bonded tapes comprised of cationic [Pt(nicotinamide)$_4$]$^{2+}$, with chloride ions in channels between tapes. (Reprinted from J. C. Mareque Rivas and L. Brammer, *New J. Chem.*, **22**, 1315–8 (1998). Reproduced by permission of the Royal Society of Chemistry (RSC) and the Centre National de la Recherche Scientifique (CNRS).

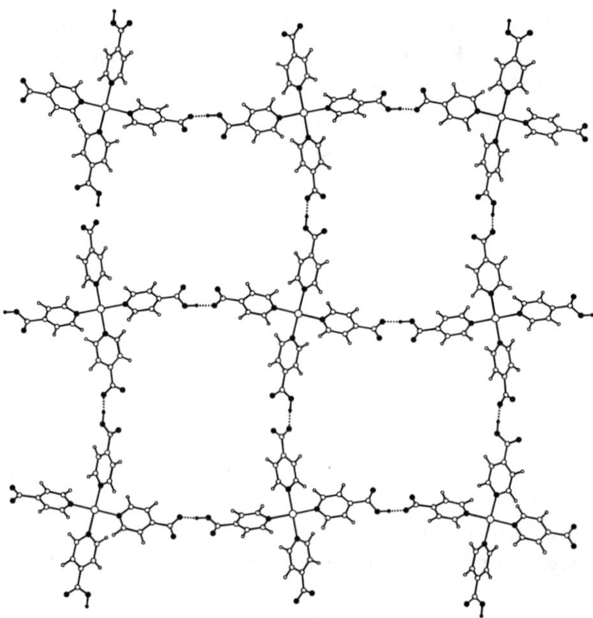

Figure 18 Section of one of the O–H···O^- hydrogen-bonded square grids found in $[Pt(L)_2(HL)_2] \cdot 2H_2O$(L = isonicotinate) (cf. **IX**) [27c]. Water molecules occupy the channels along the crystallographic *c*-axis (not shown). Pt···Pt separations are 16.2 Å. Oxygen, nitrogen and key hydrogen atoms are shaded.

topology (and also threefold interpenetrated) is found in the structure of *trans*-$[Ni(L)_4(OH_2)_2]Br_2 \cdot 2L$ (L = pyridine-4-aldoxime) [27c]. However, the use of O–H···O hydrogen bonds to link oxime groups of neighbouring building blocks rather than the expected O–H···N bonded $R_2^2(6)$ synthon leads to considerable distortion of the square grid (Figure 19). As in the previously described examples that involve amide groups in Section 4.2.1, the use of the square-planar building block $[M(L)_4]^{2+}$ (M = Ni, Pt; L = isonicotinamide) has led to a variety of different hydrogen-bonded architectures. *trans*-$[Ni(L)_4(OH_2)_2]ClO_4 \cdot 2H_2O$ leads to cationic layers in which sets of four amide groups from neighbouring units interact in an $R_4^4(16)$ hydrogen-bonded ring [27c]. $[Pt(L)_4](PF_6)_2$ [27a] and $[Pt(L)_4]Cl_2 \cdot 4L$ [27c] form complicated hydrogen-bonded networks. Both contain one-dimensional networks propagated via $R_2^2(8)$ synthons between amide groups which *trans* ligands interact (*cf.* **XIII**). In $[Pt(L)_4](PF_6)_2$ the remaining two amide groups provide cross-linking via $R_4^4(16)$ hydrogen-bonded rings involving amide groups from four different cations. In $[Pt(L)_4]Cl_2 \cdot 4L$, two sets of one-dimensional chains involving $R_2^2(8)$ amide–amide synthons run in roughly orthogonal directions and are linked by an extensive network of hydrogen bonds involving the remaining amide groups of the coordinated ligands, the isonicotinamide guest molecules and

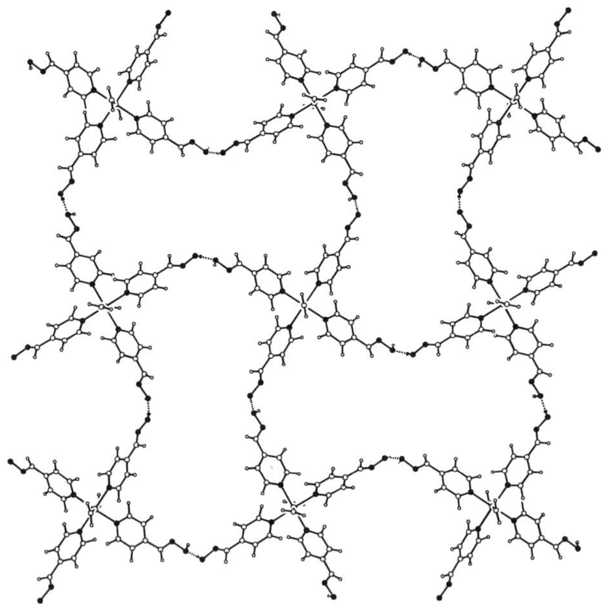

Figure 19 Distorted two-dimensional (4,4) network in *trans*-$[Ni(L)_4(OH_2)_2]Br_2 \cdot 2L$ (L = pyridine-4-aldoxime), propagated by oxime–oxime O–H···O hydrogen bonds (cf. **IX**) [27c]. Not shown are bromide anions and pyridine-4-aldoxime guest molecules, which are hydrogen-bonded to the 2D framework. Ni···Ni separations are 16.3 Å. Oxygen, nitrogen and key hydrogen atoms are shaded.

the chloride anions. Finally, $[Pt(L)_4]Cl_2 \cdot 7H_2O$, despite its chemical similarity to $[Pt(L)_4]Cl_2 \cdot 4L$, forms an entirely different structure in which the cations are not directly linked, but rather form hydrogen-bonded bridges via chloride ions and a chain of water molecules [27a].

4.2.1.4 *Tetrahedral* ML_4 *building blocks*

Munakata *et al.* used tetrahedral Cu(I) centres coordinated by four 3-cyano-6-methylpyrid-2(1*H*)-one ligands (L), CuL_4^+, as building blocks for 3D hydrogen-bonded networks [58]. When combined with PF_6^- or $CF_3SO_3^-$ anions, the amide groups link to ligands of neighbouring cations via the $R_2^2(8)$ synthon, such that each Cu(I) centre serves as a tetrahedral node in a fourfold interpenetrated diamondoid network (Figure 20). When anions BF_4^- or ClO_4^- are used instead, an alternative 3D network is obtained in which the anions occupy infinite channels parallel to one crystallographic axis. In this network, each CuL_4^+ building block is linked to eight others via catemer-type amide–amide hydrogen bonds.

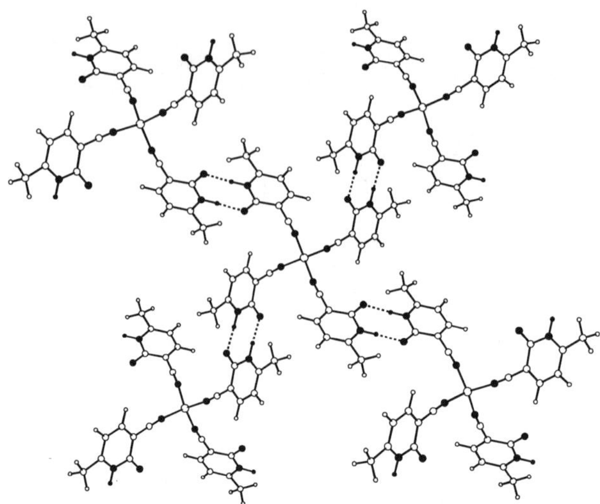

Figure 20 Part of the four-fold interpenetrated 3D diamondoid network formed by $[Cu(L)_4]^+$ [L = 3-cyano-6-methylpyrid-2(1*H*)-one] as its PF_6^- salt [58]. Hydrogen-bonded links are provided by $R_2^2(8)$ synthons involving all amide groups. Oxygen, nitrogen and key hydrogen atoms are shaded.

4.2.1.5 Heteroleptic ML_nX_m *building blocks* ($n = 2$–3)

Puddephatt and co-workers have prepared an extensive series of building blocks of the type **XII** and **XIII** in which square-planar Pt(II) or Pd(II) centres are coordinated by two carboxyl- or amide-bearing ligands of the type discussed in Section 4.2.1, namely nicotinic acid, isonicotinic acid, nicotinamide, isonicotinamide [59] and related ligands [59a,60]. The remaining ligands are either chloride (type **XIII**) or phosphine, diphosphine or 2,2′-bipyridine (type **XII**). A convenient means of preparing octahedral Pt(IV) building blocks with one set of *trans* ligands each bearing a carboxyl group has also been reported [44b]. The hydrogen-bonded networks formed by these assemblies have been investigated. In each case a 1D assembly is formed where direct hydrogen bonding takes place between the carboxyl or amide groups of neighbouring molecules (Figure 21). Some of the systems described also illustrate the potential problems that can arise owing to unintended hydrogen bond formation to solvent molecules or other ligands (e.g. chloride). However, such hydrogen bonding to other ligands can also be harnessed in the overall network formation. This is seen in the hydrogen-bonded sheet structure of *trans*-$[PtCl_2(L)_2] \cdot 2CH_3OH$ (L = pyridine-3,5-dicarboxylic acid) [60] and the related example *cis*-$[PtCl_2(L)_2] \cdot 2H_2O$ (L = nicotinic acid) from Mareque Rivas and Brammer [54d,61], which exhibits a twofold interpenetrated hydrogen-bonded ladder structure wherein water-bridged carboxyl dimers form the 'rungs' and

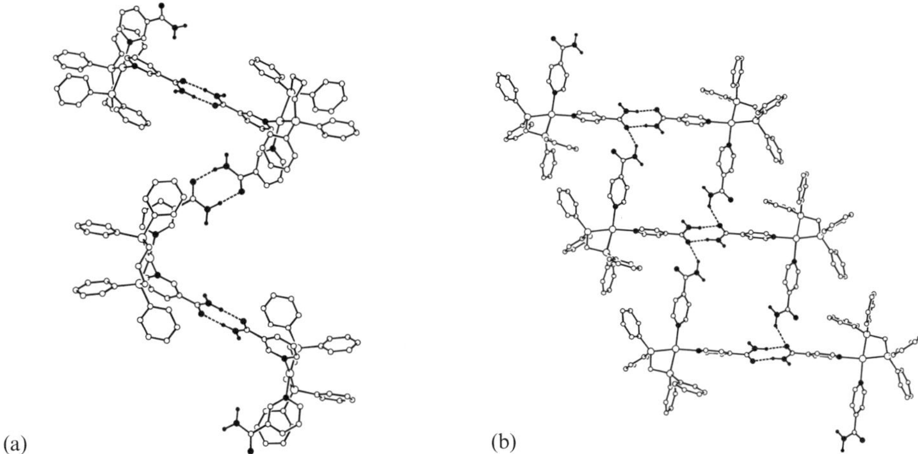

Figure 21 (a) Hydrogen-bonded tape structure of cation in $[Pd(dppp)(L)_2](CF_3SO_3)_2 \cdot 2(CH_3)_2CO$ (dppp = diphenylphosphinopropane; L = nicotinamide) [59b] showing propagation via amide–amide $R_2^2(8)$ rings. (b) Hydrogen-bonded ladder structure of cation in $[Pd(dppm)(L)_2](CF_3SO_3)_2 \cdot (CH_3)_2CO$ (dppm = diphenylphosphinomethane; L = isonicotinamide) using both ring and catemer amide–amide hydrogen bonds synthons [59b]. Oxygen, nitrogen and key hydrogen atoms are shaded.

bifurcated $O–H \cdots Cl_2Pt$ interactions, reminiscent of those described in Section 4.1.2, provide the 'uprights' (Figure 22). Recently, Stang and co-workers described the 1D networks formed by the cations of $[Pt(PPh_3)_2(L)_2](NO_3)_2 \cdot H_2O$ (type **XII**) and $[Rh(\eta^5\text{-}C_5Me_5)(L)_3](CF_3SO_3)_2 \cdot 0.5(CH_3)_2CO$ (type **XV**), L = isonicotinamide, in which all amide groups are involved in amide–amide hydrogen bonding [62]. The platinum system forms zig-zag chains analogous to those found by Puddephatt and co-workers when using chelating diphopshines (e.g. Figure 21a), whereas the rhodium system forms novel doubly stranded interwoven chains that involve both $R_2^2(8)$ and catemer linkages.

4.2.2 Chelating ligands

Chelating ligands impart greater stability to metal–ligand binding, although in the context of crystal engineering there is not necessarily a direct correspondence between the metal coordination geometry and the nodal properties of the metal site in the resulting hydrogen-bonded network (Figure 23). Thus, linear two-connected nodes can be formed from square-planar (**XXII**), tetrahedral (**XXIV**) or octahedral (**XXVIII** or **XXX**) metal centres. Hydrogen-bonded networks built up from hydrogen bonds between chelating ligands on adjacent building blocks have been less extensively explored than those using monodentate ligands. This is no doubt in part due to the greater conventional synthetic endeavour required

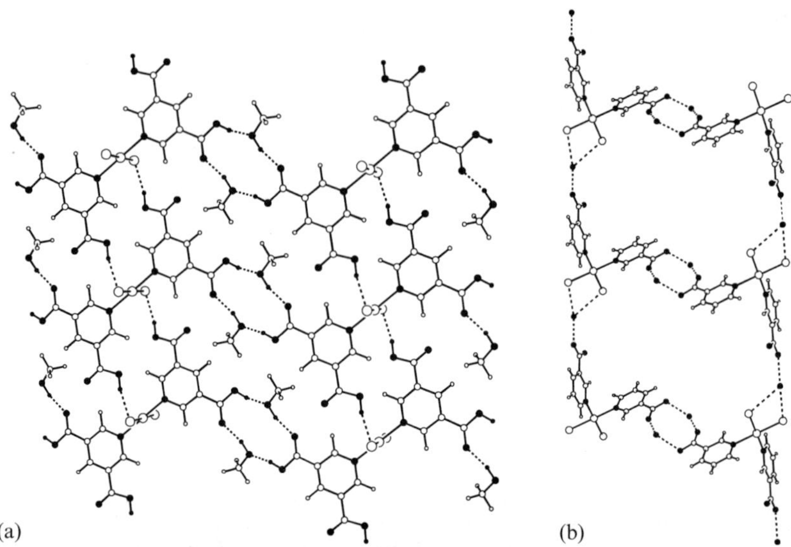

Figure 22 (a) Stepped 2D hydrogen-bonded sheet in *trans*-$[PtCl_2(L)_2] \cdot 2CH_3OH$ (L = pyridine-3,5-dicarboxylic acid), linked in one direction via $R_4^4(12)$ methanol-bridged carboxylic acid dimers and in the orthogonal direction by a chain involving O–H···Cl–Pt–Cl···H–O hydrogen bonds [60]. Oxygen, nitrogen and key hydrogen atoms are shaded. (b) Ladder structure in *cis*-$[PtCl_2(L)_2] \cdot 2H_2O$ (L = nicotinic acid) connected via water bridges between carboxyl groups ('rungs') and between carboxyl group and $PtCl_2$ acceptor ('uprights') [54d,61]. Oxygen and nitrogen atoms are shaded; amide hydrogen atoms are not shown.

to prepare suitable chelating ligands, while most of the monodentate ligands in common use (Section 4.2.1) are readily available from commercial sources.

In some of the earliest examples of combining coordination chemistry with hydrogen bonds in crystal engineering, Mingos and co-workers designed a number of chelating ligands capable of coordinating to square-planar metal centres [63]. The periphery of these ligands was functionalized so as to be able to form triple hydrogen bonds, the inspiration for the work clearly being the robust hydrogen-bonded linkages found between base pairs in DNA. This strategy permitted the construction of strongly hydrogen-bonded tapes (cf. **XXII** in which dashed lines would represent triple hydrogen bond links). Tapes were also cross-linked into sheets by additional hydrogen bonds.

Tadokoro and Nakasuji used the monoanionic 2,2′-biimidazoleate ($HBim^-$) ligand in conjunction with divalent octahedral metal centres, M(II), to prepare 2D honeycomb networks based upon $M(Hbim)_3^-$ building blocks linked by N–H···N hydrogen bonds in an $R_2^2(10)$ synthon (Figure 24) [64]. The overall crystal structure depends on the counter-cation used, and can be a layer structure or an interpenetrated network. The interligand N–H···N hydrogen bonds can

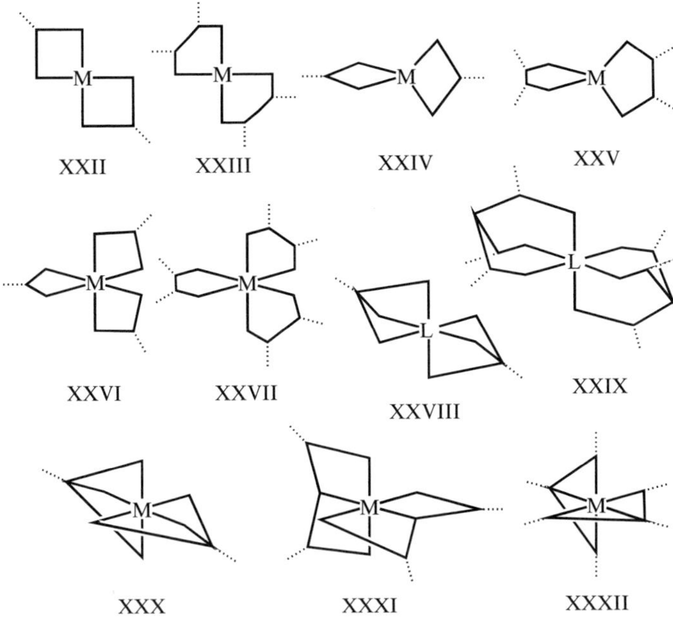

Figure 23 Schematic representation of coordination compounds with chelating ligands bearing hydrogen bonding groups. Arrangements **XXII–XXVII** show bidentate ligands with one or two hydrogen bonding sites per ligand; arrangements **XXVII–XXXII** show *facial* and *meridial* tridentate ligands with 1–3 hydrogen bonding sites per ligand.

also be interrupted by solvent molecules such as water. Importantly, the networks can be constructed through self-assembly directly from a solution of the metal ion, ligand and a suitable counter-cation. The $M(Hbim)_3^-$ building blocks [cf. **XXVI**, with the dashed line representing the $R_2^2(10)$ N–H$\cdots$N linkage] are chiral, although the resulting honeycomb network is not, since alternating *R* and *S* enantiomers are employed in its construction. Lehn and co-workers prepared nanoporous hydrogen-bonded networks based on coordination to octahedral metal centres of three chelating ligands obtained by derivatization of 1,10-phenanthroline with two urea units [65]. The building blocks resemble **XXVII**, but with hydrogen bonds to six neighbouring units in directions approximately orthogonal to phenanthroline planes. Functionalized terpyridine (terpy-NH) ligands have also been used to prepare 2D hydrogen-bonded grids from $[M(terpy\text{-}NH)_2]^{2+}$ (M = Co, Zn) building blocks similar to **XXXI** [66] (Figure 25). However, the use of different counterions, PF_6^-, BF_4^- or $CF_3SO_3^-$, results in changes from 2D interwoven grids to a 1D tape, owing to different numbers of potential hydrogen bonding sites being used in formation of the crystalline assembly.

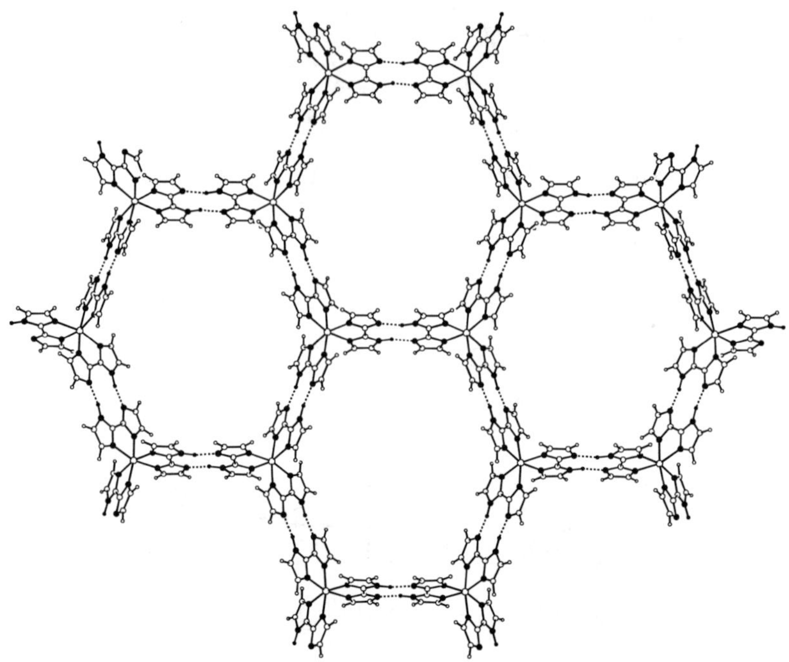

Figure 24 2D honeycomb network constructed from $Ni(Hbim)_3^-$ building blocks ($HBim^- = 2,2'$-biimidazoleate) linked by N–H···N hydrogen bonds in $R_2^2(10)$ synthons as seen in crystal structure of [K(dibenzo-18-crown-6)][$Ni(Hbim)_3$] · $3CH_3OH \cdot 2H_2O$ [64]. Nitrogen and key hydrogen atoms are shaded.

4.2.3 Metalloporphyrins

Metalloporphyrins bearing peripheral functional groups capable of hydrogen bonding have also been used as building blocks for designing inorganic crystals. In particular, the use of suitable templates can result in robust networks with large pores for potential use as molecular sieves, as illustrated by the work of Goldberg and co-workers [67] (Figure 26). The hydrogen-bonded square grid network is related to analogous coordination networks made using tetra-4-pyridylporphyrin in the same manner that assemblies **III** and **V** are related [67a]. The groups of Suslick and Aoyama have prepared metalloporphyrin-based hydrogen-bonded networks using resorcinol-substituted porphyrins that stack into infinite columns though O–H···O hydrogen bonds between hydroxyl groups [68].

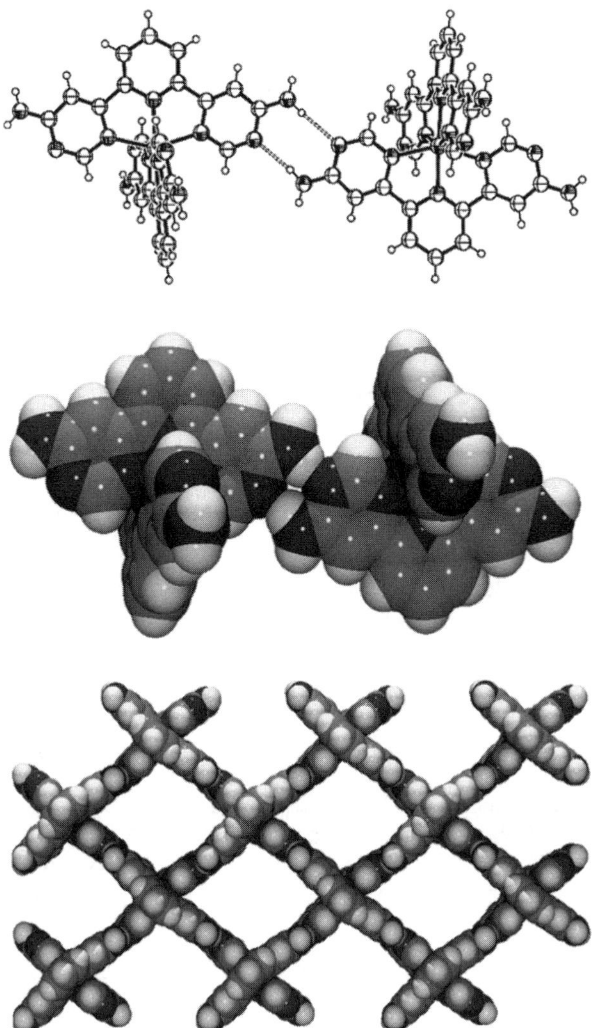

Figure 25 Structure of $[Co(terpy\text{-}NH)_2](PF_6)_2$ showing assembly of cationic cobalt complexes by self-complimentary N–H ··· N hydrogen bonds (top, middle) to give a 2D grid (bottom). Reproduced from V. Ziener, E. Brevning, J.-M. Lehn, E. Wegelius, K. Rissanen, G. Baum, D. Fenske and G. Vaughn, *Chem. Eur. J.*, **6**, 4132–9 (2000) with permission of Wiley-VCH.

4.2.4 Metal clusters and polynuclear building blocks

The building blocks considered to date in Section 4.2 have all been mononuclear, i.e. containing a single metal centre. However, dinuclear and polynuclear building blocks are also amenable to this type of inorganic crystal engineering. Thus,

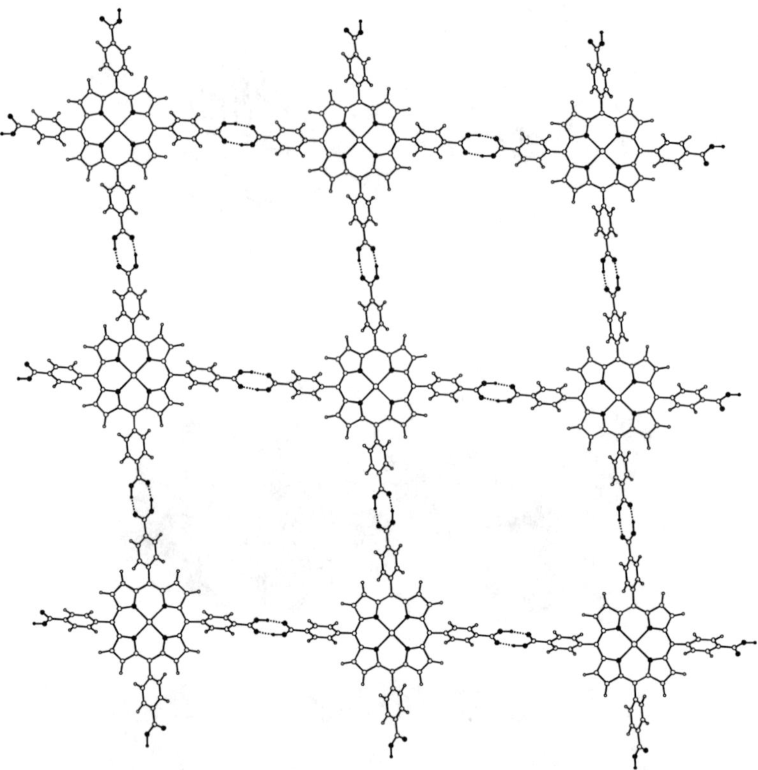

Figure 26 Hydrogen-bonded grid in [aqua(5,10,15,20-tetrakis(4-carboxyphenyl)porphyrinato)zinc (II)]·$4C_6H_5NO_2$ [67b]. Zn···Zn separations are 22.2 Å. Oxygen, nitrogen and key hydrogen atoms are shaded; coordinated water and disordered nitrobenzene solvate not shown.

Puddephatt and co-workers [44b] and Rendina and co-workers [69] each prepared dinuclear systems in which square-planar Pt(II) centres are bridged by a 4,4′-bipyridine ligand and each metal then carries a ligand with a functional group capable of self-recognition via hydrogen bonding, namely carboxyl and amide groups, respectively. In the solid state these molecular species assemble into 1D hydrogen-bonded polymers, though Rendina and co-workers reported the formation of cyclic assemblies in solution [69].

Recently, Puddephatt and co-workers have also reported Pt_3 metallatriangles in which the metal vertices are linked by the unsymmetrical bipyridine ligand 4-$NC_5H_4C(=O)NH$-4-C_5H_4N. These ligands then permit stacking of the triangles by N–H···O hydrogen bonds. However, this stacking only extends to dimerization in the examples presented and additional hydrogen bonding takes place with anions and solvent molecules.

In Section 4.1.1, Zaworotko and co-workers' use of the tetrahedral metal cluster $[Mn_4(\mu_3\text{-}OH)_4(CO)_{12}]$ to form a diamondoid network via hydrogen bonded bridges to 4,4′-bipyridine spacer was highlighted. More recently, Lehn's group has prepared a ligand (L) capable of self-assembly with suitable transition metal ions (M^{2+}) into a tetranuclear $[M_4L_4]^{8+}[2 \times 2]$ grid that is designed to undergo further self-assembly via hydrogen bonds to give a 2D 'grid of grids' (Figure 27) [70]. However, in the only example crystallized to date, 1D hydrogen-bonded tapes were formed, with π-stacking important in cohesion between tapes (Figure 28). Remarkably, in the system crystallized, rather than the intended BF_4^- anions the counterions found were chloride left over from an early part of the ligand synthesis and SiF_6^{2-}, presumably derived from adventitious reaction of free fluoride with the silica vessel in which the crystallization was conducted.

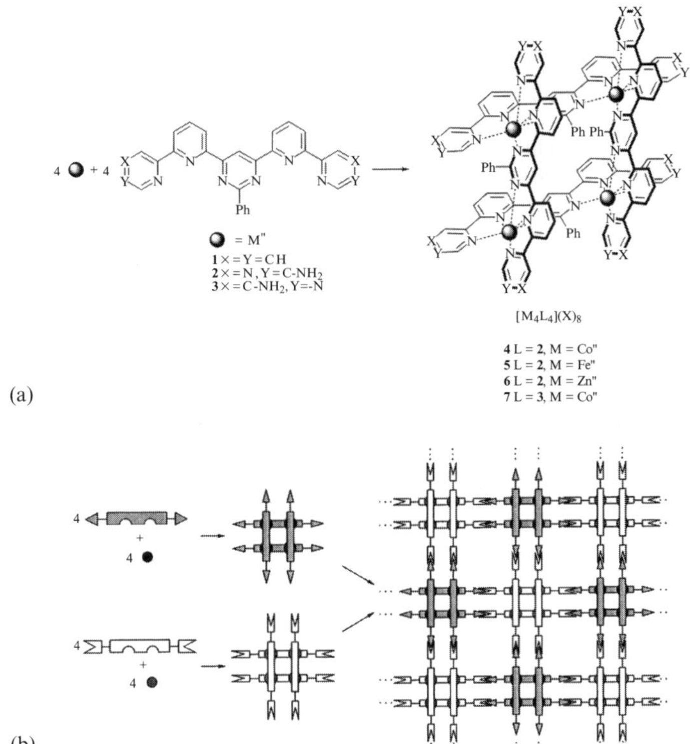

Figure 27 (a) Ligand designed to bind two adjacent transition metal centres via *meridial* tridentate coordination and illustration of its self-assembly with such metals ions to give $[M_4L_4]^{8+}$ $[2 \times 2]$ coordination grids. (b) Anticipated further self-assembly of the $[2 \times 2]$ coordination grids via hydrogen bonds to give a 2D 'grid of grids'. Reproduced from E. Brevning, V. Ziener, J.-M. Lehn, E. Wegelivs and K. Rissanen, *Eur. J. Inorg. Chem.*, 1515–21 (2001) with permission of Wiley-VCH.

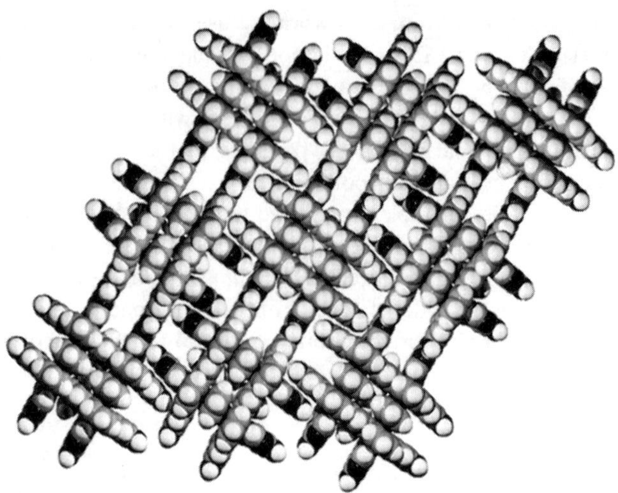

Figure 28 Crystal structure of $[Co_4L_4][Cl]_2[SiF_6]_3 \cdot 2CH_3CN \cdot 50H_2O$ showing self-assembly of the $[Co_4L_4]^{8+}$ [2 × 2] coordination grids from Figure 27a. In this case, the [2 × 2] coordination grids assemble via hydrogen bonds in only one direction; in the orthogonal direction π–π edge–face and face–face interactions are the principal cohesive forces. Reproduced from with permission of Wiley-VCH.

In exploring the chemistry of octahedral $[Mo_6Cl_8X_6]^{2-}$ clusters, Shriver and co-workers prepared the cluster in which $X^- = p\text{-}OC_6H_4CONH_2^-$, each carboxamide-bearing ligand being terminally bound to one of the metal centres [71]. The two crystal structures reported as different solvates, $[(crypt)Na]_2[Mo_6Cl_8X_6] \cdot 2DMF \cdot 6H_2O \cdot 4CH_3OH$ and $[(crypt)Na]_2[Mo_6Cl_8X_6] \cdot 6DMF$, adopt lamella structures, although with different cluster-to-cluster separations within and between layers. Within the layers, four of the six carboxamide ligands are involved in $R_2^2(8)$ amide–amide hydrogen-bonded links between neighbouring clusters. In the first of the two structures the remaining carboxamides form hydrogen-bonded links between layers to give a 3D hydrogen-bonded network (Figure 29), whereas in the second structure such cross-linking involving the other carboxamides does not occur.

4.3 Networks Formed by Hydrogen-bonded Links via Anion, Cation or Neutral Molecule Bridges Between Coordination Compound Building Blocks

A major problem in inorganic crystal engineering when using hydrogen-bonded networks is that counterions, especially anions that are in common use, are often capable of engaging in hydrogen bonding. Thus, intended hydrogen-bonded networks can be disrupted by interactions between the designed building blocks and counterions. Even when neutral building blocks are used, neutral molecules, often solvents [72], can be incorporated into the final crystal structure and modify or

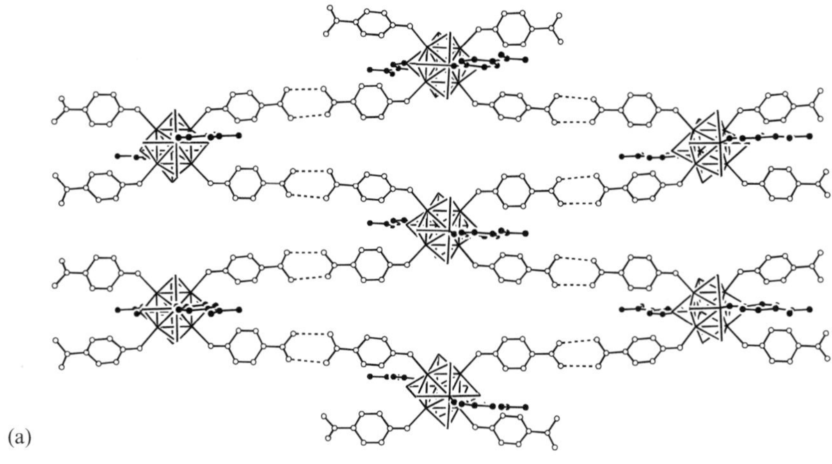

(a)

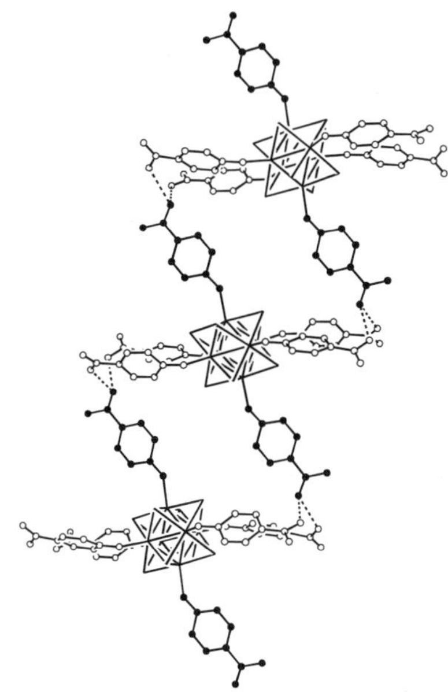

(b)

Figure 29 Views of the anionic hydrogen-bonded network found in the crystal structure of $[(crypt)Na]_2[Mo_6Cl_8X_6] \cdot 2DMF \cdot 6H_2O \cdot 4CH_3OH$ ($X = 4\text{-}OC_6H_4CONH_2$; crypt = 4,7,13,16,21,24-hexaoxo-1,10-diazabicyclo[8.8.8]hexacosane; DMF = dimethylformamide) [71]. Cation and solvent molecules are not shown; $[Mo_6Cl_8]^{4+}$ cluster core shown in wireframe style. (a) Lamella structure constructed via $R_2^2(8)$ amide–amide hydrogen bonding involving four of the six amide groups (open circles). (b) Hydrogen bonding involving two remaining (shaded) amide groups that provides the links between the layers. Hydrogen atoms are not shown.

disrupt the intended hydrogen-bonded network. This section deals not with such cases, but rather cases in which a counterion or neutral molecule guest has been used effectively, or at least intentionally, to provide a hydrogen-bonded link between the functional groups on the periphery of the building blocks.

4.3.1 Anionic bridges

Since the most common combination in coordination chemistry, that of metal cation and neutral ligands, gives rise to cationic coordination complexes, it is no surprise that the choice of anion should be of utmost important in designing hydrogen-bonded networks based upon coordination compounds as building blocks. This can be seen in Section 4.2, where anions were often selected to minimize the chance of their forming (accepting) hydrogen bonds and thus maximize the chance of direct hydrogen bonding between the cationic building blocks. However, a few groups have adopted a different approach by selecting anions that can be incorporated deliberately into the hydrogen-bonded network, forming a bridge between the metal-containing building blocks. Burrows and co-workers employed metal complexes containing two *N,S*-chelating thiosemicarbazide ligands that serve as multiple hydrogen bond donors. When combined with dicarboxylate anions such as *trans*-fumarate or terephthalate, each thiosemicarbazide ligand forms a double hydrogen bond with one carboxylate group of an anion. The anion thus forms a bridge between the cationic hydrogen bond donor building blocks (Figure 30) [73]. Hubberstey and co-workers [74] designed a family of tetradentate C_2^- or C_3^- linked bis(amidino-*O*-alkylurea) ligands which provide eight peripheral N–H hydrogen bond donor sites. These have been linked into hydrogen-bonded tape structures using four of the eight donor sites to 'recognize' a variety of anions in a common manner (Figure 31). Further hydrogen bonds involving the other donor sites, and anion as well as solvent water acceptors, lead to overall hydrogen-bonded aggregation in two dimensions. Jones and co-workers have reported a number of aminogold(I) compounds in which hydrogen bonding and aurophilic Au(I)$\cdots$Au(I) interactions are important in the solid state [75]. A particularly compelling example is the structure of bis(pyrrolidine)gold(I) chloride as its dichloromethane solvate (1.33CH_2Cl_2). The $Au(NC_4H_5)_2^+$ cations form hydrogen-bonded chains via Cl^- anion bridges. The chains, arranged in parallel fashion form layers, each layer with its chain direction rotated 90° with respect to neighbouring ones. Au$\cdots$Au interactions then provide a direct link between cations in adjacent layers leading to a three-dimensional network [75a]. An interesting system has also been reported by Chen *et al.* in which *trans*-$[Ni(OH_2)(SCN)_4]^{2-}$ metallate anions are linked via nitrate anions and 4,4′-bipyridinium cations to give a (6,3)-type honeycomb network sustained entirely by hydrogen bonds [$R_{10}^{12}(64)$ rings] [30e] (Figure 32). Nitrate anions serve as the three connected nodes by accepting three hydrogen bonds. Each six-node ring

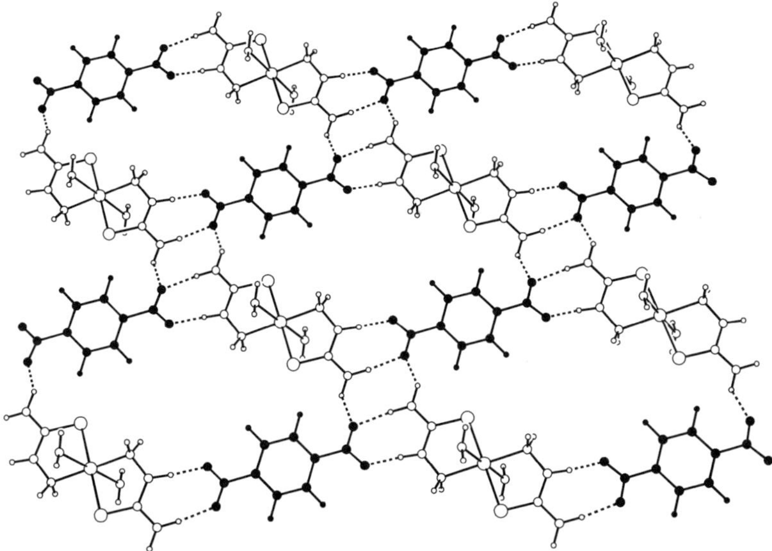

Figure 30 Hydrogen-bonded sheet in crystal structure of $[Zn(SC(NH_2)NHNH_2)_2(OH_2)_2][1,4\text{-}O_2CC_6H_4CO_2] \cdot 2H_2O$ [73a]. Cationic metal complexes $[Zn(SC(NH_2)NHNH_2)_2(OH_2)_2]^{2+}$ (open circles) serve as hydrogen bond donors to anionic terephthalate bridges (shaded). Layers are linked via N–H···O and O–H···O hydrogen bonds to solvent water molecules (not shown).

(internal dimensions 20 × 24 Å, also consists of four 4,4′-bipyridinium cations, each of which donate two hydrogen bonds, and two metallate anions. Layers of the honeycomb networks are stacked in the third dimension via dimeric $R_2^2(12)$ O–H···S hydrogen bonding interactions involving the remaining hydrogen atoms of the co-ordinated water molecules and the sulfur atoms of two of the four anions. Filling of the internal channels arises from twofold interpenetration of the network.

4.3.2 Cationic bridges

Cationic hydrogen-bonded bridges can be used to link metallate building blocks. An earlier example of such an approach involving the ligand domain was provided in the work of Brammer and Orpen wherein perhalometallates were linked via cationic bridges containing ammonium or pyridinium hydrogen bond donors (Section 4.1.2). A splendid example of the use of cationic bridges can be found in the work of MacDonald *et al.* [76], who reported an extensive study of a series of essentially isostructural layered hydrogen-bonded salts. The metallate anions contain divalent first row transition metals that are octahedrally coordinated by two *meridial* tridentate 2,6-pyridinedicarboxylate ligands. Salts formed using imidazolium cations lead to grid-like hydrogen-bonded layers involving

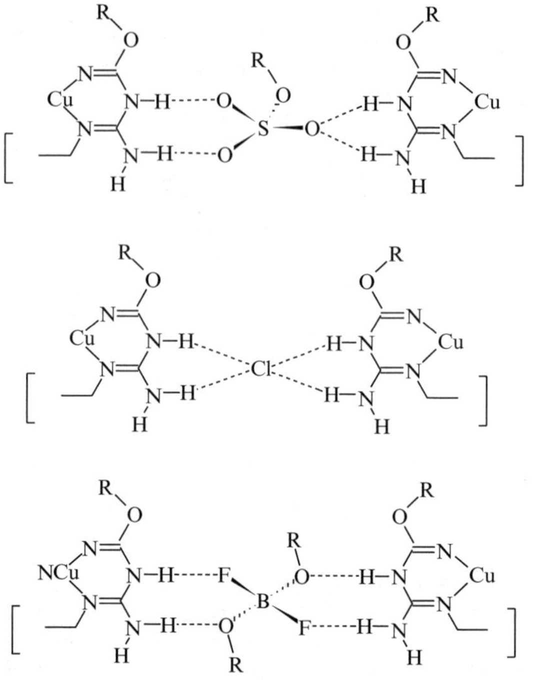

Figure 31 Network propagation via anion recognition by hydrogen bonding. Reprinted from A. J. Blake, P. Hubberstey, V. Suksanypanya and C. L. Wilson, *J. Chem. Soc., Dalton Trans.*, 3873–80 (2000) Reproduced by permission of the Royal Society of Chemistry.

change-assisted $N^{+}{-}H \cdots O^{-}$ hydrogen bonds, with a hydrogen bonding pattern at the metallate building blocks resembling that of **XXXI**. The reproducibility of this network arrangement and its metrics led the authors to investigate the co-crystallization of mixed metal systems and, using epitaxial crystal growth, the preparation of mixed metal composite layered materials (see Section 7).

4.3.3 Neutral molecule bridges

Water and short-chain alcohols are commonly used solvents in the synthesis of crystals containing hydrogen bonding building blocks owing to the inherent polar nature of such molecular units. Thus, given the capability of these solvent molecules both to accept and to donate hydrogen bonds, their (typically inadvertent) incorporation into the final hydrogen bonded network should be no surprise. Indeed, a variety of other solvent molecules have been observed to be incorporated into crystal structures through strong or weak hydrogen bonds [72], but the deliberate use of such solvent molecule incorporation in crystal design is a difficult

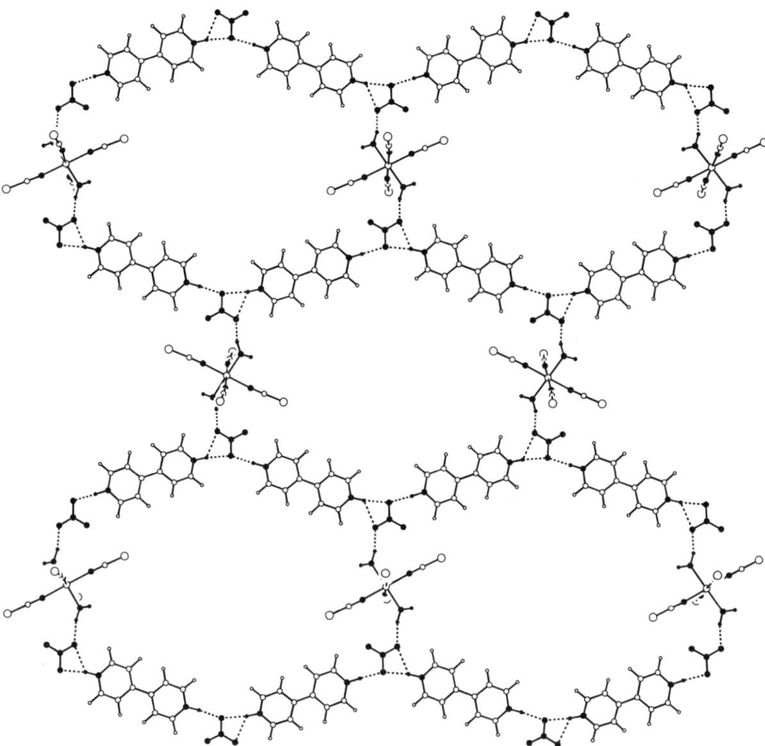

Figure 32 Hydrogen-bonded honeycomb network found in crystal structure of $[4,4'\text{-bipy}]_2$-trans-$[Ni(OH_2)(SCN)_4][NO_3]_2$ [30e]. Oxygen, nitrogen and key hydrogen atoms are shaded.

proposition. However, other neutral molecule bridges have been used effectively in hydrogen-bonded networks, the most common examples involving di- and polyamines and -pyridyls. These have been widely exploited in organic crystal synthesis through co-crystallization with molecules presenting O–H donor groups [77]. In analogous fashion, 4,4′-bipyridine has been used as a neutral molecule bridge between metal-containing building blocks, particularly those which provide hydrogen bond donors that lie in the *Ligand Domain*, for example in hydroxo [29] and aqua [30] ligands (Sections 4.1.1 and 4.4.1). A particularly nice example is provided by Chen and co-workers [30a], in which the compound $[Mn(4,4'\text{-bipy})(OH_2)_4]^{2+}$ is cross-linked in two dimensions through use of *all eight* water hydrogen atoms in forming O–H $\cdots$ N hydrogen bonds to additional 4,4′-bipy spacers.

4.4 Combining Hydrogen Bonds with Coordination Networks

Section 4 up to this point has focused upon linking discrete mono- , di- or polynuclear metal complexes via hydrogen bonds. This section will deal with metal centres that are already linked into 1D or 2D networks by coordination bonds, and consider their further cross-linking by hydrogen bonds resulting in networks of higher dimensionality.

4.4.1 1D coordination polymers with hydrogen-bonded cross-links

There are a number of examples of 1D coordination polymers propagated by 4,4′-bipyridyl (4,4′-bipy) links in which further cross-linking has been accomplished by $M(OH_2)\cdots(4,4'\text{-bipy})\cdots(H_2O)M$ hydrogen-bonded links. Such systems typically use octahedrally coordinated metal centres, in which *trans* coordination of 4,4′-bipy leads to the 1D coordination polymer. The additional coordination sites are then occupied by either two or four water molecules providing O–H donor sites suitably positioned to form hydrogen-bonded links to additional 4,4′-bipy molecules in directions orthogonal to the polymer backbone. Such systems are exemplified by the systems $[M(NCE)_2(OH_2)_2(4,4'\text{-bipy})]\cdot(4,4'\text{-bipy})$ (M = Co, E = S [30c]; M = Fe, E = S or Se [78]). Ciani and co-workers have also explored this theme, illustrating that both single and double $M(OH_2)\cdots(4,4'\text{-bipy})\cdots(H_2O)M$ hydrogen-bonded links between metal centres can be achieved in related systems [30d]. In one system, $[Fe(OH_2)_3(ClO_4)(4,4'\text{-bipy})]ClO_4\cdot 4,4'\text{-bipy}$, alternate parallel $\{Fe(4,4'\text{-bipy})\}_n^{2n+}$ coordination polymers are linked by single then double hydrogen-bonded 4,4′-bipy bridges.

Direct hydrogen bonding between the ligands of metals contained within adjacent coordination polymers appears to be less common. The coordination polymer $\{[Cd(\mu\text{-}3,3'\text{-pytz})(NO_3)_2(MeOH)_2]\}_n$ [3,3′-pytz = 3,6-bis(pyrid-3-yl)-1,2,4,5-tetrazine] exhibits interchain O–H···O hydrogen bonding between the coordinated methanol and nitrato ligands [31c]. In recent reports, the groups of Aakeröy [79] and Nishikiori [80] have cross-linked neutral coordination polymers formed via small bridging anionic ligands by using pyridine-based ligands (L) that carry functional groups (carboxyl, amide or oxime) capable of self-recognition through hydrogen bonds. These result in 2D sheet structures. The copper(I) halide coordination polymers $[Cu(\mu_3\text{-}X)L]_n$ (X = Cl, I) have a repeat distance along the polymer backbone of ca 3.7–4.1 Å that is suitable for π–π stacking between pyridine rings (Figure 33). The Ni(II) thiocyanate coordination polymers, $[Ni(\mu\text{-}SCN)_2L_2]$, by virtue of the longer bridging anions, give rise to a larger spacing (ca 5.5 Å) between the metal sites to which the pyridine-based hydrogen bond linkers are coordinated. This spacing is small enough to prevent interpenetration of the resulting 2D network, but permits inclusion of large aromatic guest molecules by expansion and compression of alternate cavities due to the flexibility of the hydrogen-bonded linkage (Figure 34). The authors also noted that isostructural host can be prepared using Cu(II) or Cd(II) in place of Ni(II). Zubieta and co-workers used a similar approach to link oxomolybdate polymers using 4-pyridyl-4′-pyridiniumamine (Hbpa) ligands that are coordinated to Mo sites on one

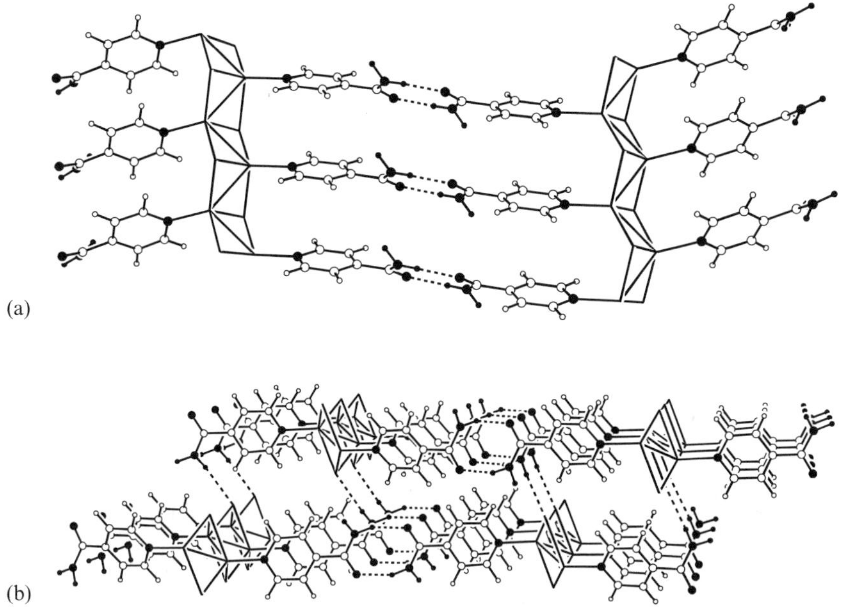

Figure 33 (a) Cross-linking of $\{Cu(L)I\}_n$ coordination polymer via amide–amide hydrogen bonds between isonicotinamide ligands (L) to give a sheet structure [79a]. (b) Linking of 2D sheets in third dimensions via short N–H···I(Cu) hydrogen bonds (H···I = 2.75 Å). Oxygen, nitrogen and key hydrogen atoms are shaded; copper iodide polymer shown in wireframe style.

chain and form bifurcated N–H···O_2Mo hydrogen bonds to a neighbouring chain, leading to 2D sheets of composition $[Mo_4O_{13}(Hbpa)_2]$ [81]. Orpen and co-workers reported the formation of $[\{MCl_4\}_n]^{2n-}$ chains (M = Mn, Cd) supported by hydrogen-bonded cross-linking of H_2(4,4′-bipy) cations that form bifurcated N–H···Cl_2M hydrogen bonds [82].

Recently, Prior and Rosseinsky [83] prepared by hydrothermal means a coordination polymer $[Ni(\mu\text{-}1,4\text{-}H_2NC_6H_4NH_2)(H_2TMA)(OH_2)_2]_n$ [H_3TMA = benzene-1,3,5-tricarboxylic acid (trimesic acid)], which has an all-*trans* geometry at the nickel centres. The carboxyl functional groups and the water molecules engage in extensive hydrogen bonding leading to infinite hydrogen-bonded sheets that lie approximately orthogonal to the direction of propagation of the coordination polymer. A 3D network, $\{(NH_4)[Yb(DMF)_4][Pt(CN)_4]_2\}_n$ (DMF = dimethylformamide), in which coordination bonds give rise to propagation in one direction while hydrogen bonds provide links in the other two has also been reported by Shore and co-workers [84]. The system consists of infinite square columns of composition $\{[Yb(DMF)_4][Pt(CN)_4]_2\}_n{}^{n-}$ in which $Yb(DMF)_4$ and two $[Pt(CN)_4]^{2-}$ moieties provide the corners of the Yb_2Pt_2 squares that are then linked into columns by further bridging $[Pt(CN)_4]^{2-}$ units. These 1D ‘coordination

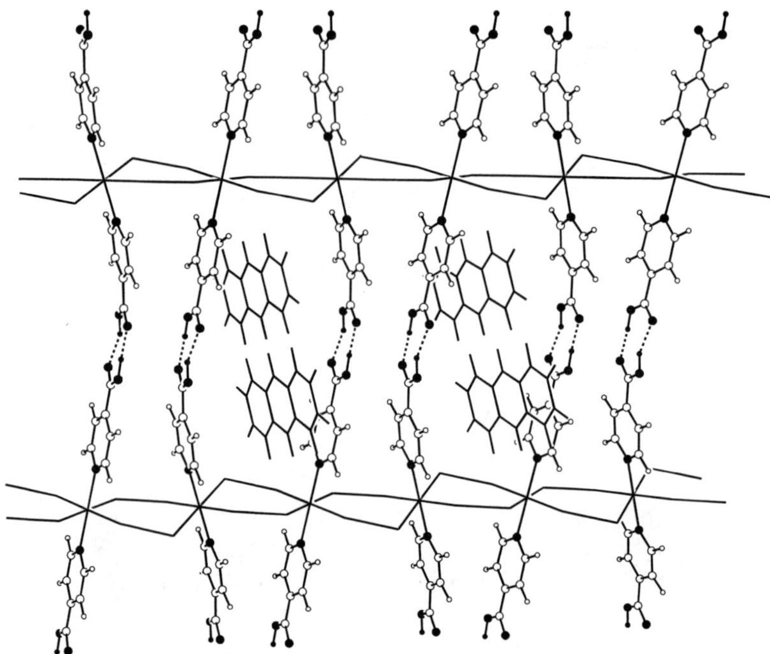

Figure 34 Cross-linking of $\{Ni(\mu\text{-}SCN)_2L_2\}_n$ coordination polymer via carboxyl–carboxyl hydrogen bonds between isonicotinic acid ligands (L) to give a sheet structure which stacks providing channels that can accommodate polycyclic aromatic hydrocarbon guest molecules, here anthracene [80]. Oxygen, nitrogen and key hydrogen atoms are shaded; nickel thiocyanate polymer and anthracene guest molecules shown in wireframe style.

columns' are then linked in the other two directions by N–H···NC(Pt) hydrogen bonds from the ammonium cations, in which *all* ammonium hydrogens and *all* terminal cyano ligands participate.

4.4.2 2D coordination networks with hydrogen-bonded cross-links

There are many examples of two-dimensional coordination networks involving organic bridging ligands, often using pyridyl groups [8c,g]. However, examples of such networks, which are also cross-linked in the third dimension by hydrogen bonds, are surprisingly rare. A particularly nice example is provided by Lauher and co-workers [85]. Pyridyl-derivatized urea and oxalamide ligands when combined with silver(I) salts lead to coordination networks in which each metal ion serves as a four-connected node, with its local geometry being that of a flattened tetrahedron. The resulting 2D grid coordination network is linked in the third dimension by a well-established pattern of hydrogen bonds between the urea or oxalamide functionalities.

While coordination networks are typically considered to involve bridging ligands consisting of at least two atoms, it is useful to include halometallate and oxometallate layer structures as a form of 2D coordination network for the purposes of the present section. Mitzi has described and reviewed the technologically important area of inorganic–organic perovskites [43], with particular attention being paid to layered halometallate perovskites in which the anionic layers are bridged by organic alkylammonium cations, R–NH_3^+ or $(H_3N{-}R{-}NH_3)^{2+}$, that form N–H $\cdots$ X hydrogen bonds (X = halogen), as illustrated schematically in Figure 35. Figure 36 indicates the most commonly observed hydrogen bonding interactions between the ammonium group and the halogens. The applications associated with this class of compounds are discussed in Section 7. Oxometallate structures are also a widely studied class of materials, particularly oxomolybdates (Chapter 6). However, in their 1999 review, Zubieta and co-workers noted [8c] that only one example is known of a layered oxomolybdate in which the layers are linked through hydrogen bonding to organic cations. $H_2(4,4'\text{-bipy})[Mo_7O_{22}] \cdot H_2O$ exhibits an oxide layer structure similar to those of MoO_3, with the interlamellar region containing the $[H_2(4,4'\text{-bipy})]^{2+}$ cations and water molecules, which are hydrogen bonded to each other and through multi-point hydrogen bonding to the oxygen atoms of the anionic layer [86].

4.5 Periphery Domain: Organometallic Building Blocks

This section deals with organic ligands bound to metals via metal–carbon bonds, but bearing substituents capable of forming hydrogen bonds with neighbouring molecules. Most examples involve ligands bonded to metals via their π-electrons, e.g. alkenes, alkynes, arenes and cyclopentadienyls, and this will be the principal focus here. Examination of hydrogen bonding groups will also be restricted to

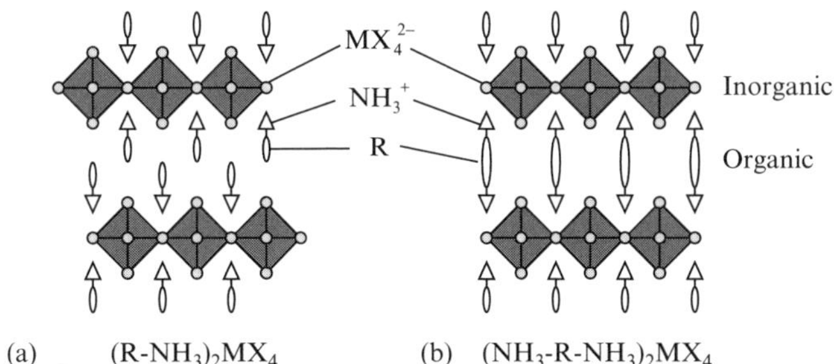

Figure 35 Schematic representation of single layer halometallate perovskites $[\{MX_4\}_n]^{2n-}$ with (a) monoammonium (RNH_3^+) and (b) diammonium ($^+H_3NRNH_3^+$) organic cations as spacer units linked to inorganic layers via N–H $\cdots$ X hydrogen bonds. Reprinted from D. B. Mitzi, *J. Chem. Soc. Dalton Trans.*, 1–12 (2001). Reproduced by permission of the Royal Society of Chemistry.

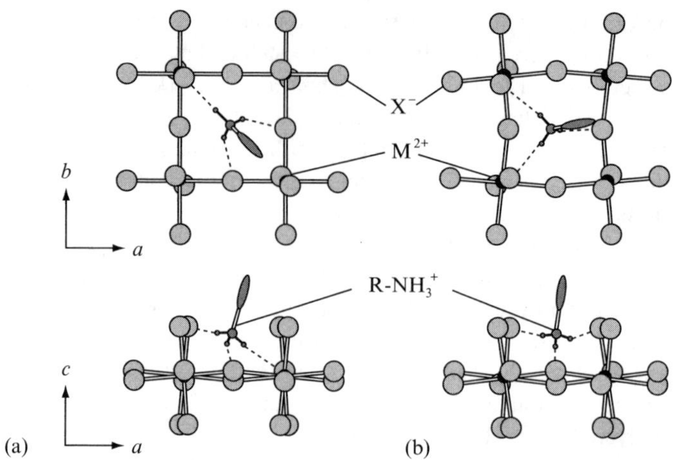

Figure 36 Two hydrogen bonding schemes typically observed in inorganic–organic perovskite structures of empirical formula $(RNH_3)_2[MX_4]$ and $(H_3NRNH_3)[MX_4]$. (a) NH_3^+ group forms two N–H···X hydrogen bonds to bridging halides and one to a terminal halide. (b) NH_3^+ group forms two N–H···X hydrogen bonds to terminal halides and one to a bridging halide. (See discussion in Ref. 43b). Reprinted from D. B. Mitzi, *J. Chem. Soc., Dalton Trans.*, 1–12 (2001). Reproduced by permission of the Royal Society of Chemistry.

alcohols, carboxylic acids and amides. These organometallic building blocks differ from many of their coordination compound analogues involved in hydrogen bonding in the periphery domain in that the metal exerts little if any directional influence upon the hydrogen bond propagation. There may be some electronic influence of the metal upon the hydrogen bonds, as was noted for the case with the corresponding class of coordination compounds. However, in a number of cases the metal (and often the remainder of its ligands) can be thought of as a substituent or appendage on an organic hydrogen-bonded network (see **IV** and **VI** in Figure 12).

4.5.1 Organic ligands with alcohol functional groups

Using the CSD, Braga *et al.* examined hydrogen bonding patterns involving alcohol functional groups present in organometallic compounds [28a]. Typical patterns of hydrogen bonds are analogous to those formed by organic alcohols [87]. Thus, hydrogen-bonded chains of graph set C(2) were reported, as well as cyclic $R_4^4(8)$ tetramers [88] and the two dimensional hydrogen-bonded sheet formed by $[Fe(CO)(\mu\text{-}CO)(\eta^5\text{-}C_5H_4CH_2CH_2OH)]_2$ via self-association of all O–H groups into cyclic $R_6^6(12)$ hexamers [89] (Figure 37). The size of the group (R) to which the alcohol (R–OH) is being bonded is one of the principal determining factors regarding the type of network formed, with smaller groups leading to chain formation and larger ones to ring formation. The participation of ubiquitous carbonyl ligands

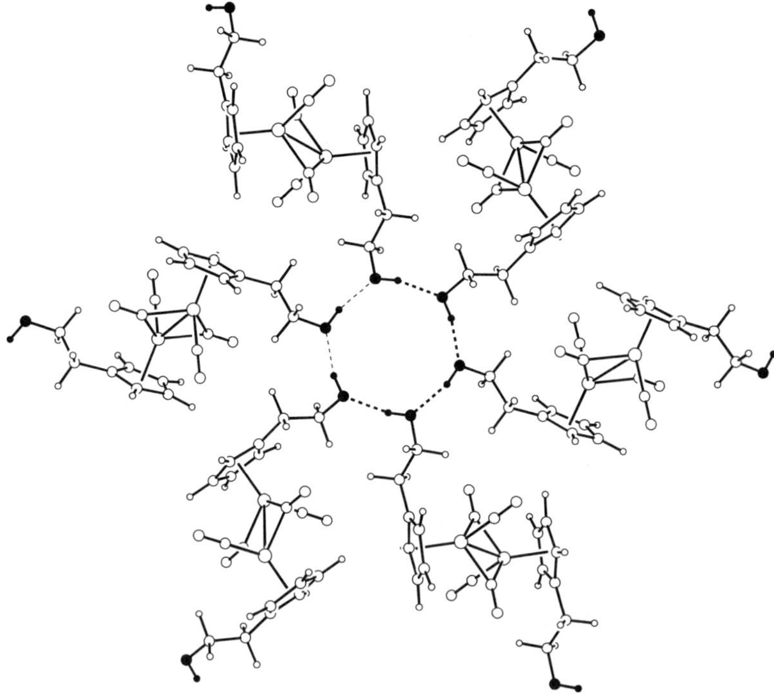

Figure 37 Part of hydrogen-bonded sheet formed by $[Fe_2(\eta^5\text{-}C_5H_4CH_2CH_2OH)_2(CO)_2(\mu\text{-}CO)_2]$ [89] showing $R_6^6(12)$ O–H···O hydrogen-bonded rings. Adapted with permission from Figure 11c in D. Braga, F. Grepioni, P. Sabatino and G. R. Desiraju, *Organo metallics*, **13**, 3532–43 (1994). Copyright 1994 American Chemical Society.

in hydrogen-bonded networks involving O–H···O≡C(M) hydrogen bonds is also documented [28a].

In an effort to use organometallic alcohols in crystal synthesis, Braga and Grepioni in collaboration with Walther and co-workers have reported the synthesis and structural characterization of a series of prop-2-yn-1-ol complexes of Pt and Ni [90]. The hydrogen bonding patterns in the structures of these compounds again resemble the aforementioned chain and ring supramolecular synthons known for alcohols in general. These patterns are then compared with patterns observed for the corresponding organic prop-2-yn-1-ols [90]. A number of 1,1′-ferrocenediols have been prepared and the hydrogen bonding patterns of their crystal structures analysed. These include the structures of $Fe(\eta^5\text{-}C_5H_4CH_2OH)_2$ [91] and $Fe(\eta^5\text{-}C_5Me_4CH_2OH)_2$ [92], which form a 1D C(2) network and a 3D network, respectively, through O–H···O hydrogen bonds. The predominant work involving such diol systems is due to Glidewell and co-workers [93,94], who examined the series $Fe(\eta^5\text{-}C_5H_4C(R)(R')OH)_2$, R = R′ = Ph [93a]; R = R′ = Me [93b]; R = Ph, R′ = Me (racemic) [93c] and R = Me, R′ = H (racemic) [93c], all of which form discrete dimers via $R_4^4(8)$ hydrogen-bonded rings. A more recent report by Braga *et al.* describes the structure of enantiomerically pure $Fe(\eta^5\text{-}C_5H_4CH(CH_3)OH)_2$, which forms a 1D hydrogen-bonded chain [95]. Co-crystallization of $Fe(\eta^5\text{-}C_5H_4CPh_2OH)_2$ can be achieved with a variety of diamine and polyamine hydrogen bond acceptors resulting in O–H···N hydrogen-bonded adducts, although rarely in extended networks [94].

4.5.2 Carboxyl- and carboxylate-bearing ligands

Patterns of hydrogen bond formation for organometallic complexes with carboxyl-substituted ligands have been compared with those of organic carboxylic acids showing that the $R_2^2(8)$ rings and C(4) chains prevalent in the latter are also common in the former [28a]. In cases where more than one ligand carboxyl group is present, hydrogen-bonded networks are found. Most interesting are the strong similarities between the networks formed by maleic acid [96] and $Fe(\eta^2$-maleic acid)$(CO)_4$ [97] and similarly between those of fumaric acid (two polymorphs) [98] and $Fe(\eta^2$-fumaric acid)$(CO)_4$ (two polymorphs) [97,99]. In a similar vein to these organic/organometallic pairings, Brammer *et al.* have prepared and crystallized the $(\eta^6\text{-arene})Cr(CO)_3$ complexes of the well-known series of benzene carboxylic acids: benzoic, terephthalic, isophthalic and trimesic acid (**XXXIII** – **XXXVI**, Figure 38) [27a] . For **XXXIII**, a centrosymmetric carboxylic acid dimer is formed by association of $(\eta^6\text{-arene})Cr(CO)_3$ molecules. The dimers then associate via C–H···O hydrogen bonds [100]. Efforts to prepare solvent free crystals of **XXXIV** – **XXXVI** have to date been unsuccessful, owing to the poor solubility of these compounds in media other than polar solvents that can compete for hydrogen bond formation. Thus, **XXXIV** and **XXXV** have been crystallized as their dimethyl sulfoxide (DMSO) and dimethylformamide (DMF) solvates, **XXXIV**·2DMSO and **XXXV** · 2DMF, respectively. Hydrogen bonding between the carboxyl groups and the solvent molecules prevents direct association between carboxyl groups. However, the disolvated organometallic molecules serve as three-component building blocks for further association via a network of C–H···O hydrogen bonds.

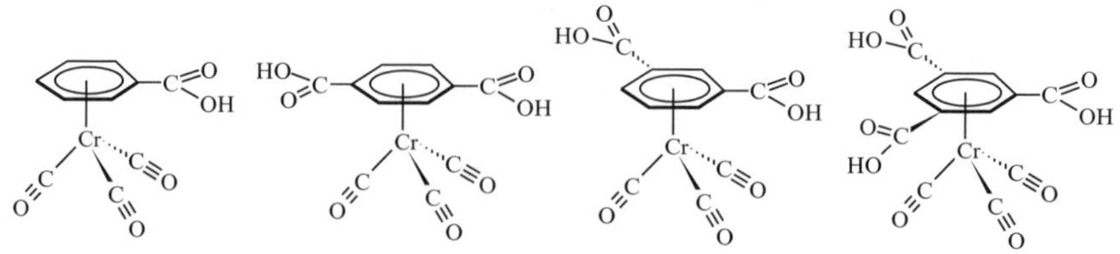

XXXIII XXXIV XXXV XXXVI

Figure 38 Organometallic benzene carboxylic acid building blocks: chromium tricarbonyl complexes [27a].

Most interesting in this series is the structure of **XXXVI** · $^{n}Bu_2O$, which exhibits a zig-zag chain owing to hydrogen-bonded association of two of the three carboxyl groups on each organometallic molecule. The third carboxyl group is occupied in forming an O–H···O hydrogen bond to the ether (Figure 39). Of note is the fact that all $Cr(CO)_3$ moieties are appended on the same side of a given hydrogen-bonded tape. The tapes associate in a bilayer arrangement involving a number of C–H···O hydrogen bonds, with both carboxyl and carbonyl oxygens participating.

One potential disadvantage of the di- and tricarboxylic acid metal–arene building blocks **XXXIV** – **XXXVI** for use in crystal engineering is that the electron-withdrawing nature of the carboxyl groups makes the arene ligands relatively poor electron donors to the metal centre, resulting in a decomplexation of the arene that is rather more facile than one might like. By contrast, cyclopentadienyl ligands offer a more strongly bound alternative to arenes. Braga and co-workers have studied extensively the organometallic bis(benzoic acid) system $[Cr(\eta^6\text{-}C_6H_5COOH)_2]$ (**XXXVII**-H_2), the bis(cyclopentadienylcarboxylic acid) systems $[Fe(\eta^5\text{-}C_5H_4COOH)_2]$ (**XXXVIII**-H_2) and $[Co(\eta^5\text{-}C_5H_4COOH)_2]^+$([**XXXIX**-$H_2$]$^+$) and some of their deprotonation products (Figure 40) [11g,101]. The structural aspects of these building blocks will be considered in this section and to some extent in Section 5.4, while aspects of the work on these systems that is pertinent to polymorphism [101d,f] and to solid–gas reaction chemistry and sensing [101h] will be mentioned briefly in Sections 7 and 8, and are covered in more detail elsewhere in this volume (see Chapter 8).

While the structures of the neutral **XXXVII**-H_2 (two polymorphs) [102] and **XXXVIII**-H_2 [101f] comprise simple dimers with a *syn* conformation of carboxyl groups (Figure 41a), neutral $[Fe(\eta^5\text{-}1,3\text{-}C_5H_3(Me)COOH)_2]$ [95] and the cationic [**XXXIX**-H_2]$^+$ (two polymorphs) [101d] and oxidized [**XXXVII**-H_2]$^+$ (two

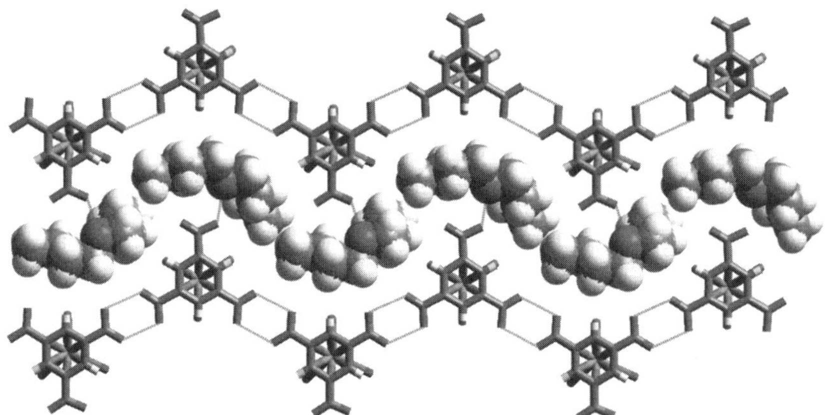

Figure 39 Hydrogen-bonded network in the crystal structure of $[Cr\{\eta^6\text{-}1,3,5\text{-}C_6H_3(CO_2H)_3\}(CO)_3]\cdot{}^nBu_2O$ [27a]. (see also **Plate 10**).

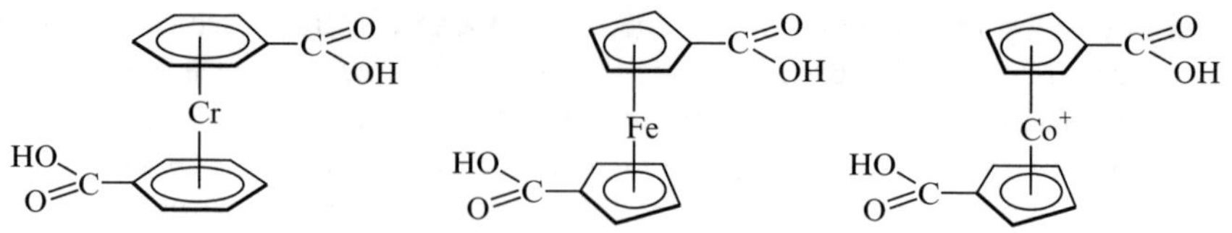

$XXXVII\text{-}H_2$ $XXXVIII\text{-}H_2$ $[XXXIX\text{-}H_2]^+$

Figure 40 Metallocene-1,1′-dicarboxylic acid and related bis(arene) building blocks, shown here in the *anti* conformation.

polymorphs) [101f] all form hydrogen-bonded chains with carboxyl groups adopting an *anti* conformation and all interacting via $R_2^2(8)$ synthons (Figure 41b). The chain arrangement is presumably adopted to accommodate the ring methyl groups or PF_6^- anions. In the presence of $[Cp_2Co]^+$ cations, the anionic carboxyl–carboxylate species [**XXVIII**-H]$^-$ forms a chain similar to the dicarboxylic acid cations, but sustained by O–H···O$^-$ hydrogen bonds (Figure 41c) [101a]. However, in the presence of the $[(\eta^6\text{-}C_6H_6)_2Cr]^+$ cation, a different arrangement is adopted in which [**XXVIII**-H]$^-$ units form dimers via O–H···O$^-$ hydrogen bonds and are then linked into a chain via bridging neutral **XXXVIII**-H_2 molecules [101a]. The neutral, zwitterionic **XXXIX**-H, crystallized with some difficulty by dehydration of its trihydrate and then seeding, forms a twisted version of the ID chain seen for [**XXVIII**-H]$^-$ (Figure 41d) [101c]. Other 1:1 mixed systems, e.g. [**XXXIX**-H_2]$^+$/**XXXIX**-H [101b], are known as well as the unusual adduct [**XXXIX**-H] [K^+] [PF_6^-], which contains potassium ions that are encapsulated within cages comprised of four $[CO(\eta^5\text{-}C_5H_4COOH)(\eta^5\text{-}C_5H_4COO)]$ (**XXXIX**-H) zwitterions linked via O–H···O$^-$ and C–H···O hydrogen bonds [101b]. The cages in turn form a 1D tape via additional C–H···O hydrogen bonds.

A few other organometallic dicarboxyl moieties are known. Particularly noteworthy are a series of organometallic compounds prepared by Štepnička and co-workers involving the phosphorus-bound 1′-(diphenylphosphino)ferrocenecarboxylic acid ligand [103]. These form rings and 1D hydrogen-bonded polymers when two such ligands are used [103a], but show the potential to form networks of higher dimensionality where more than two ligands are introduced [103c].

4.5.3 Ligands with amido groups

Despite their use in organic crystal engineering and in building blocks based upon coordination compounds (Section 4.2), amides have seen little deliberate use in organometallic crystal engineering. Their involvement in the packing in organometallic crystal structures suggests that they should behave as they do in other organic and coordination compound systems [28c]. There are some examples of

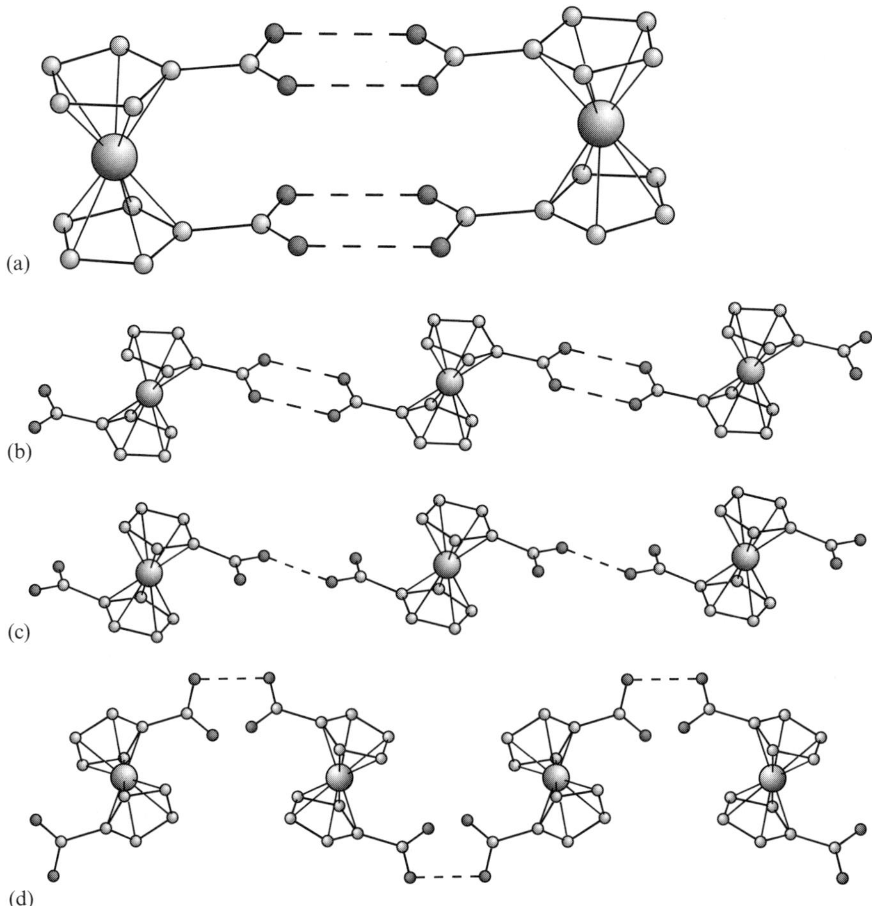

Figure 41 Hydrogen bonding patterns observed in structures involving (a) **XXXVIII**-H_2, (b) [**XXXIX**-H_2]$^+$, (c) [**XXXVIII**-H]$^-$ and (d) [**XXXIX**-H]. Reprinted from D. Braga, L. Maini, M. Polito, L. Scaccianoce, G. Cojazzi and F. Grepioni, *Coord. Chem. Rev.*, **216–217**, 225–48 (2001), Copyright 2001, with permission of Elsevier Science.

π-arene and π-cyclopentadienyl amides [104]. $Cr(\eta^6\text{-}C_6H_5CONH_2)(CO)_3$ and the related compound $Cr(\eta^6\text{-}C_6H_5CH_2CONH_2)(CO)_3$ crystallize as dimers linked via $R_2^2(8)$ N–H···O hydrogen-bonded synthons, with the second hydrogen atom from each amide group forming a bifurcated hydrogen bond to carbonyl oxygens of two neigbouring molecules, resulting in a 3D hydrogen-bonded network [104a]. Somewhat differently, $Mn(\eta^5\text{-}C_5Cl_4CONH_2)(CO)_3$ crystallizes as a 1D hydrogen-bonded tape consisting of $R_2^2(8)$ and C(4) N–H···O synthons, as is typical of many organic primary amides.

5 WEAK HYDROGEN BONDS

As weak hydrogen bonds by definition do not contribute substantially to overall crystal cohesion on an *individual* basis, their primary importance lies in their abundance. Thus, their combined energetic contribution in a given crystal may be substantial, as can be the case with other weak interactions, which at first consideration may seem to be in the realm of 'background noise' compared with stronger interactions that are (rightfully) given greater consideration in crystal design and synthesis. However, unlike van der Waals interactions, for example, weak hydrogen bonds, the C–H···O hydrogen bond being the prime example [105], are directional in nature [106] and exhibit attractive forces at fairly long distances, including beyond the nominal van der Waals contact distance [14e]. It is now recognized that hydrogen bonding is not confined to the electronegative elements in the upper right-hand corner of the Periodic Table, but that a wide range of elements can participate in hydrogen bonds, many of these forming relatively weak interactions [14e]. However, in this section the discussion of weak hydrogen bond donors will be confined to C–H groups, particularly the ubiquitous C–H···OC(M) hydrogen bonds of organometallic compounds. Weak acceptors, including phenyl, cyclopentadienyl and alkyne groups, that use filled π-orbitals rather than lone pairs to participate in hydrogen bonds, will also be discussed. It should be stated that in many if not all of the systems discussed in Section 4, weak hydrogen bonds are present in conjunction with the stronger counterparts. However, as in organic crystal engineering, the use of weak hydrogen bonds as the *primary* organizing force in inorganic crystal design is somewhat difficult to implement and in most cases will not be the preferred strategy. Nevertheless, weak hydrogen bonds should not be ignored as they will frequently be present whether by design or not. Ideally they should be used to work in harmony with stronger forces in the overall crystal design. As Aakeröy and Leinen have observed with weak hydrogen bonds [19], 'Typically, the directional preferences of these bonds are satisfied within the geometrical constraints of stronger hydrogen bonds... Furthermore, C–H···X interactions can tilt the balance between several options of stronger bonded networks, thus acting as an important "steering force" in solid-state assembly'.

5.1 $M_2(\mu_2\text{-}CH_2)$ and $M_3(\mu_3\text{-}CH)$ Groups in C–H···O Hydrogen Bonds

In Section 4.1.1, (M)N–H and (M)O–H hydrogen bond donors were discussed in the context of the influence of metal coordination upon donor capability of such groups (e.g. OH_2 vs M–OH_2). This discussion is now extended to (M)C–H donors. A useful comparison between the hydrogen bonding capabilities of $M_2(\mu_2\text{-}CH_2)$ and $M_3(\mu_3\text{-}CH)$ groups has been carried out by Braga *et al.* using the CSD to examine the metrics of C–H···O hydrogen bonds, in most cases involving carbonyl ligands as acceptors [28e]. Both methylidene (μ_2-CH_2) and methylidyne (μ_3-CH) ligands are found frequently to engage in C–H···O hydrogen bond formation, the latter generally forming shorter interactions. The authors discuss these findings in the context that these groups resemble alkenes ($C_{sp}{}^2$–H) and alkynes (C_{sp}–H), respectively, which form C–H···O hydrogen bonds with similar mean metrics to methylidene and methylidyne ligands. However, an alternative viewpoint arises if one considers the carbon atoms of these ligands to be sp^3 hydridized. It then seems apparent that coordination of these C–H donors to a metal center has the effect of enhancing the polarization of the $C^{\delta-}$–$H^{\delta+}$ bond relative to their organic counterparts, i.e. $M_2(\mu_2$-$CH_2)$ vs R_2CH_2 and $M_3(\mu_3$-CH) vs R_3CH (R = hydrocarbon group), the latter being much poorer hydrogen bond donors in each case. Thus, it seems reasonable to conclude that the polarizing effect of coordination to *three* metal centres upon one C–H donor group in $M_3(\mu_3$-CH), will be greater than that of *two* metal centres upon two C–H donor groups in $M_2(\mu_2$-CH). This is consistent with the observed C–H···O hydrogen bond lengths.

5.2 C–H···OC(M) Hydrogen Bonds

In an important paper, Braga, Grepioni, Desiraju and co-workers highlighted the abundance of C–H···OC(M) hydrogen bonds in crystals of organometallic compounds [28b]. The use of the CSD to analyse such hydrogen bonds across the series of d-block metals Ti–Ni led to a number of specific findings:

1. Terminal, doubly- and triply-bridging CO ligands frequently participate in C–H···O hydrogen bonds.
2. Bifurcated acceptors are common, since most organometallic crystals have an excess of C–H donors over OC(M) acceptors.
3. The effect of different metal atoms or the nuclearity of the complex is not detectable.
4. The mean H···O separation decreases in the order C–H···OC(M) (2.62 Å) > C–H···OC(M_2)(2.57 Å) > C–H···OC(M_3)(2.44 Å), consistent with the greater π-back-donation from the metal centres experienced by the bridging carbonyl ligands leading to increased oxygen basicity and stronger hydrogen bond formation.
5. Intermolecular hydrogen bonds involving C–H···OC(M_n), $n = 1$–3, are directional with corrected [107] H···O–C angles most commonly having values in the range 140–160°, regardless of the mode of coordination of the carbonyl ligand.

These results not only underlie the importance of considering such hydrogen bonds as part of the overall crystal design process for inorganic (organometallic) crystals, but also point the way towards the *proactive* use of C–H···OC(M) hydrogen bonds in crystal design.

Another illustrative example of the importance of C–H···OC(M) hydrogen bonds and their influence upon stronger hydrogen bonds comes from a detailed investigation of hydrogen bonding [11d,24e] in a series of salts $R_3NH^+Co(CO)_3L^-$ (L = CO, PR_3). These systems were prepared not with crystal engineering in mind, but rather to explore the N–H···Co hydrogen bonds formed [24b–f,25a]. A detailed examination of the structures of the $Co(CO)_4^-$ salts provides some useful insights into organometallic crystal engineering and in particular demonstrates the overall importance of C–H···OC(M) hydrogen bonds, which far outnumber the N–H···Co hydrogen bonds that were the focus of the original studies [11d]. This point is further accentuated in the crystal structures of the series $(DABCO)H^+Co(CO)_3L^-$; L = CO (**XL**), PPh_3 (**XLI**), PPh_2(*p*-tol) (**XLII**) and P{*p*-tol}$_3$ (**XLIII**) (*p*-tol = *para*-tolyl), in which the basicity of the hydrogen bond acceptor (Co) is steadily increased from **XL** to **XLIII** [24e]. However, the crystal structures reveal that in the solid state the N–H···Co hydrogen bond decreases in length not in parallel with the increase in basicity of the acceptor, but in the sequence **XL**, **XLIII**, **XLII**, **XLI**. As each N–H···Co hydrogen-bonded ion pair is present in a different landscape of weak interactions, including many weak hydrogen bonds, it seems clear that the strongest hydrogen bond, N^+–H···Co^-, is significantly influenced by accommodation of this multitude of weaker interactions required to minimize the overall packing energy. This example provides an interesting adjunct to the statement quoted above from Aakeröy and Leinen [19] concerning the interplay of strong and weak hydrogen bonds in crystal structures.

5.3 D–H···π Hydrogen Bonds (D = C, N, O)

Although the ability of the filled π-orbitals of multiple bonds, including aromatic rings, to serve as hydrogen bond acceptors was first observed in the 1930s [108] (as were C–H···O [109] and C–H···N [110] hydrogen bonds), it was not until the 1990s that this class of weak interactions was generally accepted as a type of hydrogen bond. On this topic, Desiraju and Steiner state in their monograph on weak hydrogen bonds, 'There is no good reason to suppose that electron-rich π-systems like aromatic rings and C≡C triple bonds would not participate in hydrogen bonding interactions. Indeed they constitute the most common class of non-conventional acceptors' [14e]. The fact that they go on to devote 80 out of 440 pages in the monograph to D–H···π hydrogen bonds gives some indication of the widespread nature of such interactions.

That having been said, their deliberate application in crystal design and specifically in inorganic crystal design has been much more limited. There are nevertheless some interesting examples of D–H···π hydrogen bonds in inorganic systems and the occurrence of such hydrogen bonds in organometallic crystals has been surveyed using the CSD [111]. Mingos and co-workers prepared the 1:2 and 1:4 adducts of the electron-rich dimetallated alkyne $NpPh_2PAu–C{\equiv}C–AuPPh_2Np$ (Np = naphthyl) with chloroform in which the acidic C–H bonds of the solvent molecules form short C–H ··· π(C≡C) hydrogen bonds [H ··· C≡C(midpoint) as short as 2.30 Å] [112]. Cyclopentadienyl C–H groups have also been observed to engage in C–H ··· π(RC≡CML_n) hydrogen bonds [113]. Hydrogen-bonded networks of this type have attracted less attention than their organic counterparts, though a cyclic hexamer constructed through C–H ··· π(C≡C) hydrogen bonds has been observed in ethynylferrocene [114].

The prevalence of phenyl rings in organometallic compounds, in no small part due to the use of the PPh_3 ligand, leads to stabilizing interactions between such phenyl groups [115], referred to by Dance and Scudder in terms of 'the phenyl factor' [21] and the 'phenyl embrace' [116]. Some of these interactions involve so-called edge-to-face contacts and might be considered alongside other weak C–H ··· π(Ph) hydrogen bonds. Cyclopentadienyl rings can also serve as acceptors for D–H ··· π(Cp) hydrogen bonds. An interesting case in which a 1D network is propagated via O–H ··· π(Cp) hydrogen bonds can be found in (η^5-1-{*p*-tolyl}-2-hydroxyindenyl)(η^5-cyclopentadienyl)ruthenium(II) (Figure 42) [117]. Carbonyl ligands are also observed to form D–H ··· π(OC) hydrogen bonds, though these are less common than D–H ··· OC(M) hydrogen bonds discussed above [22b,118].

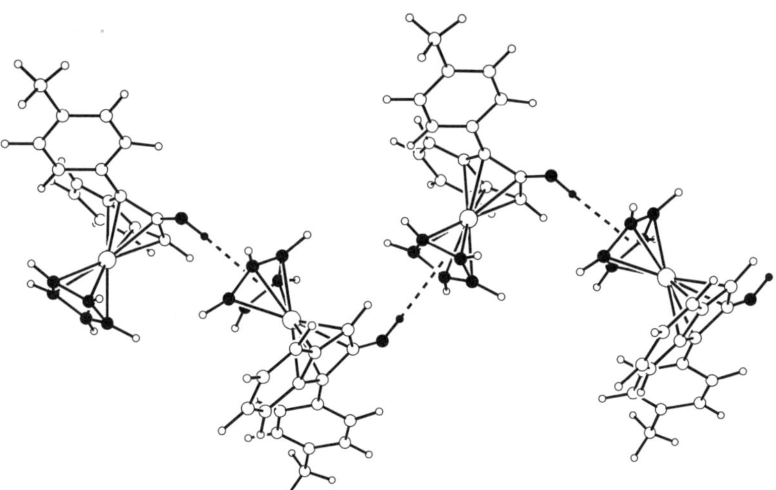

Figure 42 Zig-zag 1D network propagated by O–H ··· π(Cp) hydrogen bonds in the crystal structure of [Ru(η^5-Cp)(η^5-1-*p*-tolyl-2-hydroxyindenyl)] (Cp = C_5H_5) [117].

5.4 Application of Charge-assisted C–H···O Hydrogen Bonds in Organic–Organometallic Crystal Engineering: Carboxyl–Carboxylate Hosts with Cationic Metallocene Guests

An extensive series of crystalline systems have been prepared by Braga, Grepioni and co-workers in which $[Cr(\eta^6\text{-}C_6H_6)_2]^+$ or $[Co(\eta^5\text{-}C_5H_5)_2]^+$ cations are encapsulated in hosts formed by O–H···O hydrogen-bonded networks of organic di- and polycarboxylic acids, partial deprotonation of which provides the overall charge balance [11c,119]. The initial inspiration was an effort to mimic the remarkable system reported by Etter *et al.* in which cyclohexanedione forms O–H···O hydrogen-bonded cyclic hexamers in the presence of benzene as a template (Figure 43a), owing to stabilization of this motif by C–H···O hydrogen bonds [120]. Recognizing the similarities between the peripheries of many organometallic molecules and their organic counterparts, the aim was to use what was known from organic crystal packing to develop organometallic crystal engineering. An effort to combine $[Cr(\eta^6\text{-}C_6H_6)_2]$, as an organometallic analogue of benzene, with cyclohexanedione (CHD) led to the crystalline system $[Cr(\eta^6\text{-}C_6H_6)_2]^+[(CHD)_2]^-\cdot 2CHD$ (Figure 43b) owing to facile oxidation of the $[Cr(\eta^6\text{-}C_6H_6)_2]$ [121]. This strongly resembles Etter's *et al.* system in that a templated organic host network supported by strong O–H···O hydrogen bonds encapsulates a guest via C–H···O hydrogen bonds. However, in this case the cationic nature of the guest leads to strengthened charge-assisted $C\text{–}H^{\delta+}\cdots O^{\delta-}$ hydrogen bonds. Subsequent exploitation of these observations led to a general synthetic route to organic–organometallic host–guest networks by preparing crystals from the combination of $[Cr(\eta^6\text{-}C_6H_6)_2][OH]$ or $[Co(\eta^5\text{-}C_5H_5)_2][OH]$ with a variety of organic carboxylic acids. Deprotonation of an acid moiety by the hydroxide ion then leads to the crystalline host–guest system in which the organometallic cations interact with the organic host network via $C\text{–}H^{\delta+}\cdots O^{\delta-}$ hydrogen bonds. Prediction of the final crystal architecture is necessarily difficult in these composite systems, as is the extent of hydration, since water is a convenient polar solvent for crystal synthesis. Nevertheless, a number of guiding principles have been established [11c]. Thus, for example, it is clear that the organic host network is wrapped around the organometallic guests maximizing $C\text{–}H^{\delta+}\cdots O^{\delta-}$ hydrogen bond formation and that the highly basic carboxylate $(\text{–}CO_2^{\,-})$ sites seek strong hydrogen bond donors, which are only available on the carboxyl groups of the neutral or partially deprotonated polycarboxylic acids. The organic host network often assembles into a channel structure in which the organometallic cations are arranged, although layer structures have also been prepared when using the small planar acid squaric acid [119e,h,i]. Chirality can readily be introduced through the application of chiral carboxylic acids such are tartaric acid [119c] and or even chiral alcohols such as binaphthol [122]. More recently, mixed metal systems have also been prepared by using organometallic the carboxylic acid $Fe(\eta^5\text{-}C_5H_5COOH)_2$ (**XXXVIII**-H_2) to form the host network [101a,123].

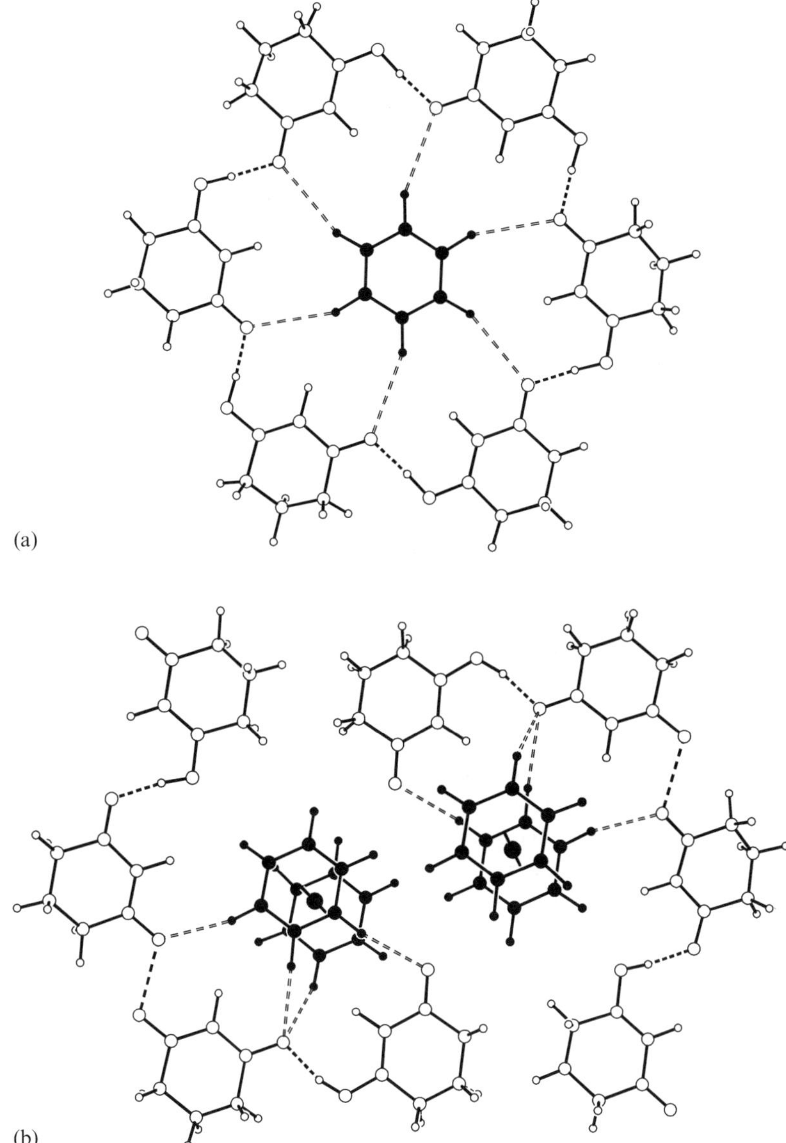

Figure 43 (a) Etter's *et al.* iconic example of self-assembly in $(C_6H_8O_2)_6 \cdot C_6H_6$ [120] (b) A close organometallic analogue, $[(C_6H_8O_2)_6(C_6H_7O_2)]^- \, [Cr(C_6H_6)_2]^+$ [121]. Note the importance of C–H···O hydrogen bonds emanating from the arenes in directing the self-assembly of the O–H···O hydrogen-bonded cyclohexenedione network.

6 DIRECT INVOLVEMENT OF METALS IN HYDROGEN BONDS

As noted in Section 3.3, spectroscopic and crystallographic evidence, principally in the past decade, has led to the recognition that metals can participate directly in hydrogen bonds as either the donor, M–H···A [23], or acceptor, D–H···M [13,24,25]. These interactions are of importance in the study of protonation of metal centres and the deprotonation of metal hydrides, but do they have a potential role to play in inorganic crystal design? Surveys of the CSD suggest that there may be many examples of such interactions [23e,25k]. M–H···O hydrogen bonds are generally regarded as being fairly weak, similar to C–H···O hydrogen bonds, and a 1996 CSD survey revealed only 17 crystal structures with (M)H···O distances of less than 2.8 Å (eight such interactions at less than 2.6 Å) [23e]. Hence, unlike C–H···O hydrogen bonds, their weakness is not compensated for by their abundance. Hence their role in inorganic crystal design is likely to be negligible. The prospects are slightly better for D–H···M hydrogen bonds. These are stronger, reaching 7 kcal/mol in solution for O–H···M hydrogen bonds between neutral molecules and undoubtedly more when charge-assisted. Hence strength is not the issue. Although they are also more abundant than their M–H···O counterparts [25 k], the question of the applicability of D–H···M hydrogen bonds in crystal design does rest upon a realistic assessment of the *availability* of such hydrogen bonds. At present D–H···M hydrogen bonds have been observed for a limited number of electron-rich transition metal centres [25] and require, in addition, sterically accessible (d-orbital) lone pairs on the metal atom at the *centre* of a metal complex. Their use in crystal design is, therefore, likely to be at best very limited compared with hydrogen bonds involving the *Ligand* and *Periphery Domains*. An added deterrent to the use of *Metal Domain* hydrogen bonds is the synthesis of the metal complexes involved. While crystal engineering has evolved from what is sometimes referred to as 'Aldrich chemistry', i.e. the use of commercially available compounds as building blocks, to the design and (conventional) synthesis of building blocks, the organometallic synthesis and potential air sensitivity of the many metal complexes able to participate in M–H···A and D–H···M hydrogen bonds is unlikely to provide an incentive to seek this approach to crystal design. These comments do not undermine the importance of such interactions in organometallic chemistry, but merely put a realistic face upon their potential use in inorganic crystal design.

Another interaction sometimes discussed alongside hydrogen bonds unique to inorganic or organometallic crystals is intermolecular X–H σ-bond complexation, sometimes confusingly so since these interactions are fundamentally different from hydrogen bonds [12a]. The interactions of X–H σ-bonds with electron-deficient transition metal centres has been widely studied and is important in the activation of such bonds by transition metals [124]. The most common among these are for X = B, C, Si, the emphasis in most studies being on intramolecular interactions. An indication that C–H σ-bonds form weak intermolecular interactions with metal centres in a number of crystal structures of organometallic compounds has been established by a survey of the CSD [25k]. However, for analogous reasons to those given in the above discussion of M–H···A and D–H···M hydrogen bonds, the role of intermolecular X–H σ-bond complexation is not of general significance in crystal design.

7 FUNCTION AND PROPERTIES

A primary goal in crystal engineering is the ability to design and synthesize crystalline materials with a high degree of predictability. This goal is being approached for certain well-defined classes of compounds for which the structural variables can be well defined and are well understood. However, reaching this goal more broadly for all classes of compounds, and all types of building blocks, still lies well in the future. The purpose of the desired structural control, of course, returns us to the intimate link between structure (and composition) and properties. Microscopic structure can in principle be controlled in order to tune the physical and chemical properties of a material. However, we cannot afford to wait until full control of structure of a wide range of materials is achieved before considering their properties; the two must be developed side by side to some extent. A number of approaches can be taken.

One can take building blocks with known properties/function and endeavour to design a crystalline solid that incorporates these units with the aim of instilling these molecular properties or some cooperative version thereof into the new material. An example of such a case is the incorporation of transition metal centres with unpaired electrons into well-defined networks in which the spins of these electrons can interact to yield desirable magnetic properties, which may then be further refined by tuning the overall structure. Similarly, one might envisage taking a catalytically active molecule and modifying its periphery so as to incorporate it into a porous crystalline material that would permit control of entry and exit sites to the catalyst, but still preserve the molecular catalytic function.

An alternative approach involves designing networks to exhibit a property or perform a function not inherent in the building blocks used. One might consider the design of porous solids or lamellar solids for host–guest chemistry, separations, etc., to fall into this category, since these properties are not inherent in the building blocks used. However, such properties of the final solid may be enhanced by the appropriate choice of building blocks, such as those with functionality capable of serving as effective binding sites.

A third approach arises when pursuing primarily structural objectives is coupled with keen observation, and examination and testing of the properties of materials synthesized. A simple example involves the common situation in which solvent molecules are incorporated into cavities in network solids. Further examination to see if the solvent can be removed or replaced by other guests is often warranted. More quantitative studies, e.g. using differential scanning calorimetry (DSC) or thermogravimetric analysis (TGA), are also useful in such circumstances.

While crystal engineering is beginning to have its successes in terms of delivering properties in designed crystalline materials, this aspect of the field is very much in its infancy. The area of crystal design covered in this chapter, that of hydrogen-bonded inorganic materials, is less well developed than either its organic counterpart or that of inorganic coordination networks, and can benefit from aspects of each of these approaches. Nevertheless, there are some interesting and promising developments in the properties of inorganic hydrogen-bonded materials, as highlighted in the remainder of this section.

Returning to the topochemical roots of crystal engineering, recent examples have illustrated that N–H···H–E hydrogen bonds (E = B, Ga) in crystalline solids can be used to form N–E bonds in the solid state by elimination of gaseous H_2. This has been illustrated by Gladfelter and co-workers in the preparation of nanocrystalline gallium nitride via N–H···H–Ga hydrogen bonds in cyclotriagallazane, $(H_2GaNH_2)_3$ [46h,i], and by Jackson and co-workers in preparing B–N-bonded networks unattainable by other means [46c–e].

The work of Mitzi has shown that inorganic–organic hybrid perovskites in which inorganic layers consisting of corner-sharing metal halide octahedra and linked by cationic organic layers are capable of exhibiting a variety of potentially useful magnetic, electrical and optical properties [43]. The organic–inorganic interface consists of many N–H···X(M) hydrogen bonds (X = halogen), leading to well organized crystalline materials, which can be characterized by single-crystal diffraction methods (see Section 4.4.2). Properties identified in such materials include semiconductor and high electron mobilities in Sn(II) halide-based hybrids [125]. Novel optical properties have also been observed for such hybrid materials based upon other divalent Group 14 halides [126] and the first-row d-block metals, M(II), have provided useful model systems for studying low-dimensional magnetic properties [127].

Neve and co-workers have also made use of hydrogen-bonded perhalometallate ions, but in this case using salts mononuclear ions linked by weak charge-assisted C–H···X(M) hydrogen bonds from alkylpyridinium cations [128]. The use of cations with long-chain alkyl substituents leads to liquid crystalline behaviour in these salts [128a–c].

Recent reports have also shown the potential for applications in sensing shown by metal-containing hydrogen-bonded crystalline solids. Albrecht, van Koten and co-workers have reported a remarkable crystalline organoplatinum ‘pincer-ligand’ complex **XLIV** [44a] that is capable of reversibly binding SO_2 [129] (Chapter 9). The complex assembles into parallel 1D tapes via O–H···Cl(Pt) hydrogen bonds such that all the organometallic building blocks within one tape are coplanar (Figure 44). The tapes are then arranged in a buckled sheet structure. Absorption of SO_2 is accompanied by a colour change from colourless **XLIV** to orange **XLV**,

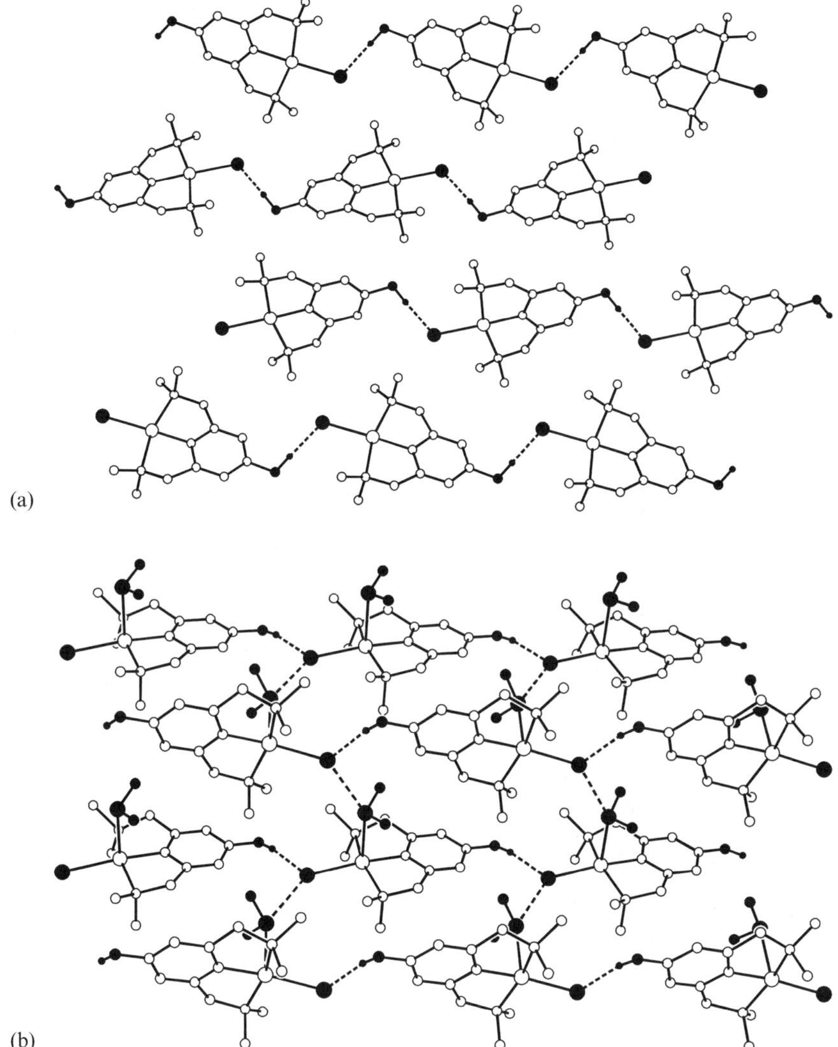

Figure 44 Crystal structures (a) [Pt(NCN-OH)Cl] XLIV [44a] and (b) [Pt(SO_2)(NCN-OH)Cl] XLV [129a,b] (NCN = tridentate 'pincer' ligand $\{C_6H_2\text{-}5\text{-}(OH)\text{-}2,4\text{-}(CH_2NMe_2)_2\}^-$). O–H · · · Cl hydrogen-bonded chains are maintained upon uptake of SO_2. Hydroxyl group, chloride ligand and SO_2 ligand are shaded. Hydrogen bond network and Cl · · · S donor–acceptor interactions are shown as dashed bonds.

which contains square pyramidal Pt(II) centres due to coordination of SO_2 [130]. A single-crystal structure determination of **XLV** shows that the hydrogen-bonded tapes remain intact and essentially unchanged in geometry relative to **XLIV**. Remarkably, **XLIV** is not inherently porous, but rather the absorption takes place through a substantial adjustment of the relative positions of the tapes, including a reversal of the direction of half of the tapes. This permits the more than 15% increase in unit cell volume needed to accommodate the absorbed gas. The absorption process in the solid state was monitored by time-resolved powder diffraction. The platinum pincer compounds that provide the molecular building blocks of **XLIV** have also been shown to be able to bind SO_2 in solution and when constructed into dendrimers [129b]. In subsequent work it was shown that removal of the aryl OH group led to crystalline solids with different packing arrangement absent of strong hydrogen bonds, but these were also able to absorb and bind SO_2 [129c]. Thus, the O–H $\cdots$ Cl(Pt) hydrogen bonds in **XLIV** provide a valuable organizing interaction but are not essential for SO_2 absorption. Importantly, the potential for development of gas sensors and optical switches in these systems is readily apparent.

Braga, Grepioni and co-workers have also reported an interesting absorption phenomenon exhibited by the crystalline zwitterion [Co(η^5-C_5H_4COOH)(η^5-C_5H_4COO)] (**XXXIX**-H), which undergoes heterogeneous reactions with a variety of hydrated acids (HCl, CF_3CO_2H, HBF_4) and bases (NH_3, NH_2Me, NMe_3) [101h]. The products of these reactions involve absorption of the acid or base and water, as exemplified by the cases for reaction with HCl and NH_3. Thus, reaction with aqueous HCl in the gas phase leads to [**XXXIX**-H_2]$^+Cl^- \cdot H_2O$ in which the O–H $\cdots$ O^- hydrogen bonds of **XXXIX**-H are replaced by O^+–H $\cdots$ Cl^- hydrogen bonds. Reaction of **XXXIX**-H with aqueous NH_3 in the gas phase leads to the formation of $NH_4{}^+$[**XXXIX**]$^- \cdot 3H_2O$, in which N^+–H $\cdots$ O^- hydrogen bonds are now present. The zwitterion **XXXIX**-H can be regenerated by heating under reduced pressure in both cases. The absorption–deabsorption process is reversible for a number of cycles and has been monitored by powder diffraction, solid-state IR spectroscopy and TGA. Such phenomena are known for crystals of some organic compounds [131], but this appears to be the first example of amphoteric absorption behaviour involving an organometallic compound.

8 DESIGN STRATEGIES: 'FAILURES' AND PROBLEMS. WHAT CAN BE LEARNED?

Design strategies for crystalline inorganic hydrogen-bonded materials necessarily rely heavily of the use of reproducible hydrogen bonding interactions, not only in terms of which donor and acceptor groups will interact with which, but also in terms of geometric reliability. The concept of the supramolecular synthon [16] is tremendously valuable in this respect, but the reliability of the synthons that one intends to use needs to be known or investigated [18], and competition from other interactions, particularly stronger ones, should be considered at the outset. Etter's studies led her to conclude that potential hydrogen bonding interactions between molecular building blocks used in crystal synthesis can typically be assessed in a hierarchical manner [1c]. Thus, it is most probable that the best hydrogen bond donor and the best acceptor will interact. Remaining donor and acceptor groups can then be considered in terms of the next best donors and acceptors. Analyses of hydrogen bond geometries also serve as important guides to likely hydrogen bonding behaviour [26,28]. Mean distances and angles are useful for summarizing hydrogen bond geometries and for making comparisons between different types of hydrogen bonds. However, examination of the overall distribution of these geometries provides information on the how readily specific hydrogen bonds may be deformed. Theoretical modelling of hydrogen bonding interactions can also provide a valuable guide to (experimental) synthetic strategies [39].

None of these approaches is foolproof, but they provide good initial guidance. However, while the literature contains predominantly reports of 'successful' syntheses, in that hydrogen-bonded networks are formed and can be analysed in terms their supramolecular synthons, etc., there are surely many unpublished examples in many laboratories of 'unsuccessful' crystal syntheses. These include products that were not or could not be satisfactorily analysed, and products in which the intended product did not form and that which formed in its place was deemed uninteresting (e.g. no network formation). Within these 'failures', these 'unsuccessful' syntheses, lies a potential wealth of information of value to crystal design. However, this information is often difficult to assess systematically and likely to be even more difficult to publish [132], given the competitive emphasis on 'success'. Nevertheless, it would be valuable to catalogue this information and one should think carefully before discarding it as merely a failed attempt.

Some of the issues that lead to 'unsuccessful' or at least unintended outcomes in crystal syntheses of inorganic hydrogen-bonded materials have been examined in this chapter. An important issue is the balance between strong and weak hydrogen bonds in determining the final structural arrangement. An assessment of potential weak hydrogen bonds can in some instances permit these to be used to augment the desired strong hydrogen bonds in the final network. Since ionic metal complexes are common, the use of ionic building blocks is more prevalent in inorganic crystal synthesis than in its organic counterpart. Counterions, frequently anions, can prevent the formation of the intended hydrogen-bonded network through their own participation in hydrogen bonding. Two approaches might be taken to address this issue. Weakly basic anions [133] could be used, although the most weakly interacting anions necessarily tend to be large in order to 'dilute' their charge, and this is clearly a limitation in crystal design. An alternative approach is to make the anion part of the solution rather than part of the problem, i.e. take advantage of the interactions made by the anion and include these in the design strategy [134]. Assessment of the role of the anion and the reliability of cationic hydrogen-bonded networks can, of course, be assessed by repeating the crystal synthesis in the presence of different anions. Solvent incorporation presents a similar problem, but perhaps more difficult to deal with. Since building blocks used in the formation of hydrogen-bonded networks are often most soluble in polar solvents (owing to hydrogen bonding interactions with the solvent), it is common for such solvents to be used in crystal syntheses from the solution phase. Thus, it is not uncommon to observe their inclusion in the crystalline product, sometimes preventing the intended association of building blocks [27a]. The torsional flexibility of ligands in the metal-containing building blocks is also a factor that should be taken into consideration. In general, the more degrees of freedom introduced, as found in torsionally flexible groups or molecules, the more possible outcomes there may be in the crystalline product.

Another design question relates directly to the issue of function. The question is whether first to establish reliable networks and synthesize crystals based upon such networks, and then to modify the building blocks to introduce a desired function while maintaining the reliability of the network. Alternatively, one can take molecules with 'functionality' built-in and then try to develop them, perhaps by minor synthetic modification into building blocks for a suitable network that will permit a functional crystalline material to be prepared. These issues have been examined in Section 7.

Finally, there is the issue of polymorphism. When a substance can exist in more than one crystalline phase, each adopting a different arrangement of molecules, ions or network, the material is described as polymorphic. Each crystalline phase is termed a polymorph. Each polymorph of a particular substance has different physical properties as a result of the differences in molecular arrangements – a dramatic demonstration of the intimate link between microscopic structure and macroscopic properties that was introduced at the outset of this chapter. The phenomenon has been widely explored in organic crystals [135] and is of considerable commercial interest, particularly in the pharmaceutical industry [135c] where different polymorphic forms of drugs may have different efficacy (bioavailablity), and require different processing and formulation. In the paint and pigment industry, polymorphic forms of pigments may result in different shades, which can be advantageous as well as presenting potential problems [136]. But what of polymorphism in hydrogen-bonded inorganic crystals? It is known, but while there is no reason to believe that it should be less widespread than in organic crystals, it has received relatively little attention to date [10]. Clearly, further exploration is needed in inorganic crystalline materials, from both the experimental side and the theoretical side (i.e. crystal structure prediction). A knowledge of the extent of polymorphic behaviour is of undeniable value in the design of any new material (Chapter 8).

9 FUTURE PROSPECTS

What can we expect from hydrogen-bonded inorganic materials in the future? Predictions are treacherous territory, but examining the current state of research in this area and developments in closely related areas can serve as a guide if viewed through the lens of likely feasibility when employing hydrogen bonds. The earlier sections of this chapter, Sections 4 and 5 in particular, illustrate the wide variety of inorganic systems in which hydrogen bonding has an important role to play. This is a diverse collection of systems.

To date fewer applications of inorganic hydrogen-bonded materials (Section 7) have been developed than in organic hydrogen-bonded systems or with inorganic coordination polymers. This is partly a chronological issue. Organic crystal engineering was sufficiently well developed by the end of the 1980s for one of its major proponents, Gautam Desiraju, to write a book to be devoted to the topic [1a], while the study of coordination polymers took off around 1990 following the pioneering work of Robson's group [137]. By contrast, few systematic efforts to study or pursue inorganic crystal design using hydrogen bonding appeared before the mid-1990s. Some examples from the organic and coordination polymer areas of crystal engineering suggest viable goals for advancements that might be expected in the near future using inorganic hydrogen-bonded materials.

Ward and co-workers have developed an extensive series of materials based upon a lamellar hydrogen-bonded motif, with layers separated by pillars that lead to void spaces for guest occupation [138]. The materials are constructed from guanidinium bisulfonates and enable remarkable metric control [138a,c], permitting applications in separation chemistry [138d]. Recently, these materials have also been designed to contain aligned polar cavities enforcing adoption of noncentric space groups and permitting nonlinear optical (NLO) activity by encapsulation of appropriate guest molecules [138e,f]. These materials represent a class of *flexible* porous solids. A very promising class of *rigid* coordination-based porous solids capable of guest encapsulation and exchange has been developed by Yaghi and co-workers using metal–carboxylate linkages [8h,i,139].

Looking further towards applications of coordination polymers, Lin and co-workers have demonstrated that polar solids (with substantial NLO response) can be designed through an appropriate choice of polar bridging ligands combined with tetrahedral metal centres that provide acentric nodes [140]. Chiral separations [141] and modest catalytic activity [141,142] have also been demonstrated in coordination polymers.

In all of these applications there appears to be scope for contributions from inorganic materials in which hydrogen bonds play an important role. Hydrogen bonding has also seen extensive use in the design of molecular receptors, and extension of this approach to the design of network solids with embedded hydrogen bonding recognition sites would appear to be an attractive proposition.

Thus, hydrogen-bonded materials offer greater flexibility than coordination-based systems. This can be advantageous or disadvantageous depending upon the desired application. Inorganic hydrogen-bonded materials have the potential to offer similar design possibilities to organic hydrogen-bonded materials, with additional properties arising from the presence of metals (optical, magnetic [142], catalytic) that might also be incorporated.

10 CONCLUSIONS AND SUMMARY

In just a few years, the application of hydrogen bonding to inorganic crystal design has developed tremendously. Systematic studies of hydrogen bonding patterns and the design and synthesis of a wide variety of systems have been conducted, incorporating many different metals and propagated via many different supramolecular synthons. Despite this output, the field is clearly in its infancy and has much yet to offer. Further work on developing *reliable* and *versatile* network designs is essential as is the need for advancement of the field in terms of development of genuinely functional materials with interesting and useful properties.

ACKNOWLEDGEMENTS

I am grateful for important contributions to our own work in this area from Reinaldo Atencio, Eric Bruton, Juan Mareque Rivas and John Swearingen, and to an ongoing collaboration with Paul Sherwood (CLRC Daresbury Laboratory, UK). Support for our work in this area has come from the National Science Foundation (CHE-9988184), the Missouri Research Board, NATO (CRG-920164) and the University of Missouri–St. Louis in the USA, and the University of Sheffield and the EPSRC (GR/R68733/01) in the UK.

REFERENCES

1. (a) G. R. Desiraju, *Crystal Engineering: The Design of Organic Solids*, Elsevier, Amsterdam (1989); (b) C. B. Aakeröy, *Acta Crystallogr.*, **B53**, 569–86 (1997); (c) M. C. Etter, *Acc. Chem. Res.* **23**, 120–6 (1990).
2. Examples most pertinent to this review are highlighted in Section 7.
3. E. G. Tulsky and J. R. Long, *Chem. Mater.*, **13**, 1149–66 (2001), and references therein.
4. C. B. Aakeröy, A. M. Beatty and B. A. Helfrich, *Angew. Chem., Int. Ed. Engl.*, **40**, 3240–2 (2001).
5. G. M. J. Schmidt, *Pure Appl. Chem.*, **27**, 647–78 (1971).
6. *J. Chem. Soc., Dalton Trans*, 3705–3998 (2000): Special Issue on Inorganic Crystal Engineering.
7. (a) D. Braga, F. Grepioni and G. R. Desiraju, *Chem. Rev.*, **98**, 1375–1405 (1998); (b) D. Braga and F. Grepioni, *Coord. Chem. Rev.*, **183**, 19–41 (1999); (c) D. Braga and F. Grepioni, *Acc. Chem. Res.*, **33**, 601–8 (2000).
8. (a) R. Robson, B. F. Abrahams, S. R. Batten, R. W. Gable, B. F. Hoskins and J. Liu, in *Supramolecular Architectures*, ed. T. Bein, *ACS Symp. Ser.*, **499**, 256–73 (1992); (b) S. R. Batten and R. Robson, *Angew Chem., Int. Ed. Engl.*, **37**, 1461–94 (1998); (c) P. J. Hagrman, D. Hagrman and J. Zubieta, *Angew. Chem., Int. Ed. Engl.*, **38**, 2638–82 (1999); (d) A. J. Blake, N. R. Champness, P. Hubberstey, W.-S. Li, M. A. Withersby and M. Schröder, *Coord. Chem. Rev.*, **183**, 117–38 (1999); (e) A. N. Khlobystov, A. J. Blake, N. R. Champness, D. A. Lemenovskii, A. G. Majouga, N. V. Zyk and M. Schröder, *Coord. Chem. Rev.*, **220**, 155–92 (2001); (f) S. R. Batten, *CrystEngComm*,

3, 67–74 (2001); (g) B. Moulton and M. J. Zaworotko, *Chem. Rev.*, **101**, 1629–1658 (2001); (h) M. Eddaoudi, D. B. Moler, H. Li, B. Chen, T. M. Reineke, M. O'Keefe and O. M. Yaghi, *Acc. Chem. Res.*, **34**, 319–30 (2001); (i) O. M. Yaghi, H. Li, C. Davis, D. Richardson and T. L. Groy, *Acc. Chem. Res.*, **31**, 474–84 (1998).

9. F. H. Allen and O. Kennard, *Chem. Des. Autom. News*, **8**, 1, 31–7 (1993).
10. For discussions of polymorphism in organometallic chemistry, see (a) D. Braga and F. Grepioni, *Chem. Soc. Rev.*, **29**, 229–38 (2000) and (b) Chapter 8 in this volume.
11. (a) D. Braga and F. Grepioni, *Acc. Chem. Res.*, **30**, 81–7 (1997); D. Braga, F. Grepioni and G. R. Desiraju, *J. Organomet. Chem.*, **548**, 33–43 (1997); (c) D. Braga and F. Grepioni, *J. Chem. Soc., Dalton Trans.*, 1–8 (1999); (d) J. C. Mareque Rivas and L. Brammer, *Coord. Chem. Rev.*, **183**, 43–80 (1999); (e) G. R. Desiraju, *J. Chem. Soc., Dalton Trans.* 3745–51 (2000); (f) A. M. Beatty, *CrystEngComm*, **3**, 243–54 (2001); (g) D. Braga, L. Maini, M. Polito, L. Scaccianoce, G. Cojazzi and F. Grepioni, *Coord. Chem. Rev.*, **216–217**, 225–48 (2001).
12. (a) L. Brammer, D. Zhao, F. T. Ladipo and J. Braddock-Wilking, *Acta Crystallogr.*, **B51** 632–40 (1995); (b) A. J. Canty and G. van Koten, *Acc. Chem. Res.*, **28**, 406–13 (1995); (c) L. M. Epstein and E. S. Shubina, *Metallorg. Khim.*, **5**, 61–74 (1992).
13. (a) R. H. Crabtree, P. E. M. Siegbahn, O. Eisenstein, A. L. Rheingold and T. F. Koetzle, *Acc. Chem. Res.*, **29**, 348–54 (1996); (b) R. Custelcean and J. E. Jackson, *Chem. Rev.*, **101**, 1963–80 (2001).
14. (a) G. C. Pimental and A. L. McClellan, *The Hydrogen Bond*, W. H. Freeman, San Francisco (1960); (b) P. Schuster, G. Zundel, and C. Sandorfy (eds), *The Hydrogen Bond: Recent Developments in Theory and Experiment*, North-Holland, Amsterdam (1976); (c) G. A. Jeffrey and W. Saenger, *Hydrogen Bonding in Biological Structures*, Springer, Berlin (1991); (d) G. A. Jeffrey, *An Introduction to Hydrogen Bonding*, Oxford University Press, Oxford (1997); (e) G. R. Desiraju and T. Steiner, *The Weak Hydrogen Bond*, IUCr Monographs on Crystallography 9, Oxford University Press, Oxford (1999).
15. (a) C. B. Aakeröy and K. R. Seddon, *Chem. Soc. Rev.*, **22**, 397–407 (1993); (b) C. V. K Sharma and G. R. Desiraju, in *Perspectives in Supramolecular Chemistry, Vol. II. The Crystal as a Supramolecular Entity*, ed. G. R. Desiraju, Wiley, New York, pp. 31–61 (1996).
16. G. R. Desiraju, *Angew. Chem., Int. Ed. Engl.*, **34**, 2311–27 (1995).
17. E. J. Corey, *Pure Appl. Chem.*, **14**, 19–37 (1967).
18. A survey of the frequency of occurrence in organic crystal structures of 75 hydrogen-bonded ring-type supramolecular synthons can be found in F. H. Allen, W. D. S. Motherwell, P. R. Raithby, G. P. Shields and R. Taylor, *New J. Chem.*, **23**, 25–34 (1999).
19. C. B. Aakeröy and D. S. Leinen, in *Crystal Engineering: From Molecules and Crystals to Materials*, ed. D. Braga, F. Grepioni and A. G. Orpen, Kluwer, Dordrecht, pp. 89–106 (1999).
20. (a) V. R. Thalladi, H.-C. Weiss, D. Bläser, R. Boese, A. Nangia, G. R. Desiraju, *J. Am. Chem. Soc.*, **120**, 8702–10 (1998); (b) V. R. Thalladi, S. Brasselet, H.-C. Weiss, D. Bläser, A. K. Katz, H. L. Carrell, R. Boese, J. Zyss, A. Nangia and G. R. Desiraju, *J. Am. Chem. Soc.*, **120**, 2563–77 (1998); (c) P. J. Langley, J. Hulliger, R. Thaimattam and G. R. Desiraju, *New J. Chem.*, **22**, 1307–9 (1998); (d) V. S. S. Kumar and A. Nangia, *Chem. Commun.* 2392–3 (2001); (e) A. Nangia, *Cryst. Eng.*, **4**, 49–59 (2001).
21. I. Dance, in *Perspectives in Supramolecular Chemistry: The Crystal as a Supramolecular Entity*, ed. G. R. Desiraju, Wiley, New York, pp. 137–233 (1996).
22. (a) Use of the domain model in the context of hydrogen bonding involving metal complexes is also discussed in Ref. 22b; (b) L. Brammer, in *Implications of Molecular and Materials Structure for New Technologies*, ed. J. A. K. Howard, F. H. Allen and G. P. Shields, Kluwer, Dordecht, pp. 197–210 (1999).

23. (a) M. A. Adams, K. Folting, J. C. Huffman and K. G. Caulton, *Inorg. Chem.*, **18**, 3020–3 (1979); (b) L. M. Epstein, E. S. Shubina, A. N. Krylov, A. Z. Kreindlin and M. I. Ribinskaya, *J. Organomet. Chem.*, **447**, 277–80 (1993); (c) S. A. Fairhurst, R. A. Henderson, D. L. Hughes, S. K. Ibrahim and C. J. Pickett, *J. Chem. Soc., Chem. Commun.*, 1569–70 (1995); (d) E. Peris and R. H. Crabtree, *J. Chem. Soc., Chem. Commun.*, 2179–80 (1995); (e) D. Braga, F. Grepioni, E. Tedesco, K. Biradha and G. R. Desiraju, *Organometallics*, **15**, 2692–99 (1996).
24. (a) L. Brammer, J. M. Charnock, P. L. Goggin, R. J. Goodfellow, A. G. Orpen and T. F. Koetzle, *J. Chem. Soc., Dalton Trans.*, 1789–98 (1991); (b) L. Brammer, M. C. McCann, R. M. Bullock, R. K. McMullan and P. Sherwood, *Organometallics*, **11**, 2339–41 (1992); (c) L. Brammer and D. Zhao, *Organometallics*, **13**, 1545–7 (1994); (d) D. Zhao, F. T. Ladipo, J. Braddock-Wilking, L. Brammer and P. Sherwood, *Organometallics*, **15**, 1441–5 (1996); (e) L. Brammer, J. C. Mareque Rivas and D. Zhao, *Inorg. Chem.*, **37**, 5512–8 (1998); (f) L. Brammer, J. C. Mareque Rivas and C. D. Spilling, *J. Organomet. Chem.*, **609**, 36–43 (2000).
25. (a) F. Calderazzo, G. Fachinetti, F. Marchetti and P. F. Zanazzi, *J. Chem. Soc., Chem. Commun.*, 181–3 (1981); (b) F. Cecconi, C. A. Ghilardi, P. Innocenti, C. Mealli, S. Midollini and A. Orlandini, *Inorg. Chem.*, **23**, 922–9 (1984); (c) I. C. M. Wehman-Ooyevaar, D. M. Grove, H. Kooijman, P. van der Sluis, A. L. Spek and G. van Koten, *J. Am. Chem. Soc.*, **114**, 9916–24 (1992); (d) I. C. M. Wehman-Ooyevaar, D. M. Grove, P. de Vaal, A. Dedieu and G. van Koten, *Inorg. Chem.*, **31**, 5484–93 (1992); (e) S. G. Kazarian, P. A. Hamley and M. Poliakoff, *J. Am. Chem. Soc.*, **115**, 9069–79 (1993); (f) A. Albinati, F. Lianza, B. Müller and P. S. Pregosin, *Inorg. Chim. Acta*, **208**, 119–22 (1993); (g) A. Albinati, F. Lianza, P. S. Pregosin and B. Müller, *Inorg. Chem.*, **33**, 2522–6 (1994); (h) E. S. Shubina and L. M. Epstein, *J. Mol. Struct.*, **265**, 367–84 (1992); (i) E. S. Shubina, A. N. Krylov, A. Z. Kreindlin, M. I. Rybinskaya and L. M. Epstein, *J. Mol. Struct.*, **301**, 1–5 (1993); (j) D. Braga, F. Grepioni, E. Tedesco, K. Biradha and G. R. Desiraju, *Organometallics*, **16**, 1846–56 (1997); (k) Y. Gao, O. Eisenstein and R. H. Crabtree, *Inorg. Chim. Acta*, **254**, 105–11 (1997); (l) G. Orlova and S. Scheiner, *Organometallics*, **17**, 4362–7 (1998); (m) J. Kozelka, J. Bergès, R. Attias and J. Fraitag, *Angew. Chem., Int. Ed. Engl.*, **39**, 198–201 (2000).
26. (a) G. Aullón, D. Bellamy, L. Brammer, E. A. Bruton and A. G. Orpen, *Chem. Commun.*, 653–4 (1998); (b) L. Brammer, E. A. Bruton and P. Sherwood, *New J. Chem.*, **23**, 965–8 (1999); (c) L. Brammer, E. A. Bruton and P. Sherwood, *Cryst. Growth Des.*, **1**, 277–90 (2001).
27. (a) L. Brammer, J. C. Mareque Rivas, R. Atencio, S. Fang and F. C. Pigge, *J. Chem. Soc., Dalton Trans.*, 3855–67 (2000); (b) C. B. Aakeröy and A. M. Beatty, *Chem. Commun.*, 1067–8 (1998); (c) C. B Aakeröy, A. M. Beatty and D. S. Leinen, *Angew. Chem., Int. Ed. Engl.*, **38**, 1815–9 (1999).
28. (a) D. Braga, F. Grepioni, P. Sabatino and G. R. Desiraju, *Organometallics*, **13**, 3532–43 (1994); (b) D. Braga, K. Biradha, F. Grepioni, V. R. Pedireddi and G. R. Desiraju, *J. Am. Chem. Soc.*, **117**, 3156–66 (1995); (c) K. Biradha, G. R. Desiraju, D. Braga and F. Grepioni, *Organometallics*, **15**, 1284–95 (1996); (d) D. Braga, F. Grepioni, K. Biradha and G. R. Desiraju, *J. Chem. Soc., Dalton Trans.* 3925–30 (1996); (e) D. Braga, F. Grepioni, E. Tedesco, H. Wadepohl, and S. Gebert, *J. Chem. Soc., Dalton Trans.*, 1727–32 (1997). (f) D. Braga, P. de Leonardis, F. Grepioni, E. Tedesco, M. J. Calhorda, *Inorg. Chem.*, **37**, 3337–48 (1998).
29. S. B. Copp, K. T. Holman, J. O. S. Sangster, S. Subramaniam and M. J. Zaworotko, *J. Chem. Soc., Dalton Trans.*, 2233–43 (1995).
30. (a) M.-L. Tong, H. K. Lee, X.-M. Chen, R.-B. Huang and T. C. W. Mak, *J. Chem. Soc., Dalton Trans.*, 3657–9 (1999); (b) A. J. Blake, S. J. Hill, P. Hubberstey and W.-S.

Li, *J. Chem. Soc., Dalton Trans.*, 913–4 (1997); (c) J. Lu, T. Paliwala, S. C. Lim, C. Yu, T. Niu and A. Jacobson, *Inorg. Chem.*, **36**, 923–9 (1997); (d) L. Carlucci, G. Ciani, D. Prosperio and A. Sironi, *J. Chem. Soc., Dalton Trans.*, 1801–3 (1997); (e) H.-J. Chen, M.-L. Tong and X.-M. Chen, *Inorg. Chem. Commun.*, **4**, 76–78 (2001).

31. (a) C. Pettinari, F. Marchetti, A. Cingolani, R. Pettinari, S. R. Troyanov and A. Drozdov, *J. Chem. Soc., Dalton Trans.*, 831–6 (2000); (b) F. A. Cotton and T. R. Felthouse, *Inorg. Chem.*, **21**, 2667–75 (1982). (c) M. A. Withersby, A. J. Blake, N. R. Champness, P. A. Cooke, P. A. Hubberstey, W.-S. Li and M. Schröder, *Inorg. Chem.*, **38**, 2259–66 (1999).
32. (a) J. Klunker, M. Bidermann, W. Schafer and H. Hartung, *Z. Anorg. Allg. Chem.*, **624**, 1503–8 (1998); (b) R. Fuchs and P. Klüfers, *J. Organomet. Chem.* **424**, 353–70 (1992); (c) H. Hoberg, G. Heger, C. Krüger, Y.-H. Tsay, *J. Organomet. Chem.*, **348**, 261–78 (1988).
33. For example, see (a) M. C. Navarro-Ranninger, M. J. Carnazon, A. Alvares Validez, J. R. Masaguer, S. Martinez Carrera and S. Garcia Blanco, *Polyhedron*, **6**, 1059–64 (1987); (b) R. Melanson, F. D. Rochon, *Acta Crystallogr.*, **C40**, 793–5 (1984); (c) M.-C. Corbeil and A. L. BeauChamp, *Can. J. Chem.*, **66**, 1379–85 (1988).
34. For example, see (a) S. F. Paloponi, S. J. Geib, A. L. Rheingold and T. B. Brill, *Inorg. Chem.*, **27**, 2963–71 (1988); (b) C. K. Prout, V. S. B. Mtetwa and F. J. C. Rossotti, *Acta Crystallogr.*, **B49**, 73–9 (1993).
35. (a) F. M. Raymo and J. F. Stoddart, *Chem. Ber.* **129**, 981–90 (1996). (b) S. J. Loeb, in *Comprehensive Supramolecular Chemistry*, Vol. 1, volume ed. G. W. Gokel; Series eds J. L. Atwood, J. E. D. Davies, D. D. MacNicol and F. Vögtle, Pergamon Press, Oxford, pp. 733–53 (1996); (c) H. M. Colquhoun, J. F. Stoddart and D. J. Williams, *J. Chem. Soc., Chem. Commun.* 847–8 (1981); (d) H. M. Colquhoun, S. M. Doughty, S. M. Maud, J. F. Stoddart, D. J. Williams and J. B. Wolstenholme, *Isr. J. Chem.* **25**, 15–26 (1985).
36. (a) J. D. Dunitz and R. Taylor, *Chem. Eur. J.*, **3**, 89–98 (1997); (b) J. A. K Howard, V. J. Hoy, D. O'Hagan and G. T. Smith, *Tetrahedron*, **52**, 12613–22 (1996); (c) P. Murray-Rust, W. C. Stallings, C. T. Monti, R. K. Preston and J. P. Glusker, *J. Am. Chem. Soc.* **105**, 3206–14 (1983); (d) L. Shimoni, J. P. Glusker, *Struct. Chem.*, **5**, 383–97 (1994); (e) P. K. Thallapally and A. Nangia, *CrystEngComm*, **3**, 114–9 (2001).
37. (a) The normalized distance follows the definition of Allen and co-workers [37b]. Thus, $R_{HX} = d(H \cdots X)/(r_H + r_X)$, where r_H and r_X are the van der Waals radii [37c] of the hydrogen and halogen atoms, respectively; (b) J. P. M. Lommerse, A. J. Stone, R. Taylor and F. H. Allen, *J. Am. Chem. Soc.*, **118**, 3108–16 (1996); (c) A. J. Bondi, *J. Chem. Phys.*, **68**, 441–51 (1964).
38. E. Peris, J. C. Lee, J. R. Rambo, O. Eisenstein and R. H. Crabtree, *J. Am. Chem. Soc.*, **117**, 3485–91 (1995).
39. L. Brammer, J. K. Swearingen, E. A. Bruton and P. Sherwood, *Proc. Natl. Acad. Sci. USA*, **99**, 4956–61 (2002).
40. L. Brammer, E. A. Bruton and P. Sherwood, unpublished work.
41. (a) J. C. Mareque Rivas and L. Brammer, *Inorg. Chem.*, **37**, 4756–7 (1998); (b) G. R. Lewis and A. G. Orpen, *Chem. Commun.*, 1873–4 (1998); (c) for further comments on bifurcated $N–H \cdots Cl_2M$ hydrogen bonds, see also G. A. Yap, A. L. Rheingold, P. Das and R. H. Crabtree, *Inorg. Chem.*, **34**, 3474–6 (1995); H. Dadon and J. Bernstein, *Inorg. Chem.*, **36**, 2898–2900 (1997).
42. (a) A. L. Gillon, G. R. Lewis, A. G. Orpen, S. Rotter, J. Starbuck, X.-M. Wang, Y. Rodríguez-Martín and C. Ruiz-Pérez, *J. Chem. Soc., Dalton Trans.*, 3897–905 (2000); (b) A. Angeloni and A. G. Orpen, *Chem. Commun.*, 343–4 (2001); (c) B. Dolling, A. L. Gillon, A. G. Orpen, J. Starbuck and X.-M. Wang, *Chem. Commun.*, 567–8 (2001).

43. (a) D. B. Mitzi, *J. Chem. Soc., Dalton Trans.*, 1–12 (2001); (b) D. B. Mitzi, *Prog. Inorg. Chem.*, **48**, 1–121 (1999).
44. (a) P. J. Davies, N. Veldman, D. M. Grove, A. L. Spek, B. T. G. Lutz and G. van Koten, *Angew. Chem., Int. Ed. Engl.*, **35**, 1959–61 (1996); (b) C. S. A. Fraser, H. A. Jenkins, M. C. Jennings and R. J. Puddephatt, *Organometallics*, **19**, 1635–42 (2000).
45. (a) J. C. Lee, Jr, A. L. Rheingold, B. Müller, P. S. Pregosin and R. H. Crabtree, *J. Chem. Soc., Chem. Commun.*, 1021–2 (1994); (b) S. Park, R. Ramachandran, A. J. Lough and R. H. Morris, *J. Chem. Soc., Chem. Commun.*, 2201–2 (1994); R. H. Crabtree, P. E. M. Siegbahn, O. Eisenstein, A. L. Rheingold and T. F. Koetzle, *Acc. Chem. Res.*, **29**, 348–54 (1996); (d) an earlier report also suggested weak O–H···H–M hydrogen bonding, see R. C. Stevens, R. Bau, D. Milstein, O. Blum and T. F. Koetzle, *J. Chem. Soc., Dalton Trans.*, 1429–32 (1990).
46. (a) T. B. Richardson, S. de Gala, R. H. Crabtree amd P. E. M. Siegbahn, *J. Am. Chem. Soc.*, **117**, 12875–6 (1995); (b) W. T. Klooster, T. F. Koetzle, P. E. M. Siegbahn, T. B. Richardson and R. H. Crabtree, *J. Am. Chem. Soc.*, **121**, 6337–43 (1999); (c) R. Custelcean and J. E. Jackson, *J. Am. Chem. Soc.*, **120**, 12935–41 (1998); (d) R. Custelcean and J. E. Jackson, *Angew. Chem., Int. Ed. Engl.*, **38**, 1661–3 (1999); (e) R. Custelcean, M. Vlassa and J. E. Jackson, *Angew. Chem., Int. Ed. Engl.*, **39**, 3299–3302 (2000); (f) J. L. Atwood, G. A. Koutsantonis, F.-C. Lee and C. L. Raston, *J. Chem. Soc., Chem. Commun.*, 91–2 (1994); (g) Z. Yu, J. M. Wittbrodt, A. Xia, M. J. Heeg, H. B. Schlegel and C. H. Winter, *Organometallics*, **20**, 4301–3 (2001); (h) J.-W. Hwang, J. P. Campbell, J. Kozubowski, S. A. Hanson, J. F. Evans and W. L. Gladfelter, *Chem. Mater.*, **7**, 517–25 (1995); (i) J. P. Campbell, J.-W. Hwang, V. G. Young, Jr, R. B. Von Dreele, C. J. Cramer and W. L. Gladfelter, *J. Am. Chem. Soc.*, **120**, 521–31 (1998).
47. (a) J. C. Lee, Jr, E. Peris, A. L. Rheingold and R. H. Crabtree, *J. Am. Chem. Soc.*, **116**, 11014–9 (1994); (b) S. C. Gatling and J. E. Jackson, *J. Am. Chem. Soc.*, **121**, 8655–6 (1999).
48. (a) K. Abdur-Rashid, D. G. Gusev, S. E. Landau, A. J. Lough and R. H. Morris, *J. Am. Chem. Soc.*, **120**, 11826–7 (1998); (b) D. G. Gusev, A. J. Lough and R. H. Morris, *J. Am. Chem. Soc.*, **120**, 13138–47 (1998); (c) K. Abdur-Rashid, D. G. Gusev, A. J. Lough and R. H. Morris, *Organometallics*, **19**, 834–43 (2000); (d) K. Abdur-Rashid, A. J. Lough and R. H. Morris, *Can. J. Chem.*, **79**, 964–76 (2001).
49. (a) F. A. Cotton, G. Wilkinson, C. A. Murillo and M. Bochmann, *Advanced Inorganic Chemistry*, 6th edn, Wiley, New York (1999); (b) J. E. Huheey, E. A. Keiter and R. L. Keiter, *Inorganic Chemistry*, 4th edn, Harper Collins, New York, 1993.
50. (a) C. B. Aakeröy and A. M. Beatty, *Cryst. Eng.*, **1**, 39–49 (1998); (b) C. B. Aakeröy, A. M. Beatty and D. S. Leinen, *J. Am. Chem. Soc.*, **120**, 7383–4 (1998).
51. (a) C. B. Aakeröy and A. M. Beatty, *J. Mol. Struct.*, **474**, 91–101 (1999); (b) B. Cova, A. Briceño and R. Atencio, *New J. Chem.*, **25**, 1516–9 (2001).
52. B.-C. Tzeng, A. Schier and H. Schmidbaur, *Inorg. Chem.*, **38**, 3978–84 (1999).
53. C. B. Aakeröy, A. M. Beatty and B. A. Helfrich, *J. Chem. Soc., Dalton Trans.*, 1943–5 (1998).
54. (a) J. C. Mareque Rivas and L. Brammer, *New J. Chem.*, **22**, 1315–8 (1998); (b) the chloride ions in [Pt(nicotinamide)$_4$]Cl$_2$ are held by hydrogen bonds in channels that lie between the cation layers. The anions lie at intervals of ca 4 Å along the channels; (c) a second polymorph of [Pt(nicotinamide)$_4$](PF$_6$)$_2$·H$_2$O was also prepared in which all four amide groups lie on he same side of the metal coordination plane (cf. **XX**) leading to a cation bilayer arrangement sustained by amide–amide and amide–water hydrogen bonds [54d]; (d) J. C. Mareque Rivas, PhD Thesis, University of Missouri–St. Louis (1999).
55. For an explanation of graph set notation used in characterizing hydrogen bonding patterns, see Ref. 1c and J. Bernstein, R. E. Davis, L. Shimoni and N.-L. Chang, *Angew. Chem., Int. Ed. Engl.*, **34**, 1555–73 (1995).

56. L. P. Wu, M. Yamamoto, T. Kuroda-Sowa, M. Maekawa, Y. Suenaga and M. Munakata, *J. Chem. Soc., Dalton Trans.*, 2031–7 (1996).
57. The structure is a dihydrate, i.e. $[Pt(L)_2(HL)_2] \cdot 2H_2O$(L = isonicotinate).
58. M. Munakata, L. P. Wu, M. Yamamoto, T. Kuroda-Sowa and M. Maekawa, *J. Am. Chem. Soc.*, **118**, 3117–24 (1996).
59. (a) Z. Qin, H. A. Jenkins, S. J. Coles, K. W. Muir and R. J. Puddephatt, *Can. J. Chem.*, **77**, 155–7 (1999); (b) Z. Qin, M. C. Jennings and R. J. Puddephatt, *Inorg. Chem.*, **40**, 6220–8 (2001).
60. Z. Qin, M. C. Jennings, R. J. Puddephatt and K. W. Muir, *CrystEngComm*, **2**, 73–6 (2000).
61. J. C. Mareque Rivas and L. Brammer, unpublished work.
62. (a) C. J. Kuehl, F. M. Tabellion, A. M. Arif and P. J. Stang, *Organometallics*, **20**, 1956–9 (2001); (b) the solvent of crystallization present in $[Rh(\eta^5\text{-}C_5Me_5)(L)_3](CF_3SO_3)_2$ is not clearly specified. The crystal structure data suggest ca $0.5(CH_3)_2CO$, whereas elemental analysis suggests one molecule of H_2O per rhodium complex.
63. (a) A. D. Burrows, C.-W. Chan, M. M. Chowdhry, J. E. McGrady and D. M. P. Mingos, *Chem. Soc. Rev.*, **24**, 329–39 (1995); (b) M. M. Chowdhry, A. D. Burrows, D. M. P. Mingos, A. J. P. White and D. J. Williams, *J. Chem. Soc., Chem. Commun.*, 1521–2 (1995); (c) A. Houlton, D. M. P. Mingos and D. J. P. Williams, *J. Chem. Soc., Chem. Commun.*, 503–4 (1994); A. D. Burrows, D. M. P. Mingos, A. J. P. White and D. J. Williams, *J. Chem. Soc., Dalton Trans.*, 3805–12 (1996)
64. M. Tadokoro and K. Nakasuji, *Coord. Chem. Rev.*, **198**, 205–18 (2000), and references therein.
65. D. G. Kurth, K. M. Fromm and J.-M. Lehn, *Eur. J. Inorg. Chem.*, 1523–6 (2001).
66. U. Ziener, E. Breuning, J.-M. Lehn, E. Wegelius, K. Rissanen, G. Baum, D. Fenske and G. Vaughan, *Chem. Eur. J.*, **6**, 4132–9 (2000).
67. (a) I. Goldberg, *Chem. Eur. J.*, **6**, 3863–70 (2000); (b) R. Krishna Kumar, S. Balasubramanian and I. Goldberg, *Chem. Commun.*, 1435–6 (1999); (c) Y. Diskin-Posner, R. Krishna Kumar and I. Goldberg, *New J. Chem.*, **23**, 885–90 (1999); (d) S. Dahal and I. Goldberg, *J. Phys. Org. Chem.*, **13**, 382–7 (2000).
68. (a) P. Bhyrappa, S. R. Wilson and K. S. Suslick, *J. Am. Chem. Soc.*, **119**, 8492–8502 (1997); (b) K. Kobayashi, M. Koyanagi, K. Endo, H. Masuda and Y. Aoyama, *Chem. Eur. J.*, **4**, 417–24 (1998).
69. N. C. Gianneschi, E. R. T. Tiekink and L. M. Rendina, *J. Am. Chem. Soc.*, **122**, 8474–9 (2000).
70. E. Breuning, U. Ziener, J.-M. Lehn, E. Wegelius and K. Rissanen, *Eur. J. Inorg. Chem.*, 1515–21 (2001).
71. N. Prokopuk, C. S. Weinert, D. P. Siska, C. L. Stern and D. F. Shriver, *Angew. Chem., Int. Ed. Engl.*, **39**, 3312–5 (2000).
72. A. Nangia and G. R. Desiraju, *Chem. Commun.*, 605–6 (1999).
73. (a) A. D. Burrows, S. Menzer, D. M. P. Mingos, A. J. White and D. J. Williams, *J. Chem. Soc., Dalton Trans.*, 4237–40 (1997); (b) M. T. Allen, A. D. Burrows and M. F. Mahon, *J. Chem. Soc., Dalton Trans.*, 215–21 (1999).
74. (a) A. J. Blake, P. Hubberstey, U. Suksangpanya and C. L. Wilson, *J. Chem. Soc., Dalton Trans.*, 3873–80 (2000); (b) A. S. Batsanov, M. J. Begley, M. W. George, P. Hubberstey, M. Munakata, C. E. Russell and P. H. Walton, *J. Chem. Soc., Dalton Trans.*, 4251–9 (1999).
75. (a) B. Ahrens, P. G. Jones and A. K. Fischer, *Eur. J. Inorg. Chem.*, 1103–10 (1999); (b) B. Ahrens, S. Friedrichs, R. Herbst-Irmer and P. G. Jones, *Eur. J. Inorg. Chem.*, 2017–29 (2000).
76. J. C. MacDonald, P. C. Dorrenstein, M. M. Pilley, M. M. Foote, J. L. Lundberg, R. W. Henning, A. J. Schultz and J. L. Manson, *J. Am. Chem. Soc.*, **122**, 11692–702 (2000).

77. (a) C. V. K. Sharma and M. J. Zaworotko, *Chem. Commun.*, 2655–6 (1996); (b) G. Ferguson, C. Glidewell, R. M. Gregson and P. R. Meehan, *Acta Crystallogr.*, **B54**, 129–38 (1998); (c) G. Ferguson, C. Glidewell, R. M. Gregson and P. R. Meehan, *Acta Crystallogr.*, **B54**, 139–50 (1998); (d) G. Ferguson, C. Glidewell, R. M. Gregson, P. R. Meehan and I. L. J. Patterson, *Acta Crystallogr.*, **B54**, 151–61 (1998); (e) G. Ferguson, C. Glidewell, R. M. Gregson and E. S. Lavendar, *Acta Crystallogr.*, **B55**, 573–90 (1999); (f) G. Ferguson, C. Glidewell and E. S. Lavendar, *Acta Crystallogr.*, **B55**, 591–600 (1999).
78. N. Moliner, M. C. Muñoz and J. A. Real, *Inorg. Chem. Commun.*, **2**, 25–27 (1999).
79. (a) C. B. Aakeröy, A. M. Beatty, D. S. Leinen and K. R. Lorimer, *Chem. Commun.*, 935–6 (2000); (b) C. B. Aakeröy, A. M. Beatty and K. R. Lorimer, *J. Chem. Soc., Dalton Trans.*, 3869–72 (2000); (c) see also M. A. S. Goher and T. C. W. Mak, *Inorg. Chim. Acta*, **101**, L27–L30 (1985).
80. R. Sekiya and S. Nishikiori, *Chem. Commun.*, 2612–3 (2001).
81. P. J. Zapf, R. L. LaDuca, Jr, R. S. Rarig, Jr, K. M. Johnson, III and J. Zubieta, *Inorg. Chem.*, **37**, 3411–4 (1998).
82. A. L. Gillon, A. G. Orpen, J. Starbuck, X.-M. Wang, Y. Rodríguez-Martín and C. Ruiz-Pérez, *Chem. Commun.*, 2287–8 (1999).
83. T. J. Prior and M. J. Rosseinsky, *Chem. Commun.*, 1222–3 (2001).
84. B. Du, E. A. Meyers and S. G. Shore, *Inorg. Chem.*, **40**, 4353–60 (2001).
85. C. L. Schauer, E. Matwey, F. W. Fowler and J. W. Lauher, *J. Am. Chem. Soc.*, **119**, 10245–6 (1997).
86. P. J. Zapf, R. C. Haushalter and J. Zubieta, *Chem. Commun.*, 321–2 (1997).
87. C. P. Brock and L. L. Duncan, *Chem. Mater.*, **6**, 1307–12 (1994).
88. (a) Y. Duscausoy, J. Protas, J. Bensancon and S. Top, *J. Organomet. Chem.*, **94**, 47–53 (1975); (b) A. J. Pearson, S. Mallik, R. Mortezaei, M. W. D. Perry, R. J. Shively and W. J. Youngs, *J. Am. Chem. Soc.*, **112**, 8034–42 (1990); (c) see also 1,1‴-bis(hydroxymethyl)biferrocene in T.-Y. Dong, C.-K. Chang, S.-H. Lee, L.-L. Lai, M. Y.-N. Chang and K.-J. Lin, *Organometallics*, **16**, 5816–25 (1997).
89. S. C. Tenhaeff, D. R. Tyler and T. J. R. Weakley, *Acta Crystallogr.*, **C48**, 162–3 (1992).
90. D. Braga, F. Grepioni, D. Walther, K. Heubach, A. Schmidt, W. Imhof, H. Görls and T. Klettke, *Organometallics*, **16**, 4910–9 (1997).
91. R. A. Bartsch, P. Kus, R. A. Holwerda, B. P. Czech, X. Kou and N. K. Dalley, *J. Organomet. Chem.*, **522**, 9–19 (1996).
92. M. Hobi, O. Ruppert, V. Gramlich and A. Togni, *Organometallics*, **16**, 1384–91 (1997).
93. (a) G. Ferguson, J. F. Gallagher, C. Glidewell and C. M. Zakaria, *Acta Crystallogr.*, **C49**, 967–71 (1993); (b) Y. Li, G. Ferguson, C. Glidewell and C. M. Zakaria, *Acta Crystallogr.*, **C50**, 857–61 (1994); (c) J. F. Gallagher, G. Ferguson, C. Glidewell and C. M. Zakaria, *Acta Crystallogr.*, **C50**, 18–23 (1994).
94. C. Glidewell, G. Ferguson, A. J. Lough and C. M. Zakaria, *J. Chem. Soc., Dalton Trans.*, 1971–82 (1994).
95. D. Braga, L. Maini, F. Paganelli, E. Tagliavini, S. Casolari and F. Grepioni, *J. Organomet. Chem.*, **637–639**, 609–615 (2001).
96. M. N. G. James and G. J. B. Williams, *Acta Crystallogr.*, **B30**, 1249–57 (1974).
97. Y. Hsiou, Y. Wang and L.-K. Liu, *Acta Crystallogr.*, **C45**, 721–4 (1989).
98. (a) C. Brown, *Acta Crystallogr.*, **21**, 1–5 (1966); (b) A. L. Bednowitz and B. Post, *Acta Crystallogr.*, **21**, 566–71 (1966).
99. C. Pedone and A. Sirigu, *Inorg. Chem.*, **7**, 2614–8 (1968).
100. (a) Benzoic acid itself associates via carboxyl dimer formation [95b], as do (η^6-4-^{t}BuC_6H_4COOH)Cr$(CO)_3$ [95c] and 4-^{t}BuC_6H_4COOH [95d]; (b) R. Feld, M. S.

Lehman, K. W. Muir and J. C. Speakman, *Z. Kristallogr.*, **157**, 215–31 (1981); (c) H. van Koningsveld, *Cryst. Struct. Commun.*, **11**, 1423–33 (1982); (d) F. van Meurs and H. van Koningsveld, *J. Organomet. Chem.*, **78**, 229–35 (1974).

101. (a) D. Braga, L. Miani and F. Grepioni, *Angew. Chem., Int. Ed. Engl.*, **37**, 2240–2 (1998); (b) D. Braga, L. Maini, M. Polito and F. Grepioni, *Organometallics*, **18**, 2577–9 (1999); (c) D. Braga, G. Cojazzi, L. Maini, M. Polito and F. Grepioni, *Chem. Commun.*, 1949–50 (1999); (d) D. Braga, L. Maini, M. Polito, M. Rossini and F. Grepioni, *Chem. Eur. J.*, **6**, 4227–35 (2000); (e) D. Braga, L. Maini, F. Grepioni, A. De Cian, O. Félix, J. Fischer and M. W. Hosseini, *New J. Chem.*, **24**, 547–53 (2000); (f) D. Braga, L. Maini, F. Grepioni, C. Elschenbroich, and O. Schiemann, *Organometallics*, **20**, 1875–81 (2001); (g) F. Grepioni, M. Rossini and D. Braga, *CrystEngComm*, **3**, 36–40 (2001); (h) D. Braga, G. Cojazzi, D. Emiliani, L. Maini and F. Grepioni, *Chem. Commun.*, 2272–3 (2001).
102. (a) G. J. Palenik, *Inorg. Chem.*, **8**, 2744–9 (1969); (b) F. Tawusagawa and T. F. Koetzle, *Acta Crystallogr.*, **B35**, 2888–96 (1979).
103. (a) P. Štěpnička, J. Podlaha, R. Gyepes and M. Polášek, *J. Organomet. Chem.*, **552**, 293–301 (1998); (b) P. Štěpnička, I. Císařová, J. Podlaha, J. Ludvík and M. Nejezchleba, *J. Organomet. Chem.*, **582**, 319–27 (1999); (c) P. Štepnička, R. Gyepes and J. Podlaha, *Collect. Czech. Chem. Commun.*, **63**, 64–74 (1998).
104. (a) M. M. Kubicki, P. Richard, B. Gautheron, M. Viotte, S. Toma, M. Hudecek and V. Gajda, *J. Organomet. Chem.*, **476**, 55–62 (1994); (b) K. Sunkel and D. Steiner, *J. Organomet. Chem.*, 3**68**, 67–76 (1989).
105. (a) R. Taylor and O. Kennard, *J. Am. Chem. Soc.*, **104**, 5063–70 (1982); (b) G. R. Desiraju, *Acc. Chem. Res.*, **24**, 270–6 (1991); (c) G. R. Desiraju, *Acc. Chem. Res.*, **29**, 441–9 (1996); T. Steiner, *Crystallogr. Rev.*, **6**, 1–57 (1996).
106. T. Steiner and G. R. Desiraju, *Chem. Commun.*, 891–2 (1998).
107. J. Kroon and J. A. Kanters, *Nature*, **248**, 667–9 (1974).
108. O. R. Wulf, U. Liddell and S. B Hendricks, *J. Am. Chem. Soc.*, **58**, 2287–93 (1936).
109. W. D. Kumler, *J. Am. Chem. Soc.*, **57**, 600–5 (1935).
110. S. Glasstone, *Trans. Faraday Soc.*, 200–14 (1937).
111. D. Braga, F. Grepioni and E. Tedesco, *Organometallics*, **17**, 2669–72 (1998).
112. T. E. Müller, D. M. P. Mingos and D. J. Williams, *J. Chem. Soc., Chem. Commun.*, 174–5 (1994).
113. T. Steiner and M. Tamm, *J. Organomet. Chem.*, **570**, 235–9 (1998).
114. T. Steiner, M. Tamm., A. Grzegorzewski, N. Schulte, N. Veldman and A. M. M. Schreurs, *J. Chem. Soc., Perkin Trans. 2*, 2441–6 (1996).
115. For a detailed analysis of interactions between aromatic rings, see (a) C. A. Hunter, *Chem. Soc. Rev.*, **23**, 101–9, (1994); see also (b) W. B. Jennings, B. M. Farrell and J. F. Malone, *Acc. Chem. Res.*, **34**, 885–94 (2001); (c) C. A. Hunter and J. K. M. Sanders, *J. Am. Chem. Soc.*, **112**, 5525–34 (1990); C. Janiak, *J. Chem. Soc., Dalton Trans.*, 3885–96 (2000).
116. (a) I. G. Dance and M. Scudder, *Chem. Eur. J.*, **2**, 481–6 (1996); (b) I. G. Dance and M. Scudder, *J. Chem. Soc., Dalton Trans.*, 2755–69 (1996); (c) M. Scudder and I. G. Dance, *J. Chem. Soc., Dalton Trans.*, 329–44 (1998); (d) M. Scudder and I. G. Dance, *J. Chem. Soc., Dalton Trans.*, 2909–15 (2000); (d) M. Scudder and I. G. Dance, *CrystEngComm*, **3**, 46–9 (2001); (e) I. G. Dance and M. Scudder, *New J. Chem.*, **25**, 1510–5 (2001).
117. J. Yang, J. Yin, K. A. Abhoud and W. M. Jones, *Organometallics*, **13**, 971–8 (1994). Hydrogen bonding interpreted by G. R. Desiraju and T. Steiner [14e].
118. J. C. Mareque Rivas and L. Brammer, unpublished results.
119. (a) D. Braga, A. L. Costa, F. Grepioni, L. Scaccianoce and L. Tagliavini, *Organometallics*, **16**, 2070–9 (1997); (b) D. Braga, A. Angeloni, F. Grepioni and E. Tagliavini,

Organometallics, **16**, 5478–85 (1997); (c) D. Braga, A. Angeloni, F. Grepioni and E. Tagliavini, *Chem. Commun.*, 1447–8 (1997); (d) D. Braga, A. Angeloni, F. Grepioni and A. Tagliavini, *J. Chem. Soc., Dalton Trans.*, 1961–8 (1998); (e) D. Braga and F. Grepioni, *Chem. Commun.*, 911–2 (1998); (f) D. Braga, A. Angeloni, L. Maini, A. W. Götz and F. Grepioni, *New J. Chem.*, **23**, 17–24 (1999); (g) D. Braga, O. Benedi, L. Maini and F. Grepioni, *J. Chem. Soc., Dalton Trans.*, 2611–8 (1999); (h) D. Braga, C. Bazzi, L. Maini and F. Grepioni, *CrystEngComm*, **1**, 15–20 (1999); (i) D. Braga, L. Maini, L. Prodi, A. Caneschi, R. Sessoli and F. Grepioni, *Chem. Eur. J.*, **6**, 1310–7 (2000).

120. M. C. Etter, Z. Urbanczyck-Lipowska, D. A. Jahn and J. S. Frye, *J. Am. Chem. Soc.*, **108**, 5871–6 (1986).
121. D. Braga, F. Grepioni, J. J. Byrne and A. Wolf, *J. Chem. Soc., Chem. Commun.*, 1023–4 (1995).
122. F. Grepioni, S. Gladiali, L. Scaccianoce, P. Ribeiro and D. Braga, *New. J. Chem.*, **25**, 690–5 (2001).
123. D. Braga, L. Maini and F. Grepioni, *J. Organomet. Chem.*, **593–594**, 101–8 (2000).
124. (a) R. H. Crabtree, *Angew. Chem., Int. Ed. Engl.*, **32**, 789–805 (1993); (b) M. Brookhart, M. L. H. Green and L.-L. Wong, *Prog. Inorg. Chem.*, **36**, 1–124 (1998); (c) J. Y. Corey, J. Braddock-Wilking, *Chem. Rev.* **99**, 175–292 (1999).
125. (a) D. B. Mitzi, C. A. Feild, W. T. A. Harrison and A. M. Guloy, *Nature*, **369**, 467–9 (1994); (b) D. B. Mitzi, S. Wang, C. A. Feild, C. A. Chess and A. M. Guloy, *Science*, **267**, 1473–6 (1995); (c) D. B. Mitzi, C. A. Feild, Z. Schlesinger and R. B. Laibowitz, *J. Solid State Chem.*, **114**, 159–63 (1995).
126. (a) D. B. Mitzi, *Chem. Mater.*, **8**, 791–800 (1996); (b) G. C. Papavassiliou and I. B. Koutselas, *Synth. Met.*, **71**, 1713–4, (1995); (c) J. Calabrese, N. L. Jones, R. L. Harlow, N. Herron, D. L. Thorn and Y. Wang, *J. Am. Chem. Soc.*, **113**, 2328–30 (1991).
127. (a) R. D. Willett, H. Place and M. Middleton, *J. Am. Chem. Soc.*, **110**, 8639–50 (1988); (b) G. V. Rubenacker, D. N. Haines, J. E. Drumheller, K. Emerson, *J. Magn. Magn. Mater.*, **43**, 238–42 (1984).
128. (a) F. Neve, A. Crispini, S. Armentano and O. Francescangeli, *Chem. Mater.*, **10**, 1904–3 (1998); (b) F. Neve, A. Crispini and O. Francescangeli, *Inorg. Chem.*, **39**, 1187–94 (2000); (c) F. Neve, O. Francescangeli, A. Crispini and J. Charmant, *Chem. Mater.*, **13**, 2032–41 (2001); (d) F. Neve and A. Crispini, *Cryst. Growth Des.*, **1**, 387–93 (2001).
129. (a) M. Albrecht, M. Lutz, A. L. Spek and G. van Koten, *Nature*, **406**, 970–4 (2000); (b) M. Albrecht, R. A. Gossage, M. Lutz, A. L. Spek and G. van Koten, *Chem. Eur. J.*, **6**, 1431–45 (2000); (c) M. Albrecht, M. Lutz, A. M. M. Scheurs, E. T. H. Lutz, A. L. Spek and G. van Koten, *J. Chem. Soc., Dalton Trans.*, 3797–3804 (2000); (d) see also the *Nature* News and Views commentary, J. W. Steed, *Nature*, **406**, 943–4 (2000).
130. Coordination of SO_2 to the Pt(II) centre involves the behaviour of the metal as a Lewis base using its filled d_{z^2} orbital to donate electron density to the Lewis acidic SO_2. This resembles the behaviour of square planar Pt(II) centres as hydrogen bond acceptors [12a, 24a].
131. (a) I. C. Paul and D. Y. Curtin, *Science.*, **187**, 19–26 (1975); (b) R. S. Miller, D. Y. Curtin and I. C. Paul, *J. Am. Chem. Soc.*, **96**, 6329–34 (1974); (c) R. S. Miller, D. Y. Curtin and I. C. Paul, *J. Am. Chem. Soc.*, **96**, 6334–9 (1974); (d) R. S. Miller, D. Y. Curtin and I. C. Paul, *J. Am. Chem. Soc.*, **96**, 6340–9 (1974).
132. A valuable example of the assessment of a 'failed' or unreliable strategy for crystal synthesis can be found in Ref. 51a. The issue is also raised in Ref. 27a.
133. For reviews on so-called weakly coordinated anions, see (a) S. H. Strauss, *Chem. Rev.*, **93**, 927–42 (1993); (b) C. A. Reed, *Acc. Chem. Res.*, **31**, 133–9 (1998).

134. (a) For examples of deliberate incorporation of counteranions into hydrogen bonding networks, see Refs 73 and 74; (b) a recent example of this sort of approach applied to Ag(I) coordination polymers, which are typically cationic, can be found in L. Brammer, M. D. Burgard, C. S. Rodger, J. K. Swearingen and N. P. Rath, *Chem. Commun.*, 2468–9 (2001).
135. (a) W. C. McCrone, in *Physics and Chemistry of the Organic Solid State*, ed. D. Fox, M. M. Labes and A. Wessenberg, Interscience, New York, vol. 2, pp. 725–67 (1965); (b) J. Bernstein, in *Organic Solid State Chemistry*, ed. G. R. Desiraju Elsevier, Amsterdam, pp. 471–518 (1987); (c) T. Threllfall, *Analyst*, **120**, 2435–60 (1995); (d) J. D. Dunitz, *Pure Appl. Chem.*, **63**, 177–85 (1991); (e) J. Berstein, R. J. Davey and J.-O. Henck, *Angew Chem., Int. Ed. Engl.*, **38**, 3440–61 (1999).
136. M. U. Schmidt, in *Crystal Engineering: From Molecules and Crystals to Materials*, ed. D. Braga, F. Grepioni and A. G. Orpen, Kluwer, Dordrecht, pp. 331–48 (1999).
137. (a) B. F. Hoskins and R. Robson, *J. Am. Chem. Soc.*, **111**, 5962–4 (1989); (b) B. F. Hoskins and R. Robson, *J. Am. Chem. Soc.*, **112**, 1546–54 (1990); (c) R. W. Gable, B. F. Hoskins and R. Robson, *J. Chem. Soc., Chem. Commun.*, 1677–8 (1990); (d) B. F. Abrahams, B. F. Hoskins, J. Liu and R. Robson, *J. Am. Chem. Soc.*, **113**, 3045–51 (1991); (e) B. F. Abrahams, B. F. Hoskins, D. M. Michall and R. Robson, *Nature*, **369**, 727–9 (1994).
138. (a) K. T. Holman, A. M. Pivovar, J. A. Swift and M. D. Ward, *Acc. Chem. Res.*, **34**, 107–18 (2001); (b) V. A. Russell, C. C. Evans, W. Li and M. D. Ward, *Science*, **276**, 575–9 (1997); (c) K. T. Holman and M. D. Ward, *Angew. Chem., Int. Ed. Engl.*, **39**, 1653–6 (2001); (d) A. M. Pivovar, K. T. Holman and M. D. Ward, *Chem. Mater.*, **13**, 3018–31 (2001); (e) J. A. Swift and M. D. Ward, *Chem. Mater.*, **12**, 1501–4 (2000); (f) K. T. Holman, A. M. Pivovar and M. D. Ward, *Science*, **294**, 1907–1911 (2001).
139. (a) O. M. Yaghi, G. Li and H. Li, *Nature*, **378**, 703–6 (1995); (b) H. Li, M. Eddaoudi, M. O'Keefe and O. M. Yaghi, *Nature*, **402**, 276–9 (1999); (c) H. Li, M. Eddaoudi, T. L. Groy and O. M. Yaghi, *J. Am. Chem. Soc.*, **120**, 8571–2 ((1998); (d) B. Chen, M. Eddaoudi, S. T. Hyde, M. O'Keefe and O. M. Yaghi, *Science*, **291**, 1021–3 (2001); (e) M. Eddaoudi, J. Kim, N. Rosi, D. Vodak, J. Wachter, M. O'Keefe and O. M. Yaghi, *Science*, **295**, 469–72 (2002).
140. (a) O. R. Evans and W. Lin, *Chem. Mater.*, **13**, 3009–17 (2001); (b) O. R. Evans, R.-G. Xiong, Z. Wang, G. K. Wong and W. Lin, *Angew. Chem., Int. Ed. Engl.*, **38**, 536–8 (1999); (c) W. Lin, O. R. Evans, R.-G. Xiong and Z. Wang, *J. Am. Chem. Soc.*, **120**, 13272–3 (1998); (d) W. Lin., Z. Wang and L. Ma, *J. Am. Chem. Soc.*, **121**, 11249–50 (1999).
141. O. R. Evans, H. L. Ngo and W. Lin, *J. Am. Chem. Soc.*, **123**, 10395–6 (2001).
142. J. S. Seo, D. Whang, H. Lee, S. I. Jun, J. Oh, Y. J. Jeon and K. Kim, *Nature*, **404**, 982–6 (2000).

Chapter 2

Molecular Recognition and Self-Assembly Between Amines and Alcohols (Supraminols)

Raffaele Saladino

Università Ďegli Studi Ďella Tuscia, Viterbo, Italy

Stephen Hanessian

Université de Montréal, Montréal, Canada

1 INTRODUCTION

The role of the H-bonds involving neutral amidic-type NH and carbonyl-type O atoms (amidic H-bonding) in the controlled assembly of biologically relevant architectures in Nature is of paramount importance [1]. Not surprisingly, amidic-H-bonding has been widely exploited for the design and preparation of a variety of stable, higher-order solid-state supramolecular assemblies [2]. The preparation of these organized molecular frameworks through self-assembly of complementary molecules, has in fact, become a most challenging research endeavour [3]. For example, there is an increasing interest in the field of nonlinear optical materials (NLO), where supramolecular assemblies may play a central role [4]. In this context, the prediction of the physical and chemical properties and the structure and aesthetic quality of the supramolecular product by rational design of the self-assembly by complementary molecules (tectons) [1f,5] is relevant for a novel engineering of the solid state [3d,6]. Studies of self-assembling processes based on amidic H-bonding showed the possibility of forming different supramolecular constructs. Two- and three-dimensional molecular networks [7], extended sheet structures [8], columnar networks [9], molecular tapes [10], and helical aggregates [11] are only

Crystal Design: Structure and Function, edited by G. R. Desiraju

some of the different architectures observed. Despite the numerous reports dealing with the role of amidic H-bonding on the processes of molecular recognition and self-assembly, less attention has been devoted to nonamidic H-bonding, and the prospects of designing new supramolecular architectures. Spectra of the ammonia–water [12] and trimethylamine–water [13] complexes provided first examples of nonamidic H-bonding between amino and hydroxyl groups. In the case of trimethylamine–water, the spectra of all isotopomers were indicative of a structure with an essentially linear single H-bond (1.82 Å). H-bonding between alcohols and amines can play an important role in different physical and chemical phenomena, as revealed by the study of the interaction between native cellulose I or regenerated cellulose II with liquid ammonia [14], alkylamines [15], and other amines [16]. These transformations have been the subject of a number of studies, and complexes of cellulose with amines showed distinct X-ray diffraction patterns [17]. In these complexes, a unit cell containing two almost identical chains with a twofold screw axis symmetry, in which a conformational change for the hydroxymethyl group of the anhydroglucose repeating unit is evident, has been suggested by X-ray and ^{13}C NMR analyses [18]. H-bonding interactions may influence the glass transition temperature (T_g) in polyamine–polyalcohol mixtures [19], where clear maxima are observed in contrast to those of isolated polyalcohol mixtures. The origin of this excess of configurational entropies has been ascribed to the structural change of the H-bonding networks in the polyamine–polyalcohol systems compared to the polyalcohol alone. Moreover, nonamidic H-bonding interactions can effectively compete with metal–ligand coordination in determining the structure of a supramolecular aggregate, as observed during the study of the crystal engineering of organogallium alkoxides derived from primary amines and alcohols [20]. In this case, conformational changes in the metal ligand complex and increased steric bulk on the carbon atom adjacent to the amine group can generate a different H-bonding network.

This chapter is concerned with the assembly of supramolecular motifs based on H-bonding interactions between alcohols and amines. We shall name these 'supraminols' (see Section 1.3). Some useful insights about the effect of the structural and physical properties (as for example the chirality) of the component molecules on the structure and properties of the supramolecular product will be made whenever possible, even if it is widely accepted that predictions of crystal structures from knowledge of chemical composition of the components are far from being generalized [21].

2 SUPRAMOLECULAR ARCHITECTURES BASED ON ALCOHOL–AMINE (OH–NH) H-BONDING

2.1 General Principles

H-bonding interactions between simple amines and alcohols in solution have been demonstrated spectroscopically [13,22] and analytically [23]. The effect of the addition of different concentrations of triethylamine on the IR absorption spectra (in the 3333 cm^{-1} region) of methanol and ethanol in different solvents was described by Baker and co-workers over 50 years ago [24]. They showed that a mixture of alcohol and amine gives rise to a new band at 3226 cm^{-1} which was ascribed to the formation of the alcohol–amine complex. Many years later, the effect of steric parameters on the formation of H-bonds between tertiary amines and linear chain alcohols [25] or branched chain alcohols [26] was reported. The steric bulk of the alcohol decreases the value of the association constant, and with branched chain alcohols, different conformational complexes were observed even though they were present in a 1:1 ratio [26]. In the early 1990s, Kawashima and Hirayama reported the resolution of racemic diols by complex formation with enantiopure diamines [27], but they did not directly discuss the formation of supramolecular assemblies. In 1994, Ermer and Eling described for the first time the general principles for molecular recognition among alcohols and amines utilizing aromatic substrates [28]. This was based on the consideration that the hydroxyl group in alcohols and the amino group in primary amines are complementary in H-bonding, both stoichiometrically and geometrically. The alcoholic hydroxyl group is characterized by one H-bond donor (the hydrogen atom) and two H-bond acceptors (two lone-pairs on the oxygen atom). A primary amine group has two donors (two hydrogens) and one acceptor (one nitrogen lone-pair) (Figure 1).

The structural compatibility of H-bond donors and acceptors in alcohols and primary amines determines a favorable geometry for the formation of a 1:1 alcohol–amine complex in which all the donors and acceptors on each group could participate in H-bonding with a tetrahedral geometry. In this tetrahedral coordination (Figure 1), three H-bonds emanate from every oxygen and nitrogen atom, compared with only two in the free alcohol and amine, respectively, providing the energetic advantage of the process. While isolated crystalline alcohols or amines can form only one H-bond per hydroxy and amino group, respectively, generating H-bonded motifs involving free lone pairs between 1:1 complexes of alcohols and amines could result in various structural possibilities with a full or partial tetrahedral H-bonding geometry. In these complexes there is a general tendency to avoid N(H)N bonds, because they are comparatively long and of low polarity with respect to N(H)O and O(H)N bonds (Figure 2) [28]. Therefore, molecular recognition between alcohols and amines might be expected to be an efficient process, since complementarity is a crucial requisite for the interaction of molecular partners interactions. Moreover, it is clear that the congruence of steric compatibility with

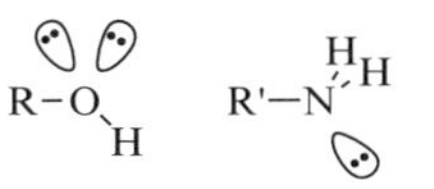

R–O–H   R'–N(H)H

Figure 1 Hypothetical tetrahedral geometry in H-bond donor–acceptor complementarity of alcohols and amines [28].

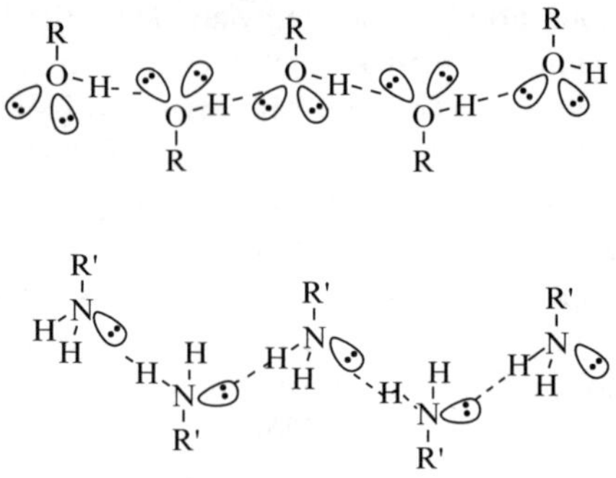

Figure 2 Usual chain-like pattern of H-bonding in isolated crystalline alcohols and amines according to Ermer and Eling [28].

favorable geometry during self-assembly might be responsible for the generation of a high degree of order on a macromolecular scale. Evidence in favor of this consideration may be obtained from the analysis of melting-point diagrams of binary alcohol–amine mixtures, which usually show dystectic maxima (congruent melting points) at 1:1 hydroxy:amino stoichiometry in a complex (Figure 3) [29].

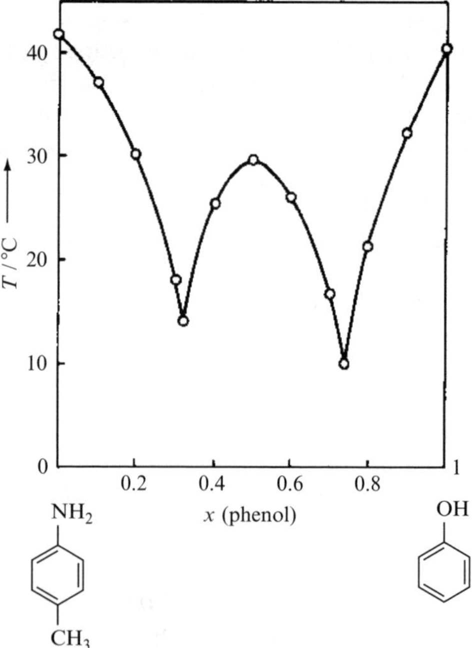

Figure 3 Melting-point diagram of the binary system phenol–*p*-toluidine (adapted from Ref. 29) as cited in Ref. 28.

2.2 Super-tetrahedral H-bonded Alcohol–Amine Architectures

In a 1:1 complex between alcohols and primary amines, the usual zig-zag chains of H-bonds may be joined together alternately by N(H)O interactions to generate sheets of *trans*-fused super-cyclohexane rings with chair-like conformations. From a topological point of view, these super-structures resemble the sheet structure of the stable allotrope of arsenic as super-arsenic sheets in which O(H)O, N(H)N, and N(H)O interactions are involved (Figure 4) [28].

Alternatively, N(H)N bonds may be completely avoided by considering the same 'super-arsenic' motif which consist of N(H)O bonds only (Figure 5) [28].

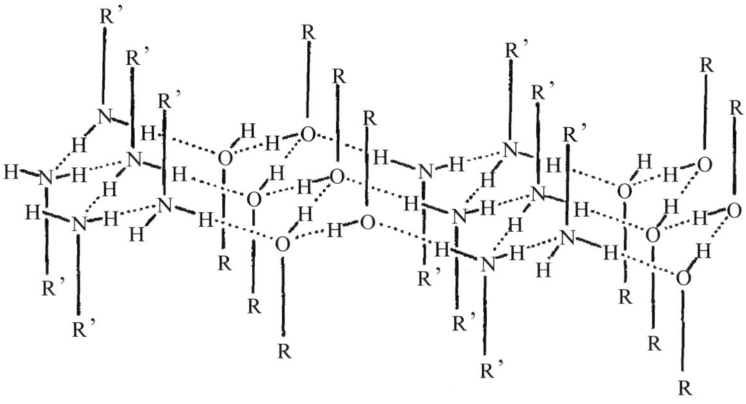

Figure 4 H-bonded structure of a 1:1 alcohol–amine complex of the super-arsenic type [28] involving 1/3 O(H)O, N(H)N, and N(H)O H-bonds each.

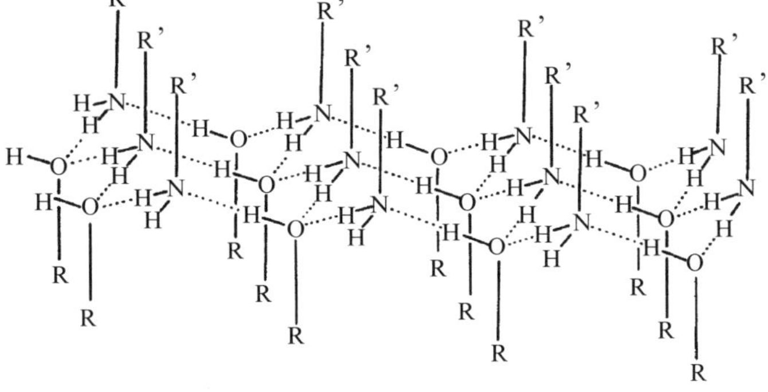

Figure 5 H-bonded structure of 1:1 alcohol–amine complex of the super-arsenic type involving N(H)O bonds exclusively [28].

H-bonded sheets analogous to those of black phosphorus are possible when two thirds of the *trans*-fusion between the super-cyclohexane chairs is replaced by *cis*-fusions. Such a super-black phosphorus type sheet structure may be obtained by cutting through a cubic diamond lattice (Figure 6) rather than a hexagonal one [28].

In the cubic diamond lattice, every oxygen atom is coordinated to three nitrogen atoms and one oxygen atom, and every nitrogen atom to three oxygen atoms and one nitrogen atom. On the other hand, in the hexagonal diamond lattice every oxygen atom is tetrahedrally coordinated to four nitrogen atoms and every nitrogen atom to four oxygen atoms. The tetrahedrons are all oriented in one direction to produce the hexagonal (sixfold rotational) symmetry. In principle, when monofunctional alcohols and amines are replaced by linear dialcohols and diamines of general formula HO–R–OH and NH_2–R–NH_2, respectively, the two-dimensional super-arsenic sheets may be joined to form a three-dimensional super-tetrahedral and, in particular, super-zincblende or super-wurtzite networks (Figure 7).

Other super-structures for 1:1 alcohol–amine complexes from polyhedral H-bond clusters with the shape of a tetrahedron, cube, pentagonal dodecahedron, trigonal-prismatic, truncated-icosahedral, or even extremely hollow truncated-icosahedral are theoretically possible [28] (Figure 8).

2.3 Assembly of Super-tetrahedral H-bonded Alcohol–Amine Architectures [30]

Probably the first published structure of an H-bonded super network between amines and alcohols is that of 4-aminophenol (AMPHOL) described by Brown, although in this study the positions of the H-atoms were not unambiguously determined [31]. The crystal structure of hydrazine monohydrate, $N_2H_4{\bullet}H_2O$, has been reported by Liminga and Olovsson [32]. Single crystals suitable for

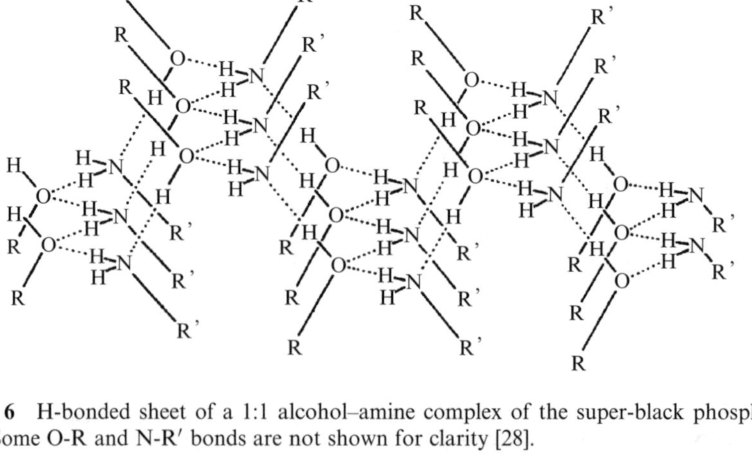

Figure 6 H-bonded sheet of a 1:1 alcohol–amine complex of the super-black phosphorus type. Some O-R and N-R′ bonds are not shown for clarity [28].

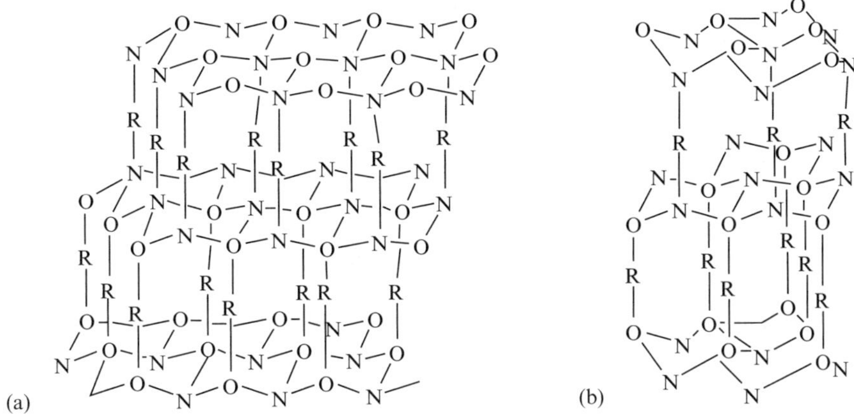

Figure 7 3D super-tetrahedral H-bonded architectures of 1:1 complexes between linear dialcohols and linear diamines. (a) Super-diamond network with a super-adamantane type building block. (b) Super-wurtzite network with combined super-iceane and super-bicyclo [2.2.2]octane-type building blocks [28].

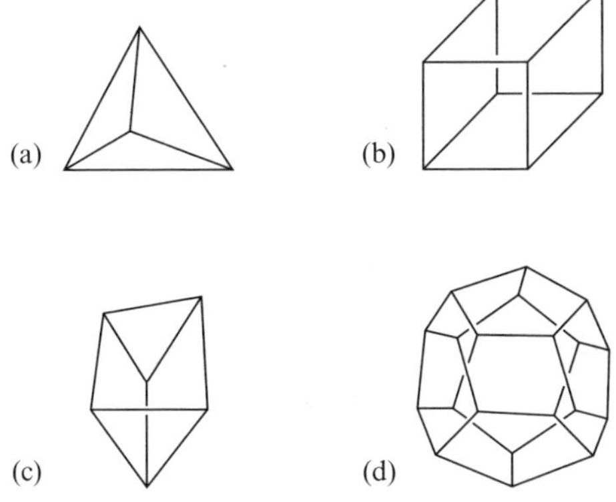

Figure 8 Some examples of hypothetical polyhedral 1:1 alcohol–amine complexes. (a) Tetrahedron; (b) cube; (c) trigonal prism; (d) pentagonal dodecahedron according to Ermer and Eling [28].

X-ray analysis were obtained working at −165 °C. Crystals are trigonal (space group $P3_121$) with three molecules per hexagonal unit cell, where each oxygen atom is H-bonded to six nitrogen atoms of six different molecules of hydrazine. The reverse is true for the water molecules thus the system forms a three-dimensional network. Two ranges of values were found for distances between the oxygen and nitrogen atoms, and the nitrogen atoms did not form H-bonds to each other. Subsequently, Liminga [33] described the crystal structure of hydrazine bis(ethanol). As evidenced by the corresponding freezing-point diagram [34], only one complex, of elemental formula $N_2H_4{\bullet}2C_2H_5OH$, was deduced for the hydrazine **1**–bis(ethanol) **2** system **1•2** with a melting point of −188 °C (Scheme 1).

$$\underset{\mathbf{1}}{H_2N{-}NH_2} + 2\ \underset{\mathbf{2}}{CH_3CH_2OH} \longrightarrow \underset{\mathbf{1{:}2}}{H_2N{-}NH_2 \cdot 2\ CH_3CH_2OH}$$

Scheme 1 Preparation of the complex between hydrazine and ethanol.

Crystals are described as orthorhombic, of the *Pbcn* space group with parameters $a = 18.470$, $b = 4.889$, $c = 8.755$, and $Z = 4$. The structure is characterized by the presence of infinite layers of H-bonded amine and alcohol molecules that are parallel to the *yz* plane. On the basis of the full complementarity between the partners, each nitrogen and oxygen atom is involved in three H-bonds. In particular, the nitrogen atom is acceptor in one H-bond and donor in two. The reverse is true for the oxygen atom. The nitrogen atom binds three oxygen atoms and each oxygen atom is bonded to three nitrogen atoms in a tetrahedral geometry. The bond angles formed by the nitrogen and oxygen atoms deviate less than 7° from the theoretical tetrahedral angle with the only exception of the longer N–O bond (3.06 Å). In the latter case, large deviations from the mean value are observed. The H-bonded layers are connected by van der Waals interactions and the value of the shortest distance between two methyl carbon atoms of two different layers was found to be 4.0 Å (Figure 9).

Ermer and Eling [28] reported the synthesis of various 1:1 alcohol–amine complexes, **3•5**, **3•6**, **4•5**, and **4•6**, starting from hydroquinone (**3**), 4,4′- dihydroxybiphenyl (**4**), 4-phenylenediamine (**5**), and benzidine (**6**), respectively (Scheme 2). These compounds were prepared by dissolving equimolar amounts of the appropriate components in acetone at room temperature with subsequent slow evaporation of the solvent. After elaboration of the X-ray data, the four complexes **3•5**, **3•6**, **4•5**, and **4•6**, showed an H-bonded super-architecture in which every oxygen and nitrogen atoms is engaged in three H-bonds of the O(H)N type, with approximately regular tetrahedral coordination. These three-dimensional architectures are formed by the connection of the two-dimensional super-arsenic type sheets. The H-bonds are arranged in all-*trans* fused super-cyclohexane chairs, held together by the covalent aromatic bridges in a cubic diamond lattice of the zincblende type. As reported by the authors [28], the puckered H-bonded sheets for **3•5**, **3•6**, **4•5**, and **4•6** complexes are very similar, and in all cases there are three crystallographically independent O(H)N bonds, one O–H···N and two O···H–N bonds, as required stoichiometrically.

In Figures 10–13 are illustrated the stereopacking diagrams of complexes **3•5**, **3•6**, **4•5**, and **4•6**. In these cases, a super-adamantane unit may be cut out as a characteristic building block of the super-diamond architecture. The space group of the former super-structure was found to be $P2_1/c$, which is a subgroup of that of cubic diamond, $Fd3m$.

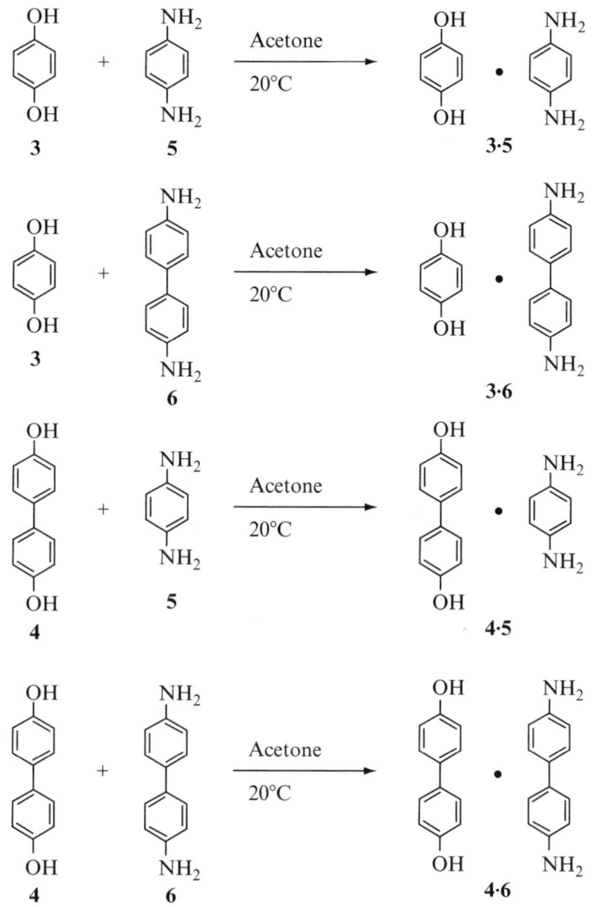

OH NH$_2$ Acetone OH NH$_2$
+ 20°C •
OH NH$_2$ OH NH$_2$
3 5 3·5

NH$_2$ NH$_2$
OH Acetone OH
+ 20°C •
OH OH
3 NH$_2$ NH$_2$
6 3·6

OH OH
NH$_2$ NH$_2$
Acetone
+ 20°C •
NH$_2$ NH$_2$
OH 5 OH
4 4·5

OH NH$_2$ OH NH$_2$
Acetone
+ 20°C •
OH NH$_2$ OH NH$_2$
4 6 4·6

Scheme 2 Preparation of alcohol–amine complexes **3•5**, **3•6**, **4•5**, and **4•6** from the appropriate components [28].

Complexes between linear amino alcohols that contain hydroxy and amino groups within the same molecule such as 4-aminophenol (**7**) and 4-aminobiphenyl-4-ol (**8**) also revealed a super-tetrahedral H-bonded architecture of the wurtzite type. In Figures 14–16 are reported the stereopacking diagrams of **7** and **8**. The building block representing the super-wurtzite network is now the fused super-iceane/super-bicyclo[2.2.2]octane unit (where iceane is used to describe a tetracyclo[5.3.1.1^{2,6}.0^{4,9}]dodecane). In this case, a *Pna*2$_1$ space group was observed.

The question arises as to why the complexes **3•5**, **3•6**, **4•5**, and **4•6** have a super-diamond architecture, whereas **7** and **8** prefer the super-wurtzite alternative. As suggested by the authors [28], the explanation is not obvious since 'both

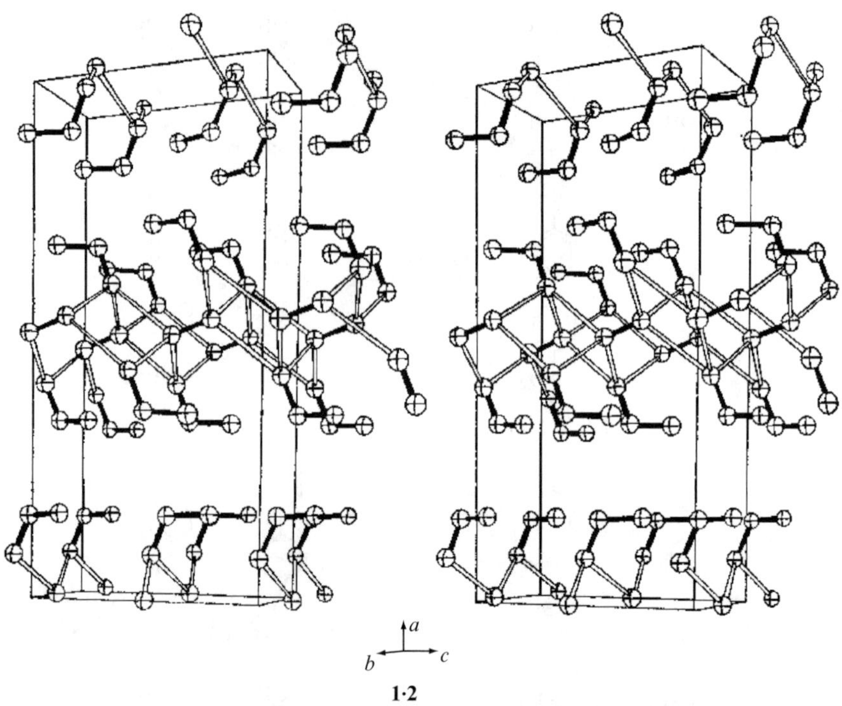

Figure 9 ORTEP representation of the adduct between hydrazine and ethanol. Note the ribbon-like or ladder-like motif of the H-bonding network [33].

packing modes are expected to be energetically very similar' (the hexagonal diamond is only about 1.5 kcal/mol less favored than cubic diamond). A partial answer might be proposed on the basis of the structural properties of the aromatic spacers. In fact, the aromatic spacers in **3•5**, **3•6**, **4•5**, and **4•6** complexes experience a tilt with respect to the normal of the planes of the H-bonded sheets to improve the stacking within the aromatic layers. This tilt was found to be in all cases opposite at oxygen and nitrogen and is satisfied by a diamond-type lattice. In the case of aminophenols **7** and **8**, a wurtzite-type lattice may satisfy this condition, since 'the tilting distortion is antisymmetric with respect to the mean plane through the aromatic layers' [28]. Thus, the authors concluded that the opposite tilting preferences at oxygen and nitrogen determine whether a super-zincblende or a super-wurtzite network is built up. Notably, Desiraju and co-workers [35] showed that regioisomers of 4-aminophenol, such as 2-aminophenol (**9**) and 3-aminophenol (**10**), do not have the same encoded information for the molecular recognition process. The structures of these two isomers, determined by low-temperature neutron diffraction, showed the presence of unexpected N–H $\cdots\pi$ and C–H $\cdots$ O bonding interactions. For example, in the case of 2-aminophenol (**9**), each hydroxy moiety donates one H-bond to the nitrogen atom, and accepts

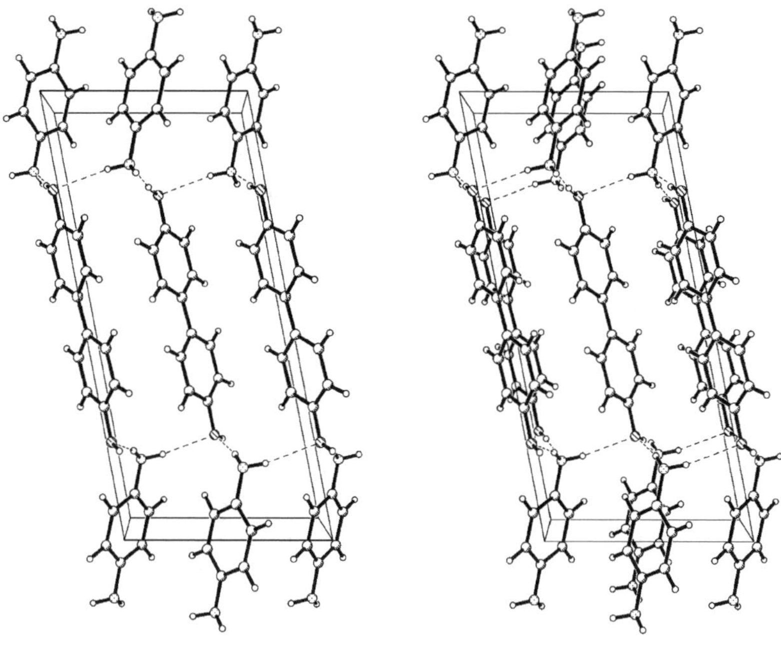

3·5

Figure 10 Stereopacking of diphenol–diamine complexes **3•5** and **3•6** with cell edges outlined. For all left-handed members of stereopairs, the view is along the short cell edge *b* (ca 5.3 Å) defined by a 1,3-distance of the puckered H-bonded super-arsenic sheets [28].

N–H···O and C–H···O bonds with three other molecules. The amino group behaves similarly and participates in the formation of the N–H···π interaction (Figure 17).

The authors [35] suggested that the reason for these anomalous structures could be rationalized 'in terms of the need to attain a herringbone or T-shaped geometry of the phenyl rings' in accordance with data previously reported for the correlation of the structure of fused-ring aromatic hydrocarbons with the corresponding crystal structures [36]. Again, the equilibrium between H-bonding and stacking interactions define the shape of the supramolecular architecture. Crystal and molecular structures of two polymorphs of 3-amino-4-nitrobenzyl alcohols (**11**) have also been reported by X-ray diffraction methods [37]. In both crystals one of the amine hydrogens participates in the intramolecular and intermolecular H-bonds. The second amine and also the hydrogens of the hydroxy group participate in intermolecular H-bonds in both crystals forming a two-dimensional H-bonding network, parallel to the plane *bc* in the first polymorph (form I), and parallel to the plane *ac* in the second polymorph (form II). In form I the hydroxy moiety is both an acceptor and a donor of the H-bond, whereas in form II it is only an H-bond donor in intermolecular H-bonds. Mootz and

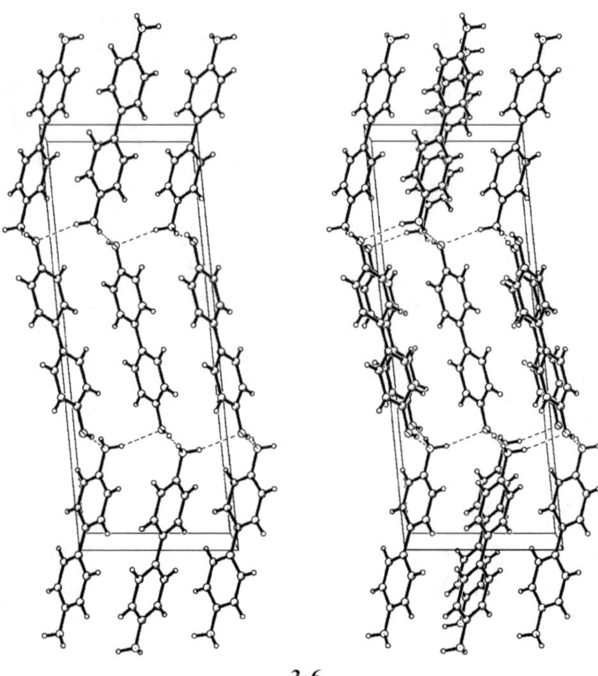

3·6

Figure 11 Stereopacking of diphenol–diamine complexes **3•5** and **3•6** with cell edges outlined. For all left-handed members of stereopairs, the view is along the short cell edge *b* (ca 5.3 Å) defined by a 1,3-distance of the puckered H-bonded super-arsenic sheets [28].

co-workers [38] reported the crystal structure of ethanolamine (**11**), $HOCH_2$–CH_2NH_2, in which the hydroxy and amino groups are separated by a two-carbon ethano spacer. A detailed analysis of this structure reveals a distorted super-tetrahedral architecture formed by strong N(H)O bonds in the diagonal of the *ac* plane. Weaker N(H)O and O(H)N bonds are formed between the main chains. In the three-dimensional network all molecules are H-bonded to six neighbouring ones. These do not generate a super-arsenic network as previously observed for aromatic amino phenols [28]; rather, a super-black phosphorus sheet connected in a three-dimensional network with a pentacyclo $[8.4.0.0^{3.8} \cdot 0^{5.14} \cdot 0^{7.12}]$ tetradecane (isodiamantane, *D*2 symmetry) as a building block is observed (Figure 18a). The structure of diethanolamine is depicted in Figure 18b. In this case two molecules of amino alcohol form a dimer by intermolecular H-bonding of the O(H)O type. The dimers are further linked by strong O(H)N bonds in a one-dimensional tubular super-architecture in which the NH groups point to the inner side.

Meyers and Lipscomb [39] and Donohue [40] reported the X-ray analysis of hydroxylamine (**12**). A detailed revision of these data revealed the presence of a single and characteristic 3D connected network formed by undulating H-bonded sheets with a repetitive covalent O–N bond (Figure 19).

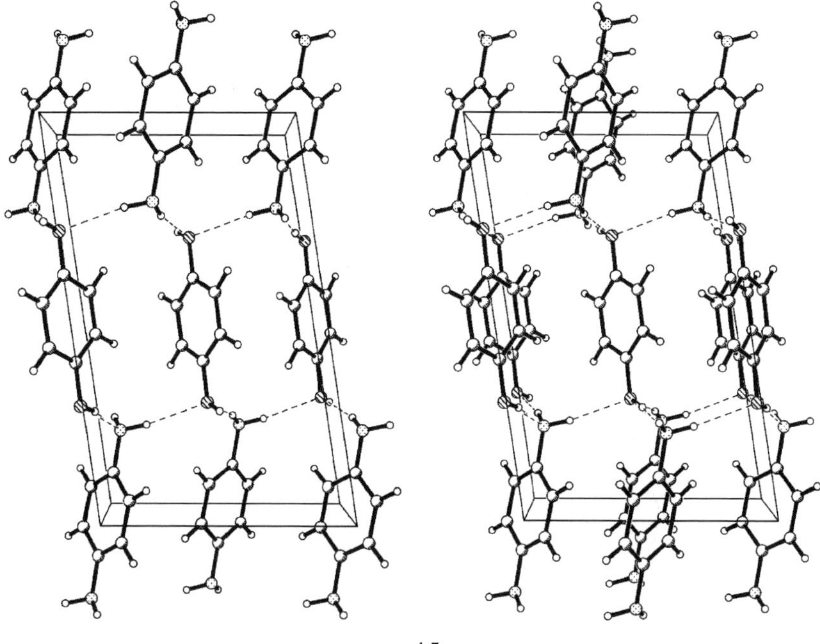

4.5

Figure 12 Stereopacking of diphenol–diamine complexes **4•5** and **4•6** with cell edges outlined. For all left-handed members of stereopairs, the view is along the short cell edge *b* (ca 5.3 Å) defined by a 1,3-distance of the puckered H-bonded super-arsenic sheets.

The H-bonded sheets create distorted super-cyclohexane boats (and not chairs!) that are neither zincblende-nor wurtzite-like super-structures. A careful revision of the literature reveals other interesting examples of super-structures formed by alcohols and amines. For example, during their studies dealing with the preparation of a stable form of hydrazine, Toda and co-workers [41] reported the formation of the supramolecular 1:1 amine–alcohol complex **1•3** between hydrazine (**1**) and hydroquinone (**3**) (Scheme 3).

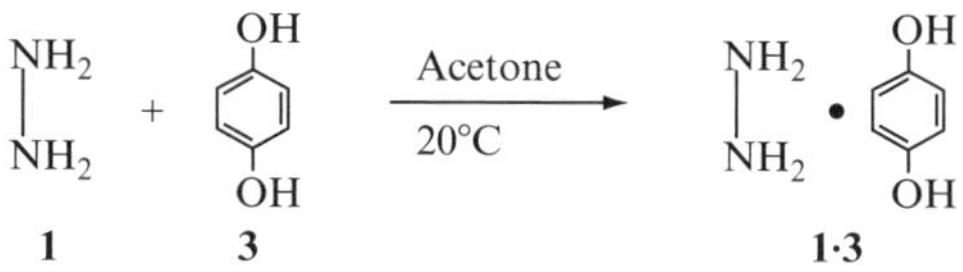

NH_2–NH_2 (**1**) + hydroquinone (**3**, OH/OH) $\xrightarrow[20°C]{Acetone}$ NH_2–NH_2 • hydroquinone (**1·3**)

Scheme 3 Preparation of the complex **1•3** between hydrazine **1** and hydroquinone **3**.

In the crystal of complex **1•3**, hydroquinone and hydrazine molecules form an extensive H-bonding network in which all oxygen and nitrogen atoms are fully coordinated in a super-tetrahedral architecture. In this aggregate, the amino and

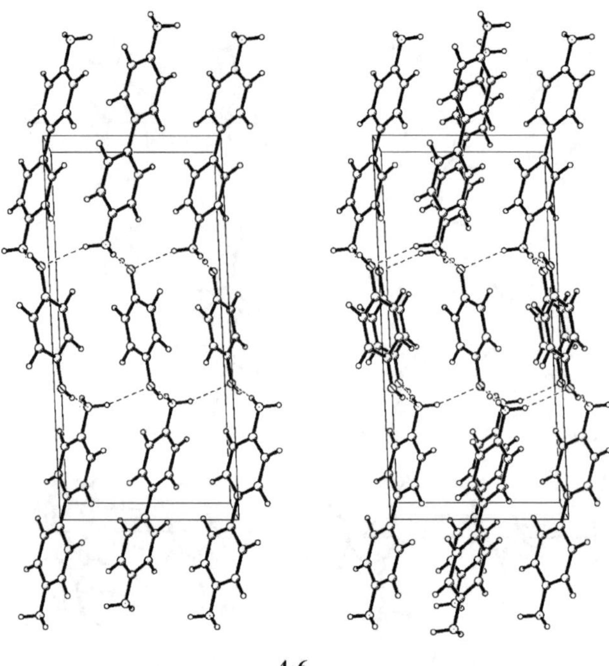

4·6

Figure 13 Stereopacking of diphenol–diamine complexes **4•5** and **4•6** with cell edges outlined. For all left-handed members of stereopairs, the view is along the short cell edge *b* (ca 5.3 Å) defined by a 1,3-distance of the puckered H-bonded super-arsenic sheets.

hydroxy groups generate a layered packing of the molecules in the unit cell, with a super-black phosphorus sheet-like structure, stacked perpendicular to the crystallographic *b* axis with alternating hydrophobic (benzene rings) and hydrophilic (OH and NH_2) regions (Figure 20). It is interesting that the introduction of an aromatic spacer in the hydrazine molecule changes the shape of the aggregate, and affords complexes with a cubic diamond lattice network of the zincblende type.

The same authors [42] reported the synthesis of 1:1 complexes between 2,2′-di(4-hydroxyphenyl)propane (**13**) and several alcohols, primary amines, hydrazine, and methylhydrazine (**14**). In this context the structure of the complex **13•14**, has been elucidated by means of X-ray crystallography (Scheme 4) [43].

The space group observed for this complex was $P2_1/a$. As illustrated in Figure 21, two molecular components are connected exclusively by O(H)N H-bonds, the first connecting one hydroxy group to one amino nitrogen atom and the second connecting the hydroxy group to the methylamine nitrogen atom. The resulting structure is a centrosymmetric cyclic dimer in which the two 2,2′-di(4-hydroxyphenyl) moieties are partially stacked.

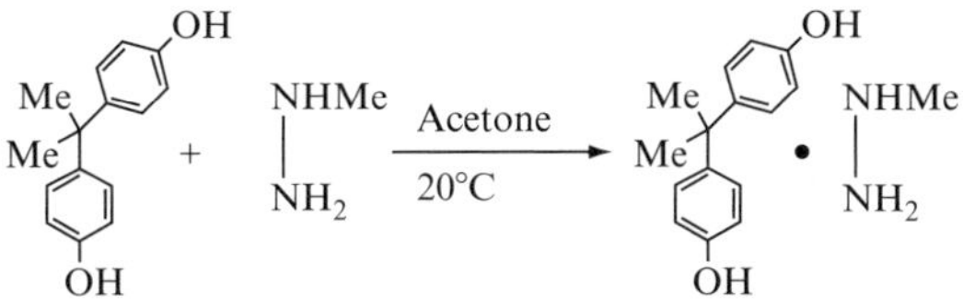

OH Me Me OH + NHMe | NH_2 —Acetone, 20°C→ OH Me Me OH • NHMe | NH_2

13 **14** **13·14**

Scheme 4 Preparation of the complex **13•14** between 2,2′-di(4-hydroxyphenyl)propane (**11**) and methylhydrazine (**12**).

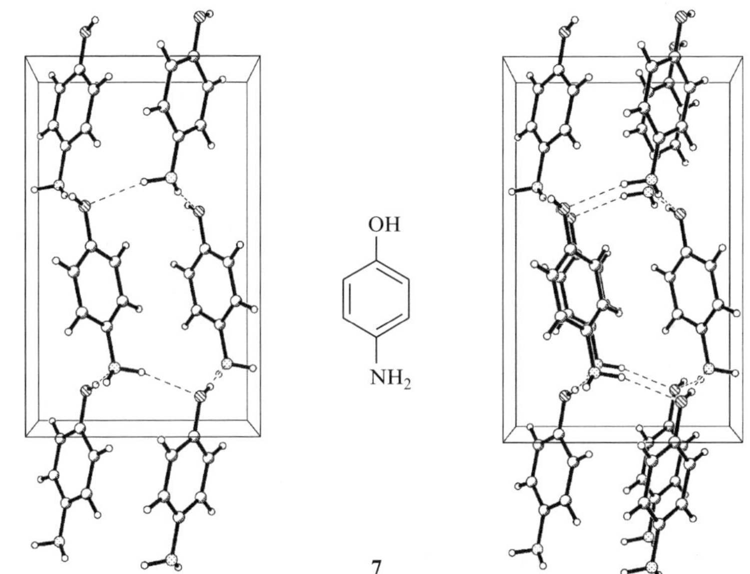

Figures 14 Stereopacking of **7** and **8** with cell edges outlined: O, white; N, grey; C, black; H-bonds, dotted line. For all left-handed members of stereopairs, the view is along the short cell edge *b* (ca 5.3 Å) defined by a 1,3-distance of the puckered H-bonded super-arsenic sheets [28].

Although the cyclic dimers are connected by additional H-bonds, the authors did not observe a super-tetrahedral network. Notably, a 2:1 alcohol–amine complex **14•15**, was also observed [41] between 4-methoxyphenol (**15**) and methylhydrazine (**14**) (Scheme 5). Thus, a dimer formed by a primary amine and a secondary amine in such cases does not have the general requisites to generate a supramolecular architecture.

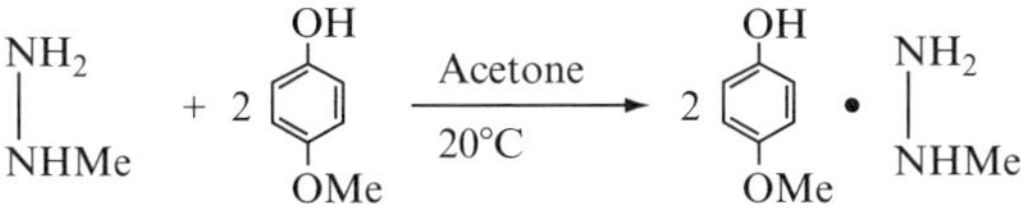

NH_2 | NHMe + 2 OH OMe —Acetone, 20°C→ 2 OH OMe • NH_2 | NHMe

14 **15** **14·15**

Scheme 5 Preparation of the complex **14•15** between 4-methoxyphenol (**15**) and methylhydrazine (**14**).

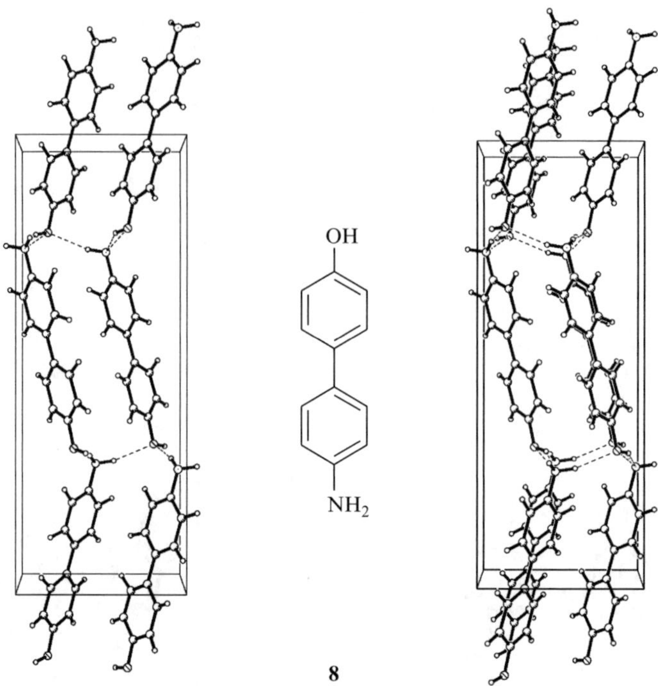

Figure 15 Stereopacking of **7** and **8** with cell edges outlined: O, white; N, grey; C, black; H-bonds, dotted line. For all left-handed members of stereopairs, the view is along the short cell edge *b* (ca 5.3 Å) defined by a 1,3-distance of the puckered H-bonded super-arsenic sheets [28].

The structure found for **14•15** appears to be closely related to that of a previously described complex between hydrazine and hydroquinone, as evidenced by the similar constant lattice and space group. The authors [41] considered the complex **14•15** as formally generated by the replacement of one of the hydrophilic regions located on both sides of the complex between hydrazine and hydroquinone by the 4-methoxy groups (Figure 22). Crystallographic twofold axes characterize the H-bonding network around the hydrazine molecules.

In 1998, Loehlin and co-workers [44] reported the structures of five new crystals of the super-tetrahedral type formed by complementary amines and alcohols. They described the H-bonding networks of three complexes of the diamine type with monoalcohols: 4-phenylenediamine (**5**)–phenol (**16**) (ratio 1:2, as in **5•16**), 4-phenylenediamine (**5**)–4-phenylphenol (**17**) (ratio 1:2, as in **5•17**), and 4-phenylenediamine (**5**)–4-chlorophenol (**18**) (ratio 1:2, as in **5•18**) (Scheme 6). Two complexes of 4-phenylenediamine with diols were also reported: 4-phenylenediamine (**5**)–2,6-dihydroxynaphthalene (**19**) (ratio 1:1, as in **5•19**), and 4-phenylenediamine (**5**)–1,6-hexanediol (**20**) (ratio 1:1, as in **5•20**). The H-bonded networks were

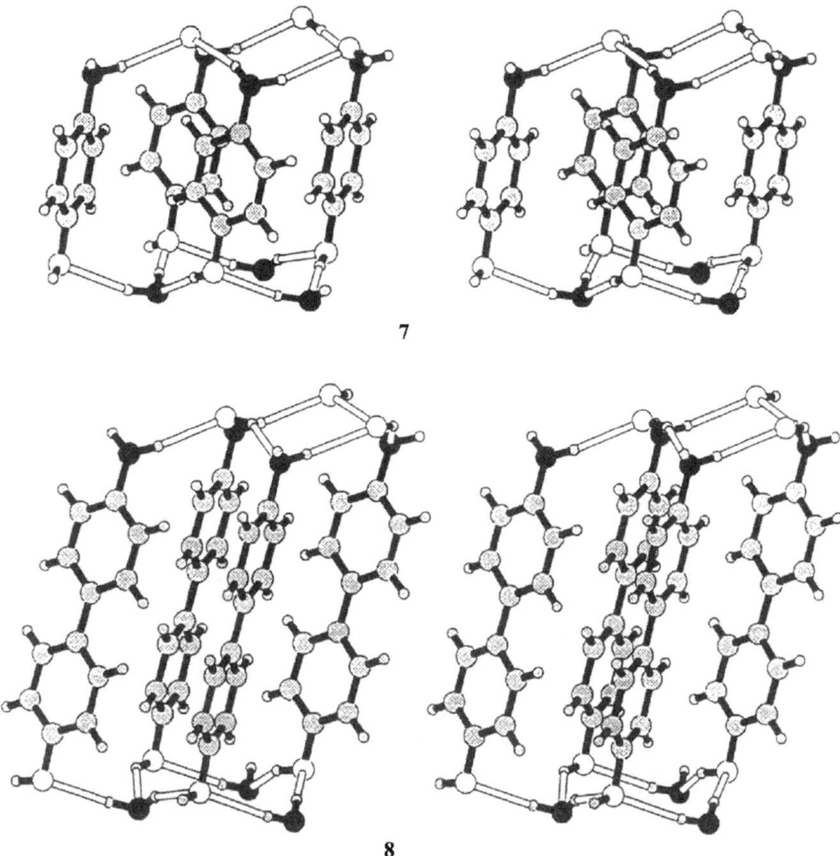

Figures 16 Stereopacking of **7** and **8** with cell edges outlined: O, white; N, grey; C, black; H-bonds, dotted line. For all left-handed members of stereopairs, the view is along the short cell edge *b* (ca 5.3 Å) defined by a 1,3-distance of the puckered H-bonded super-arsenic sheets [28].

defined by the application of the graph-set analysis based on the procedures previously reported by Etter and co-workers [45] and Bernstein and co-workers [46]. In these complexes the donor and acceptor oxygen and nitrogen atoms show a partially distorted tetrahedral geometry, and in the presence of polyfunctional molecules three-dimensional H-bonding networks are formed. In all cases studied, the alcohol acts as donor in the shortest H-bond.

Complexes **5•16**, **5•17**, **5•19**, and **5•20** show the hexagonal-like layer structure that is characteristic of the majority of the known 'saturated H-bonded' complexes between amines and alcohols. ORTEP representations of complexes **5•16**, **5•17**, **5•19**, and **5•20** are shown in Figures 23–26.

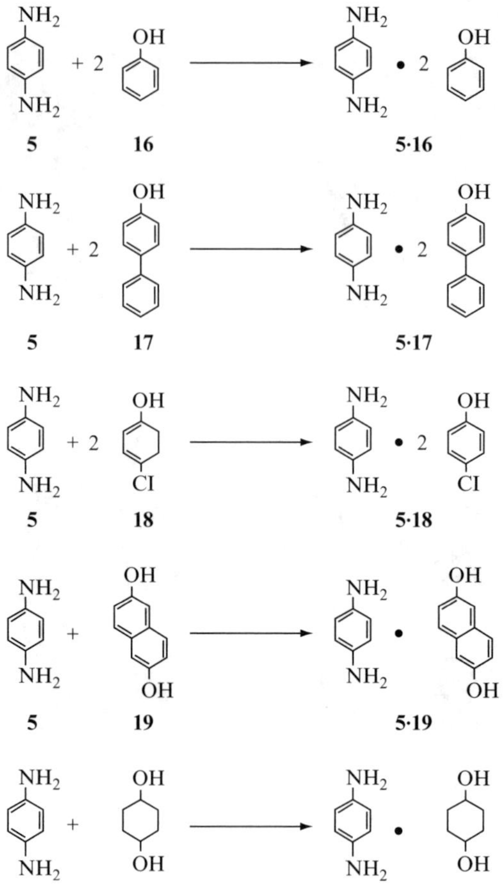

Scheme 6 Preparation of complexes **5•16**, **5•17**, **5•18**, **5•19**, and **5•20**.

Instead, the complex **5•18**, between 4-phenylenediamine and 4-chlorophenol, shows a ladder-like pattern of the H-bond network, that is similar to that previously described by Liminga [32,33] for the complex between hydrazine and ethanol (Figure 27). In this case, the ladder-like pattern did not allow the interactions between chlorine atoms on adjacent layers, losing the extra stabilization observed by Desiraju for many others complexes [47]. The authors described some interesting features for the diol moiety arrangement in the complex 4-phenylenediamine–1,6-hexane diol, **5•20** [44].

In this case, the alcohol moiety is not disposed parallel to the plane, as usually observed for other complexes, but has two ends in different unit cells. Moreover,

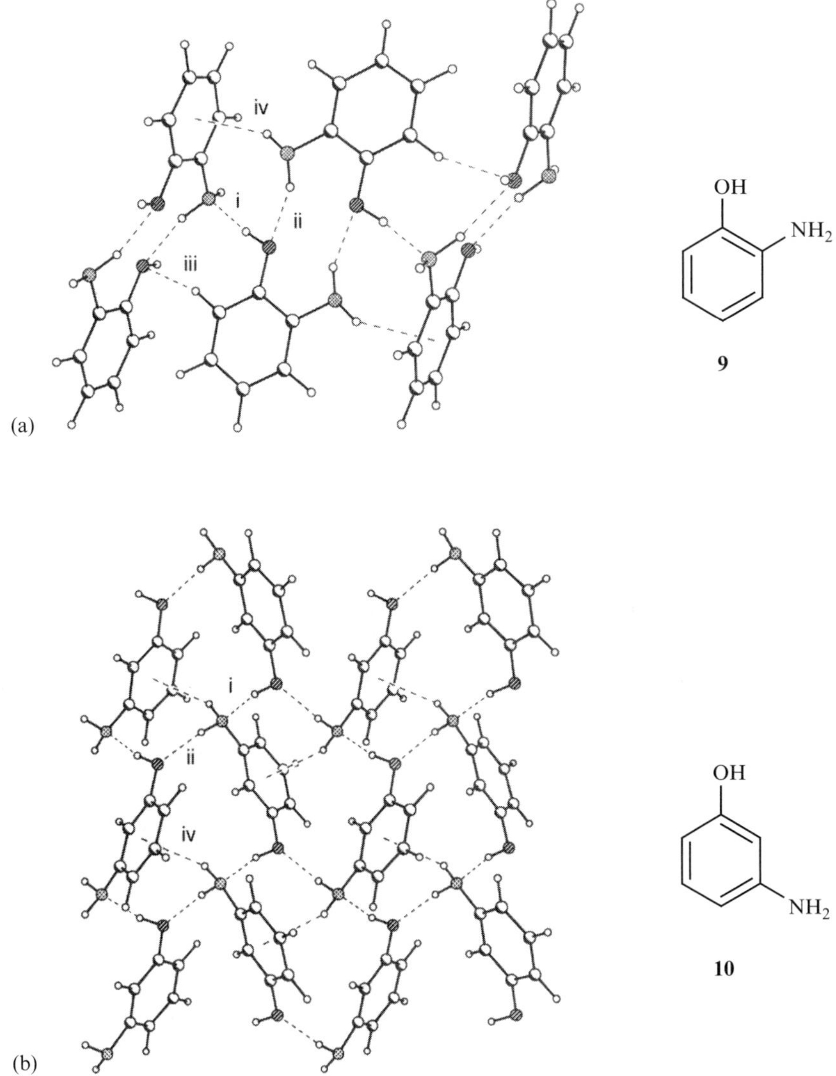

Figure 17 (a) H-bonding network in the crystal structure of 2-aminophenol (**9**). (b) H-bonding network in the crystal structure of 3-aminophenol (**10**) [35].

the diol molecules are inclined by at least 30° with respect to the *a* axis, and by 70° with respect to the *b* axis. Further examples of supramolecular architectures based on H-bonding between the hydroxyl and amino moieties have been reported with oligosiloxanediols [48]. For example, when 1,1,3,3,5,5-

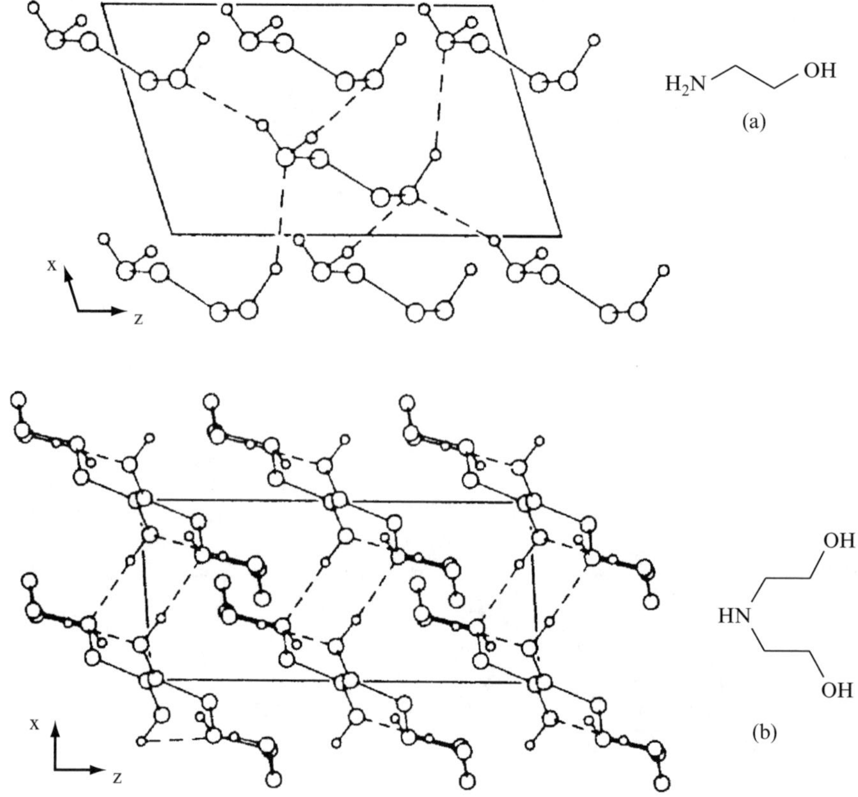

Figure 18 (a) Super-tetrahedral H-bonded 3D network of crystalline ethanolamine. (b) Super-tetrahedral H-bonded 3D network of crystalline diethanolamine. The O(H)N bonds are drawn as thin lines and covalent alkyl linkers as thick black rods [38].

hexaphenyltrisiloxane-1,5-diol (**21**) was treated with pyrazine (**22**), crystals of the complex **21•22** (4:1 ratio) were obtained and analysed by X-ray diffraction (Scheme 7). In the aggregate, the diol units form centrosymmetric dimers by means of ordered OH $\cdots$ H bonds. Pairs of such dimers are linked to the pyrazine by means of OH $\cdots$ N bonds.

$$\mathrm{HO{-}Si(Ph)_2{-}O{-}Si(Ph)_2{-}O{-}Si(Ph)_2{-}OH} \; + \; \text{pyrazine} \longrightarrow 4\,\mathrm{HO{-}Si(Ph)_2{-}O{-}Si(Ph)_2{-}O{-}Si(Ph)_2{-}OH} \bullet \text{pyrazine}$$

21 **22** **21·22**

Scheme 7 Preparation of the complex between 1,1,3,3,5,5-hexaphenyltrisiloxane-1, 5-diol (**21**) and pyrazine (**22**).

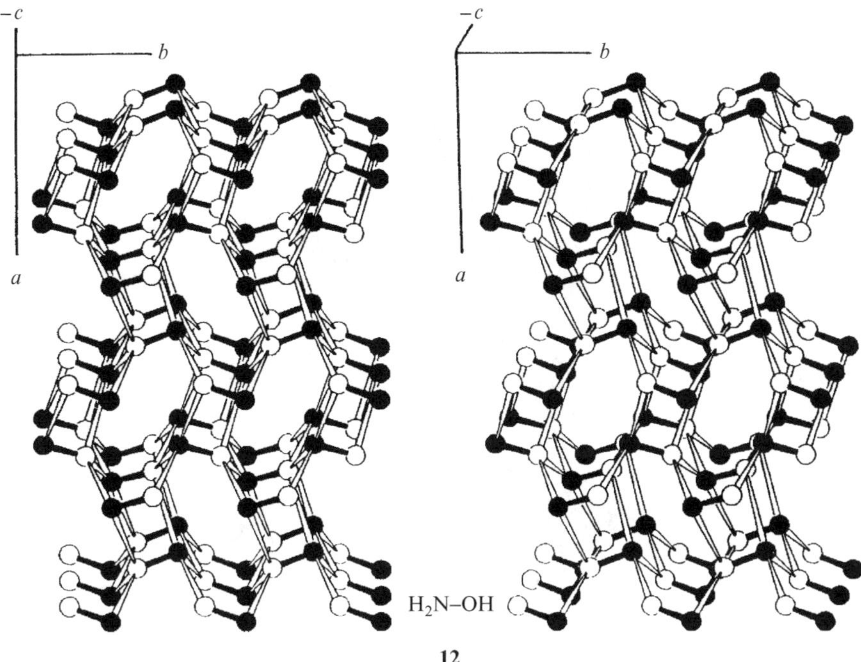

Figure 19 Stereoview of the super-tetrahedral H-bonded 3D network of hydroxylamine (**12**). O(H)N H-bonds are drawn as open white bonds and covalent O–N bonds as filled black sticks [39,40].

1,1,3,3,5,5-Hexaphenyltrisiloxane-1,5-diol (**21**) forms aggregates with pyridine (**23**) in the ratio of 2:3 (Scheme 8). In these aggregates, diol units are again linked to the diol dimer, but only disordered OH···H bonds were present. Two pyridines are linked to the diol dimer by means of ordered OH···N bonds, while the

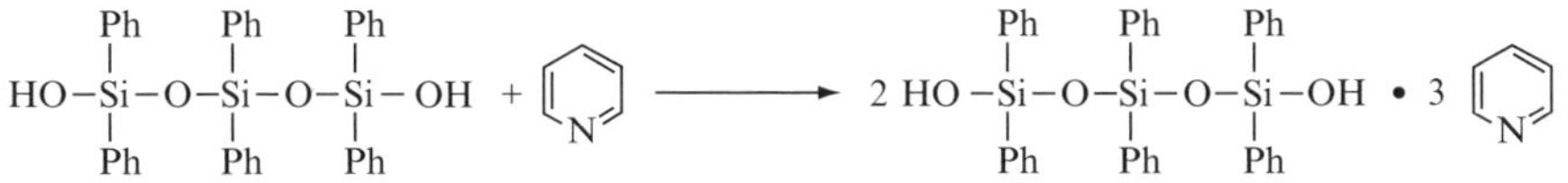

HO–Si(Ph)₂–O–Si(Ph)₂–O–Si(Ph)₂–OH (**21**) + pyridine (**23**) ⟶ 2 HO–Si(Ph)₂–O–Si(Ph)₂–O–Si(Ph)₂–OH • 3 pyridine (**21·23**)

Scheme 8 Preparation of the complex between 1,1,3,3,5,5-hexaphenyltrisiloxane-1,5-diol (**21**) and pyridine (**23**).

third pyridine unit, which is disordered across a centre of inversion, links the diol dimers into a $C3^3(9)$ chain by means of OH···N and CH···O bonds.

A different H-bonding network was obtained with 1,1,3,3-tetraphenyldisiloxane-1,3-diol (**24**) [20]. In the molecular recognition process with hexamethylenetetramine (**25**), the diol acts as a double donor and the diamine as a double

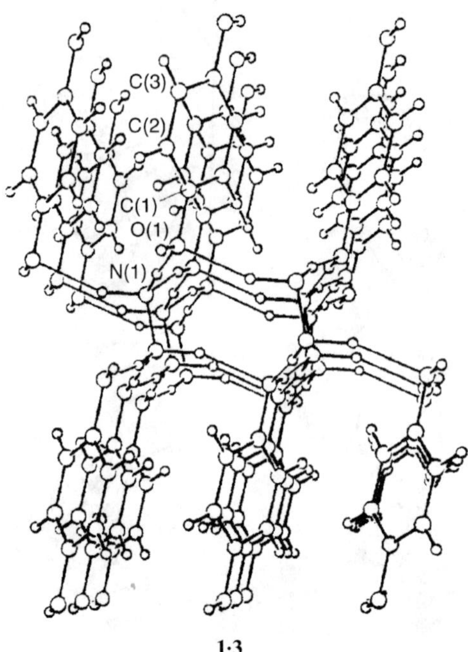

1·3

Figure 20 Crystal packing of the complex **1•3** between hydrazine (**1**) and hydroquinone (**3**). Thin lines illustrate H-bonds [41].

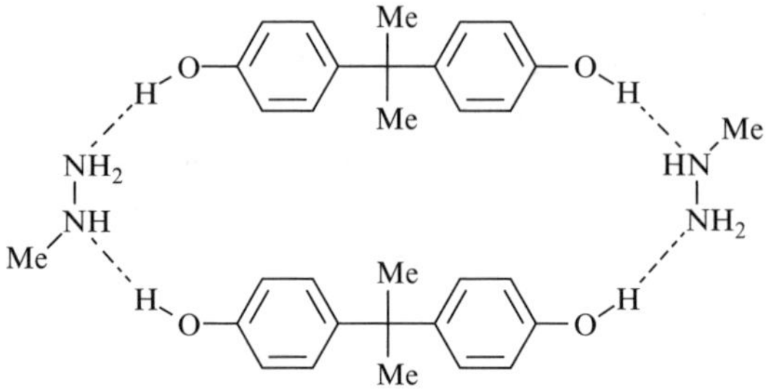

13·14

Figure 21 Schematic representation of the centrosymmetric cyclic dimer **13•14** [42].

acceptor in ordered OH···N bonds to give a 1:1 complex. The structure consists of $C_2^2(10)$ chains of alternating diol and diamine units. Moreover, in the complex between **24** and bipyridyl (**26**) (diol:diamine ratio 1:1), there are two independent molecules, both lying across centres of inversion and therefore both containing linear Si–O–Si groups. Each diol acts as a double donor of H-bonds, and the 2,2′-bipyridyl molecule acts as a double acceptor, thus forming $C_2^2(11)$ chains of alternating diol and amine units (Scheme 9).

$HO{-}Si(Ph)_2{-}O{-}Si(Ph)_2{-}OH$ + 2,2'-bipyridyl ⟶ $HO{-}Si(Ph)_2{-}O{-}Si(Ph)_2{-}OH$ • 2,2'-bipyridyl

24 **26** **24·26**

Scheme 9 Preparation of the complex between 1,1,3,3-tetraphenyldisiloxane-1,3-diol (**24**) and 2,2′-bipyridyl (**26**).

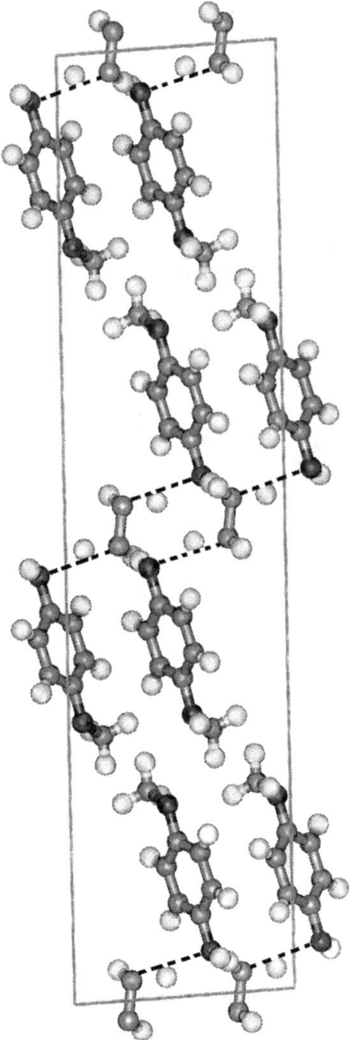

Figure 22 Crystal packing of the complex **14•15** between methylhydrazine (**14**) and 4-methoxyphenol (**15**). Thin lines illustrate H-bonds [41].

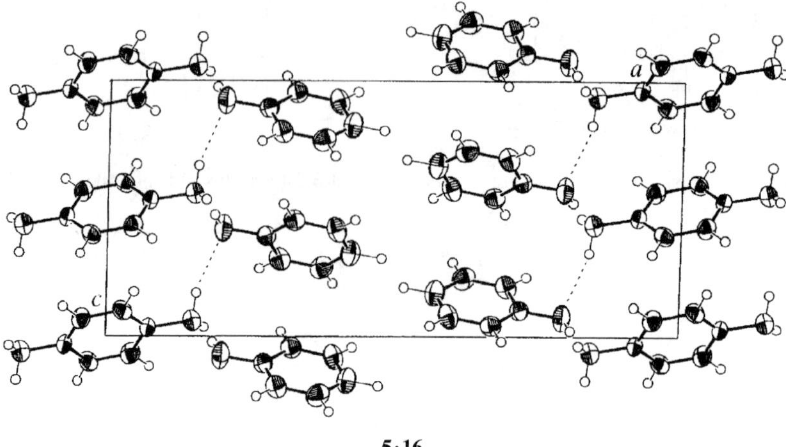

5·16

Figure 23 ORTEP representation of adducts **5•16**. View in the *bc* plane. Note the presence of van der Waals interactions between phenol molecules of separate layers [44].

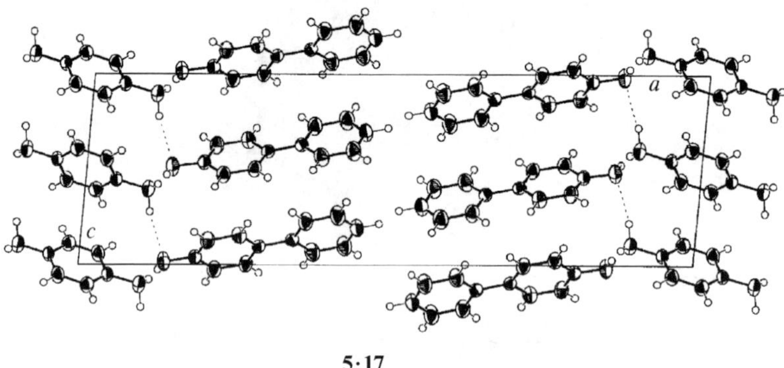

5·17

Figure 24 ORTEP representation of the adduct **5•17** [44].

The structural motif observed in the complex 1,1,3,3-tetraphenyldisiloxane-1,3-diol (**24**) and pyrazine (**22**) (diol:diamine ratio 2:1) (Scheme 10) is a chain of rings in which pairs of diol molecules are linked by OH ··· O bonds into centrosymmetric

$$HO{-}Si(Ph)_2{-}O{-}Si(Ph)_2{-}OH \; + \; C_4H_4N_2 \longrightarrow HO{-}Si(Ph)_2{-}O{-}Si(Ph)_2{-}OH \bullet C_4H_4N_2$$

24 **22** **24·22**

Scheme 10 Formation of the complex between 1,1,3,3-tetraphenyltrisiloxane-1,3-diol (**24**) and pyrazine (**22**).

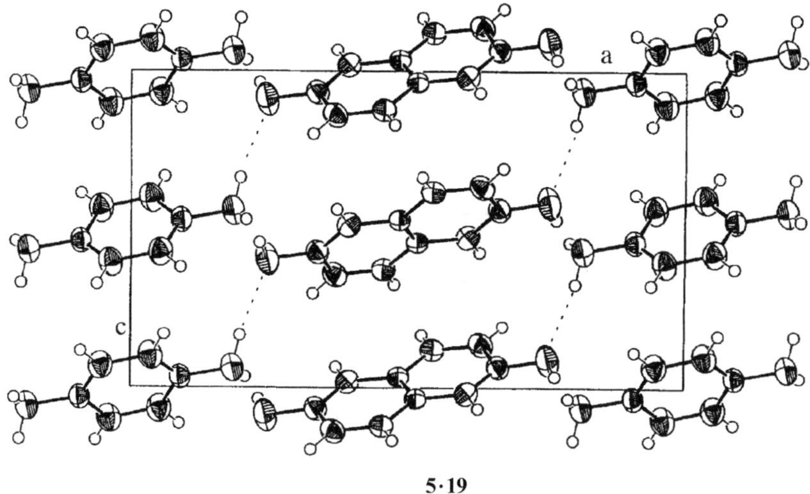

Figure 25 ORTEP representation of the adduct **5•19**. The diol moieties are approximately parallel to the projection [44].

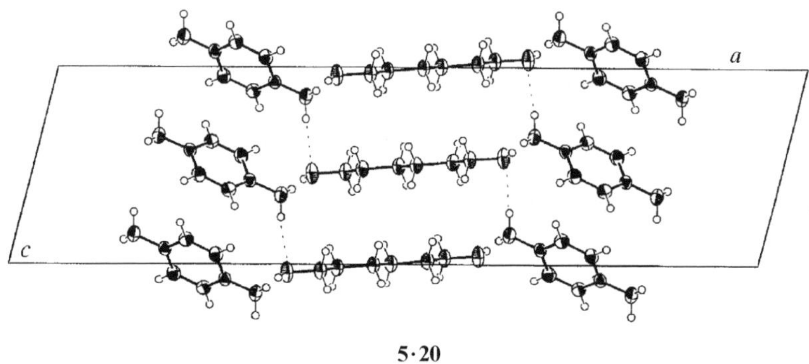

Figure 26 ORTEP representation of the adduct **5•20**. Note that the alcohol moiety is not disposed parallel to the plane and has two ends in different unit cells [44].

$R_2^2(12)$ dimers. These dimers are further linked into $C_2^2(13)$ chains by means of OH · · · N bonds to the pyrazine units.

3 HELICAL H-BONDED MOLECULAR NETWORKS FORMED BY ACHIRAL ALCOHOLS AND ACHIRAL AMINES

Hosseini and co-workers [49] described the synthesis of an infinite single strand helical supramolecular network **27•4** by mixing equimolar amounts of the calix[4]-arene derivative **27**, bearing four pyridine groups as H-bond acceptors, with

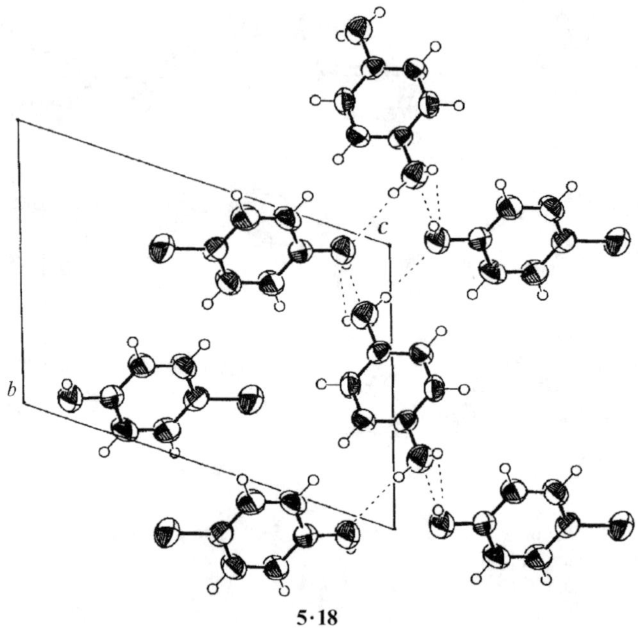

Figure 27 ORTEP representation of the adduct **5•8**. The ladder-like network is orthogonal to the projection. Chlorine atoms on adjacent layers are largely separated.

4,4′-biphenol (**4**) in nitromethane ($MeNO_2$). Single crystals consisting of compounds **27•4** and $MeNO_2$ in a 1:1:1 ratio were formed upon slow cooling of the solution (Scheme 11). See also Chapter 4.

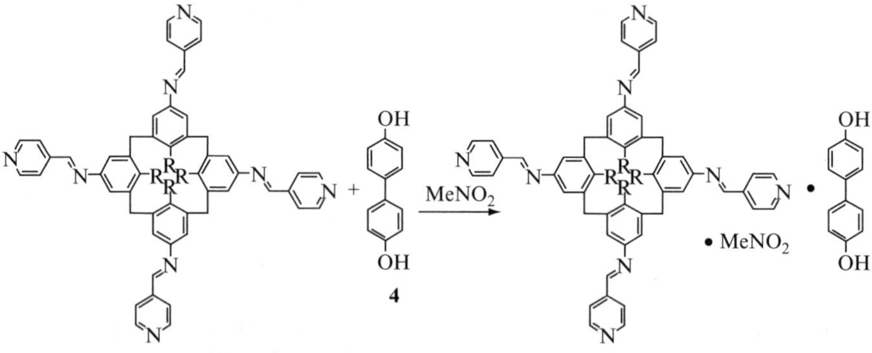

OH
OH
4
$MeNO_2$
• $MeNO_2$
R= $O(CH_2)_2Me$
27
R= $O(CH_2)_2Me$
27·4

Scheme 11 Preparation of the 1:1:1 complex **27•4**, between the calix[4]arene derivative **27**, 4,4′-biphenol (**4**), and $MeNO_2$.

A single-stranded helical network whose pitch along the z axis is composed of four molecules of **27**, four molecules of **4**, and four molecules of $MeNO_2$ was observed. Two of the four pyrimidines are located on opposite sides of the supramolecular architecture and participate in the formation of the helical arrangement through strong H-bonds with O(H)N distances of 2.70 and 2.77 Å, respectively. The others two pyridines are located in close proximity of the two $MeNO_2$ molecules with C···N distances of 3.29 and 3.31 Å, respectively. It was suggested that the observed helical structure is derived from the primary structure of **27**, which possesses an S_4 axis (Figure 28) [49].

Five helical strands associate laterally leading to a quintuple helical braid (Figure 29). The authors [49] suggested that the formation of this supramolecular assembly may be due to edge-to-face interactions between the pyridine units and the phenyl groups in conjunction with van der Waals interactions. Finally, the quintuple helices associate laterally to form the crystal. In this supramolecular synthesis H-bonding plays a role only in the first level of the assembly process via a single-stranded H-bonded helical network.

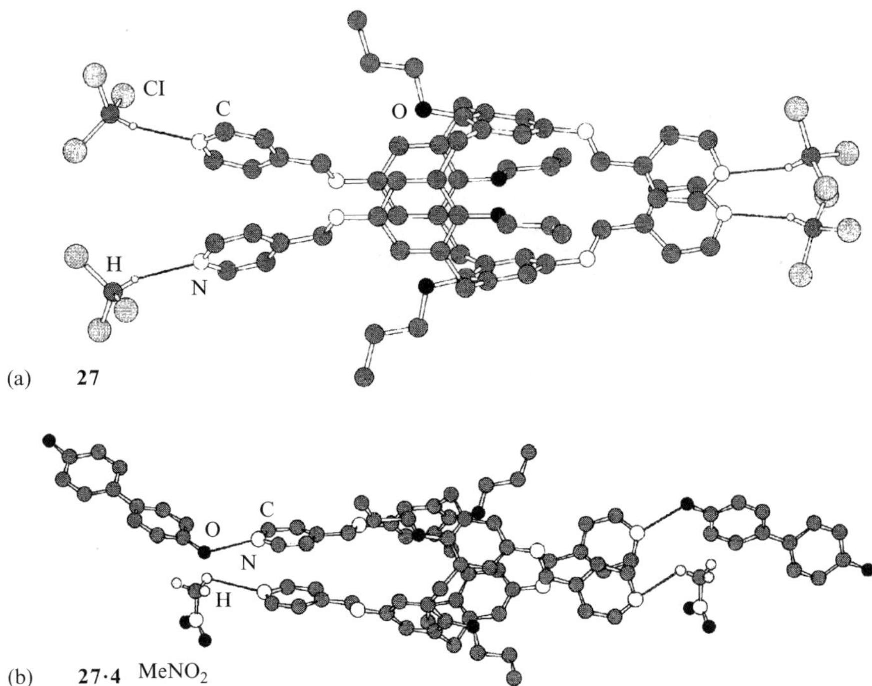

Figure 28 (a) X-ray structure of **27**. (b) X-ray structure of complex **27•4•**$MeNO_2$. Hydrogen atoms except for $MeNO_2$ are omitted for clarity [49].

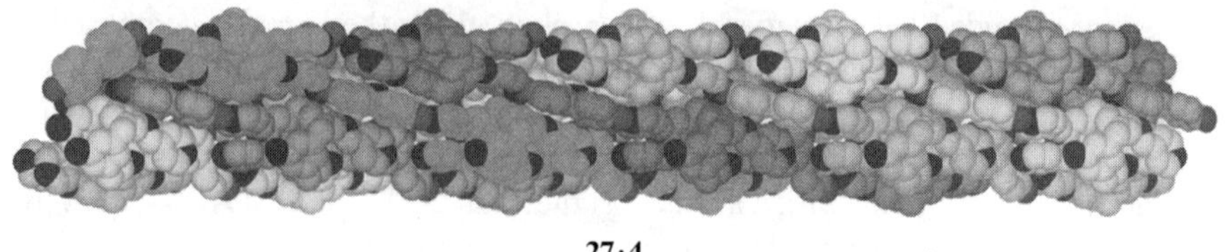

27·4

Figure 29 Quintuple helical supramolecular assembly of different complexes **27•4** [49] (see also **Plate 10**).

4 HELICAL H-BONDED MOLECULAR NETWORKS FORMED BY CHIRAL ALCOHOLS AND CHIRAL AMINES (SUPRAMINOLS)

4.1 Neutral Supraminols Based on *trans*-1,2-Diaminocyclohexane

In 1994, Hanessian and co-workers [50] reported the first examples of metal-free three-dimensional triple-stranded helicates through spontaneous self-assembly of chiral C_2-symmetrical diols and chiral C_2-symmetrical diamines. The initial observation resulted from the utilization of enantiopure C_2-symmetrical vicinal *trans*-1,2-diaminocyclohexane [51,52] as ligands in the asymmetric dihydroxylations of olefins [53] and as reagents for asymmetric synthesis [54]. When equimolar amounts of (*S,S*)-*trans*-1,2-diaminocyclohexane (**28**) and its (*R,R*)-enantiomer (**29**) were individually mixed with (*S,S*)-*trans*-1,2-cyclohexanediol and heated in refluxing benzene, crystals of the respective supraminol complexes **28•30** and **29•30** were formed quantitatively (Scheme 12). This was the physical basis for the separation of racemic diols with *trans*-1,2-diaminocyclohexane [27].

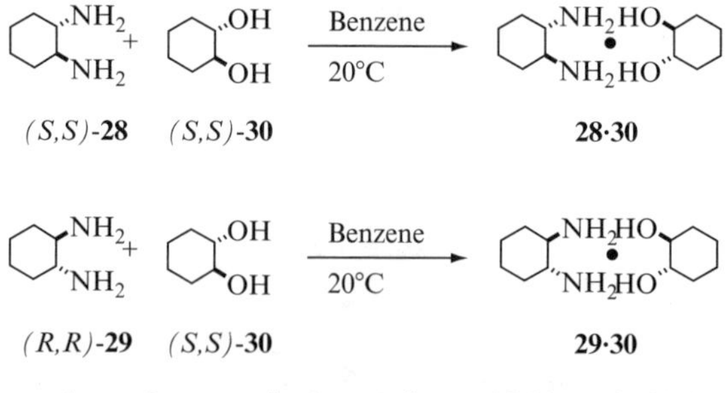

NH_2 + OH —Benzene, 20°C→ NH_2HO • NH_2HO

(*S,S*)-**28** (*S,S*)-**30** **28·30**

NH_2 + OH —Benzene, 20°C→ NH_2HO • NH_2HO

(*R,R*)-**29** (*S,S*)-**30** **29·30**

Scheme 12 Preparation of supraminol complexes **28•30** and **29•30** from equimolar amounts of enantiomerically pure *trans*-1,2-diaminocyclohexanes (**28** and **29**) and (*S,S*)-*trans*-1,2-cyclohexanediol (**30**).

X-ray analyses of complexes **28•30** and **29•30** show well-defined and extremely ordered supramolecular architectures in which the cyclohexane rings of both reagents are stacked into four vertical columns with the amino and alcohol polar groups facing inwards. A pleated sheet-like central core resembling a staircase and consisting of eight-membered square-planar, H-bonded units, with fully tetracoordinated heteroatoms and involving one functional group for each molecule, is formed. The remaining amino and alcohol moieties form two symmetrical side-rows of H-bonds, which flank opposing sides of the central core and join the amine and alcohol groups in a zig-zag pattern (Figure 30). Infrared spectra of complexes **28•30** and **29•30** in KBr also showed extensive H-bonding networks compared with the diamine or diol progenitors alone [55]. The entire set of H-bond distances ranges from 1.87 to 2.47 Å for **28•30** and from 1.88 to 2.42 Å for **29•30**.

Unlike the central core, the side-row H-bonding occurs through tricoordinated oxygen atoms on alternating diol units. The nitrogen atom of the intervening diamine unit is also tricoordinated because of the steric inaccessibility of a second hydrogen atom for coordination. Notably, even if supraminol complexes **28•30** and **29•30** have identical staircase-like cores and side-rows with alternating H-bonding, the relative sense of donation in the core is opposite to that observed in the side-rows [51]. H-bonds of the side-strands in **28•30** are formed between oxygen donors and nitrogen acceptors only. This motif is reversed in **29•30**, where the side-strands consist of H-bonds between nitrogen atoms as donors and oxygen atoms as acceptors. A remarkable consequence of the extensive H-bonding network is the emergence in these three-dimensional supraminols of a secondary structure in which the diamine-diol components wrap around the central core in a triple-stranded helicate. The sense of chirality for both homochiral and heterochiral supraminols is determined by the chirality of the diamine. (*S,S*)-*trans*-1,2-Diaminocyclohexane (**28**) leads to a left-handed helicate, where as (*R,R*)-*trans*-1,2-diaminocyclohexane (**29**) leads to right-handed helicate with the same diol. In each case, a full turn comprises four units, involving a pair each of alternating diamines and diols. Three strands run parallel in the helicate at fixed distances so that the groove of the helix is one-third of the pitch (the pitch of the helix is 15.724 Å in **28•30** and 15.048 Å in **29•30**). In Figure 31 are shown the CPK representations of the triple-stranded helical structures of **28•30** and **29•30**.

The complexing affinities of (*R,R*)-*trans*-1,2-diaminocyclohexane (**29**) to (*R,R*)-*trans*-1,2-cyclohexanediol (**31**) and (*S,S*)-*trans*-1,2-cyclohexanediol (**30**), and to the corresponding cyclopentanediols, have been measured by van Gunsteren and coworkers [56] in benzene and CCl_4 at 298 K utilizing microcalorimetry. Interestingly, the two complexes show significant differences (Figure 32).

Diol (*S,S*)-**30** binds to diamine (*R,R*)-**29** in benzene with an enthalpy change (ΔH_b) about 20 kJ/mol lower, and an entropy change (ΔS_b) about 60 J/(K mol) lower than its enantiomer [57]. The cyclopentanediols (not shown) behave similarly. When the solvent is changed from benzene to CCl_4, the enthalpy and entropy changes are quantitatively the same. It is of interest that the heterochiral complex (*S,S*)/(*R,R*)-**29•30** appears to be energetically twice as favorable, compared with

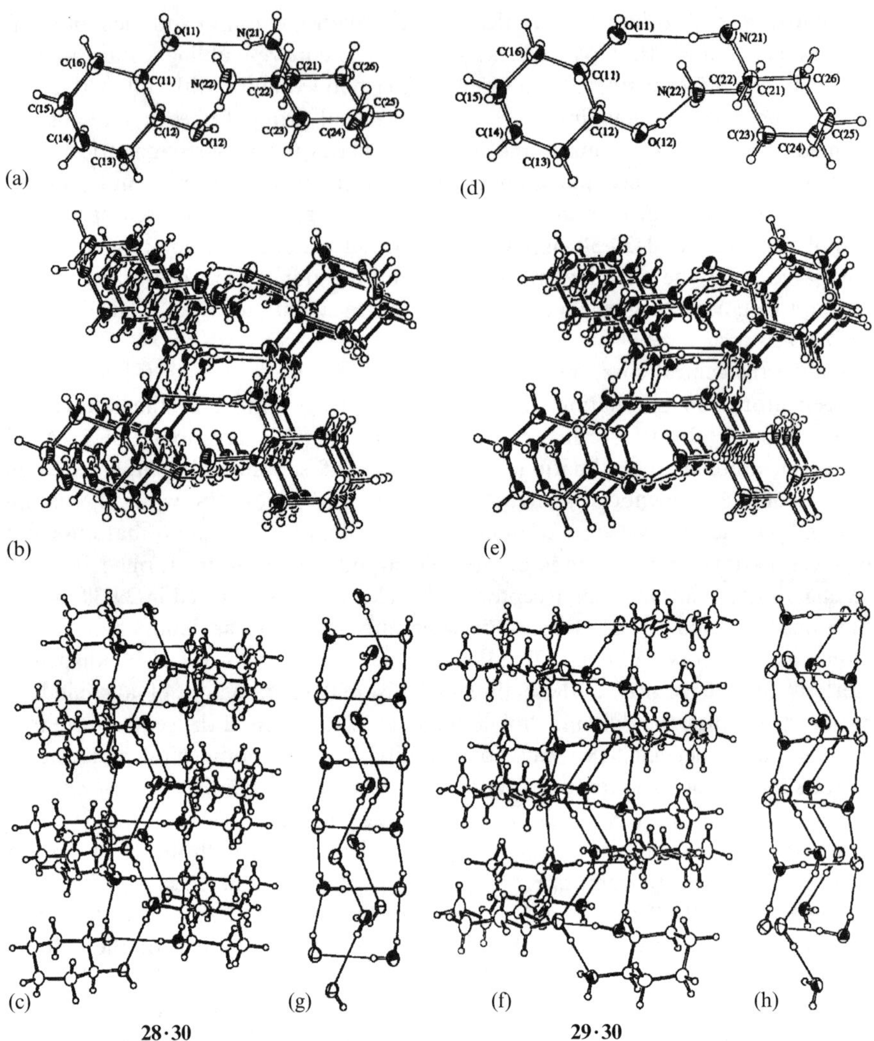

Figure 30 (a and d) ORTEP views of the molecular structure of complexes **28•30** and **29•30**, respectively (H-bonds are represented by thin lines). (b, e) H-bonding network for **28•30** and **29•30** (top view down the *a* axis). (c, f) Side view of the H-bonding network of **28•30** and **29•30**. (g, h) Simplified depiction of the views in (c) and (f) showing the pleated sheet motif of the core and the side-rows of H-bonds [50,51].

the homochiral complex, but entropically less favorable by a factor of 3–4 depending on the solvent. To explain these data, the authors [56] performed some long-time-scale (0.1 μs) molecular-dynamics simulations in a simplified solvent model [58], but discrepancies were found with respect to the experimental data,

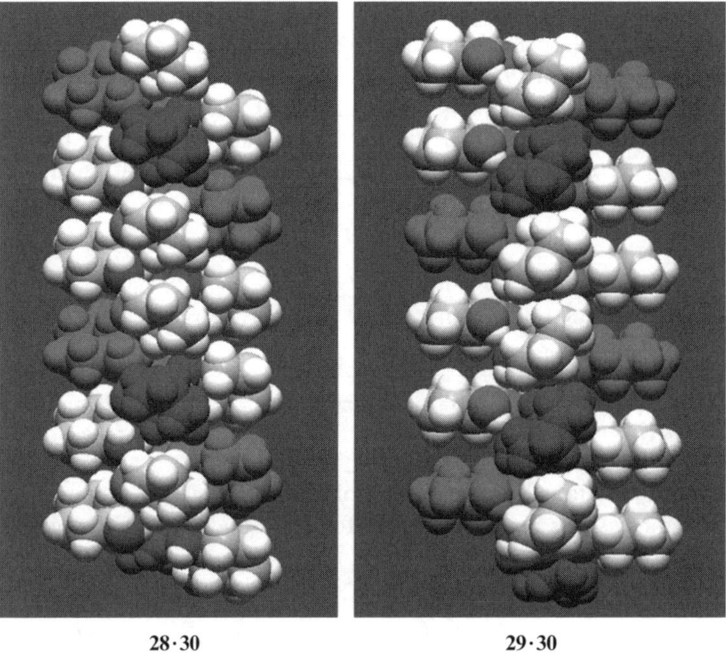

Figure 31 CPK representation of the triple-stranded helical structures of **28•30** (left) and **29•30** (right) [50,51] (see also **Plate 11**).

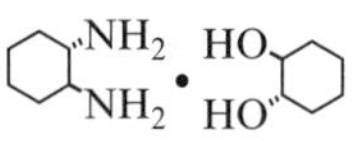

Complex	Parameter		
28·30	k_b [M^{-1}]	19	53
	ΔG_b[KJ/mol]	−7.4	−10.0
	ΔH_b[KJ/mol]	−21.8	−15.9
	ΔS_b[J/(K/mol)]	−47.9	−19.8
29·30	k_b [M^{-1}]	14	38
	ΔG_b[KJ/mol]	−6.6	−9.1
	ΔH_b[KJ/mol]	−40.6	−36.0
	ΔS_b[J/(K/mol)]	−113.2	−89.6

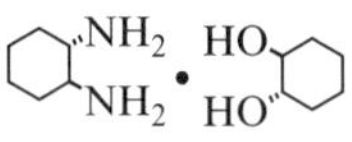

Figure 32 Experimental results for the thermodynamic parameters in benzene and in CCl_4 at 298 K. K_b is the equilibrium constant for binding; ΔG_b is the binding (Gibbs) free energy; ΔG_b is the binding enthalpy; ΔS_b is the entropy change upon binding [56].

even after a systematic analysis of the dependence of simulations on model parameters. To explain this, the authors claimed as possible causes specific solute–solvent or solvent–solvent interactions or the formation of supramolecular architectures of more than two molecules.

Maintaining (*R*,*R*)-*trans*-1,2-diaminocyclohexane (**29**) as the diamine of choice, other supraminols were successively synthesized with a number of readily available

C_2-symmetric diols [59]. The first example of a self-assembled H-bonded coil or ribbon involving exclusively amino and hydroxy groups was obtained by reaction of equimolar amounts of **29** and (*R*,*R*)-*trans*-2,3-butanediol (**32**). In contrast to the previously described supraminols **28•30** and **29•30**, the melting point of product **29•32** (65–67 °C) was found to be higher than that of both isolated components (diamine, m.p. 41–43 °C; diol, liquid), suggesting a stronger H-bonding network (Scheme 13).

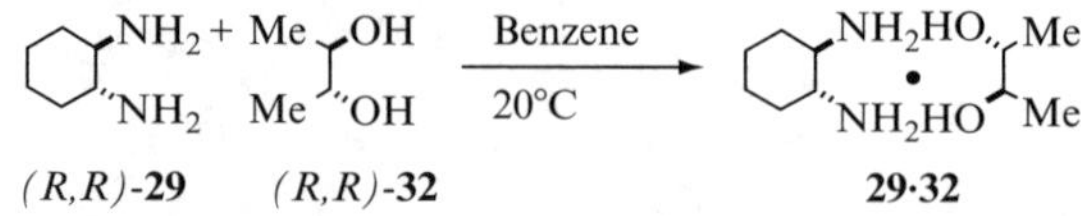

NH_2 + Me OH / NH_2 Me OH —Benzene, 20°C→ NH_2HO Me • NH_2HO Me

(*R,R*)-**29** (*R,R*)-**32** **29·32**

Scheme 13 Preparation of the supraminol complex **29•32** from (*R*,*R*)-**29** and (*R*,*R*)-*trans*-2,3-butanediol (**32**).

In the X-ray crystal structure, the cyclohexane and methyl groups are stacked in four columns as was previously found for **28•30** and **29•30**, although the structure is markedly tilted (Figure 33).

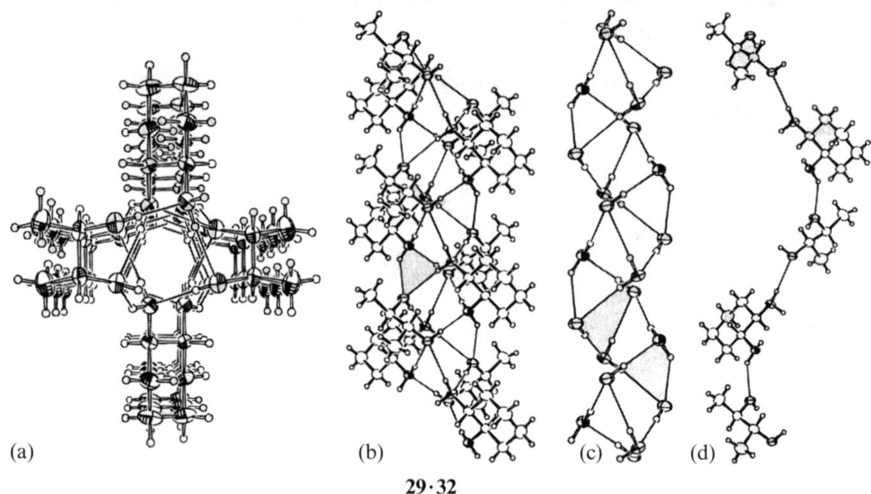

Figure 33 (a) H-bonding network of supraminol (*R,R*)-**29•**(*R,R*)-**32** (top view down the *a* axis). (b) Side view (H-bonds represented by thin lines) showing the coordination of all heteroatoms. (c) Simplified representation of the view in (b) showing the right-handed helical motif of the pleated sheet ribbon constituting the H-bonded core of the assembly. (d) Single strand of H-bonded extracted from the triple-stranded helicate [51].

The H-bonding motif of the core is a right-handed-helical ribbon, consisting of eight-membered square-planar H-bonded units characterized by an unprecedented full coordination of *all* the heteroatoms in the structure. This network defines a rigid cylindrical channel of 4.2–4.4 Å diameter along the axis. In the three-dimensional H-bonding array present in **29•32** the H-bonds constituting the edges of the ribbon run in donating directions opposite to each other (antidromic), as do face-to-face perpendicular H-bonds in the ribbon. Three antidromic H-bonded strands run parallel at a distance of 6.764 Å, generating a helicate of 20.293 Å pitch, considerably longer than those **28•30** and **29•30** (see in particular Figure 33d). While the peripheral residues adopt a left-handed helix, the ribbon-like motif within the core has an opposite helicity (Figure 33b and c). The CPK representation of the X-ray structure of **29•32** is shown in Figure 34.

A comparison of the H-bonding motif between supraminols **28•30**, **29•30**, and **29•32** suggests that the structure of the partner diol (e.g. cyclic diol versus acyclic diol) is an important factor in the control of the efficiency of the full or partial coordination in these supramolecular structures. Further insight into this aspect was obtained by the X-ray crystal analysis of heterochiral and homochiral complexes **29•33** and **29•34**, formed between the diamine (*R*,*R*)-**29** and (*S*,*S*)-hydrobenzoin (**33**) and (*R*,*R*)-hydrobenzoin (**34**), respectively [51]. After crystallization from benzene, the heterochiral complex **29•33** showed a higher melting

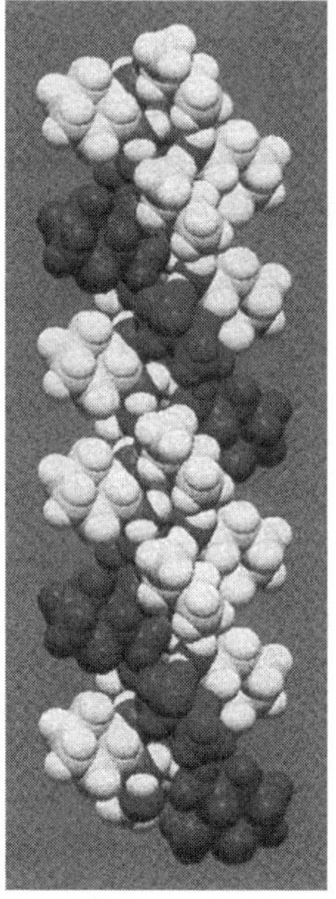

29·32

Figure 34 CPK representation of the X-ray structure of **29•32** [51] (see also **Plate 12**).

point (99–100 °C) than the homochiral diastereoisomer (75–76 °C), and the X-ray analysis revealed two different H-bonding networks (Scheme 14).

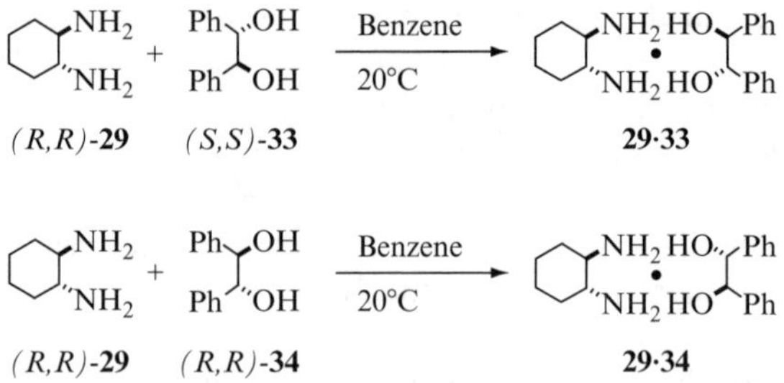

(*R,R*)-**29** + (*S,S*)-**33** → (Benzene, 20°C) **29·33**

(*R,R*)-**29** + (*R,R*)-**34** → (Benzene, 20°C) **29·34**

Scheme 14 Preparation of supraminol complexes **29•33** and **29•34** between the diamine (*R,R*)-**29** and (*S,S*)-hydrobenzoin (**33**) and (*R,R*)-hydrobenzoin (**34**).

In the first case, a typical staircase-like motif of the H-bonded core was observed showing a close similarity with **28•30** and **29•30**. Some H-bonds were longer resulting in a partially formed staircase motif with an 'S'-shaped motif (Figure 35). Thus, the complementarity between the partners is affected by increasing the steric congestion and the stacking properties of the substituents in a *trans*-1,2-diol scaffold.

A different architecture was found for the complex **29•34**. In this case the H-bonding network develops along two directions, generating a layered structure in which the diamine and diol components are bonded together by only H-bonds between one oxygen donor and one nitrogen acceptor. The top view of the network shows a new motif consisting of alternating pairs of diamines and diols, which belong to contiguous complexes, interlinked in a face-to-face arrangement by H-bonds between the remaining hydroxy and amino groups of the partners (Figure 36). The CPK representation of the triple-stranded helical structure of the complex **29•34** is shown in Figure 37.

The ability of diamine (*R,R*)-**29** to organize the secondary structures of supraminols is further confirmed by its recognition properties with diols having functional diversity, as for example (*R,R*)-4-cyclohexene-1,2-diol (**35**) and (*R,R,R,R*)-4, 5-dibromocyclohexane-1,2-diol (**37**) [59]. Thus heating equimolar amounts of **29** with **35** and of **29** with **36** in benzene gives the corresponding crystalline complexes suitable for the X-ray analyses (Scheme 15) [60].

As a common feature, the cyclohexane and cyclohexene moieties in the complex **29•35** are stacked in four columns (Figure 38b, c, and d) with all the oxygen and nitrogen atoms involved in a fully H-bonded right-handed helical ribbon motif of the core, which is similar to that previously observed for the complex **29•32** between (*R,R*)-*trans*-1,2-diaminocyclohexane and (*R,R*)-*trans*-butanediol, but different from the plated-sheet-like motif found in the complex **29•30**. Notably, while the core

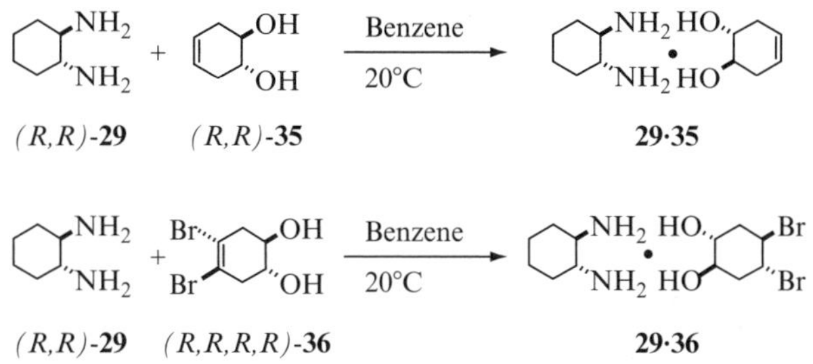

Benzene, 20°C

(R,R)-**29** *(R,R)*-**35** **29·35**

Benzene, 20°C

(R,R)-**29** *(R,R,R,R)*-**36** **29·36**

Scheme 15 Preparation of supraminol complexes **29•35** and **29•36** from the (*R*,*R*)-*trans*-1,2-diaminocyclohexane (**29**) and (*R*,*R*)-4-cyclohexene-1,2-diol (**35**) and (*R*,*R*,*R*,*R*)-4,5-dibromocyclohexane-1,2-diol (**36**).

of **29•35** adopts the right-handed helical ribbon-like H-bonded motif, the peripheral residues have an opposite helicity (Figure 38d, e, and f).

In the case of the complex **29•36**, the bromine atoms influence the recognition process as the diol moiety assumes, for the first time, two different orientations in the crystal that are rotated of about 20° in the *a–c* plane (Figure 39). A ribbon like H-bonded motif for the core in which only two amino groups are fully engaged in H-bonding is evident. The CPK representations of complexes **29•35** and **29•36** are shown in Figure 40.

The efficiency of **29** to generate supraminol architectures is further emphasized by its ability to coordinate alcohols with a lower degree of symmetry, such as (*S*)-1-phenyl-1,2-ethanediol (**37**) and *meso*-1,2-cyclohexanediol (**38**). The 1:1 complex between **29** and (*S*)-**37** has been reported [61], showing the typical array expected for the complexes based on **29** as the assembling diamine. From the X-ray data of the complex **29•37** (Scheme 16) it appears that the transition from the *C*2 symmetry of previously considered diols to the *C*1 symmetry of **37** had some appreciable effects on the recognition process, even if the helicate shape is retained. The oxygen and nitrogen atoms are engaged in only three H-bonds.

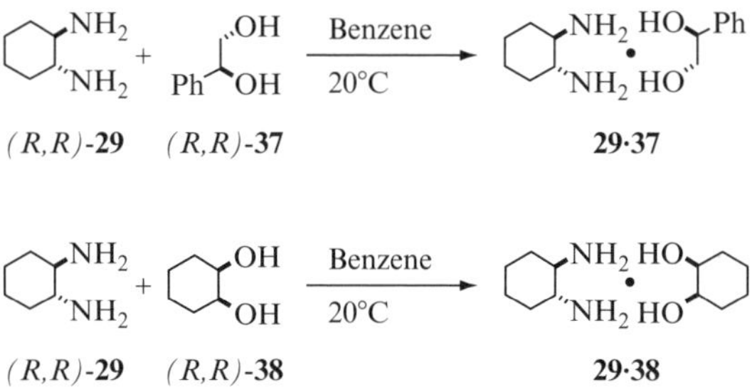

Benzene, 20°C

(R,R)-**29** *(R,R)*-**37** **29·37**

Benzene, 20°C

(R,R)-**29** *(R,R)*-**38** **29·38**

Scheme 16 Preparation of the supraminol complexes **29•**(*S*)-**37** and **29•38**.

The ORTEP representation of the complex **29•**(*S*)-**37**, and the corresponding CPK representations are shown in Figures 41 and 42, respectively.

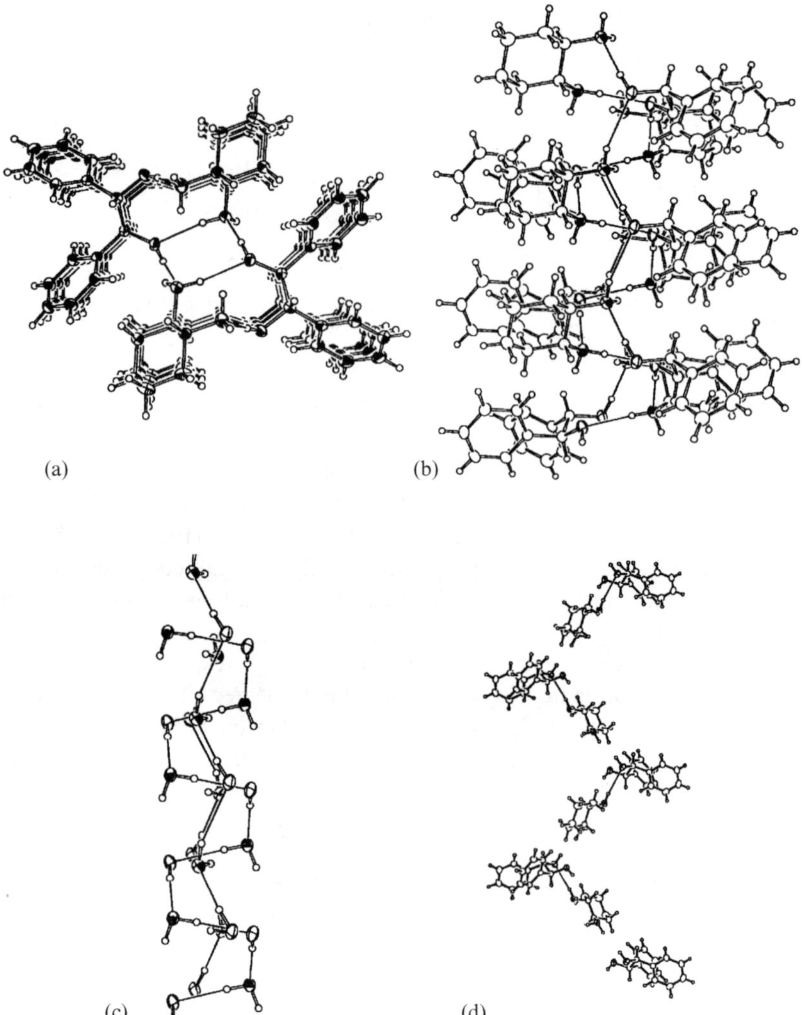

Figure 35 (a) H-bonded assembly of the adduct **20•33** (top view down the *b* axis). (b) Side view of the assembly (H-bonds represented by thin lines). (c) Simplified representation of the view in (b) showing the 'S'-shaped motif on the H-bonded core. (d) Single strand of partially H-bonded diamine and diol units. The arrangement of the units in combination with the direction of the H-bonds (thin lines) imparts the right-handed helical shape of the strand producing a pseudo-trihelicate structure.

When stoichiometric amounts of **29** and **38** were heated in benzene, the complex **29•38** was isolated in crystalline form (Scheme 16) [60]. The structure of this complex is different from that previously observed for the diols with *C*1 and *C*2 symmetries (Figure 43). In contrast to other helical complexes, where the ratio of

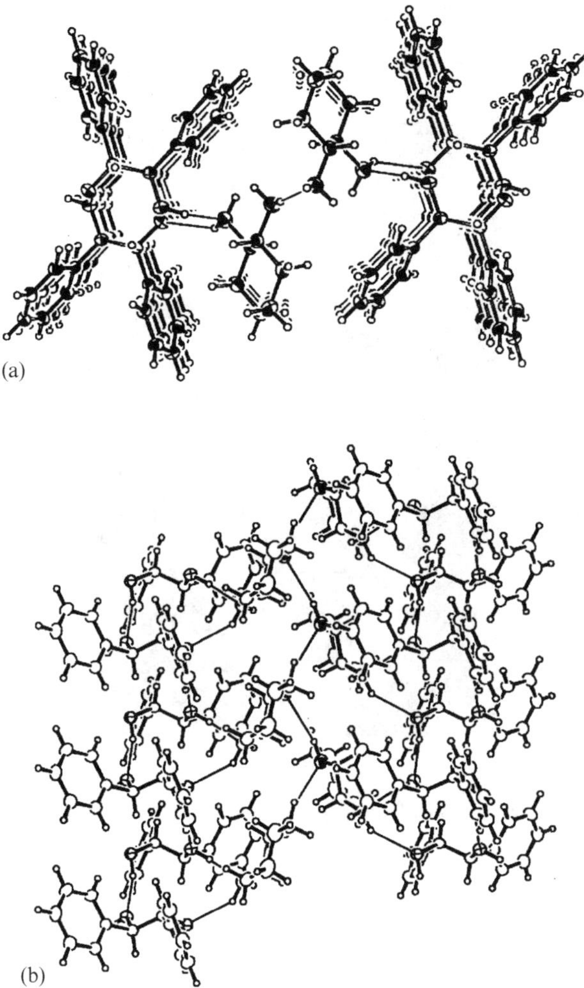

Figure 36 (a) Top view of the assembly of **29•34** down the *b* axis (the first dimension of the layered structure), showing a segment of the layer taken from the *c* axis (the second dimension). (b) Side view of the assembly down the *a* axis (H-bonds in thin lines) [51].

the diol to the diamine moiety is 1:1, the supraminol **29•38** is defined by two diamine and two diol moieties. The complex is characterized by a layered structure with an incomplete ribbon-like core motif in which two interlinking H-bonds connect any unit cell (Figure 43). Thus, even if the *C*2 symmetry of the diol is not an essential feature for the synthesis of supraminols, the partial loss of symmetry influences the order and the topology of the H-bonding network. The CPK representation of the complex **29•38** is shown in Figure 44.

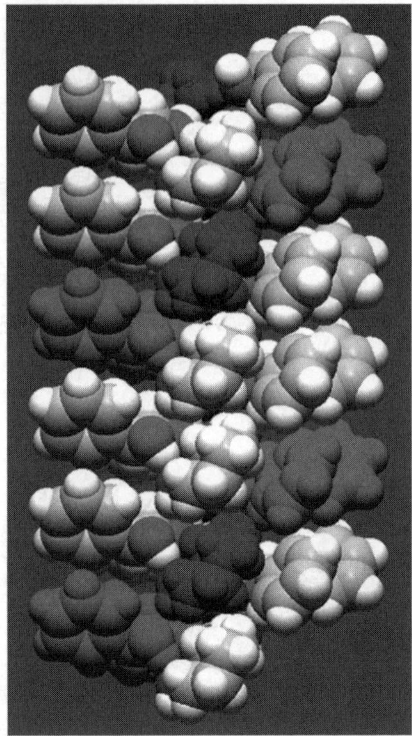

29·34

Figure 37 CPK representation of the triple-stranded helical structures of **29•34** [51] (see also **Plate 13**).

4.2 Neutral Supraminols Based on *trans*-2,3-Diaminobutane

Specific recognition and assembly instructions for the synthesis of supraminols are encoded also in other chiral diamines, as in the case of (*R*,*R*)-2,3-diaminobutane (**39**) [60]. Equimolar amounts of (*R*,*R*)-**39** and (*R*,*R*)-*trans*-1,2-cyclohexanediol (**31**) self-assemble into a helicate **39•31** (Scheme 17). This structure is characterized by a right-handed helical ribbon core and a left-handed helical strand motif that appears to be more compressed in the *a*–*c* plane and more extended along the *b* plane compared with that previously described for the complex between (*R*,*R*)-*trans*-1,2-diaminocyclohexane and (*R*,*R*)-2,3-butanediol (Figure 44) [51].

Similar results were obtained in the recognition process between (*R*,*R*)-**39** and (*R*,*R*)-4-cyclohexene-1,2-diol (**35**) [60]. The motif of the crystal core of the complex **39•35** is a right-handed helical ribbon while the outer core adopts a left-

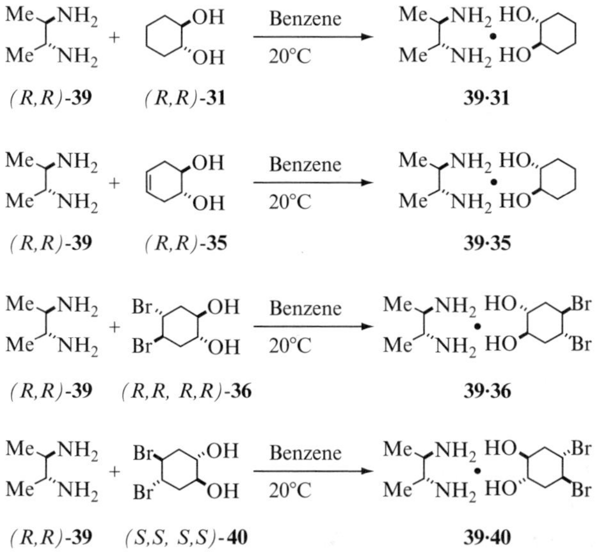

Scheme 17 Preparation of the supraminol complex **39•31** from (*R*,*R*)-2,3-diaminobutane (**39**) and (*R*,*R*)-*trans*-1,2-cyclohexanediol (**31**).

handed triple helicate as in previous structures such as **29•32** (Figure 45, Scheme 17). In this case, molecules of benzene were included in the crystal at the intersection of four columns in a plane parallel to the *a–b* plane of the unit cell. When the complex **39•35** was prepared from a 1:1 mixture of benzene and toluene, only crystals including benzene were recovered. Thus, supraminol **39•35** is capable of discriminating between benzene and toluene molecules during the self-assembly process of the crystals.

(*R*,*R*)-2,3-Diaminobutane (**39**) is able to recognize substituted 1,2-*trans*-cyclohexanediols such as (*R*,*R*,*R*,*R*)- and (*S*,*S*,*S*,*S*)-4,5-dibromocyclohexane-1,2-diol (**36** and **40**), respectively. With the matched partner the core of the complex **39•36** is characterized by an incomplete right-handed helical ribbon, which is more compressed along the *a–c* plane than all other helicates in this series (Figure 46, Scheme 17).

Unlike the complex **39•36**, the complex **39•40**, formed with the mismatched diol, consists of a vastly interlinked layered H-bonded network that is in part reminiscent of the crystal structure previously observed for (*R*,*R*)-*trans*-1,2-diaminocyclohexane and *meso*-1,2-cyclohexanediol (Figure 47, Scheme 17). The efficiency of **39** to recognize the mismatched diol suggests that acyclic diamines are the optimal candidates for the molecular recognition of sterically demanding diols with respect to the cyclic diamines. In fact, (*R*,*R*)-*trans*-1,2-diaminocyclohexane did not recognize mismatched **39•40**.

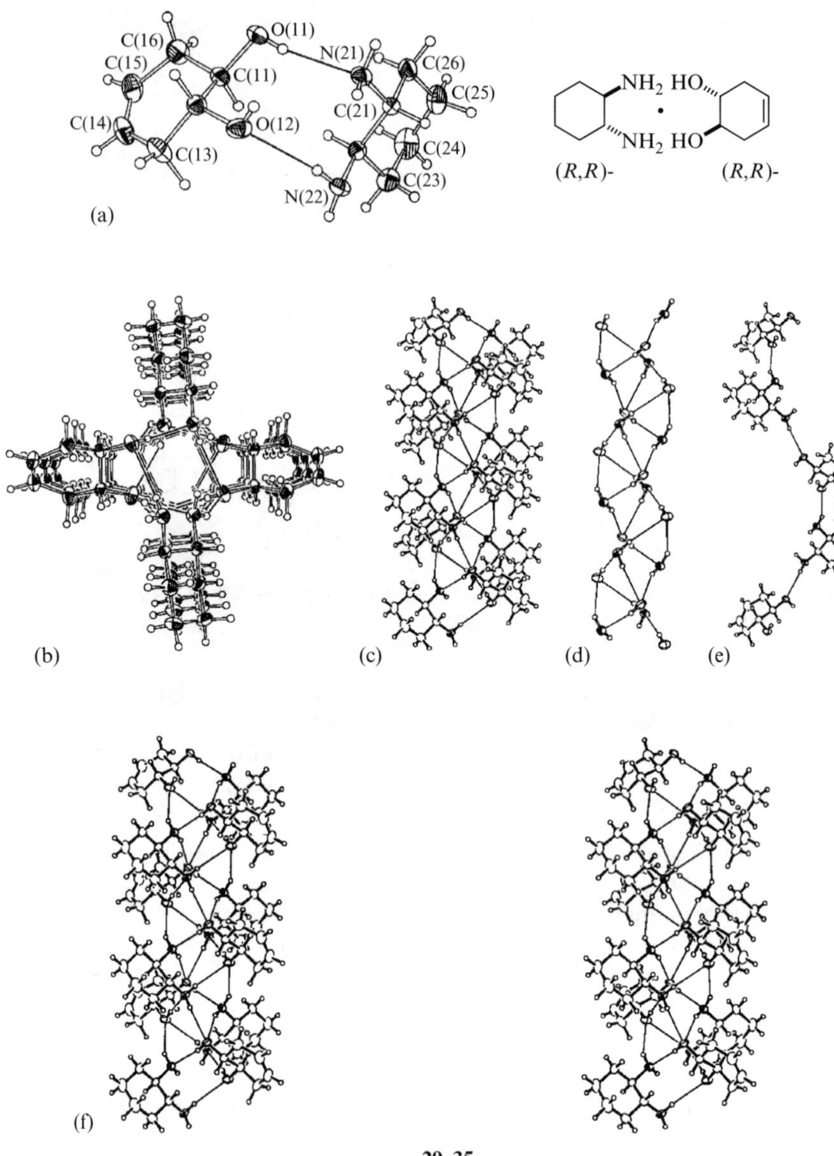

29·35

Figure 38 ORTEP views of the adduct **29•35** [60].

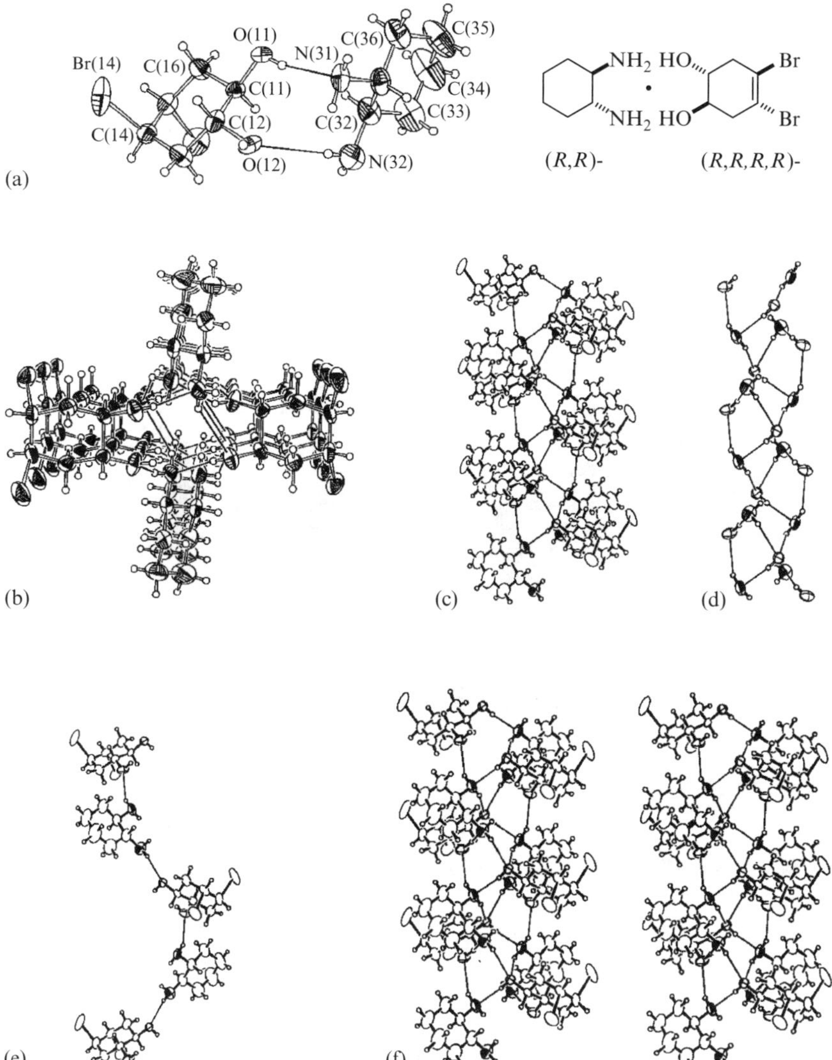

Figure 39 (a) ORTEP view of the molecular adduct **29•36**. (b) H-bonding network of **29•36** (top view down the *a* axis. (c) Side view of the H-bonding network. (d) Simplified representation of the view in (c) showing the left handed helical motif of the core. (e) Single strand for H-bonded units. (f) Stereoview of the H-bonding network (side view along the *b* axis) from the X-ray analysis. The helical motif formed by the peripheral residues is of the left-handed type [60].

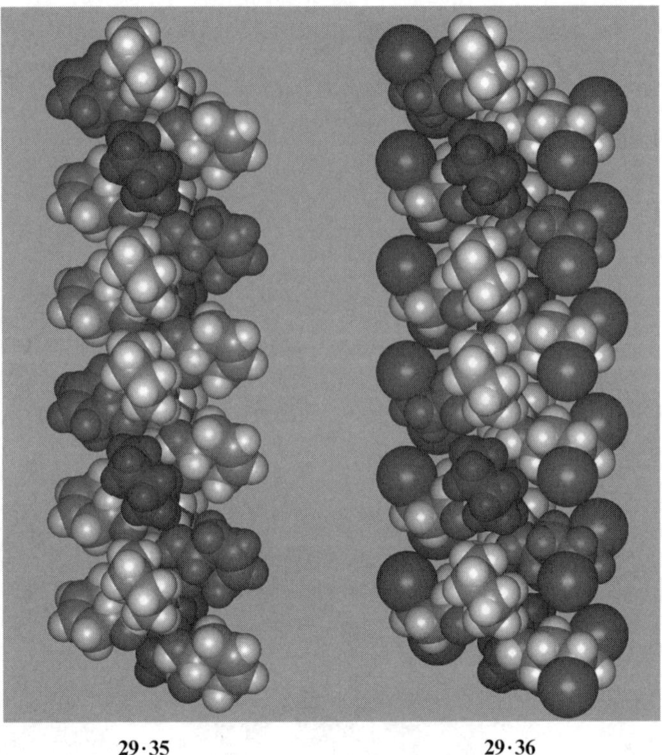

29·35 **29·36**

Figure 40 CPK representations of the adducts **29•35** and **29•36** [60] (see also **Plate 14**).

4.3 Charged Supraminols Based on *trans*-1,2-Diaminocyclohexane

The simplest example of a charged network based on *trans*-1,2-diaminocyclohexane is based on the carbon dioxide complex [51]. A forced carbonation of (*R*,*R*)-*trans*-1,2-diaminocyclohexane with carbon dioxide afforded the corresponding crystalline carbamate salt **41** (from ethanol–water), which was suitable for X-ray analysis (Scheme 18) [51,62].

A two-directional H-bonding network appears to have similarities to the previously described neutral network with staircase-shaped cores (Figure 48).

A pleated sheet-like central core formed by two opposite pairs of ammonium and carboxylate groups runs along the vertical *b* axis of the structure. The third ammonium proton and the second carboxylate oxygen are engaged in a bifurcated H-bond with the carbamic NH, and give rise to a pair of rows of H-bonds flanking the core. Four piles of carbamate molecules are joined by the core through one of the two functional groups. This arrangement describes the constitutive element (Figure 48a) of the layered structure, which develops in the second

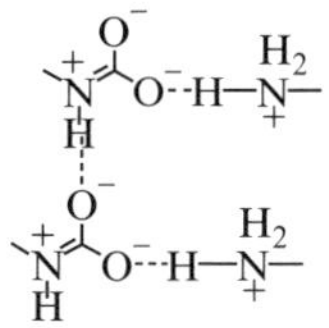

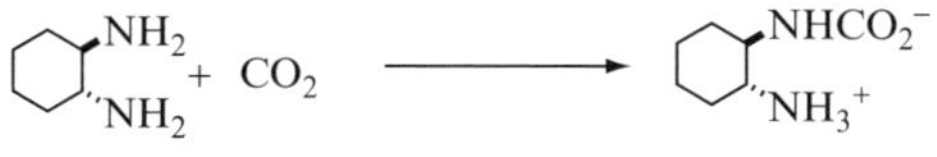

(*R,R*)-**29** **41**

Scheme 18 Preparation of the carbamate derivative **41** of (*R,R*)-*trans*-1,2-diaminocyclohexane [51].

direction (*c* axis) by H-bonding between contiguous elements. This interlink is the basic structural motif of the core rotated by 90° around the *b* axis. Along the *c* axis the network shows a double layer of nonpolar residues enclosing a layer of polar cores. In the core, all the carbonyl oxygen and ammonium nitrogen atoms are tetracoordinate, whereas those present in the side-rows are tricoordinate. The CPK representation of the complex **41** along the *a* axis is shown in Figure 49.

As in previous examples, the chirality of the superstructure of complex **41** is imposed by the configuration of the diamine, showing its relevant role in the recognition and assembly process. Based on the efficient molecular recognition between the carbonyl oxygen and ammonium nitrogen atoms, some experiments were performed treating both enantiomers of tartaric acid, (*R*,*R*)-**42** and (*S*,*S*)-**43**, with (*R*,*R*)-*trans*-1,2-diaminocyclohexane (**29**) [51]. In all cases, crystals suitable for X-ray analysis were obtained. The matched complex (*R*,*R*)-**42•**(*R*,*R*)-**29** (Scheme 19) is characterized by a tweezer-shaped motif, involving H-bonding interactions between carboxylate groups and ammonium ions in one direction and hydroxyl groups in the other. Regular cavities are located inside the core, as seen along the *a* axis (Figure 50).

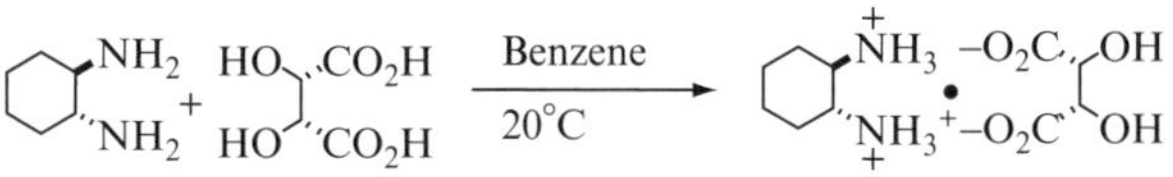

(*R,R*)-**29** (*R,R*)-**42** **29·42**

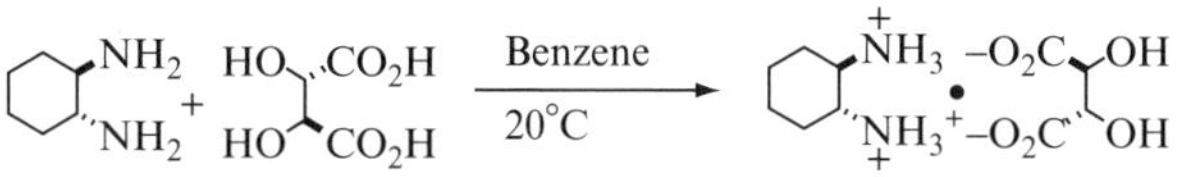

(*R,R*)-**29** (*S,S*)-**43** **29·43**

Scheme 19 Preparation of supraminol complexes (*R,R*)-**29•**(*R,R*)-**42** and (*S,S*)-**29•**(*R,R*)-**43**.

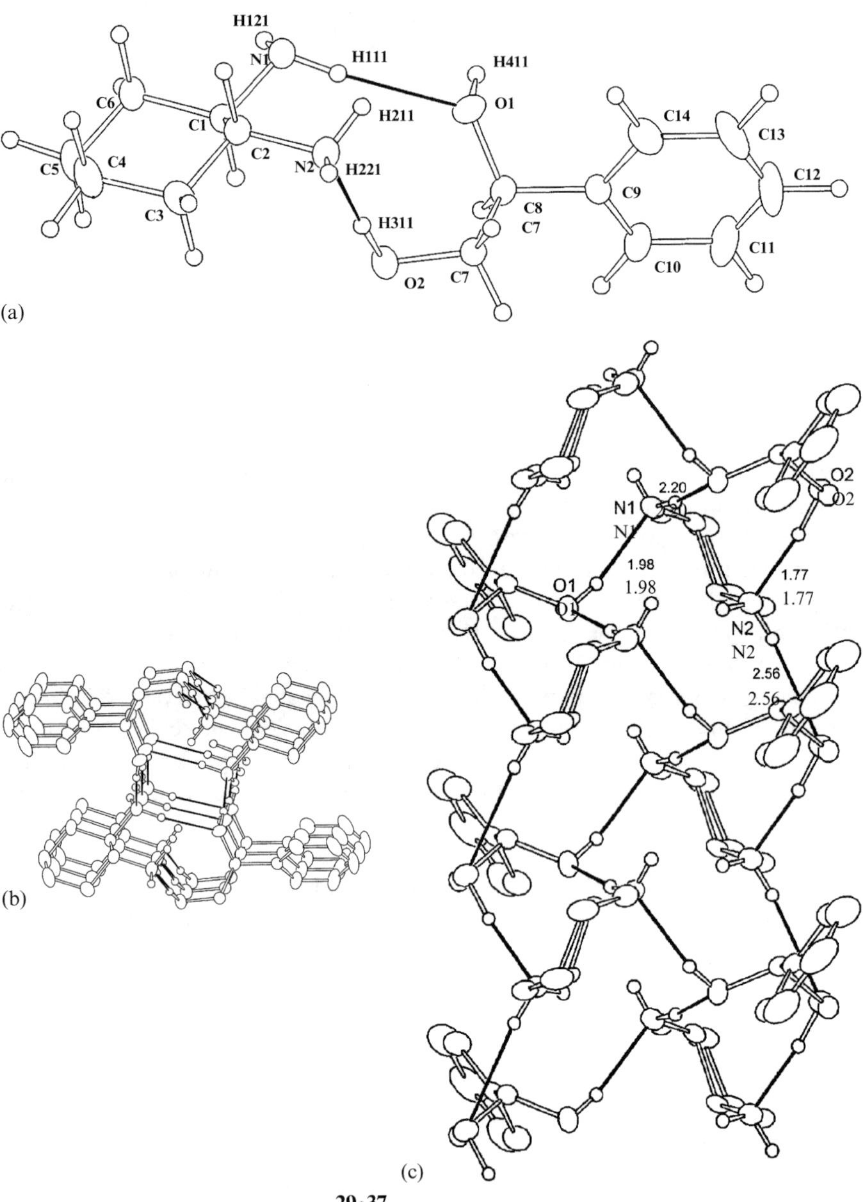

29·37

Figure 41 (a) ORTEP drawing of the adduct **29**•(*S*)-(**37**). (b) Side-view along the *a* axis with relevant H-bonding distances (Å). H-bonds are represented as black lines. H-atoms on carbon are omitted for clarity [61].

29·37

Figure 42 CPK representation of the adduct **29**•(*S*)-**37** [61].

The side-view along the *c* axis shows a 10-membered H-bonded ring as a characteristic motif of the aggregate (Figure 50c). In this network the tartrate units generate a layered structure in which the diamine units are alternatively bonded on both sides of the core. A CPK representation of the complex (*R*,*R*)-**29**•(*R*,*R*)-**42** is shown in Figure 51.

The molecular recognition between mismatched (*R*,*R*)-**29** and (*S*,*S*)-**43** is less efficient compared with the matched pair complex. Even if the complex (*R*,*R*)-**29**•(*S*,*S*)-**42** could be obtained in a crystalline form (Scheme 19), the X-ray analysis shows that one H-bond between the hydroxy and carboxylate moieties is lost and coordination has been achieved with two water molecules included in the crystal (Figure 52). The loss of such water molecules, for example after drying of the sample, determines the erosion of crystallinity.

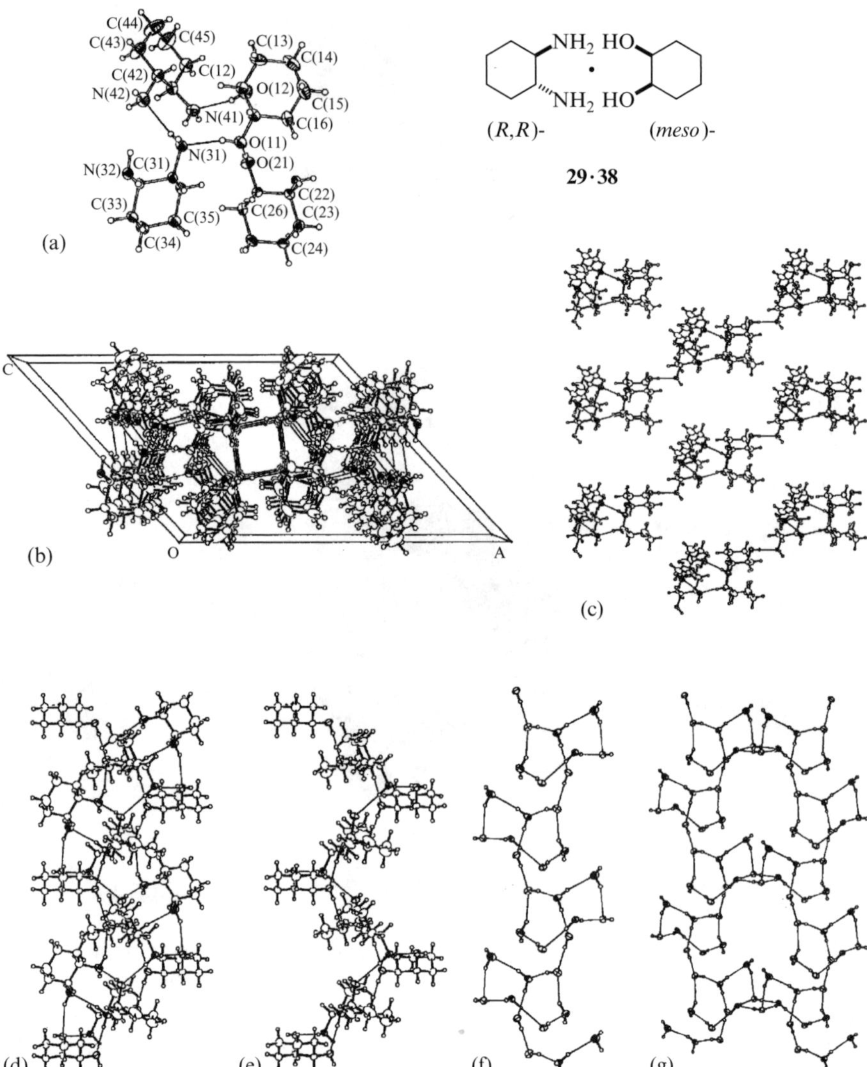

Figure 43 (a) ORTEP view of the adduct **29•38**. (b) Top view down the *b* axis. (c) H-bonding network linking columns of dimeric units. (d) Side view of one dimeric column. (e) Single strand for H-bonded units extracted from (d). (f) Simplified representation of the view in (d) showing the incomplete right-handed ribbon motif of the core of the assembly. (g) right-handed incomplete ribbon motif of two columns of dimeric units with interlinked square-planar H-bonds [60].

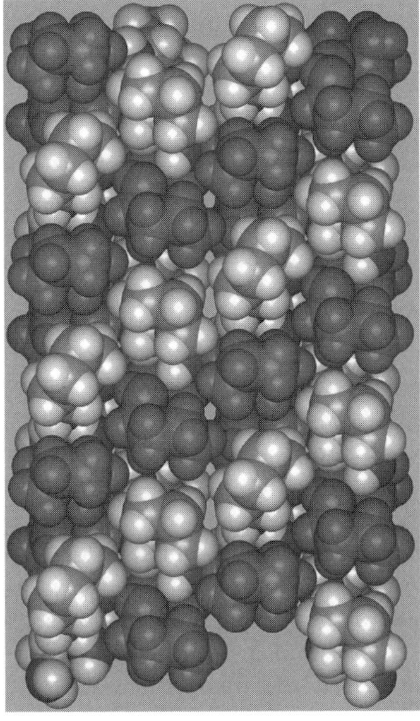

29·38

Figure 44 CPK representation of the adduct **29•38** [60] (see also **Plate 15**).

A lower level of organization in the supramolecular structure is evident even if the layered structure of the core and the stacking of the cyclohexane residues are retained. In this context, ordered H-bonded ionic triple-stranded helicates, involving (*R*,*R*)-*trans*-1,2-diaminocyclohexane (**29**), have been described [63]. In this case, mixing of (*R*,*R*)-**29** with matched (*R*,*R*)-*trans*-cyclohexane-1,2-dicarboxylic acid, (*R*,*R*)-**44**, in D_2O afforded the complex (*R*,*R*)-**29•**(*R*,*R*)-**44** as large, colourless crystals suitable for X-ray analysis (Scheme 20). The same reaction performed in the presence of the mismatched (*S*,*S*)-*trans*-cyclohexane-1,2-dicarboxylic acid (**45**) did not give any solid material. A similar result was obtained with (*S*)-methylsuccinic acid.

As reported by the authors [63], the structural motif of the complex is characterized by an infinite number of diammonium dicarboxylate units that self-assemble by a network of H-bonds. The cyclohexane groups are stacked in four columns as previously found for several neutral helical supraminols [51] (Figure 53). The three-dimensional H-bonding array present in the complex (*R*,*R*)-**29•**(*R*,*R*)-**44** revealed that the network develops through a right-handed helical ribbon which wraps

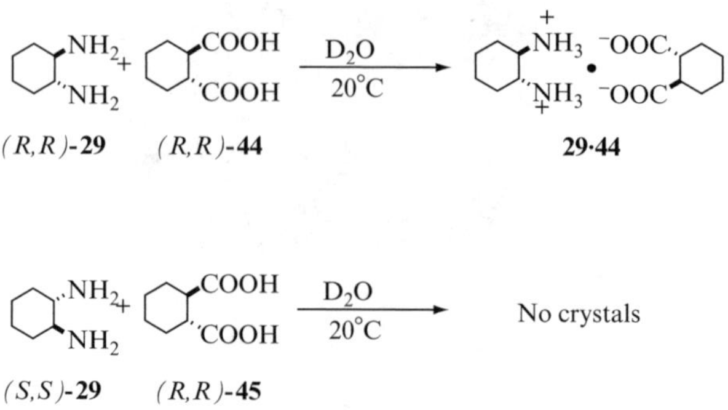

NH_2 + COOH / NH_2 COOH — D_2O, 20°C → $\overset{+}{N}H_3$ ^{-}OOC • $\underset{+}{N}H_3$ ^{-}OOC

(*R,R*)-**29** (*R,R*)-**44** **29·44**

NH_2 + COOH / NH_2 COOH — D_2O, 20°C → No crystals

(*S,S*)-**29** (*R,R*)-**45**

Scheme 20 Preparation of the supraminol complex (*R,R*)-**29**•(*R,R*)-**44**.

around the axis of the assembly. Three strands of diammonium dicarboxylate molecules run parallel in the structure to form a triple-stranded left-handed helicate.

The space-filling representation of the complex (*R,R*)-**29**•(*R,R*)-**44** in which the triple-stranded helicate super-structure is more evident is shown in Figure 54.

Notably, as was observed for neutral supraminols [51,60], the complex (*R,R*)-**29**•(*R,R*)-**44** assumes a helical motif in which chirality is determined by the absolute configuration of the *trans*-1,2-diaminocyclohexane.

5 SUPRAMINOLS FROM ACYCLIC AMINO ALCOHOLS (AMINOCAVITOLS)

An intriguing new supramolecular architecture based only on H-bonding interactions between alcohol and amine moiety has been obtained by Hanessian and co-workers [64]. Treatment of 3,4-diazido-3,4-dideoxy-1,2:5,6-di-*O*-isopropylidene-D-iditol with Dowex H^+ affords 3,4-diazido-3,4-dideoxy-D-iditol (**46**) [64], which shows extensive HO···HO H-bonding networks in its X-ray crystal structure (Figure 55). As seen in Figure 55, the diazidotetrol molecules are arranged into ordered columns with the azido and hydroxyl moieties in a regular pattern. Two vicinal OH groups present in a *trans*-configuration are involved in the formation of two H-bonding interactions with adjacent molecules. This structural motif is repeated along the crystallographic *c* axis. Interactions between the peripheral azido groups connect adjacent regular structures.

When the azido groups were reduced to give 3,4-diamino-3,4-dideoxy-D-iditol (**47**) [64], a dramatic change in the H-bonding networks was observed. In this case a new complex featuring a canal or cavity was observed which the authors named 'aminocavitol'. The monomeric unit of this aminocavitol is characterized by the presence of two H-bonded intramolecular interactions between vicinal OH and NH_2 moieties in a *cis*-configuration, as depicted in Figure 56. The values found

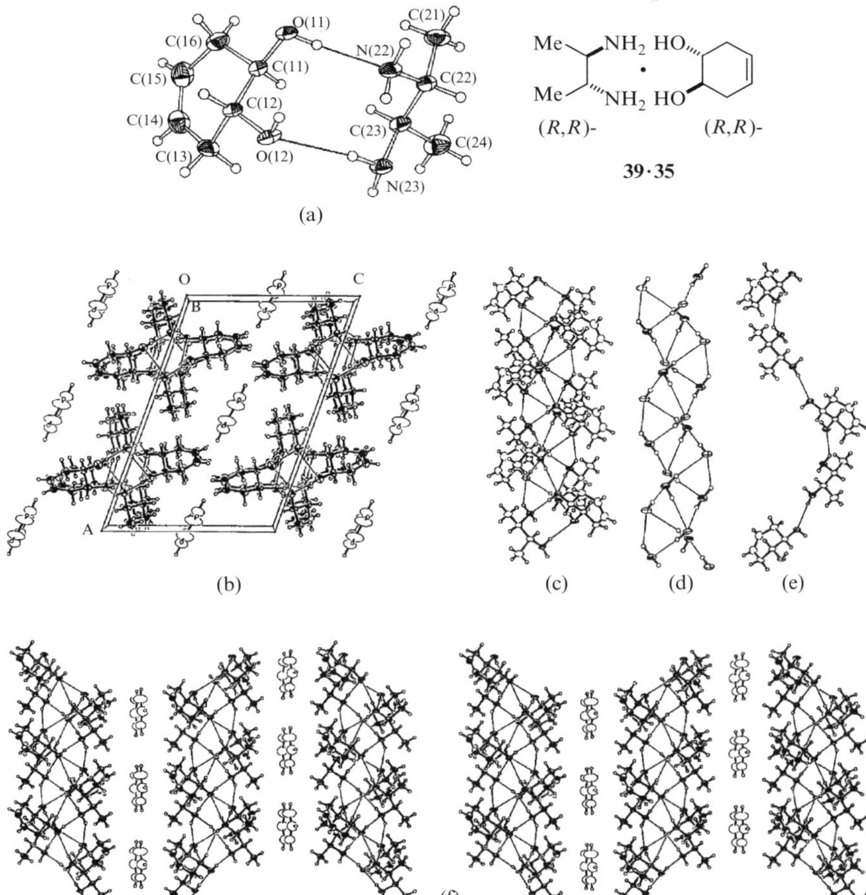

Figure 45 (a) ORTEP view of the molecular adduct **39•35** (H-bonds are represented by thin lines). (b) ORTEP view of the inclusion complex between benzene and adduct **39•35**. (c) Side view of the H-bonding network of adduct **39•35**. (d) Simplified representation of the view in (c) showing the right-handed helical motif of the ribbon like H-bonded core of the assembly. (e) Single strand for H-bonded units extracted from the triple-stranded helicate structure in **39•35** showing left-handed helicity. (f) Stereoview of the inclusion complex between benzene and adduct **39•35** [60].

for these NH···O interactions along the '*a*' crytallographic axis were 3.01 and 2.95 Å, respectively. Two such units are connected along the '*a*' axis through a pair of OH···O bonding interactions (2.79 and 2.73 Å, respectively) involving the secondary and primary alcoholic functions. In this structure, the secondary hydroxyl moiety is an H-bond acceptor from a vicinal NH_2 donor and a donor to an OH of an adjacent molecule.

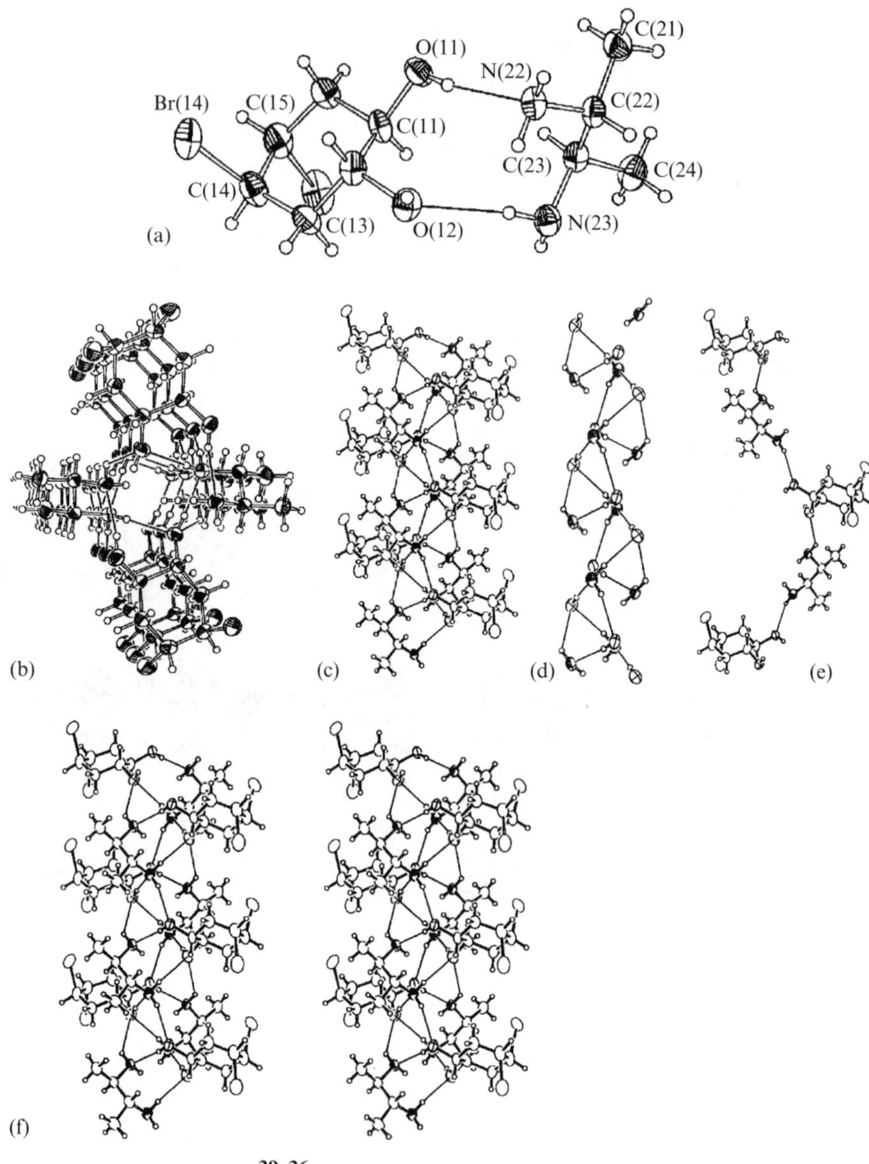

Figure 46 (a) ORTEP view of the molecular adduct **39•36** (H-bonds are represented by thin lines). (b) H-bonding network of adduct **39•36**. (c) Side view of the H-bonding network. (d) Simplified representation of the view in (c) showing the right-handed helical motif of the ribbon like H-bonded core of the assembly. (e) Single strand for H-bonded units extracted from the triple-stranded helicate structure in **39•36** showing left-handed helicity. (f) Stereoview of the H-bonding network [60].

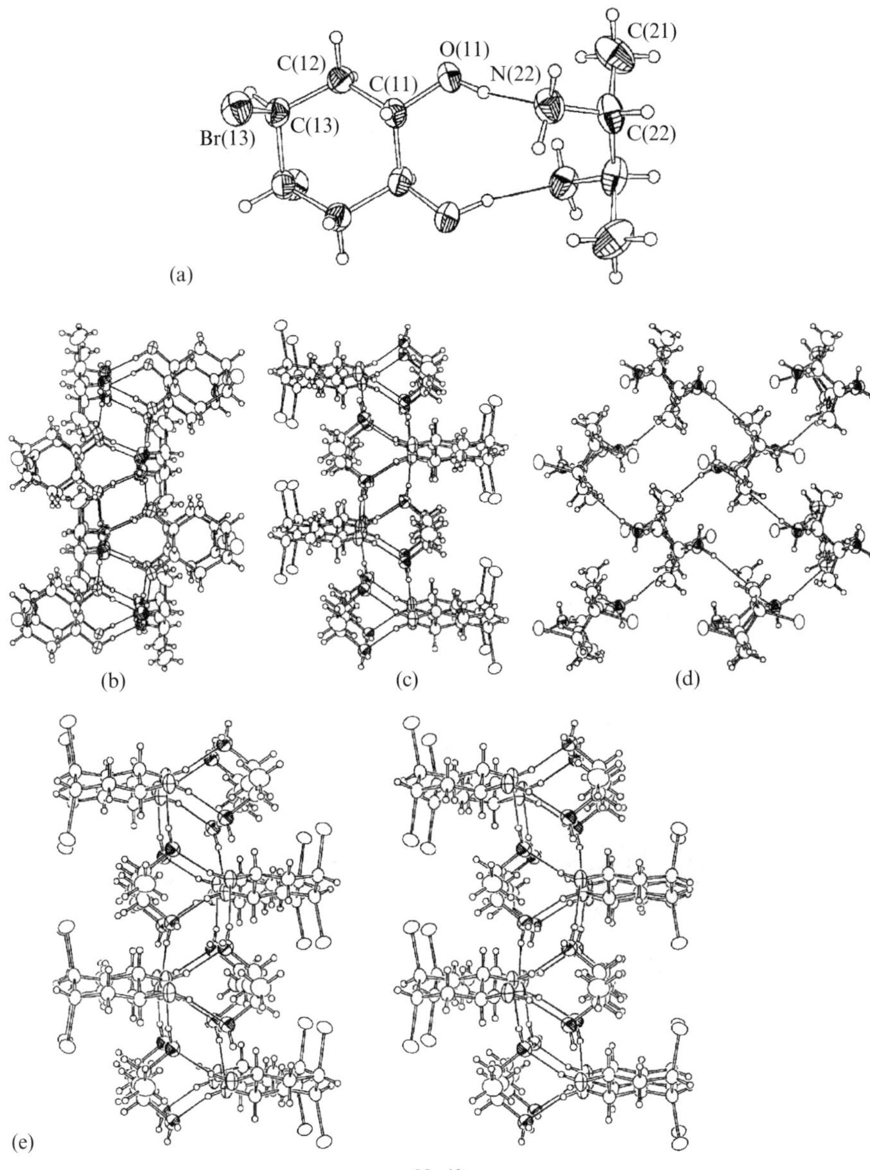

Figure 47 (a) ORTEP view of the molecular adduct **39•40** (H-bonds are represented by thin lines). (b) H-bonding network of adduct **39•40**. (c) Side view of the H-bonding network. (d) Simplified representation of the view in (c) Showing the right-handed helical motif of the ribbon like H-bonded core of the assembly. (e) Single strand for H-bonded units extracted from the triple-stranded helicate structure in **39•40** showing left-handed helicity. (f) Stereoview of the H-bonding network [60].

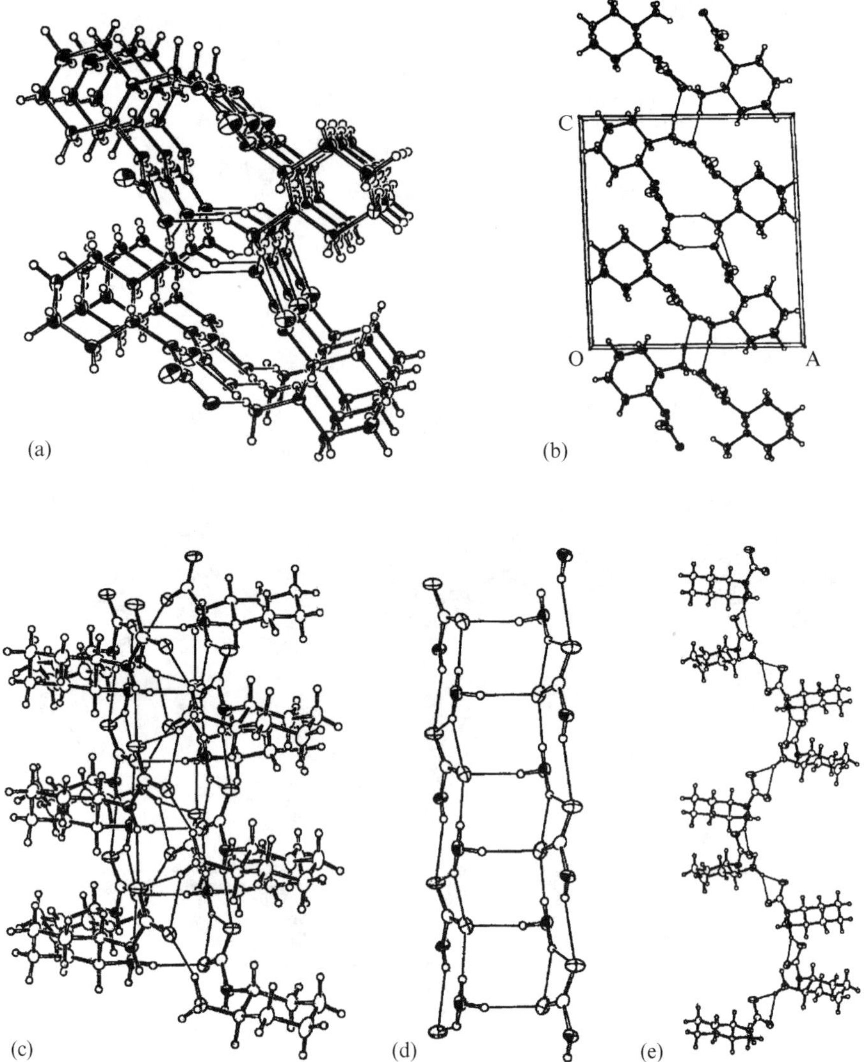

Figure 48 (a) H-bonded assembly of the carbamate **41**. (b) Interlinking H-bonds with contiguous cell units. (c) Side view of the assembly. (d) Pleated-sheet staircase-like H-bonded core with bifurcated H-bonds. (e) Single strand of diamine **29** and CO_2 separated from the triple-stranded helicate for viewing purposes. Note right-handed helicity [51].

This motif is repeated along the *a* axis generating oligomers with a pitch of 8.12 Å. These oligomers are connected in the *ac* plane by $OH \cdots N$ bonding interactions involving only the primary alcoholic moiety in an alternating mode, that is, the sequence NH_2, OH, NH_2, OH, etc. for any oligomer. The view down

41

Figure 49 CPK representation of the adduct **41** along the *a* axis [51] (see also **Plate 16**).

the *a* axis clearly shows the presence of unused NH groups. The absence of these H-bonds causes the formation of cavities and the characteristic aminocavitol supramolecular architecture appears as an infinite abyss or canal (Figure 57). In the aminocavitol superstructure the hydrogen atoms are directed toward the interior of the cavity and oxygen and nitrogen atoms are in the periphery of the cavity in a crown ether-like motif. The dimensions of the cavity are 8.3 and 1.8 Å along the *a* and *b* axes, respectively. There is not a measurable dimension along the *c* axis, hence the formation of the abyss or canal. The larger distances between two nitrogen or oxygen atoms are 4.65 and 6.48 Å, respectively.

6 OTHER ASPECTS IN THE MOLECULAR RECOGNITION PROCESS OF SUPRAMINOLS: THE 'SUPRAMOLECULAR CHIRON CONCEPT'

The concept of 'supramolecular synthon' defines specific substructural units that contain the logical code for self-assembly by noncovalent bonds (consisting mainly of H-bonding interactions) [65]. A chiral counterpart of this concept is the 'supramolecular chiron' [60,66]. It is defined as 'the minimal homo- or heterochiral molecular unit or ensemble capable of generating ordered superstructures by self-assembly through H-bonding or other noncovalent forces, and leading to

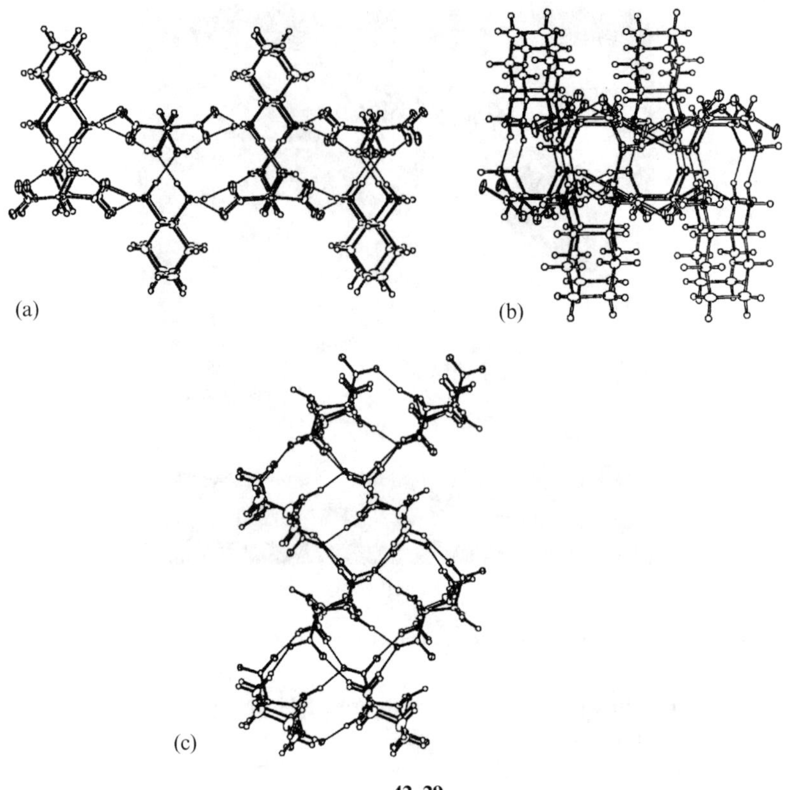

Figure 50 (a) View of the adduct (*R*,*R*)-**42**•(*R*,*R*)-**29** along the *a* axis. (b) View of the adduct (*R*,*R*)-**42**•(*R*,*R*)-**29** down the *b* axis showing the bilayered core. (c) Side view down the *c* axis [51].

topologically distinct enantio- or diastereopure architectures' [60]. The supramolecular chiron is then the ancestral assembler that encodes information for *recognition, complementarity*, and *organization* that results in ordered superstructures such as the supraminol. A compelling dilemma is to understand the molecular basis for the assembly of these supraminols. Two main scenarios are in principle possible. In the first, a preorganized arrangement of one of the components (e.g. the diamine) acts as a matrix. The other partner (e.g. the diol) could be accommodated in this matrix in the energetically most favorable terms, followed by the cascade of H-bonding interactions that leads to the observed supramolecular architectures. This mode of supraminol assembly, starting with a layered two-dimensional matrix of alternating *trans*-1,2-diaminocyclohexanes and resulting in a three-dimensional helical motif, can be considered as the 'inclusion' or 'imprinting' hypothesis. A second possibility requires an initial recognition process between the diamine and the diol to produce the 'starter unit' as a supramolecular

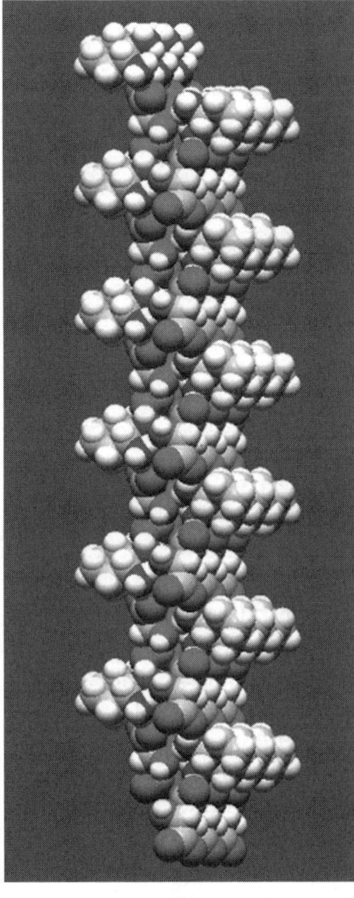

42·29

Figure 51 CPK representation of the adduct (*R*,*R*)-**42**•(*R*,*R*)-**29** [51] (see also **Plate 17**).

chiron followed by juxtaposition of other such complementary units along preferred crystallographic coordinates. In this case, the rules for the transcription of chiral information are defined by symmetry, steric, and geometric requirements. This mode of supraminol assembly can be thought of as the 'building up' or 'erector' hypothesis. It is difficult to discriminate between the two hypotheses, particularly if the assembly is also taking place in solution, or if it is manifested mostly in the solid state by well defined crystal structures. For example, interchangeable donor–acceptor interactions between diamines and diols were observed for supraminols in solution by nuclear magnetic resonance analyses [60]. In all complexes studied, the individual CH*OH* and CHNH_2 signals collapse to single broad signals, the positions of which change in proportion to the amount of

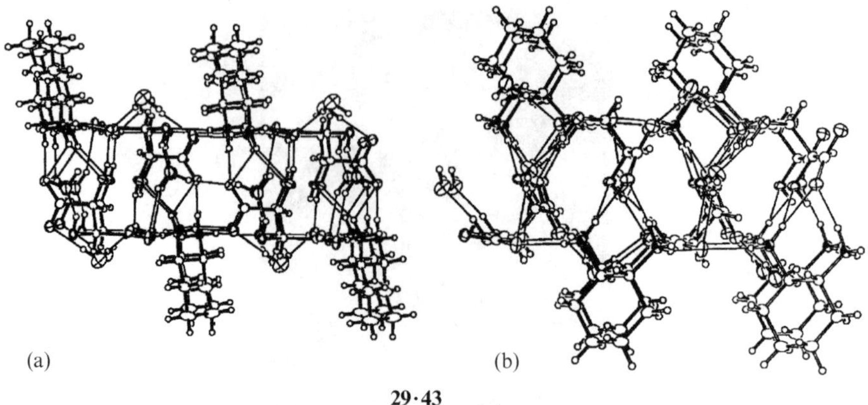

Figure 52 (a) Top view of the adduct complex (*R*,*R*)-**29**•(*S*,*S*)-**43**. Note the presence of two water molecules bound to tartrate hydroxyl and carboxylate groups, respectively. (b) Side view of the adduct complex (*R*,*R*)-**29**•(*S*,*S*)-**43** [51].

added diol. Thus, the interchange of protons between the amine and alcohol moieties is faster than the lifetime of the NMR technique even if the spectra are recorded at low temperature (−80 °C). As a general pattern, complexes with a ribbon-shaped core showed a higher downfield shift for the collapsed CH*OH* and CHNH_2 signals than that observed for the complex with a pleated-sheet motif, particularly when NMR experiments were performed in benzene-d_6. The association constant (K_b) for the matched complex between (*R*,*R*)-*trans*-1,2-diaminocyclohexane and (*R*,*R*)-*trans*-1,2-cyclohexanediol was calculated by means of ^{1}H NMR titrations in benzene-d_6 at room temperature [56]. In this experiment, the diol (from 0.2 to 2.0 equiv.) was added to the diamine assuming a simple 1:1 association in the equilibrium. A value of K_b of 18.3 M^{-1} was calculated, in accord with $K_b = 19$ previously obtained by microcalorimetry [58]. Notably, the mismatched complex, (*R*,*R*)-*trans*-1,2-diaminocyclohexane–(*S*,*S*)-*trans*-1,2-cyclohexanediol, showed a lower value of K_b(13.7 M^{-1}). Molecular recognition between matched diols and diamines was further demonstrated for the complexes (*R*,*R*)-*trans*-1,2-diaminocyclohexane–(*R*,*R*)-*trans*-1,2-cyclopentanediol and (*R*,*R*)-*trans*-1,2-diaminocyclohexane–(*R*)-1,1′-bi-2-naphthol by ^{13}C NMR experiments [60]. In fact, a 1:1 mixture (Figure 58c, left) and a 0.5:1 mixture (Figure 58d) of (*R*,*R*)-*trans*-1,2-diaminocyclohexane and 1,2-*trans*-(±)-cyclopentanediol showed a resonance for the matched (*R*,*R*)–(*R*,*R*) pair and another for the unbounded diol at lower field. In a similar way, a 1:1 mixture of racemic *trans*-1,2-diaminocyclohexane and (*R*)-1,1′-bi-2-naphthol (Figure 58, right) could be easily distinguished from an complex of two enantiopure components (Figure 58b, c and d).

Another experiment in favor of the 'building up' or 'erector' mechanism has been obtained from sublimation experiments on (*R*,*R*)-*trans*-1,2-diaminocyclohexane and (*R*,*R*)-2,3-butanediol [51,61]. The authors placed the reagents in two capillary tubes under moderate vacuum. After a few hours the crystals had filled

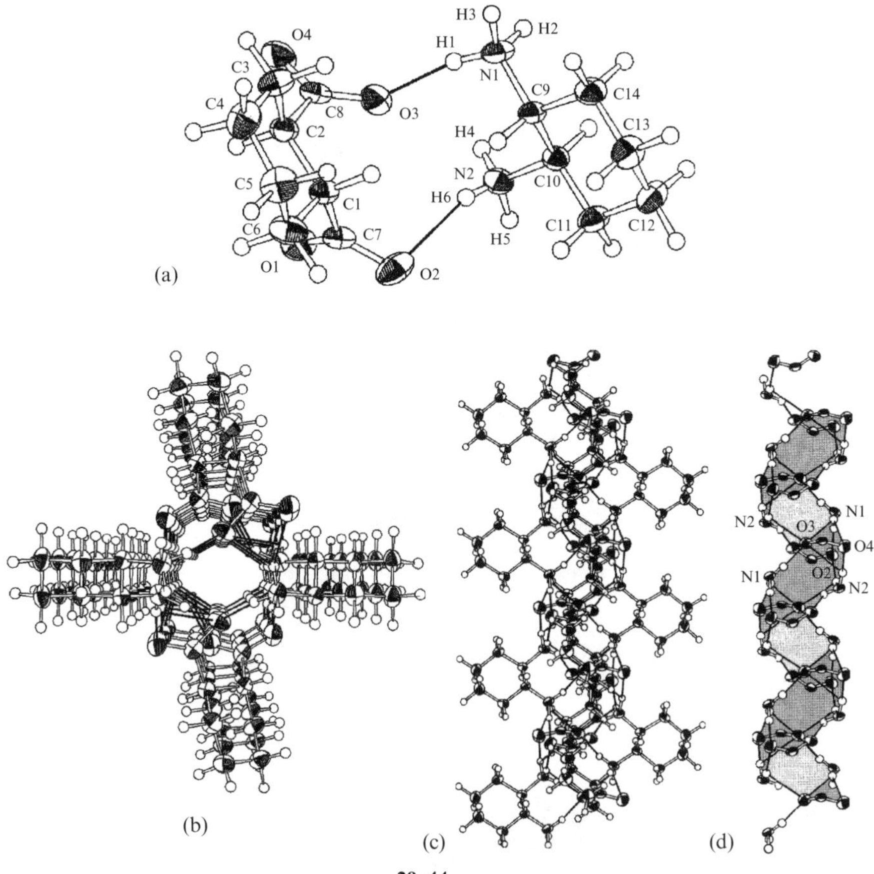

Figure 53 (a) molecular structure of (*R*,*R*)-**29**•(*R*,*R*)-**44**. (b) Top view of the adduct. (c) Side view along the *a* axis. (d) Simplified representation showing only the ammonium and the carboxylate groups involved in the H-bonding network [63].

the capillary containing the least volatile diol. The crystals so obtained showed chemical and physical properties identical with those for crystals prepared by crystallization from benzene solution. Thus, the assembly of the supramolecular structure had taken place in the gas phase, most probably by the transcription rules characteristic of recognition and complementary that are necessary for the assembly by the building up or erector hypothesis.

7 SELECTIVITY IN SUPRAMINOL ASSEMBLY [60]

The supraminols generated from *trans*-1,2-diaminocyclohexane and *trans*-1,2-diols are characterized by two distinct structural motifs. The first defines the geometric

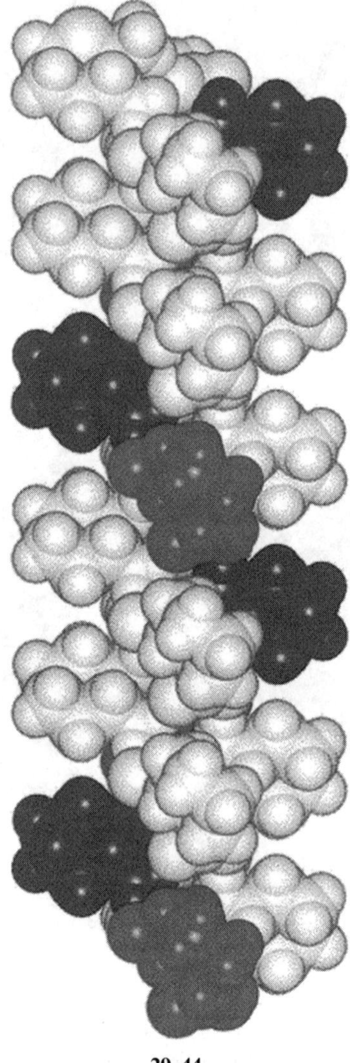

Figure 54 CPK representation of the adduct (*R*,*R*)-**29**•(*R*, *R*)-**44** [63] (see also **Plate 18**).

shape of the core of the assembly and the second characterizes the disposition of the residues carrying the functional groups. The ribbon-shaped and the pleated sheet, staircase-like motifs are the most prominent H-bonded arrays within the core of the crystalline supraminol structures in general. The disposition of the residues with respect to each other is represented by the helical outer motif displaying alternate units. The core and the residues may assume helical shapes with

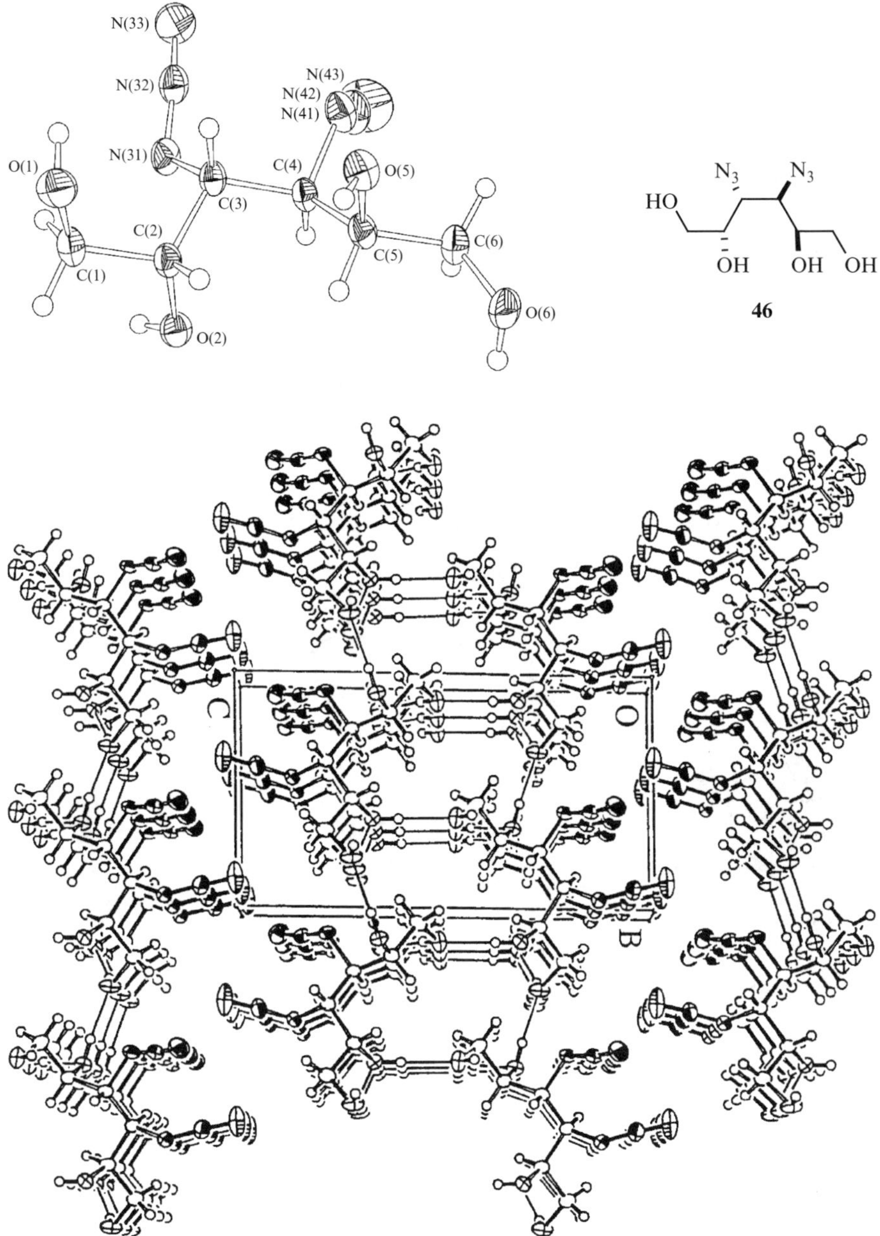

Figure 55 Supramolecular structure of 3,4-diazido-3,4-dideoxy-D-iditol (**46**) [64].

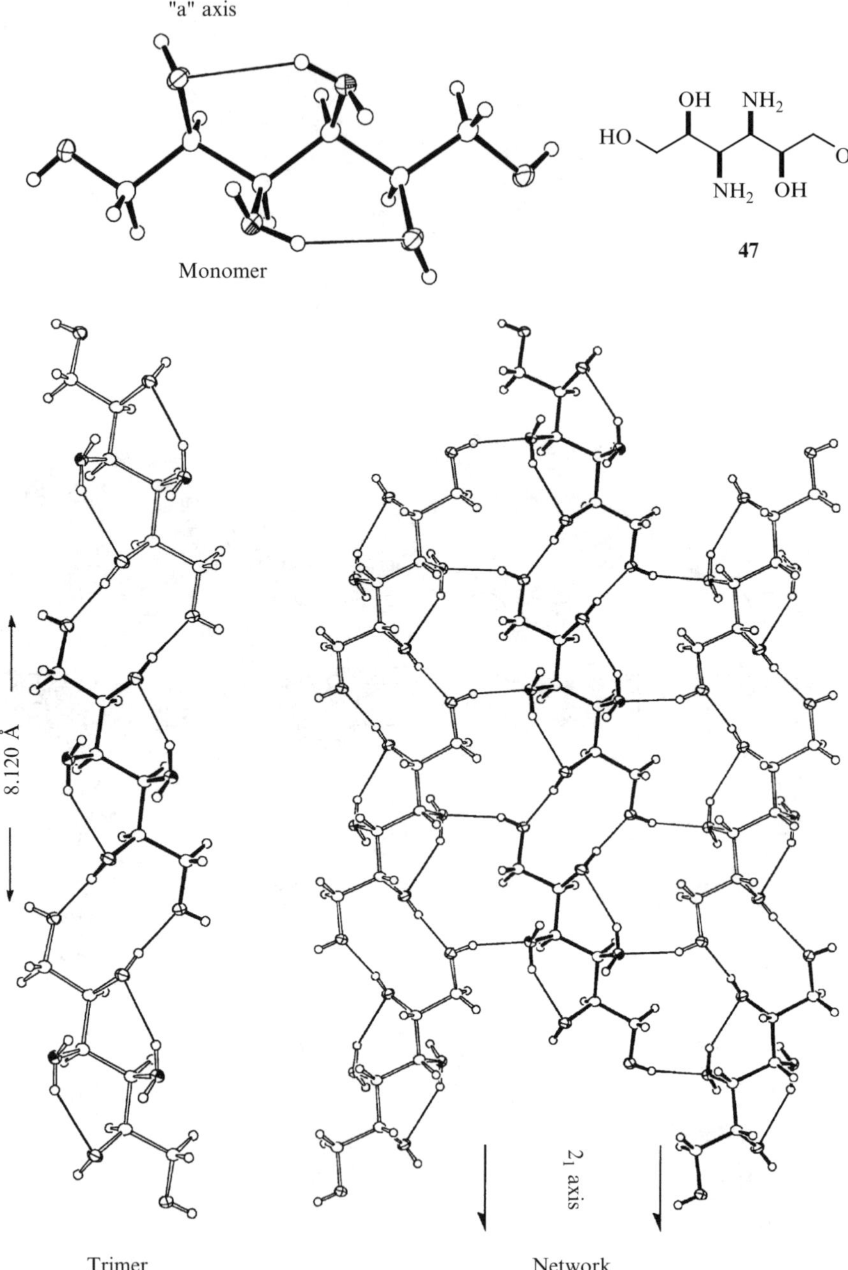

Figure 56 Aminocavitol structure of 3,4-diamino-3,4-dideoxy-D-iditol (**47**) [64].

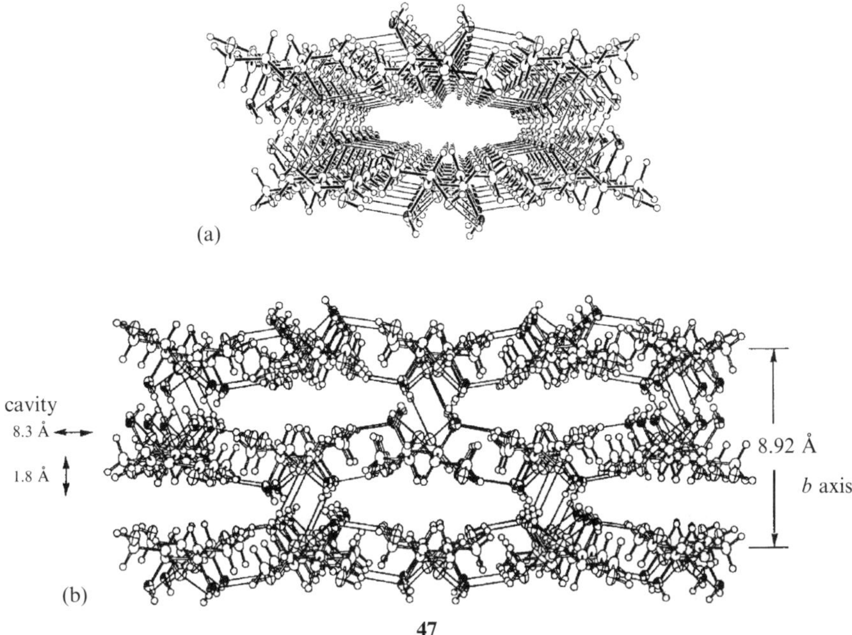

Figure 57 (a) Abyss or canal-like motif as an isolated unit. (b) Inter-unit H-bonding generating adjacent canal-like cavities [64] in aminocavitol (**47**) [64].

opposite senses of chirality. For example, the core in the (*R,R*)-diamine–diol complex **29•32** shows a right-handed helical ribbon, while the residues adopt a left-handed disposition in the outer periphery (Figure 30). When the residues in both partners consist of saturated cyclohexane rings, the resulting supraminols give a pleated sheet, staircase-like core motif as in complexes **28•30** and **29•30** (Figure 33). Another fascinating aspect of these homo- and heterochiral pairs of complexes is that the sense of the helix for the residues changes from left to right-handed simply by changing the chirality of the diamine from (*R,R*)- to (*S,S*). The origin of this reversal in the type of H-bonded assembly is unclear.

Thermodynamic studies favour the heterochiral complex over the homochiral complex [56]. On the other hand, crystallization from a mixture seems to favour the matched pairs. Thus, when the mismatched complex **29•48** between (*R,R*)-*trans*-1,2-diaminocyclohexane and (*S,S*)-1,4-cyclohexene-1,2-diol was heated (melted) with 1 equiv. of the matched (*R,R*)-diol, only crystals of the matched pair **29•35** were formed upon cooling (Scheme 21). In a similar way, when the mismatched complex (*R,R*/*S,S*)-**29•49** was melted with 1 equiv. of the matched diol (*R,R*)-**32** and the mixture was crystallized from benzene, only crystals of the matched pair (*R,R*/*R,R*)-**29•32** were obtained (Scheme 21). The corollary that 'matched diols and diamines compete better in the crystallization and assembly process than mismatched pairs' was found true also when the type of diol was modified during the competition experiment. In fact, when the mismatched complex (*R,R*/*S,S*)-**29•49** was melted with 1 equiv. of (*R,R*)-*trans*-1,2-cyclohexanediol (**31**), crystals of the matched complex (*R,R*/*R,R*)-**29•31** were again obtained as the only recovered product (Scheme 21).

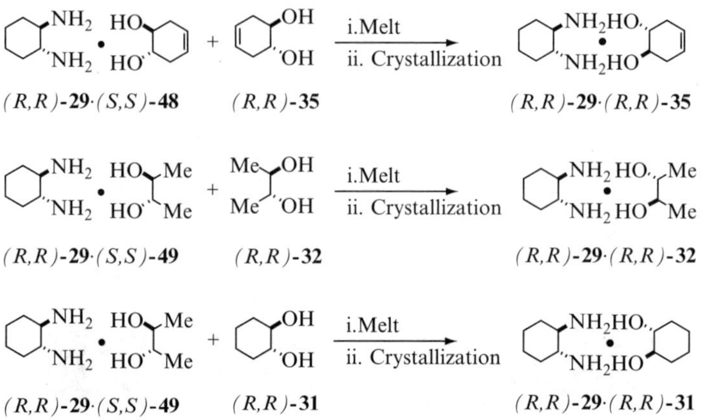

(*R,R*)-**29**·(*S,S*)-**48** + (*R,R*)-**35** → i.Melt, ii. Crystallization → (*R,R*)-**29**·(*R,R*)-**35**

(*R,R*)-**29**·(*S,S*)-**49** + (*R,R*)-**32** → i.Melt, ii. Crystallization → (*R,R*)-**29**·(*R,R*)-**32**

(*R,R*)-**29**·(*S,S*)-**49** + (*R,R*)-**31** → i.Melt, ii. Crystallization → (*R,R*)-**29**·(*R,R*)-**31**

Scheme 21 Examples of competition experiments to evaluate the relative stability of supraminols.

A set of matched and mismatched supraminols were compared with respect to their melting points, δ angles between diamino and diol groups in the crystalline complexes, and the pitch of the assembly [60]. The most compact structure among the helical motifs was that of the mismatched pair **29•30** (pitch 15.05 Å), compared with **28•30** (pitch 15.72 Å) and to all other matched complexes (pitch range 19.52–20.34 Å). This would tempt one to assume that the mismatched pair **29•30**, would also be the more stable supraminol with a shorter average H-bonded network throughout the assembly. However, when allowed to compete with the matched diamine **28**, it loses out in the reassembly–crystallization process to the higher melting, slightly more elongated helical complex **28•30**.

A second corollary that 'matched complexes with the pleated sheet staircase-like motif of the core are preferred in the reassembly–crystallization process compared with those characterized with the ribbon-like motif' was corroborated by further competition experiments. For example, when the matched complex **29•32** was melted with 1 equiv. of (*R,R*)-*trans*-1,2-cyclohexanediol (**31**) and the mixture was crystallized from benzene, crystals of the new complex **29•31** were obtained as the only recovered product (Scheme 22). On the other hand, treatment of the complex **29•31** with 1 equiv. of (*R,R*)-1,2-butanediol (**32**) did not result in any exchange of the diol partner (Scheme 22).

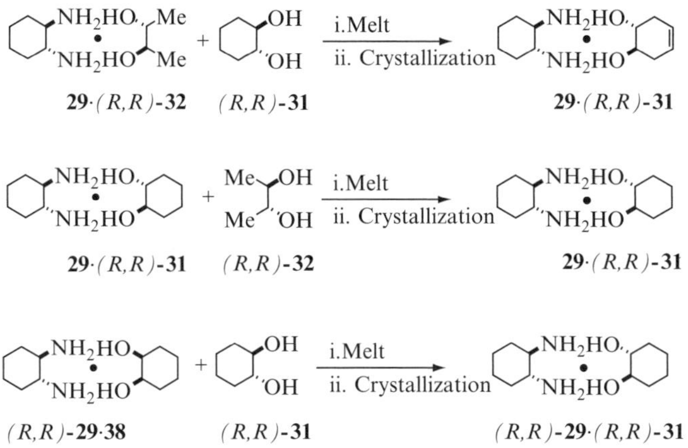

NH_2HO Me • NH_2HO Me + OH OH i. Melt ii. Crystallization NH_2HO • NH_2HO

29·(*R,R*)-**32** (*R,R*)-**31** **29**·(*R,R*)-**31**

NH_2HO • NH_2HO + Me OH Me OH i. Melt ii. Crystallization NH_2HO • NH_2HO

29·(*R,R*)-**31** (*R,R*)-**32** **29**·(*R,R*)-**31**

NH_2HO • NH_2HO + OH OH i. Melt ii. Crystallization NH_2HO • NH_2HO

(*R,R*)-**29**·**38** (*R,R*)-**31** (*R,R*)-**29**·(*R,R*)-**31**

Scheme 22 Competition experiments to evaluate the relative stability of supraminols.

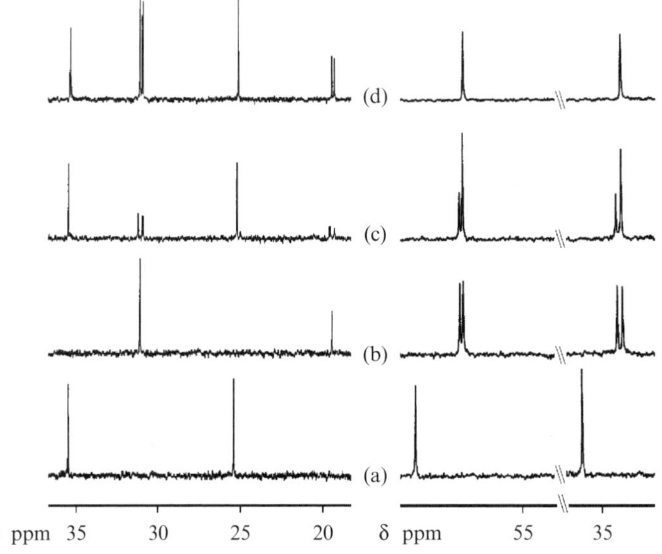

Figure 58 Partial ^{13}C NMR spectra for (*R,R*)-*trans*-1,2- diaminocyclohexane–(±)-*trans*-1, 2-cyclopentanediol and (*R*)-1,1′-bi-2-naphthol adducts. Left: (a) (*R,R*)-*trans*-1, 2-diaminocyclohexane; (b) *trans*-(±)cyclopentanediol; (c) 1:1 adduct between (*R,R*)-*trans*-1, 2-diaminocyclohexane and *trans*-(±)-cyclopentanediol; (d) 0.5:1 adduct between (*R,R*)-*trans*-1, 2-diaminocyclohexane and *trans*-(+/−)-cyclopentanediol. Right: (a) (*R,R*)-*trans*-1, 2-diaminocyclohexane; (b) 1:1 adduct between (*R*)-1,1′-bi-2-naphthol and *trans*-(±)-1, 2-diaminocyclohexane; (c) 1:1 adduct between (*R,R*)-*trans*-1, 2-diaminocyclohexane and (*R*)-1,1′ -bi-2-naphthol (*e.e.* 35%) showing molecular recognition and chiral discrimination of the diastereomeric supraminols; (d) 1:1 adduct between (*R*)-1,1′-bi-2-naphthol and (*R,R*)-*trans*-1, 2-diaminocyclohexane [60].

Assuming the pleated sheet-like motif and the ribbon-like motif as reference H-bonded core structures, a further increase in the steric demand of the residues, as for example the presence of aromatic rings as substituents on the diol or diamine moiety, or the absence of the required *trans* configuration, lead to a loss of efficiency in the recognition process with formation of incomplete ribbon-like or staircase-like structures. In the latter cases, competition experiments show that matched complexes with the pleated sheet-motif or with the ribbon-like motif of the core crystallize more easily than those characterized by incomplete or low organized core structures. Thus, when the complex **29•38** was melted with 1 equiv. of (*R,R*)-*trans*-1,2-cyclohexanediol (**31**) and the mixture was crystallized from benzene, again **29•31** was obtained as the only crystalline product (Scheme 22) [60].

It is difficult to offer definitive interpretations for these trends observed in the reassembly–crystallization experiments. Entropy and enthalpy play significant roles in the pairing of homochiral and heterochiral assemblies [56]. It is possible that the slightly more elongated structure of **28•30** is entropically more favored in the crystallization process. Clearly, many more examples of supraminol assembly are needed to appreciate fully the subtle forces involved in their stabilities and crystallization properties.

8 MISCELLANEA

8.1 Optical Resolution of Diols with Chiral Diamines

Optically active diols have been used for several asymmetric syntheses [67] and chiral resolutions [68]. In 1990, Kawashima and co-workers [69] reported the first example of direct optical resolution of racemic 2,2′-dihydroxy-1,1′-binaphthyl (**34**) with (*R,R*)-**29**. In this procedure equimolar amounts of racemic **34** and (*R,R*)-**29** were added in benzene and the mixture was heated to a homogeneous solution and then cooled to room temperature. After crystallization of the precipitate and treatment with hydrochloric acid, (*R*)-(+)-**34** was obtained with an optical purity of 94% and in a yield of 86% based on the amount of enantiomer presents in the racemate (Scheme 23).

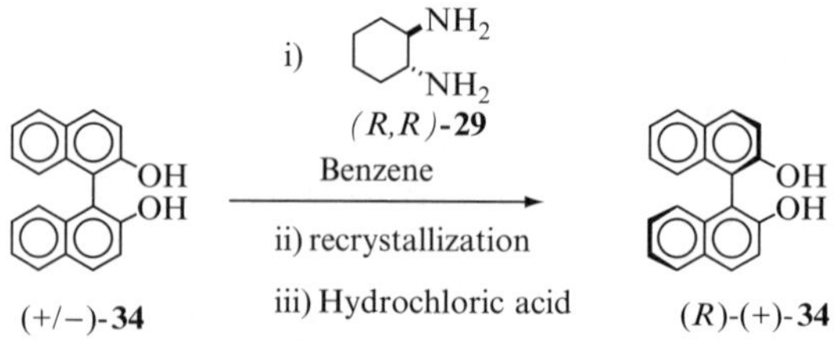

i) NH_2 NH_2 (*R,R*)-**29** Benzene

OH OH (+/−)-**34** → ii) recrystallization iii) Hydrochloric acid → OH OH (*R*)-(+)-**34**

Scheme 23 Optical resolution of 2,2′-dihydroxy-1,1′-binaphthyl (**34**) with (*R,R*)-**29**.

Evaporation of the mother liquor and treatment of the residue under similar experimental conditions afforded (*S*)-(−)-**34** in a yield of 57% with an optical purity of 62%. Optically active *threo*-1,2-diamino-1,2-diphenylethane was also an effective diamine for the optical resolution of racemic **34** that was obtained in a yield of 78% with an optical purity of 92% [70]. As previously described, the optical resolution of racemic *trans*-1,2-cyclohexanediol could be performed under similar experimental conditions [27]. Notably, this procedure was found effective also for the optical resolution of hydroxyoximes. Thus, when equimolar amounts of racemic (*E*)-1,2-diphenyl-2-(hydroxyimino)ethanol (**50**) and (*R,R*)-**29** were added to benzene and crystallized, (*R*)-(−)-(*E*)-**50** was recovered in a yield of 56% based on the enantiomer in the racemate (Scheme 24) [27].

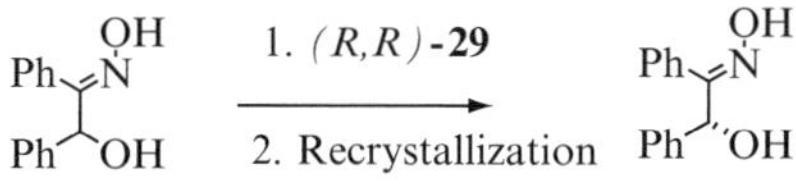

Ph, N–OH, Ph, OH — (±)-**50** → 1. (*R,R*)-**29**; 2. Recrystallization → Ph, N–OH, Ph, OH — (-)-**50**

Scheme 24 Optical resolution of (*E*)-1,2-diphenyl-2-(hydroxyimino)ethanol (**50**) with (*R,R*)-**29**.

These studies were extended to a new method of optical resolution based on the isomerization–crystallization strategy [70]. This method consists mainly of three steps: (i) formation of the diatereomeric complexes between racemic diol and optically active diamines; (ii) separation of the diastereomeric complexes; and (iii) epimerization of the undesirable complex. As an example, when the complex between (*R*)-(+)-**34** and (*R,R*)-**29** was melted at 170 °C, (*R*)-(+)-**34** included in the complex was completely racemized. The epimerized diastereoisomeric complexes were recrystallized from toluene to obtain (*R*)-(+)-**34**•(*R,R*)-**29** and (*S*)-(−)-**34**•(*R,R*)-**29**. The authors reported kinetic studies of epimerization by heating the solutions of (*S*)-(−)-**34**•(*R,R*)-**29** in different solvents (benzene, toluene, and *m*-xylene), showing that these reactions obeyed the reversible first-order kinetics of $(k_1 + k_2)t = \ln(100/OP)$, where k_1 and k_2 are the forward and reverse rate constants, respectively, and *OP* is the optical purity of **34** at time *t*. It is interesting that in some cases the chiral recognition ability between diols and diamines appears to be dependent on their molar ratio. As an example, when (−)-5,5′-dichloro-2,2′-dihydroxy-4,4′, 6,6′-tetramethylbiphenyl [(−)-**51**] was treated with (*S,S*)-**28**, a 2:1 complex was obtained. The resolution of racemic 5,5′-dichloro-2,2′-dihydroxy-4,4′, 6,6′-tetramethylbiphenyl [(±)-**51**] was accomplished only in the latter case [71]. An interesting extension of the strategy of the direct optical resolution of diols with chiral diamines consists in the chromatographic optical resolution on *trans*-1,2-diaminocyclohexane derivatives attached to a solid support [72]. Different reversed-phase chiral HPLC stationary phases (CSPs) were prepared starting from a glycidoxypropyl silica gel intermediate that was functionalized with racemic or enantiopure *trans*-1,2-diaminocyclohexane. The anchored diamine was later derivatized with different aroyl chlorides in order to obtain a broad spectrum of CPSs (Scheme 25).

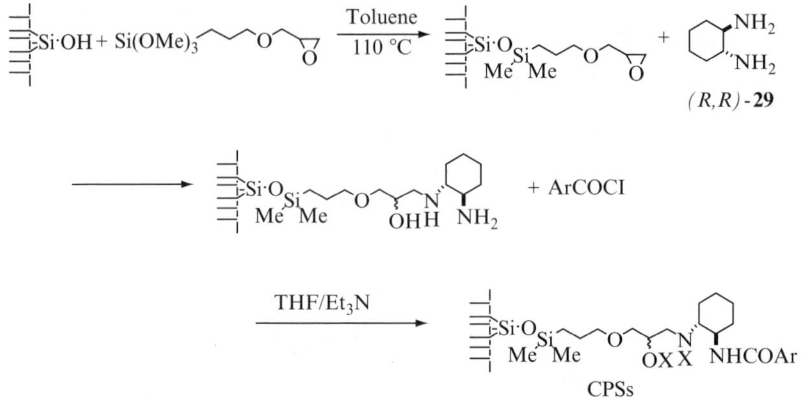

Si·OH + Si(OMe)$_3$ — Toluene, 110 °C → Si·O–Si(Me)(Me) + (R,R)-**29** → + ArCOCl — THF/Et$_3$N → CPSs

Scheme 25 Chiral stationary phases based on *trans*-1,2-diaminocyclohexane.

Others CPSs have been prepared in a similar way and characterized at different steps by physicochemical methods. For example, the structure of the chiral selector *N*-[2′-(*S*)-hydroxypropyl]-*N*, *N*′-bis(3,5- dichlorobenzoyl)-(*R*,*R*)-*trans*-1,2-diaminocyclohexane was solved by X-ray analysis and its absolute configuration confirmed unambiguously [73]. These CPS phases were used to resolve a large number of racemic mixtures belonging to different classes of organic compounds, such as α-aryloxyacetic acids, alcohols, sulfoxides, selenoxides, phosphinates, amino acids, amino alcohols, etc.

8.2 Gelling Agents Based on Chiral Diamines

Self-complementary interactions, which lead to macromolecule-like aggregates, are essential for gelation process of organic solvents [74]. In the gelation mechanism, self-complementary recognition processes, based mainly on intermolecular H-bonding and/or hydrophobic interactions, form macromolecule-like aggregates. Once formed, these aggregates are juxtaposed and interlocked to immobilize molecules of the organic solvents. In this context, a high gelation ability has been shown by alkylamides derived from *trans*-1,2-diaminocyclohexane [75]. The gelation ability of these alkylamides has been correlated to two cooperating noncovalent forces, namely the intermolecular H-bonding between the amide groups of adjacent molecules and the hydrophobic interactions of the long hydrocarbon tails. It is interesting that alkylamides synthesized starting from optically pure diamines (*S,S*)-**28** and (*R,R*)-**29** are the only derivatives that show gelation ability among several candidates tested. In fact, their racemic mixture, or the *cis*-1,2-diaminocyclohexane, did not form gels under the usual experimental conditions. Moreover, the formation of the gels failed when the chain of the lipophilic tails was shortened to four carbon atoms. The chiral structure of the aggregates in the gels was evaluated both by circular dichroism (CD) analysis and transmission electron microscopy (TEM). It is noteworthy that the helicity of the gel fibres was always right-handed for the *trans*-(*R,R*)-alkylamide derivative and left-handed for the (*S,S*)-enantiomer. Thus, the chirality of the diamine is a crucial factor for the chirality of the aggregate. On the basis of molecular modelling studies, the authors suggested that the two equatorial amide–NH atoms and the amide–CO group are antiparallel to each other and perpendicular to the cyclohexyl ring. In this way an extended molecular tape, such as that represented in Figure 59, is formed [75].

In a similar way, other derivatives of diamines (*S,S*)-**28** and (*R,R*)-**29** were evaluated for their gelation properties. As an example, *trans*-(*R,R*)-1,2-bis(dodecylureido)cyclohexane and *trans*-(*R,R*)-1,2-bis(octadecylureido)cyclohexane can cause physical gelation in a wide variety of organic solvents [76]. As an extension of these studies, novel polymerizable organogelators based on *trans*-(*R,R*)-1,2-bis(ureido)cyclohexane derivatives have been reported [77]. In these molecules, the presence of methacrylate-functionalized carboxylic acid groups determines the

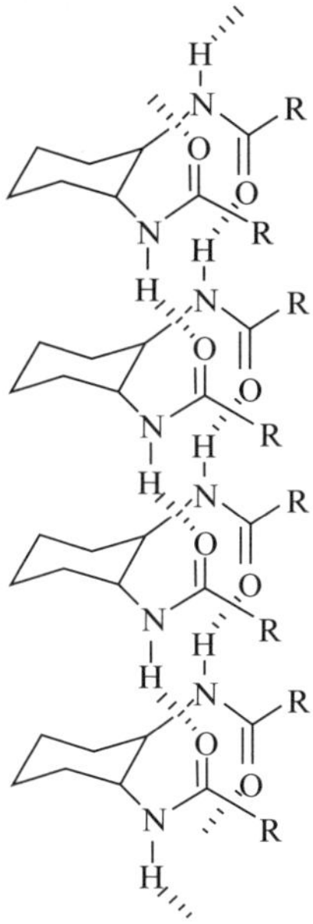

Figure 59 Schematic representation of the molecular structure of the fibers in the gelation process [75].

possibility of formation of covalent cross-links between adjacent macromolecule-like aggregates after treatment with appropriate photoinitiators. TEM studies show that the gelator molecules in the fibres are more densely packed upon polymerization, probably as the result of the formation of covalent linkages between the fibres. The gels obtained by this procedure remain stable at elevated temperatures and for prolonged periods of time. A further interesting consequence of the gelation properties of derivatives of chiral diamines (*S,S*)-**28** and (*R,R*)-**29** is the possibility of using these organized aggregates for the transcription of the chirality from organic molecules to inorganic compounds, a process that presents several analogies with the formation of fossils [78]. In particular, mixtures of neutral and cationic derivatives of diamines (*S,S*)-**28** and (*R,R*)-**29** were used as templates for a sol–gel polymerization of anionic tetraethoxysilane (TEOS). Interestingly, the authors found that right- and left-handed helical silica structures can be created by transcription of right- and left-handed structures in the organogel fibres, respectively [78]. Moreover, TEM analyses after removal of the organic gelators by calcination show that in the right-handed helical silica, a right-handed helical inner tube with a 20–60 nm inner diameter is present. In the case of the left-handed silica, the observed inner tubular structure is also left-handed.

8.3 Synthetic Receptors of Natural Products Based on Chiral Diamines

The preparation of synthetic receptors of natural products is of paramount importance to mimicking biological receptors [79]. Synthetic receptors bearing minimal structural complexity and high binding selectivities for peptide derivatives have been assembled starting from trimesic acid **52** (compound A) and (*R,R*)-**29** (compound B) (Scheme 24). When these were dissolved in THF in the presence of diisopropylethylamine, a cyclooligomer **53** of general formula A_4B_6 was obtained in acceptable yield (Scheme 26) [79].

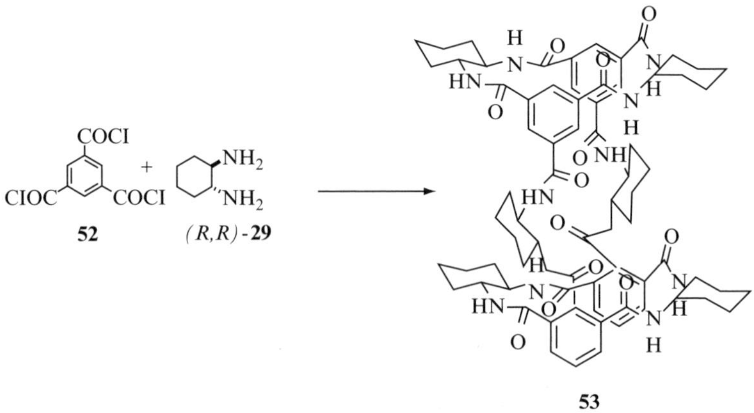

Scheme 26 Synthesis of the cyclooligomer **53** starting from trimesic acid (**52**) and (*R,R*)-**29**.

The selectivity of A_4B_6 to recognize different peptides was evaluated by ^1H NMR titrations in $CDCl_3$, showing its capability to bind amino acid residues in peptide chains with very high selectivities for chirality (up to 99+ *e.e.*) and side-chain identity (up to 3+ kcal/mol). On the basis of computational studies, the authors suggested that as a consequence of the molecular recognition process, a complex between the cyclooligomer and the peptide could be selectively formed. In this aggregate the receptor and the substrate form four intermolecular H-bonds to give a structure resembling a peptide three-strand β-sheet as schematically represented in Figure 60.

Moreover, intramolecular H-bonding between the (*R,R*)-**29** units can link the end of the receptor to produce a cavity which may be responsible for the high selectivity observed for the side-chain of the amino acids present on the peptide. Thus, while valine and phenylglycine side-chains fit well the receptor cavity, a substantial reduction in binding occurs when methylenes are added to the structure of the substrate. Different synthetic receptors have been synthesized by extension of the original concept. For example, receptors bearing more conformationally flexible acyclic diamines have been prepared under similar experimental conditions and evaluated for their ability to recognize peptides [79]. The novel receptors were sensitive to the structure of the components used to assemble them, but rigid cyclic building blocks were not necessary to obtain high binding selectivity for peptides. The influence of molecular conformation on the self-assembly of cyclohexane dicarboxylic acid diamides derived from L-phenylalanine has been studied [80]. *Ab initio* methods have been studied in relation to the self-discrimination of enantiomers in H-bonded dimers of hypothetical α-amino alcohols [81].

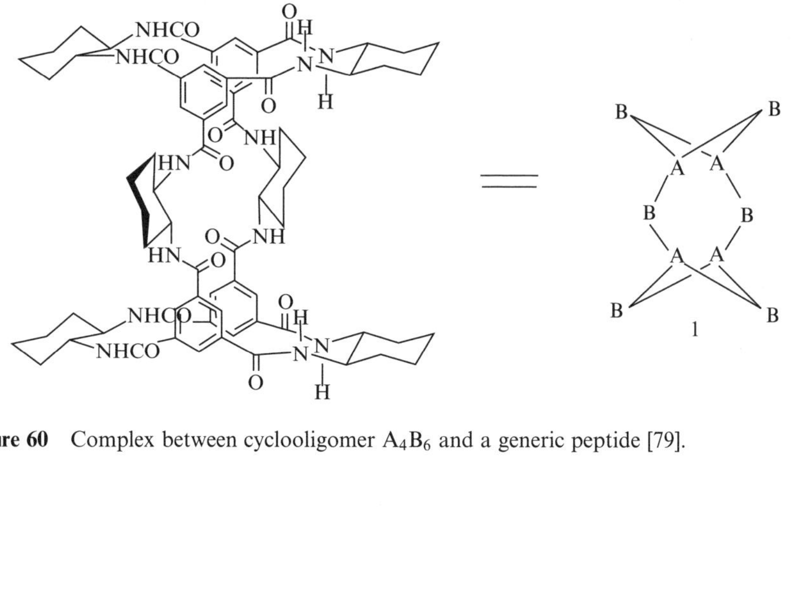

Figure 60 Complex between cyclooligomer A_4B_6 and a generic peptide [79].

9 CONCLUSION

H-bonding between amide-type NH and carbonyl-type O atoms has been known ever since the first insights into secondary and tertiary structures of proteins and polypeptides. This type of H-bonding can engage only two atom types and is one-dimensional. H-bonding between an amine and an alcohol, on the other hand, can take full advantage of the donor–acceptor characteristics of each group, resulting in a possible tetrahedral geometry. Thus, in ideal cases all hydrogens and lone pairs of an amine and an alcohol can engage in inter- or intramolecular H-bonding to produce two- and three-dimensional arrays of polyamines, polyalcohols, and, more interestingly, polyamino alcohols. Complexes between organic amines and alcohols have been known in solution and in the solid state for many years. Depending on the structures of the amines and alcohols, different supramolecular architectures are possible, and many have been characterized by X-ray analysis, revealing intriguing analogies with the structures of inorganic and organic lattices.

When the component amines and alcohols incorporate elements of chirality and symmetry, the resulting complexes are assembled by H-bonding through recognition, complementarity, and favourable organization to give three-dimensional supramolecular structures. These 'supraminols' can assemble in ordered right- or left-handed helical structures where NH and OH groups can be fully or partially coordinated (H-bonded). In this regard, *trans*-1,2-diaminocyclohexane is an exceptional supramolecular chiron or 'assembler' of supraminols in conjunction with complementary alcohols and diols. The recognition process can be extended to optical resolution and the design of artificial receptors of small molecules.

Because of the weaker nature of the H-bonds between amines and alcohols, compared with more polar amidic H-bonds, supraminols are compatible with hydrophobic environments when crystallized, although this is not a requirement. The supraminol helical *trans*-1,2-diaminocyclohexane–CO_2 complex can be recrystallized from aqueous ethanol [51].

Intra- and intermolecular H-bonding is well documented for small organic molecules, even if super-architectures are not involved [30]. We are not aware of examples H-bonding between an amino or alcohol group of a small organic molecule with complementary similar groups in the active site of an enzyme. For example, an H-bond between the C-2″ hydroxyl group in the aminoglycoside antibiotic paromomycin and a cytidine unit 1407 in 16S ribosomal RNA has been inferred from NMR studies [82] and X-ray crystallography [83]. We are also unaware of H-bonding between amino and hydroxyl groups of amino acids with proteins (as for example a serine OH and a lysine NH_2). Even if they existed, such H-bonds would be much weaker than amide types.

As for the supraminols and related superstructures covered in this chapter, entropic and enthalpic forces may combine with steric factors, recognition, and complementarity to ultimately decide the architecture, organization, dimensionality, and, in human visual terms, the aesthetic quality of the resulting motif. In contrast to normal disconnective analysis in the synthesis planning of a complex organic molecule, where simpler building blocks can be deduced, then chemically assembled to give the intended target, the assembly of amines and alcohols by H-bonding to give supraminols to superstructures of predictable shapes and architectures in the solid state is still a challenging goal.

ACKNOWLEDGEMENTS

We thank Eric Therrien for the skillful art work and Dr Michel Simard for the X-ray analyses.

REFERENCES

1. For selected references, see for example: (a) Khazanovich, N.; Granja, R. J.; McRee, D. E.; Milligan, R. A.; Ghadiri, M. R. *J. Am. Chem. Soc.* **1994**, *116*, 6011. (b) Mathias, J. P. ; Simanek, E. E.; Whitesides, G. M. *J. Am. Chem. Soc.* **1994**, *116*, 4316. (c) Yang, J.; Marendaz, J.; Geib, L. S. J.; Hamilton, A. D. *Tetrahedron Lett.* **1994**, *35*, 3665. (d) Chang, Y.; West, M.; Fowler, F. W.; Lauher, J. W. *J. Am. Chem.* Soc. **1993**, *115*, 5991. (e) Lehn, J.-M.; Mascal, M.; DeCian, A.; Fisher, J. *J. Chem. Soc., Perkin Trans. 2* **1992**, 461. (f) Simard, M.; Su, D.; Wuest, J. D. *J. Am. Chem.* Soc. **1991**, *113*, 4696. (g) For references to crystallographic studies, see: Taylor, R.; Kennard, O. *Acc. Chem. Res.* **1984**, *17*, 320.
2. For selected references, see for example: (a) Chin, D. N.; Zerkowski, J. A.; McDonald, J. C.; Whitesides, G. M. In *Organised Molecular Assemblies in the Solid State*; Whitesell, J. K., Ed.; Wiley, Chichester, **1999**, p. 185. (b) Reichert, A.; Ringsdorf, H.; Schumacher, P.; Baumeister, W.; Scheybani, T. In *Comprehensive Supramolecular Chemistry*; Atwood, J. L.; Davies, J. E. D.; McNicol, D. D.; Vogtel, F., Eds; Pergamon Press, Oxford, Vol. 9, **1996**, p. 313. (c) Fredericks, J. R.; Hamilton, A. D. In *Comprehensive Supramolecular Chemistry*; Atwood, J. L.; Davies, J. E. D.; McNicol, D. D.; Vögtle, F., Eds; Pergamon Press, Oxford, Vol. 9, **1996**, p. 565. (d) Lawrence, D. S.; Jiang, T.; Levett, M. *Chem. Rev.* **1995**, *95*, 2229. (e) Subramanian, S.; Zaworotko, M. J. *J. Coord. Chem. Rev.* **1994**, *137*, 357. (e) Aakeroy, C. B.; Seddon, K. R. *Chem. Soc. Rev.* **1993**, *22*, 397. (f) Lehn, J.-M. *Angew. Chem., Int. Ed. Engl.* **1990**, *29*, 1304. (g) Lehn, J.-M. *Angew. Chem., Int. Ed. Engl.* **1988**, *27*, 89. (h) Vögtle, F. *Supramolecular Chemistry*; Wiley, Chichester, **1991**.
3. For selected references see for example: (a) Aakeroy, C. B.; Borovik, A. S., Eds; *Coordination Chemistry Reviews*; Elsevier, Amsterdam, Vol. 183, **1999**. (b) Weber, E., Ed.; *Design of Organic Solids, Topics in Current Chemistry*; Springer, Berlin, Vol. 198, **1998**. (c) Fredericks, J. R, Hamilton, A. D. In *Comprehensive Supramolecular Chemistry*; Atwood, J. L.; Davies, J. E. D.; McNicol, D. D.; Vogtle, F., Eds; Pergamon Press, Oxford, Vols 6, 7, 9, and 10, **1996**. (d) Desiraju, G. R.; Sharma, C. V. K., in *Perspectives in Supramolecular Chemistry*; Desiraju, G. R., Ed., Wiley, Chichester, Vol. 2, p. 31, **1995**. (e) Dunitz, J. D. in *Perspectives in Supramolecular Chemistry*; Desiraju, G. R., Ed., Wiley, Chichester, Vol. 2, p. 1, **1995**. (f) Desiraju, G. R., *Crystal Engineering: The Design of Organic Solids*; Elsevier, Amsterdam, **1989**. (g) Mascal, M. *Contemp. Org. Synth.* **1994**, *1*, 31.

4. (a) Guru Row, T. N. *Coord Chem. Rev.* **1999**, *183*, 81. (b) Russel, VA. ; Ward, M. D. *Chem. Mater.* **1996**, *8*, 1654. (c) Blanchard-Desce, M. In *Comprehensive Supramolecular Chemistry*; Atwood, J. L.; Davies, J. E. D.; McNicol, D. D.; Vogtle, F., Eds; Pergamon Press, Oxford, Vol. 10, **1996**, p. 833.
5. (a) Kaes, C.; Hosseini, M. W.; Rickard, C. E. F.; Skelton, B. W.; White, A. H. *Angew. Chem., Int. Ed. Engl.* **1998**, *7*, 920;
6. Di Salvo, F. J. *Science* **1990**, *247*, 649.
7. (a) Felix, O.; Hosseini, M. W.; De Cian, A.; Fischer, J. *Angew. Chem., Int. Ed. Engl.* **1997**, *36*, 102, and references cited therein. (b) See also Ref. 1f.
8. Zhao, X.; Chang, Y.-L.; Fowler, F. W.; Lauher, J. W. *J. Am. Chem. Soc.* **1990**, *112*, 6627.
9. Kimbara, C. K.; Hashimoto, Y., Sukewaga, M., Nohira, H.; Saigo, K. *J. Am. Chem. Soc.* **1996**, *118*, 3441.
10. Zerkowski, D. J. A.; Seto, C.; Wierda, D. A.; Whitesides, G. M. *J. Am. Chem. Soc.* **1990**, *112*, 9025.
11. (a) See as examples of helical aggregates: Geib, S. J.; Vincent, C.; Fan, E.; Hamilton, A. D. *Angew. Chem., Int. Ed. Engl.* **1993**, *32*, 119. (b) Sanchez-Quesada, J.; Seel, C.; Pradas, P.; De Mendoza, J. *J. Am. Chem. Soc.* **1996**, *118*, 277. (c) Koshima, S.; Honke, S. *J. Org. Chem.* **1999**, *64*, 790.
12. Stockman, P. A., Bumgarner, R. E., Suzuki, S.; Blake, G. A. *J. Chem. Phys.* **1992**, *96*, 2496.
13. Tubergen, M. J.; Kuczwoski, R. L. *J. Am. Chem. Soc.* **1993**, *115*, 9263.
14. Barry, A. J.; Peterson, F. C.; King, A. J. *J. Am. Chem. Soc.* **1936**, *58*, 333.
15. Davis, W. E.; Barry, A. J.; Peterson, F. C.; King, A. J. *J. Am. Chem. Soc.* **1943**, *65*, 1294.
16. (a) Creely, J. J.; Wade, R. H. *J. Polym. Sci., Polym. Lett. Ed.* **1978**, *16*, 291. (b) Henrissat, B.; Marchessault, R. H.; Taylor, M. G.; Chanzy, H. *Polym. Commun.* **1987**, *28*, 113.
17. Trogus, C.; Hess, K. Z. *Phys. Chem.* **1931**, *B14*, 387.
18. Lee, D. M.; Burnfield, K. E.; Blackwell, J. *J. Biopolymers* **1972**, *11*, 89.
19. Kiyoshi, T.; Katsuo, M.; Shinsuke, Y. *J. Phys. Chem. B* **1999**, *103*, 3457.
20. Lewinski, J.; Zachara, J.; Kopec, T., Starawiesky, B. K.; Lipkowski, J.; Justyniak, I.; Kolodziejczyk. E. *Eur. J. Inorg. Chem.* **2001**, *5*, 1123.
21. For selected references on predictions of crystal structures, see: (a) Perlstein, J. *J. Am. Chem. Soc.* **1994**, *116*, 455. (b) Perlstein, J. *J. Am. Chem. Soc.* **1992**, *114*, 1955. (c) Gavezzotti, A. *J. Am. Chem.* Soc. **1991**, *113*, 4622. (d) Gavezzotti, A. *Acc. Chem. Res.* **1994**, *27*, 309. (e) Maddox, J. *Nature* **1988**, *335*, 201. (f) Leiserowitz, E.; Hagler, A. T. *Proc. R. Soc. London, Ser. A* **1983**, *388*, 133. (g) Curtin, D. Y.; Paul, I. C. *Chem. Rev.* **1981**, *81*, 525.
22. See for example: (a) Edlund, U.; Holloway, C.; Levy, G. C. *J. Am. Chem. Soc.* **1976**, *98*, 5069. (b) Joesten, M. D.; Schaad, L. J., Eds; *Hydrogen Bonding*; Marcel Dekker, New York, **1974**.
23. See, for example: Pirkle, W. H.; Pochapsky, T.-C. *Chem. Rev.* **1989**, *89*, 347. (b) Weisman, G. R. In *Asymmetric Synthesis*; Morrison, J. D., Ed.; Academic Press, New York, Vol. 1, **1983**, p. 153.
24. Baker, J. W.; Davies, M. M.; Gaunt, J. *J. Chem. Soc.* **1949**, 24.
25. Stevenson, D. P. *J. Am. Chem. Soc.* **1962**, *84*, 2849.
26. Saikia, B.; Suryanarayana, I.; Saikia, B. H.; Haque, I. *Spectrochim. Acta, Part A* **1991**, *47*, 791.
27. Kawashima, M.; Hirayama, A. *Chem. Lett.*, **1991**, 763.
28. Ermer, O.; Eling, A. *J. Chem. Soc., Perkin Trans 2* **1994**, 925.

29. (a) Kofler, A. In *Landolt–Bornstein, Zahlenwerte und Funktionen aus Physik, Chemie, Astronomie, Geophysic und Technik*, 6th edn; Schafer, K.; Lax, E., Eds; Springer, Berlin, Vol.2, Part 3, **1956**, p. 350. (b) Moelwyn-Hughes, E. A. *Physical Chemistry*, 2nd edn; Pergamon Press, Oxford, **1964**, p. 1060.
30. For examples of intermolecular H-bonding in small organic molecules, see the following Cambridge Crystallographic Data base codes: a, aromatic amino alcohols, ACBUET, BADWUX, CAGMAX, FLCTRT, IPAXP, KOSIR, MHPEAM, VANKIE, XEJLUT, BVJAJ; b, cyclic hydrocarbons, cyclitols, indanes, bicyclic compounds, steroids, BOLSOK, CBTANP, DENBIH, DOCINA, QEZSET, LATCOY, SUVNIG, WAZDEG, XESRES, TODCOE, TAYWOF, EIMCHO, EPICHO, QUDDUO; c, nucleosides, AMICET, BURNUX, COMBOV; d, amino acids, AETHRE, HIBUJ, WIRMIT; e, heterocycles, DAQYAV, HIWYIV, JUNDUL, JUNDOR. For examples of intramolecular H-bonding: a, azaspiro compound, BODKEK; b, amino sugar, BEHPOT; c, steroid, ERALK10, EPCHOL; d, piperidine, EPINUP; e, pyrrolidine, ZOGNAK; f, kanamaycin sulfate, KNMSYL; g, azithromycin, SAHWAZ; h, erythronocyclamine c, CUHPUQ10.
31. Brown, C. J. *Acta Crystallogr.* **1951**, *4*, 100.
32. Liminga R. and Olovsson, I. *Acta Crystallogr.* **1964**, *17*, 1523.
33. Liminga, R. *Acta Chem. Scand.* **1967**, *21*, 1206.
34. Corcoran, J. M.; Kruse, H. W.; Skolnik, S. *J. Phys. Chem.* **1953**, *57*, 435.
35. Allen, F. H.; Hoy, V. J.; Howard, J. A. K.; Thalladi, V. R., Desiraju, G. R., Wilson, C. C.; McIntyre, G. J. *J. Am. Chem. Soc.* **1997**, *119*, 3477.
36. (a) Robertson, J. M. *Proc. R. Soc. London Ser. A* **1951**, *207*, 101. (b) Desiraju, G. R.; Gavezzotti, A. *Acta Crystallogr.* **1989**, *B45*, 473.
37. Rusek, G.; Mazurek, J.; Lis, T.; Krajewski, K. *Pol. J. Chem.* **1997**, *71*, 119.
38. Mootz, D.; Brodalla, D.; Wiebcke, M. *Acta Crystallogr.* **1989**, *C45*, 754.
39. Meyers, E. A.; Lipscomb, W. N. *Acta Crystallogr.* **1955**, *C8*, 583.
40. Donohue, J. *Acta Crystallogr.* **1958**, *C11*, 512.
41. Toda, F.; Hyoda, S.; Okada, K.; Hirotsu, K. *J. Chem. Soc., Chem. Commun.* **1995**, 1531.
42. Goldberg, I.; Stein, Z.; Kai, A.; Toda, F. *Chem. Lett.* **1987**, 1617.
43. Toda, F.; Tanaka, K.; Hyoda, T.; Mak, T. C. W., *Chem. Lett.* **1988**, 107.
44. Loehlin, J. H.; Franz, K. J.; Gist, L.; Moore, R. H. *Acta Crystallogr.* **1998**, *B54*, 695.
45. (a) Etter, M. C. *Acc. Chem. Res.* **1990**, *23*, 120. (b) Etter, M. C.; MacDonald, J. C.; Bernstein, J. *Acta Crystallogr.* **1990**, *B46*, 256.
46. Bernstein, J.; Davis, R. E.; Shimoni, L.; Chang, N.-L. *Angew. Chem., Int. Ed. Engl.* **1995**, *34*, 1555.
47. Desiraju, G. R. *Crystal Engineering, The Design of Organic Solids*; Elsevier, New York, **1989**, chapt. 6, p. 175.
48. O'Leary, B.; Spalding, T. R.; Ferguson, G.; Glidewell, C. *Acta Crystallogr.* **2000**, *B56*, 273.
49. Jaunky, V.; Hosseini, M. W.; Planeix, J. M.; Cian, A. D.; Kyritsakas, N.; Fischer, J. *J. Chem. Soc., Chem. Commun.* **1999**, 2313.
50. Hanessian, S.; Gomtsyan, A.; Simard, M.; Roelens, S. *J. Am. Chem. Soc.* **1994**, *116*, 4495.
51. Hanessian, S.; Simard, M.; Roelens, S. *J. Am. Chem. Soc.* **1995**, *117*, 7630.
52. For references on the preparation of enantiomerically pure *trans*-1,2-diamines, see: (a) Scaros, M. G.; Yonan, P. K.; Laneman, S. A.; Fernando, P. N. *Tetrahedron: Asymmetry* **1997**, *8*, 1501. (b) Chooi, S. Y. M.; Leung, P.-H. ; Ng S.-C., Quek, G. H.; Sim, K. Y. *Tetrahedron: Asymmetry* **1991**, *2*, 981; (c) Onuma, K. ; Ito, T.; Nakamura, A. *Bull. Chem. Soc. Jpn.* **1984**, *53*, 2012. (d) Galsbøl, F.; Steenbøl, P.; Søndergaard, G.;

Sørensen, S. B. *Acta Chem. Scand.* **1972**, *26*, 3605. (e) Asperger, R. G.; Liu, C. F. *Inorg. Chem.* **1965**, *4*, 1492. (f) Jaeger, F. M.; Bijkerk, L. *Z. Anorg. Allg. Chem.* **1937**, *97*, 233.

53. Hanessian, S., Meffre, P.; Girard, M.; Beaudoin, S.; Sancéau, J.-Y.; Benanni, Y. *J. Org. Chem.* **1993**, *58*, 1991.
54. (a) For a review, see Hanessian, S.; Bennani, Y. *Chem. Rev.* **1997**, *97*, 3161. (b) Hanessian, S.; Gomtsyan, A. *Tetrahedron Lett.* **1994**, *35*, 7509. (c) Hanessian, S.; Gomtsyan, A.; Malek, N. *J. Org. Chem.* **2000**, *65*, 5623; (d) Hanessian, S.; Gomtsyan, A., Payne, A.; Hervé, Y.; Beaudoin, S. *J. Org. Chem.* **1993**, *58*, 5032. (e) Hanessian, S.; Benanni, Y.; Hervè, Y. *Synlett* **1993**, 35. (d) Hanessian, S.; Beaudoin, S. *Tetrahedron Lett.* **1992**, *33*, 7655. (f) Hanessian, S.; Beaudoin, S. *Tetrahedron Lett.* **1992**, *33*, 7659. (g) Hanessian, S.; Benanni, Y. *Tetrahedron Lett.* **1990**, *31*, 6465, and previous references. (h) Hanessian, D.; Delorme, D.; Beaudoin, S.; Leblanc, Y. *J. Am. Chem. soc.* **1984**, *106*, 5754.
55. For studies with vicinal diols, see: Huang, C.; Cabell, L. A.; Anslyn, E. V. *J. Am. Chem. Soc.* **1994**, *116*, 2778. For FT-IR studies on hydrogen bonding, see for example: Dado, G. P., Gellman, S. H. *J. Am. Chem. Soc.* **1993**, *115*, 4228. For reviews of the use of IR spectra in connection with hydrogen bonding interactions, see for example: (a) Symons, M. C. R. *Chem. Soc. Rev.* **1983**, *12*, 1. (b) Aaron, H. S. *Top. Stereochem.* **1979**, *11*, 1.
56. Hunenberg, P. H.; Granwehr, J. K.; Aebischer, J.-N.; Ghoneim, N.; Haselbach, E.; van Gunsteren, W. F. *J. Am. Chem. Soc.* **1997**, *119*, 7533.
57. Scott, W. R. P., van Gunsteren, W. F. In *Methods in Computational Chemistry: METECC-95*, Clementi, E.; Corongiu, G., Eds; STEF, Cagliari, **1995**, p. 397.
58. For another example of applications of microcalorimetric methods to complexes with similar structure, see: Haselbach, E.; Ghoneim, N.; Aebischer, J.-N. Progress Reports, Chiral 2–Workshops of the Swiss National Science Foundation, **1994**, **1995**.
59. Suemune, H.; Hasegawa, A.; Sakai, K. *Tetrahedron: Asymmetry* **1995**, *6*, 55.
60. Hanessian, S., Saladino, R.; Margarita, R.; Simard, M. *Chem. Eur. J.* **1999**, *5*, 2169.
61. Roelens, S.; Dapporto, P.; Paoli, P. *Can. J. Chem.* **2000**, *78*, 723.
62. For examples of synthesis of carbamate salts, see: (a) Von Dreele, R. B.; Bradshaw, R. L.; Burke, W. *Acta Crystallogr.* **1983**, *C39*, 253. (b) Adams, J. M.; Small, R. W. H. *J. Acta Crystallogr.* **1973**, *B29*, 2317. (c) Braibanti, A.; Manotti Lanfredi, M. L.; Pellinghelli, M. A; Tiripicchio, A. *Acta Crystallogr.* **1971**, *B27*, 2248, 2261.
63. Dapporto, P.; Paoli, P.; Roelens, S. *J. Org. Chem.* **2001**, *66*, 4930.
64. (a) Hanessian, S.; Nunez, J. C. unpublished data. (b) Hanessian, S.; Gauthier, J.-Y.; Okamoto, K.; Beauchamp, A. L.; Theophanides, T. *Can. J. Chem.* **1993**, *71*, 880.
65. Desiraju, G. R. *Angew. Chem., Int. Ed. Engl.* **1995**, *34*, 2311.
66. For references about the chiron concept, see: (a) Hanessian, S. In *Total Synthesis of Natural Products: The Chiron Approach*; Pergamon Press, New York, **1983**. (b) Nangia, A.; Desiraju, G. R. *Topics in Current Chemistry*; Springer, Berlin, Vol. 198, **1998**. (c) Nangia, A. *Curr. Sci.* **2000**, *78*, 374.
67. See, for example: (a) Seebach, D.; Beck, A. K.; Roggo, S.; Wonnacott, A. *Chem. Ber.* **1985**, *118*, 3673. (b) Trost, B. M., Murphy, D. J. *Organometallics* **1985**, *4*, 1143.
68. See, for example: (a) Kassai, C.; Juvancz, Z.; Balint, J.; Fogassy, E.; Kozma, D. *Tetrahedron* **2000**, *56*, 8355. (b) Gong, B.-Q.; Chen, W.-Y.; Hu, B.-F. *J. Org. Chem.* **1991**, *56*, 423. (c) Tamai, Y.; Heung-Cho, P.; Iizuka, K.; Okamura, A.; Miyano, S. *Synthesis* **1990**, 222. (d) Jacques, J.; Fouquey, C. *Org. Synth.* **1988**, *67*, 1. (e) Truesdale, L. K. *Org. Synth.* **1988**, *67*, 13. (f) Toda, F.; Tanaka, K. *J. Org. Chem.* **1988**, *53*, 3607. (g) Brussee, J.; Groenedijik, J. L. G.; Koppele, J. M.; Jansen, A. C. A. *Tetrahedron* **1985**, *41*, 3313.

69. (a) Kawashima, M.; Hirata, R. *Bull. Chem. Soc. Jpn.* **1993**, *66*, 2002. (b) Kawashima, M.; Hirayama. A., *Chem. Lett.* **1990**, 2299.
70. Saigo, K.; Kubota, N.; Takebayashi, S.; Hasegawa, M. *Bull. Chem. Soc. Jpn.* **1986**, *59*, 931.
71. Tanaka, K.; Moriyama, A:, Toda, F. *J. Org. Chem.* **1997**, *62*, 1192.
72. (a) Sinibaldi, M.; Carunchio, V.; Corradini, C.; Girelli, A. M. *Chromatographia* **1984**, *18*, 459. (b) Gasparrini, F.; Misiti, D.; Villani, C. *Chirality* **1992**, *4*, 447.
73. Cirilli, R.; Gasparrini, F.; Villani, C.; Gavuzzo, E.; Cirilli, M. *Acta Crystallogr.* **1997**, *C53*, 1937.
74. (a) Hanabusa, K.; Matsumoto, Y.; Miki, T.; Koyama, T.; Shirai, H. *J. Chem. Soc., Chem. Commun.* **1994**, 1401. (b) Hanbusa, K.; Naka, Y.; Koyama, T.; Shirai, H. *J. Chem. Soc., Chem. Commun.* **1994**, 2683. (c) Jokic, M.; Makarevic, J.; Zinic, M. *J. Chem. Soc., Chem. Commun.* **1995**, 1723. (d) Stocks, H. T.; Turner, N. J.; McCague, R. *J. Chem. Soc., Chem. Commun.* **1995**, 585.
75. Hanabusa, K.; Yamada, M.; Kimura, M.; Shirai, H. *Angew. Chem., Int, Ed. Engl.* **1996**, *35*, 1949.
76. Hanabusa, K.; Shimura, K.; Hirose, K.; Kimura, M.; Shirai, H. *Chem. Lett.* **1996**, 885.
77. De Loss, M.; Van Esch, J.; Stokroos, I.; Kellog, R. M.; Feringa, B. L. *J. Am. Chem. Soc.* **1997**, *119*, 12675.
78. (a) Jung, J. H.; Ono, Y., Hanabusa, K.; Shinkai, S. *J. Am. Chem. Soc.* **2000**, *122*, 5008. (b) Ono, Y.; Kanekiyo, Y.; Inoue, K.; Hijo, J.; Shinkai, s. *Chem. Lett.* **1999**, 23. (c) Ono, Y.; Nakashima, K.; Sano, M.; Hojo, J.; Shinkai, S. *Chem. Lett* **1999**, 119.
79. (a) Yoon, S. S.; Still, W. C. *J. Am. Chem. Soc.* **1993**, *115*, 823. (b) *Tetrahedron Lett.* **1994**, *35*, 2117.
80. Bergeron, R. J.; Yao, G. Y.; Erdos, G. W.; Milstein, S.; Gao, F.; Rocca, J.; Weimar, W. R.; Price, H. L.; Phanstiel, O., *Bioorg. Med. Chem.* **1997**, *5*, 2049.
81. Akorta, I.; Elguero, J. *J. Am. Chem. Soc.* **2002**, *124*, 1488.
82. (a) Fourmy, D.; Recht, M. I.; Blanchard, S. C.; Puglisi, J. D. *Science* **1996**, *274*, 1367. (b) Blanchard, S. C.; Fourmy, D.; Eason, R. G.; Puglisi, J. D. *Biochemistry* **1998**, *37*, 7716.
83. Vicens, Q.; Westhof, E. *Structure* **2001**, *9*, 647.

Chapter 3

Very Large Supramolecular Capsules Based on Hydrogen Bonding

JERRY L. ATWOOD, LEONARD J. BARBOUR, AND AGOSTON JERGA
University of Missouri–Columbia, Columbia, MO, USA

1 INTRODUCTION

An important feature in the evolution of biological systems is the encapsulation of entities within the confines of a structure [1]. Ultimately, the cell presents itself as a functional example of this enclosure of chemical space [2]. Nature has also used proteins and polypeptide chains as the building blocks for assemblies which contain and protect guests [3]. The guests may be the nucleic acid of viruses in the rhinovirus [4], poliovirus [5], cowpea chlorotic mottle virus [6], or the iron-containing core in the iron storage protein ferritin [7]. Even now, these biological capsules are beyond the synthetic grasp of the chemist because of the size and complexity of the enclosure. A further complication which has not often been addressed in the chemical literature is the very high level of organization found on the interior of enclosures of biological importance.

The encapsulation of chemical space on the scale of simple molecules has been a topic of considerable interest for more than two decades. For the formation of capsules two independent strategies have emerged: covalent synthesis and self-assembly. The groups of Cram [8], Collet [9], and Sherman [10] have synthesized capsules capable of encapsulating up to three small molecular guests [11]. Rebek and co-workers pioneered the use of self-assembly to produce a variety of assemblies held together by hydrogen bonds [12–15]. Multi-component systems have

Crystal Design: Structure and Function, edited by G. R. Desiraju

also been assembled by means of transition metal-based coordinate bonds [16–21]. The synthesis of three-dimensional hosts capable of containing a wide variety of guests has been achieved by the use of the principles of crystal engineering [22,23].

From our viewpoint, self-assembly [24], the single-step construction of molecular architecture using noncovalent forces, has provided an attractive means for constructing large, highly organized chemical entities. Owing to the reversibility of such interactions, noncovalent bonds, upon selection of appropriate chemical subunits, can facilitate error-free generation of either discrete or infinite supramolecular species.

2 SUPRAMOLECULAR CAPSULES

Typically, synthetic capsules have a contained volume in the range 200–350 Å^3. In 1997, we discovered a spherical assembly consisting of [(*C*-methylresorcin[4]arene)$_6$ $(H_2O)_8$], **1**. This assembly, Figure 1, with an enclosed volume of 1375 Å^3, was characterized by a single-crystal X-ray diffraction study and was found to be stable in nonpolar solvents [25]. Supramolecular assembly **1** ultimately led to the discovery of the link between **1** and the solid geometry principles of Plato and Archimedes [26]. Before this discussion progresses, it is useful to examine briefly Platonic and Archimedean solids.

The Platonic solids comprise a family of five convex uniform polyhedra (Figure 2, Table 1) which possess cubic symmetry and are made of the same regular polygons (equilateral triangle, square, pentagon) arranged in space such that the vertices, edges, and three coordinate directions of each solid are equivalent. That there is a finite number of such polyhedra is due to the fact that there exists a limited number of ways in which identical regular polygons may be adjoined to construct a convex corner. There are thus only five such isometric polyhedra, all of which are achiral.

In addition to the Platonic solids, there exists a family of 13 convex uniform polyhedra known as the Archimedean solids. Each member of this family is made up of at least two different regular polygons and may be derived from at least one Platonic solid through either truncation or twisting of faces (Figure 3, Table 2). In the case of the latter, two chiral members, the *snub cube* and the *snub dodecahedron*, are realized. The remaining Archimedean solids are achiral.

Returning to the discussion of **1**, it was the discovery that members of the resorcin[4]arene family, **2**, self-assemble to form the capsule shown as **1** in Figure 1 that prompted our research group to examine the topologies of related spherical hosts. Further, we sought to understand existing structures on the basis of symmetry. In addition to providing a grounds for classification, it was anticipated

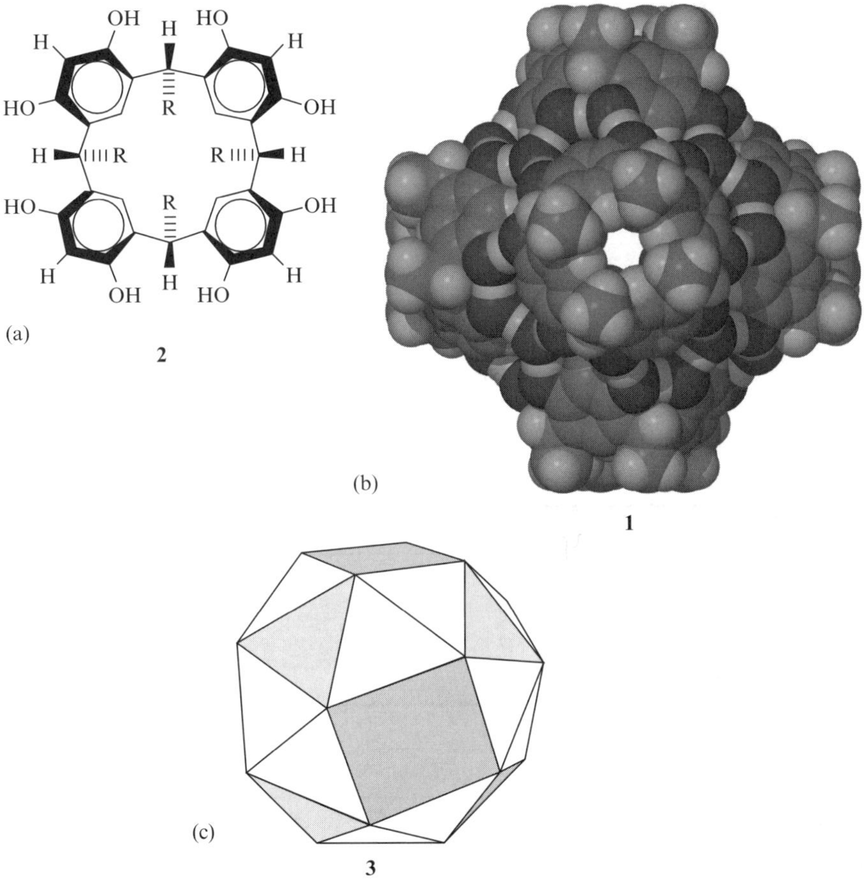

Figure 1 (a) General formula for resorcin[4]arenes, **2**; (b) hexamer of *C*-methylresorcin[4]arene, **1**; (c) the snub cube, an Archimedean solid, **3**.

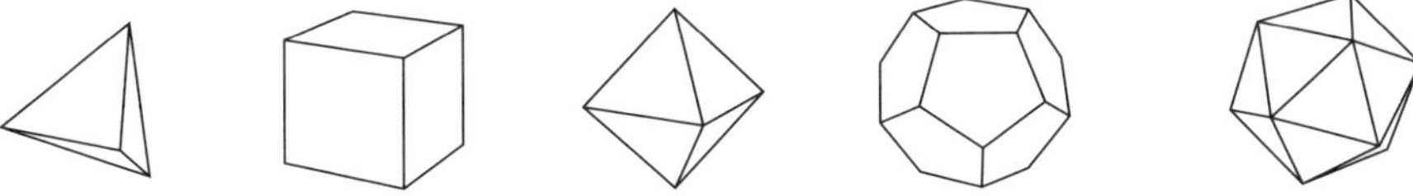

Figure 2 Geometrical representation of the Platonic solids.

that such an approach would allow one to identify similarities at the structural level, which, at the chemical level, may not seem obvious and may be used to design large, spherical host assemblies similar to **1**.

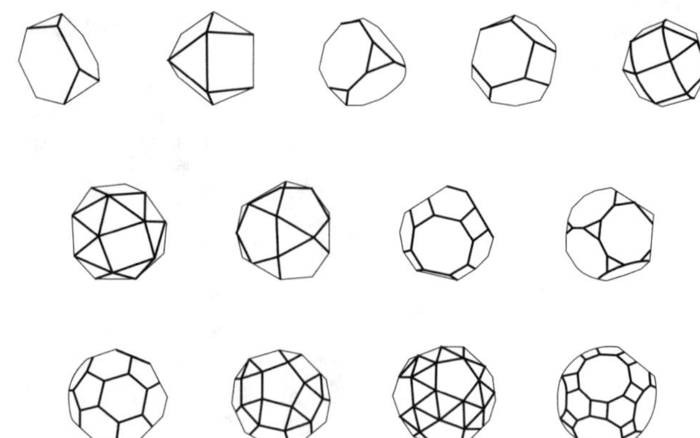

Figure 3 Geometrical representation of the Archimedean solids.

Table 1 Platonic solids.

Name	Faces (f)	Corner (c)	Edges (e)
Tetrahedron	4	4	6
Cube	6	8	12
Octahedron	8	6	12
Dodecahedron	12	20	30
Icosahedron	20	12	30

Table 2 Archimedean solids.

Name	c	e	f3	f4	f5	f6	f8	f10
Truncated tetrahedron	12	18	4	—	—	4	—	—
Truncated cube	24	36	8	—	—	—	6	—
Truncated octahedron	24	36	—	6	—	8	—	—
Cuboctahedron	12	24	8	6	—	—	—	—
Small rhombicuboctahedron	24	48	8	18	—	—	—	—
Great rhombicuboctahedron	48	72	—	12	—	8	6	—
Snub cube	24	60	32	6	—	—	—	—
Truncated dodecahedron	60	90	20	—	—	—	—	12
Truncated icosahedron	60	90	—	—	12	20	—	—
Icosidodecahedron	30	60	20	—	12	—	—	—
Small rhombicosidodecahedron	60	120	20	30	12	—	—	—
Great rhombicosidodecahedron	120	180	—	30	—	20	—	12
Snub dodecahedron	60	150	80	—	12	—	—	—

We were also able to link the spherical assembly to the Archimedean solid known as the *snub cube*, Table 2. In a recent review, we have set forth structural classifications and general principles for the design of spherical molecular hosts based, in part, on the solid geometry ideas of Plato and Archimedes [27]. Indeed, we are now using the well-known solid geometry principles embodied in Platonic and Archimedean solids to design new, large, spherical container assemblies. Based on this wide variety of work focused on the enclosure of chemical space, we confidently predict a rich and diverse array of discrete spherical molecular hosts will be assembled [27,28]. A summary of the contributions to date is presented in this chapter. However, it will be instructive to review briefly the essential portions of the geometrical analysis of exiting spherical supramolecular assemblies.

To construct a spherical host from two subunits ($n = 2$), each unit must cover half the surface of the sphere. This can only be achieved if the subunits exhibit curvature and are placed such that their centroids lie at a maximum distance from each other. These criteria place two points along the surface of a sphere separated by a distance equal to the diameter of the shell. As a consequence of this arrangement, there exist two structure types: one with two identical subunits attached at the equator and one belonging to the point group D_{nd} which is topologically equivalent to a tennis ball. In both cases, the subunits must exhibit curvature.

To construct a spherical host from three subunits ($n = 3$), each must cover one-third the surface of the sphere. Following the design conditions described previously, placing three identical subunits along the surface of a sphere results in an arrangement in which their centroids constitute the vertices of an equilateral triangle. As a result, there is only one structure type, that belonging to D_{3h}. Each of the subunits must exhibit curvature.

For $n = 4$, positioning four points along the surface of a sphere such that they lie a maximum distance from each other places the points at the vertices of a tetrahedron. This is the first case in which joining the points via line segments gives rise to a closed surface container. The container, a tetrahedron, is comprised of four identical subunits, in the form of equilateral triangles where surface curvature is supplied by edge-sharing of regular polygons rather than by the subunits themselves.

In the following discussion, the advantages of the use of the Platonic and Archimedean solids as models for supramolecular assemblies are clearly pointed out. However, the limitations in the use of these models in complex supramolecular assemblies are also revealed. Both the power and the limitations of this approach are made clear with regard to assembly of *p*-sulfonatocalix[4]arene anions, **4**, into the large spherical twelve-*p*-sulfonatocalix[4]arene entity **4** (Figure 4).

p-Sulfonatocalix[4]arene is a macrocyclic anion that has a truncated pyramidal shape and contains a hydrophobic cavity bounded by four aromatic rings. The base and apical square faces of the truncated pyramid are defined by sulfonate groups and phenolic hydroxyl groups, respectively, and the trapezoidal faces consist of the external surfaces of the aromatic rings. The bipolar amphiphilic nature of **4**, in conjunction with its truncated pyramidal shape, serves as a dominant structure-directing factor in the organization of this macrocycle in its solid-state

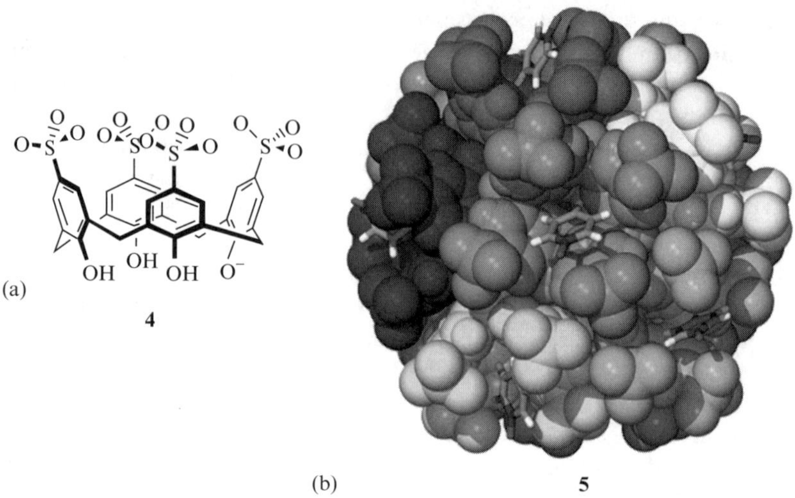

Figure 4 (a) Two representations of the 5−anion of *p*-sulfonatocalix[4]arene, **4**; (b) spherical assembly of 12 anions of *p*-sulfonatocalix[4]arene, **5**.

structures. For example, in the crystal structure of Na_5**4**, the hydrophobic cores of the truncated pyramids align to form a bilayered structure consisting of alternating organic and aqueous layers. This structure is consistent with the arrangement of a bipolar amphiphilic molecule according to the influences of hydrophobic effects. The aqueous layers are composed of the polar surfaces of the truncated pyramids, water molecules, and counterions. The organic layers consist of the π-stacked, two-dimensional bilayered grid composed of truncated pyramids arranged in an alternating up–down anti-parallel fashion with their aromatic rings in van der Waals contact with those of adjacent calixarene molecules. This aspect of the structure may be interpreted in terms of organization of the truncated pyramids according to shape complementarity.

Although the bilayer arrangement of **4** has been found in many of its structures [29,30], we anticipated that it should be possible to influence the relative orientation of the truncated pyramids with respect to each other. Indeed, based on our experience with Platonic and Archimedean solids related above, we sought to control the orientation of the truncated pyramids such that an up–up arrangement would result. Our design strategy ultimately led to a three-component system consisting of Na_5**4**, pyridine *N*-oxide, and a lanthanide metal nitrate [31]. In a representative example, aqueous solutions of Na_5**4**, pyridine *N*-oxide, and $La(NO_3)_3$ were combined in a 2:2:1 molar ratio. Crystallization and subsequent single-crystal X-ray structural analysis revealed a framework in which two *p*-sulfonatocalix[4]arene anions coordinated to a La^{3+} ion through their sulfonate groups to form a C-shaped dimeric assembly. In addition, two pyridine *N*-oxides, both of which coordinate to the La^{3+} ion, are bound within the respective calixarene cavities of the dimer. In these dimeric assemblies, the La^{3+} ion acts a hinge, and the steric requirements of the pyridine *N*-oxides binding concomitantly to the La^{3+} ion and the calixarene cavities, impart a dihedral angle of ca. 60° between the *p*-sulfonatocalix[4]arene anions. This structural feature helps **4** assemble into structures with curved surfaces, such as spherical clusters, **5** [31].

The amount of 'chemical space' enclosed by spherical assembly **5** is about 1700 $Å^3$. This space houses 30 water molecules and two sodium ions. However, the van der Waals volume of **5** is about 11 000 $Å^3$. Structure **5** is differentiated from **1** by several factors. First, the supramolecular forces used to hold **1** together are hydrogen bonds, while a combination of van der Waals forces, π-stacking forces, and metal ion coordinate covalent bonds is employed for **5**. Second, the surface which encloses the chemical space is essentially one atom thick for **1**, whereas it is the thickness of the *p*-sulfonatocalix[4]arene building block in **5** (hence the 11 000 $Å^3$ volume of the assembly with only 1700 $Å^3$ of space within). Third, the contents of the capsule are ordered for **5** (by the hydrogen bonds from the enclosed water to the phenolic oxygen atom hydrogen bond acceptors at the base of the *p*-sulfonatocalix[4]arene), but the contents are completely disordered for **1** (because of the lack of any directional bonding force connecting the skeleton of the assembly to the contents therein).

In the context of the discussion of Platonic and Archimedean solids, it is interesting that spherical assembly **5** corresponds to a *great rhombicuboctahedron*. As can be seen from Table 2, the *great rhombicuboctahedron* consists of 12 square faces, together with eight hexagons and six octagons. The square faces are the upper rims of the calix[4]arenes (the larger square faces of the truncated pyramids). Since there are 12 calix[4]arenes, there are 12 square faces, as called for by the *great rhombicuboctahedron*. However, the hexagons and the octagons are not represented by chemical compounds. Rather, the hexagons and octagons are formed by connecting three calix[4]arenes (squares) or four calix[4]arenes, respectively. That is to say, the lines connecting the squares for the *great rhombicuboctahedron* in Figure 3 do not represent chemical bonds, weak or strong. Indeed, the hexagons and octagons in **5** are not regular, but are approximations, just as the upper rim of the *p*-sulfonatocalix[4]arene is an approximation of a square. One must accept the fact that complex building blocks of large supramolecular assemblies may often only approximate the triangles, squares, pentagons, etc. called for by the Platonic or Archimedean solids.

There is a second limitation to the use of Platonic and Archimedean solids as models for supramolecular assemblies. For example, the *snub dodecahedron* is a target of some of our synthetic efforts. For the *snub dodecahedron*, a total of 60 triangles is called for, but triangles can be simply the result of hydrogen bonds from adjacent triangles or pentagons. For example, the *snub cube* in Figure 1 is composed of 32 triangles and six squares, with the triangles being represented by water molecules and the squares, resorcin[4]arenes, **2**. However, in assembly **1** only eight of the 32 triangles contain water molecules, as one can observe from the shading in

Figure 1. Therefore, in the search for the *snub dodecahedron*, utilizing calix[5]arene pentacarboxylic acid molecules as the pentagons and water as the triangles, the ratio of water to calix[5]pentacarboxylic acid cannot be higher than 60:12 (see Table 2). However, the ratio could (and probably will) be lower than 60:12.

To elucidate further the point made concerning the *snub dodecahedron*, consider the mixture of squares and triangles. The square:triangle ratio of 6:32 defines the *snub cube*, but 6:8 defines the *cuboctahedron*, and 18:8 the *small rhombicuboctahedron*. When this is considered along with the reasoning in the above paragraph, it is apparent that the Platonic and Archimedean concepts are meant to serve as a guide to the assembly of spherical supramolecular capsules. At this stage of development, the actual ratios of synthons must be determined experimentally.

An important outgrowth of the work related to the dodecamer of *p*-sulfonatocalix[4]arene, **5**, was the discovery of a method of control of molecular architecture such that spherical assembly **5** (a *great rhombicuboctahedron*, an Archimedean solid) was converted into a tubular structure [31]. When aqueous solutions of Na_5**4**, pyridine *N*-oxide, and $La(NO_3)_3$ were combined in a 2:8:1 molar ratio, the result was the tubular assembly, **6**, shown in Figure 5. The tube is ca 28 Å in diameter and consists of *p*-sulfonatocalix[4]arene anions arranged along the surface of a cylinder. The tubes are aligned with the long axis of the crystals and have lengths aproaching 1 cm in some cases. In contrast to spherical capsule **5**, the tubule has an organic shell that is not solely composed of calix[4]arene molecules. In addition to the *p*-sulfonatocalix[4]arenes, the shell contains two crystallographically unique pyridine *N*-oxide molecules intercalated between the aromatic rings of adjacent calixarenes.

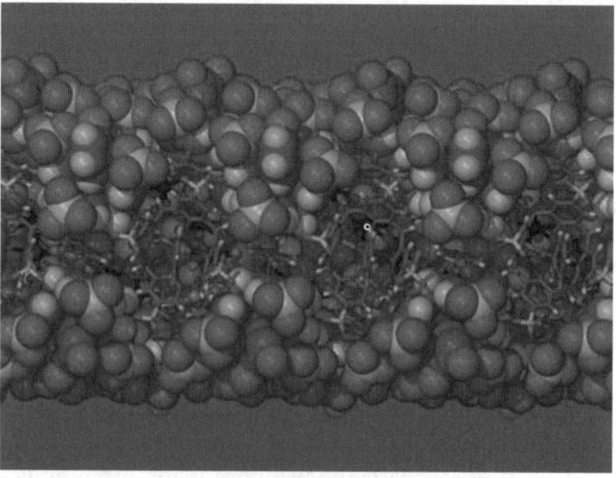

Figure 5 Structure of the tubular assembly **6** (see also **Plate 19**).

A further feature of **6** is that the *p*-sulfonatocalix[4]arene anions and the intercalated pyridine *N*-oxide molecules form a chiral helical assembly along the length of the tube. The helix consists of a single strand of alternating *p*-sulfonatocalix[4]-arene and pyridine *N*-oxide, and there are 4.5 of these units in each turn [31]. These tubular assemblies are arranged in a hexagonal array, Figure 6, in a pattern similar to the organization of cylindrical micelles.

The relationship between sphere and tube is shown in Figure 7. It is interesting that the diameters of the sphere and the tube are virtually identical. It is important to emphasize that both sphere and tubule can be formed from a wide range of metal ions, both 3+ (lanthanides) and 2+ (cadmium), and the stoichiometry of the reaction mixture controls the architecture of the final product, sphere or tubule.

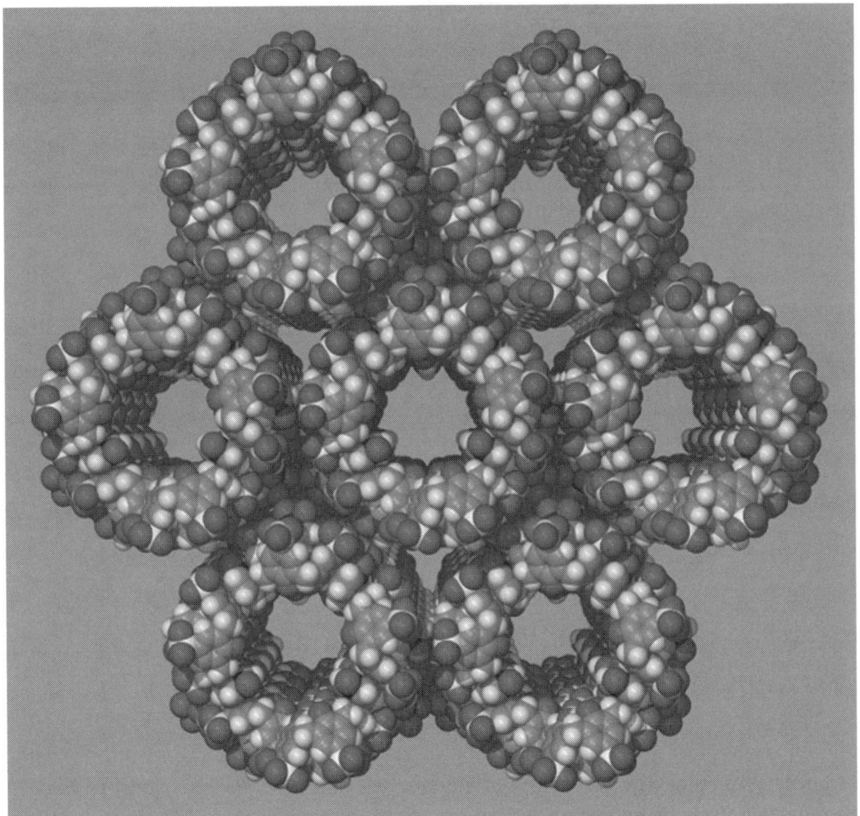

Figure 6 View of the hexagonal packing of tubules composed of *p*-sulfonatocalix[4]arene with intercalated pyridine *N*-oxide.

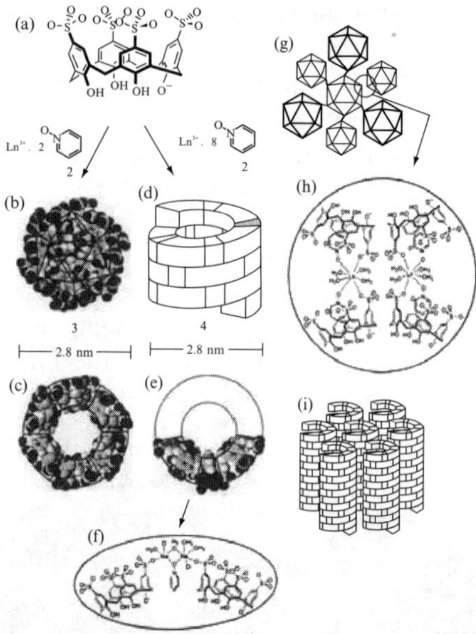

Figure 7 Schematic representation of the self-organization of *p*-sulfonatocalix[4]arene anions into structures with spherical and tubular morphologies.

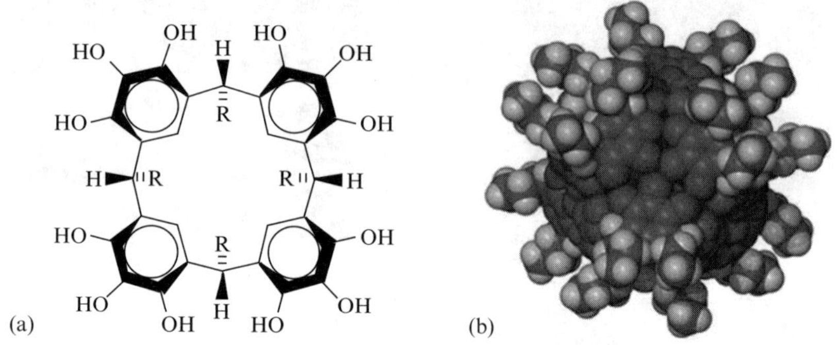

Figure 8 (a) General forumla for pyrogallol[4]arene, **7**; (b) structure of the hexamer, **8**, *C*-propylresorcin[4]arene with the oxygen atoms shown in red, (see also **Plate 20**).

There has been a recent report of a very interesting, large supramolecular assembly related to the resorcin[4]arene work. The synthesis of several pyrogallol [4]arenes, **7**, has been accomplished under mild conditions [32]. However, the authors report that they obtained the hexamer of *C*-isobutylpyrogallol[4]arene, **8**, only one time out of many attempts (Figure 8).

Further, they used this lack of reproducibility as evidence that the hexamer is unstable compared with our [(*C*-methylresorcin[4]arene)$_6$(H_2O)$_8$], **1** [25]. This did not seem credible since the hexamer appears to be held together by 72 hydrogen bonds, 48 of which are intermolecular. This means that the assembly is bound together by eight intermolecular hydrogen bonds per bonded entity. Capsule **1** is bound together by 36 intermolecular hydrogen bonds, or 2.6 per bonded molecule. For comparison, the tennis balls of Rebek are bound together by four hydrogen bonds per bonded molecule [33]. It seemed to us that the pyrogallol[4]arene hexamer should be more, not less, stable than our [*C*-methylresorcin[4]arene)$_6$(H_2O)$_8$] in the same solvents. We therefore synthesized a variety of pyrogallol[4]arenes by the acid-catalyzed condensation of pyrogallol with aldehydes. The yields are high, approaching quantitative, and we have X-ray structural data for the hexamer **8** with R = ethyl, propyl, isobutyl, and butyl. The hexamer for R = propyl is shown in Figure 9. Figure 10 displays the beautiful hydrogen bonding pattern which spans the sphere in a continuous belt. The synthesis of the large, beautiful assembly is reproducible. We have obtained the hexamer from a variety of solvents, including Et_2O with nitrobenzene, and, surprisingly, methanol. The volume available for guests is 1510 Å^3, so the range of guests available for study is practically limitless.

Conclusive proof for the existence of hexamer, **8**, in solution was obtained from an NMR spectroscopic investigation [34,35]. A representative experiment afforded the ^{1}H NMR spectrum of the hexamer in deuterochloroform, which is shown in

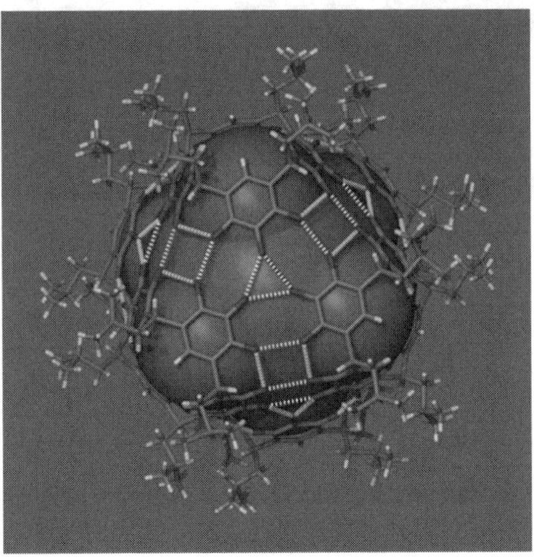

Figure 9 Stick-bond representation of hexameric capsule with the enclosed space represented in green (see also **Plate 21**).

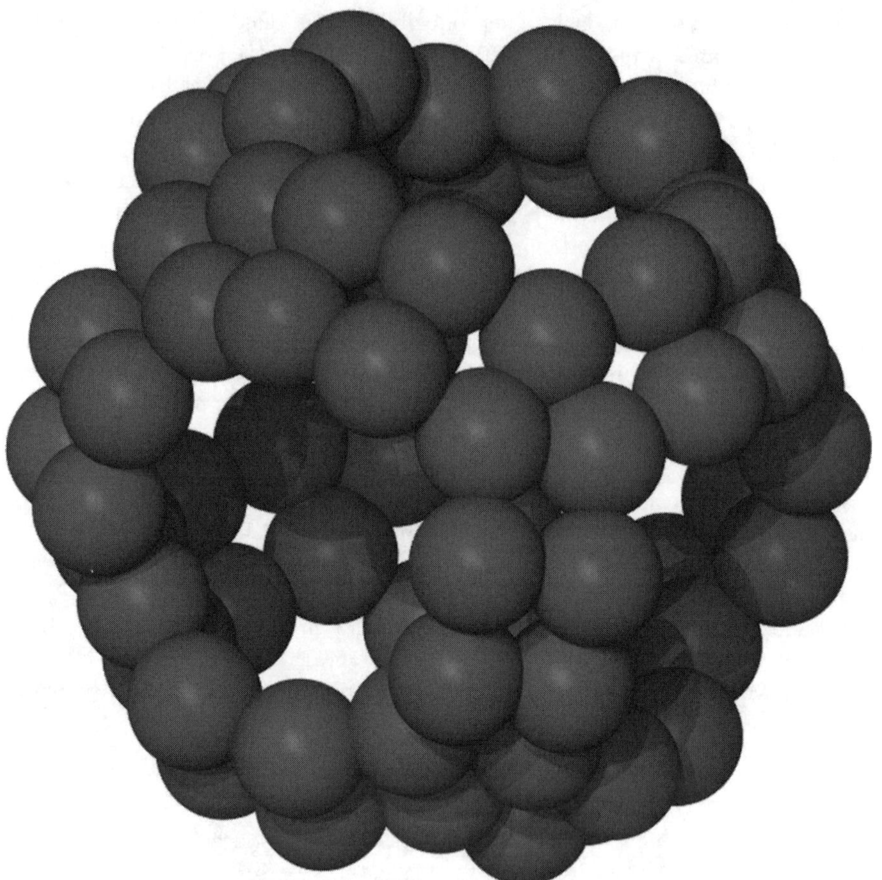

Figure 10 Structure of the hexamer of *C*-propylpyrogallol[4]arene with only the hydrogen-bonded oxygen atoms displayed.

Figure 11. For the purpose of this discussion, the most important features in the spectrum are the two methanol peaks labeled b and c, where b is the methyl proton resonance for methanol trapped within the hexameric sphere, and c is the methyl proton resonance for methanol in the bulk solvent. The identity of these peaks was proved by a spiking experiment in which additional methanol added to the NMR tube had the effect of increasing the intensity of peak c. Similar results were obtained in acetone-d_6, DMSO-d_6 and toluene-d_8. It should also be noted that NMR spectra for hexamer **8** synthesized in ethanol or 2-propanol also showed two sets of resonances for each inequivalent proton in the ethanol or 2-propanol molecules. Indeed, in a pressurized NMR sample tube we observed no change in the intensities of peaks b and c at 150 °C in acetone-d_6.

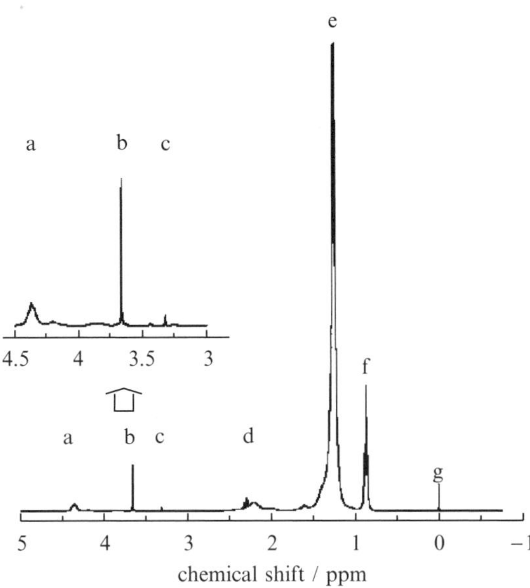

Figure 11 ^1H NMR spectrum of *C*-tridecylpyrogallol[4]arene at room temperature in $CDCl_3$ with TMS. The region of the spectrum displaying the methanol resonances is enlarged in the inset.

Using the techniques described above, stable hexameric spheres have been obtained for R = *n*-propyl up to *n*-hexyl while, for R = methyl or ethyl, there is no evidence for the existence of hexameric spheres. The solubility of *C*-alkylpyrogallol[4]arenes is low in apolar solvents when the *C*-alkyl group is a short chain (i.e. C-3, C-4). Similarly, low solubility occurs in polar solvents when the *C*-alkyl chain is long (i.e. C-10, C-13). It is therefore possible, with the appropriate choice of an R-group, to have stable hexameric spheres in polar or apolar solvents. Remarkably, these hexameric spheres are also stable in mixtures of acetone and water to the limit of their solubility. For example, the sphere for R = *n*-pentyl is stable in a 50% (v/v) mixture of acetone-d_6 and D_2O (precipitation without decomposition occurs with a higher percentage of water). These results show the extraordinary stability of these hydrogen bonded hexameric spheres in highly polar media [33,34]. We refer here to the stability of the capsule both with regard to the structural integrity imparted by the hydrogen bonding arrangement, and also the failure of the entrapped methanol to exchange with the bulk solvent.

Spherical hexamer **8** may be regarded as an example of the Archimedean solid, the *small rhombicuboctahedron*. The four-step analysis of the geometry is presented in Figure 12. The squares consist of both the pyrogallol[4]arenes and the four-center hydrogen bonding pattern, while the triangles represent the three-center hydrogen bonding pattern.

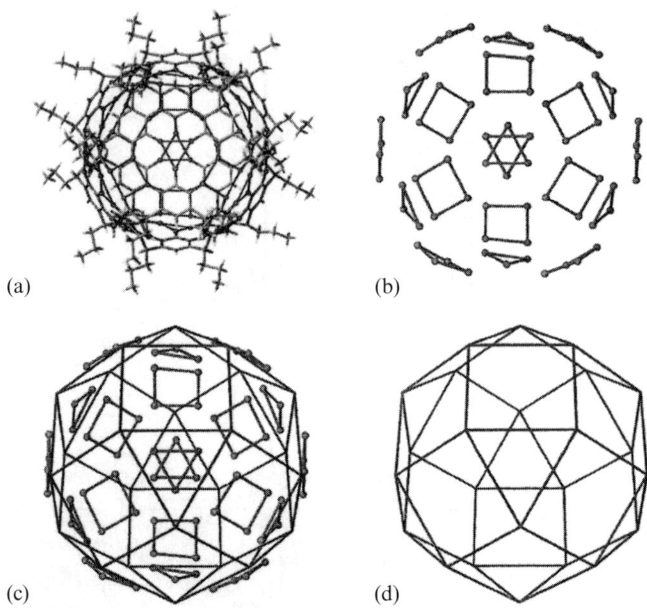

Figure 12 (a) The spherical capsule consisting of six pyrogallol[4]arene molecules shown in the capped-stick metaphor, and (b) with the carbon and hydrogen atoms removed. Hydrogen bonds are shown as thin, solid red lines. Parts (c) and (d) show the remarkable correspondence of the hydrogen bonded pattern with the Archimedean solid, the *small rhombicuboctahedron* (see also **Plate 22**).

In the above sections examples of the *snub cube*, the *great rhombicuboctahedron*, and the *small rhombicuboctahedron* have been presented. The guests are badly disordered for all of the capsules except for that made from *p*-sulfonatocalix [4]arene anions, pyridine *N*-oxide, and lanthanide ions.

3 CAVITIES OF SUPRAMOLECULAR CAPSULES

Once the enclosure of space has been accomplished, the organization of the guests contained within becomes a key issue. Rebek and co-workers used steric constraints to organize two guests within a tubular dimer [36], but for those assemblies with large enclosed volumes, both discrete and infinite, the guests, as noted in the previous section, are most often disordered [25,32,34,35]. The one example of significant order within an enclosure is the polar core consisting of 30 water molecules and two sodium ions in the twelve-*p*-sulfonatocalix[4]arene assembly [31].

4 ORGANIZATION IN CLOSED CAVITIES WITHIN INFINITE FRAMEWORKS

Our initial understanding of organization within large molecular capsules has come from the synthesis and characterization of host lattices, based upon *C*-methylresorcin[4]arene, which form hitherto unseen solid-state motifs, skewed molecular bricks (Figure 13), that possess closed cavities with internal volumes of ca 1000 Å^3. Remarkably, the cavities, although not inherently chiral, order the achiral guests such that a chiral arrangement results within their interiors [37]. Achiral symmetry breaking plays a role in a number of areas including chemistry (for example, stereochemical transformations), biology (differentiation), and materials science (ferromagnetism).

During studies aimed at determining the robustness of *C*-methylresorcin[4] arene, **2**, as a supramolecular synthon [38], we discovered that **2** can adopt a 'T-shaped' conformation in which the aromatic units lie approximately perpendicular to each other. This conformation makes **2** an eightfold hydrogen bond donor and, upon cocrystallization with 4, 4′-bipyridine, **9**, from hydrogen bond donor–acceptor solvents (for example acetone, ethanol) yields skewed molecular bricks, **2.9**·4(acetone), **10a**, and **2**·2(**9**)·3.5(ethanol)·H_2O, **10b**, that possess achiral cavities with idealized C_{2h} symmetry and internal volumes of ca 1000 Å^3 (Figure 14). In the case of **10b**, the cavities allow achiral symmetry breaking of achiral guests, giving rise to nanoscopic chiral domains within the solid.

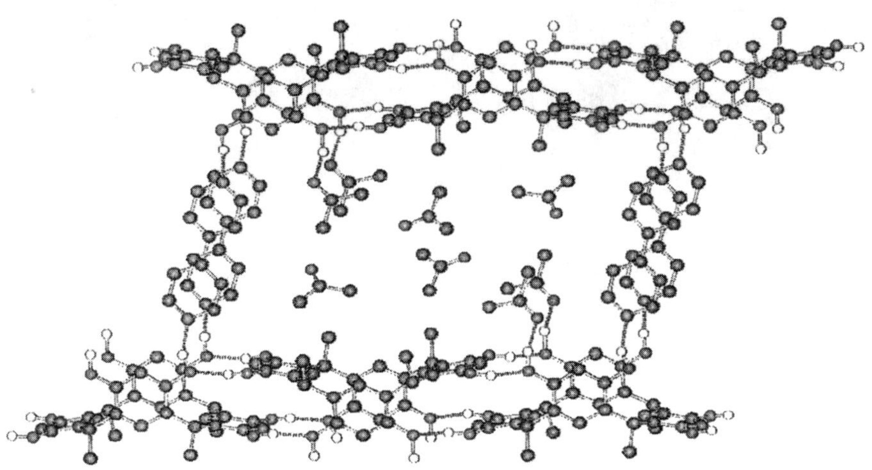

Figure 13 Skewed molecular bricks made from *C*-methylresorcin[4]arene and 4, 4′-bipyridine (see also **Plate 23**).

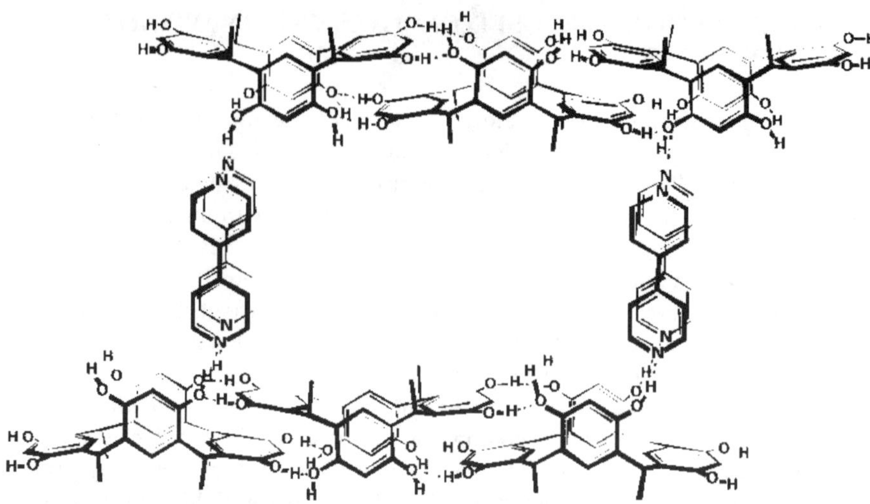

Figure 14 Schematic view of the framework of **10**. Note the four –OH hydrogen atoms oriented into the cavity pairwise at the upper left and lower right of the structure.

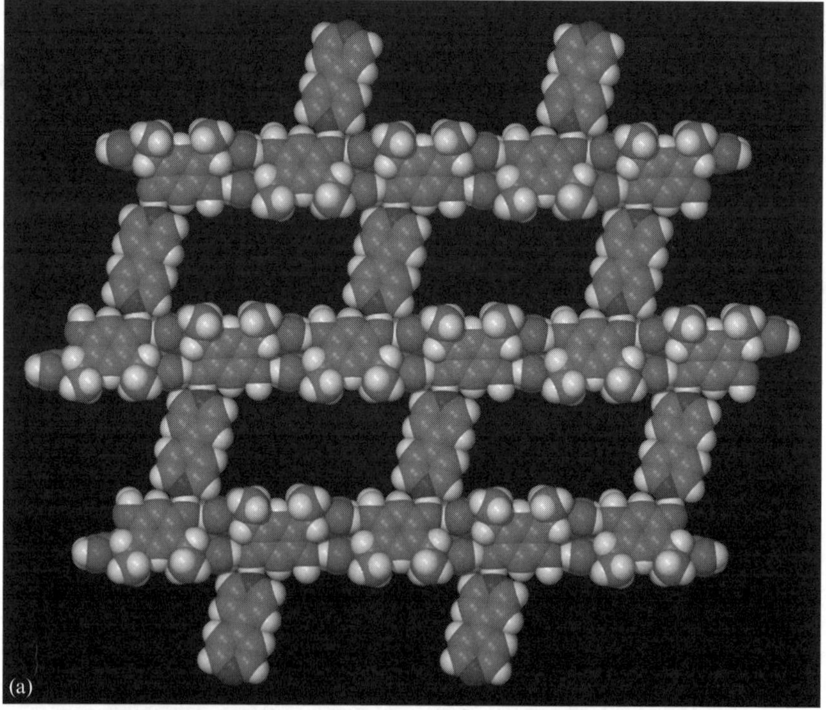

Figure 15 Structure of (a) **10a** and (b) **10b** (see also **Plate 24**).

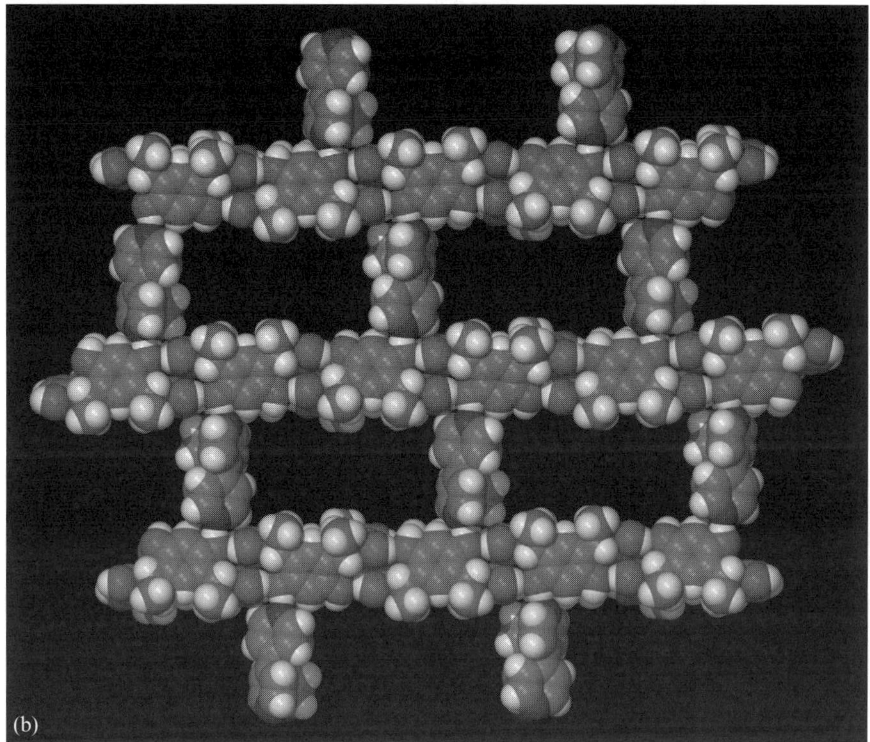

Figure 15 (*continued*)

A view depicting the framework of **10** is shown in Figure 15. Structure **2** has adopted a T-shaped conformation in which a pair of **9** forms two O–H···N hydrogen bonds to the upper rim of **2** while neighboring resorcin[4]arenes interact by way of four O–H···O hydrogen bonds yielding 1D polymeric chains. Owing to the ability of **9** to serve as a bifunctional hydrogen bond acceptor, **9** is observed to link resorcin[4]arenes of adjacent chains giving rise to 2D frameworks which conform to skewed molecular bricks. The bricks do not interpenetrate which, in turn, produces guest-filled box-like cavities (guests: eight acetones **10a**, two waters, two 4, 4′-bipyridines, seven ethanols **10b**). The internal volumes are 980 Å^3 for **10a** and 1020 Å^3 for **10b**. The cavities of **10** are spacious, being two to three times larger than those of most zeolites. Importantly, *four hydroxyl groups from two different resorcin[4]arenes line each cavity at opposite corners and form O–H···O hydrogen bonds to four guest species.*

A view depicting the solid-state packing of **10** is shown in Figure 16. The bricks of **10a** and **10b** have assembled such that they form identical... ABA... stacking patterns, the corners of the bricks lying above and below the cavities of another. As a result, the guests are completely enclosed such that there are

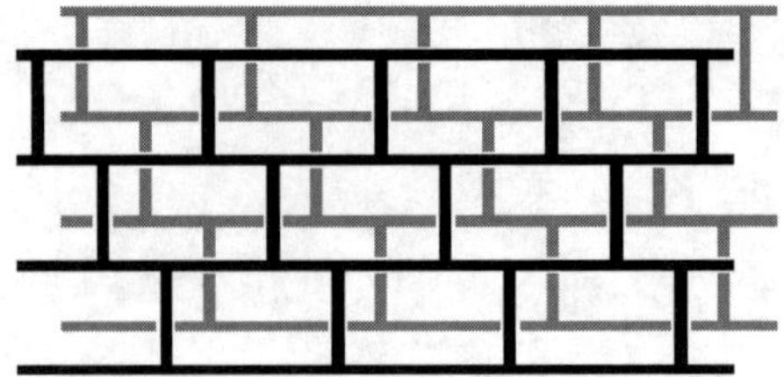

Figure 16 Representation of the ... ABAB ... packing layer structure of **10a** and **10b**.

no pores connecting adjacent cavities. The major difference between **10a** and **10b** lies in the nature of the included guests and the levels of organization the guests display within the cavities which we now address only with respect to **10b**.

A view depicting the guests of **10b** is shown in Figure 17. As stated, the cavities of **10b** contain 11 guests consisting of three different components, namely two waters, two 4, 4′-bipyridines, and seven ethanols. In the case of **10b**, the guests have assembled, by way of 15 O–H···X (X = O, N) hydrogen bonds, such that they form *a single chiral domain* in which the center of inversion of the box has been destroyed. Two waters and two ethanols participate in four O–H···O hydrogen bonds with four hydroxyl groups of **2**. Guests of neighboring boxes are

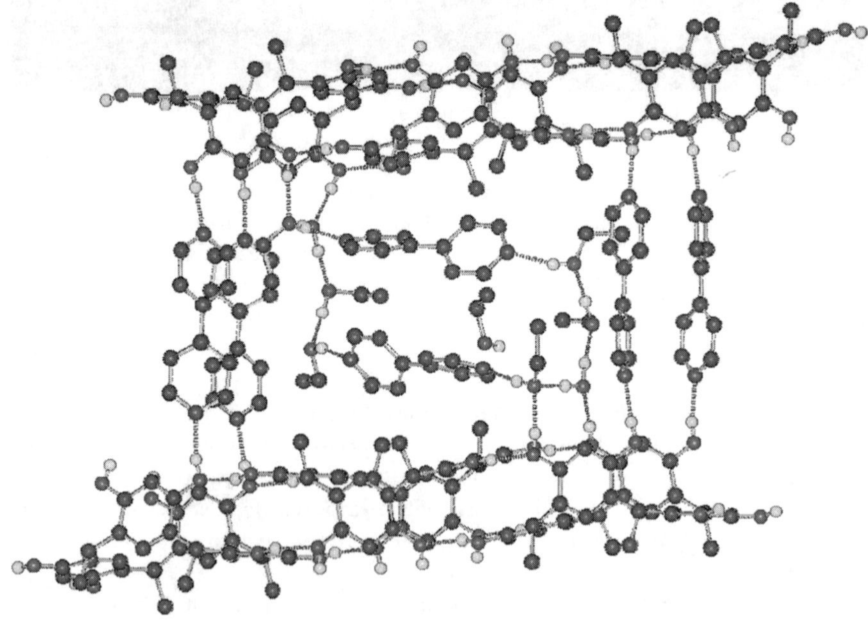

Figure 17 The 11 guests on the interior of **10b** are completely organized by hydrogen bonds emanating from the four –OH hydrogen atoms oriented into the cavity pariwise at the upper left and lower right of the structure.

related by a crystallographic center of inversion, meaning that guest configurations of adjacent cavities of **10b** are enantiomers.

It is clear that, in addition to infinite frameworks, observations made above can be applied to discrete assemblies. Specifically, the basic principle learned from the infinite frameworks, *the projection of hydrogen bond donors into a cavity can organize the guests at a remarkable level*, can be applied to the organization of guests within a molecular capsule.

5 ORGANIZATION WITHIN MOLECULAR CAPSULES

We have previously reported[25] that the capsule formulated as [(*C*-methylresorcin[4]arene)$_6$(H_2O)$_8$], **1**, possesses an excess of four hydrogen bond donors, but these donors are positioned such that they project outward from the surface of the enclosure. In this orientation they are incapable of effecting organization of the guests within the capsule. In a similar fashion, all the hydrogen bond donors in [(pyrogallol[4]arene)$_6$] are used in completing the hydrogen bond pattern that forms the capsule; the guests within are not ordered. It was reasoned that a mixed system consisting of pyrogallol and resorcinol units might form a capsule with the desired property of an excess of hydrogen bond donors, together with an orientation of some of these hydrogen donors into the capsule. The mixed possibilities are shown as **11–14**, in Figure 18. A facile synthesis of macrocycle **11** has now been discovered. (Macrocycle **11** was synthesized, along with other new related macrocycles, from the acid-catalyzed condensation of equimolar amounts of resorcinol and pyrogallol with butyraldehyde. Macrocycle **11** has also been prepared by the acid catalyzed condensation of resorcinol and pyrogallol in ratio 1:2, respectively, with isovaleraldehyde.) Upon recrystallization from Et_2O, the remarkable structure shown in Figure 19 results. The hexamer **15** takes the shape of a trigonal antiprism with the centers of **15** at the corners of the trigonal antiprism. This assembly, with six hydrogen bond donors positioned toward the interior of the capsule, possesses an internal volume of 860 Å^3. The six Et_2O molecules on the inside of the capsule are thus ordered by these six hydrogen bond donors. There are six more hydrogen bond donors oriented toward the outside, and these donors bind six additional Et_2O molecules on the outside of the hexamer, further sealing the capsule, as shown in Figure 18.

In order to understand the complicated hydrogen bonding pattern, the fulfillment of which leads to hexamer **15**, it is useful to account for the hydrogen bond donors and acceptors compared with those of the hexamer of **7**, capsule **8**. In capsule **8**, each macrocycle is rimmed with 12 hydroxyl groups, thus providing a potential of 12 hydrogen bond donors. As we have noted in the hexamer, 48 of the 72 potential hydrogen bond donors are used in intermolecular hydrogen bonds to seam the capsule together. The remaining 24 hydrogen bond donors are used in intramolecular hydrogen bonds between adjacent rings in the macrocyclic

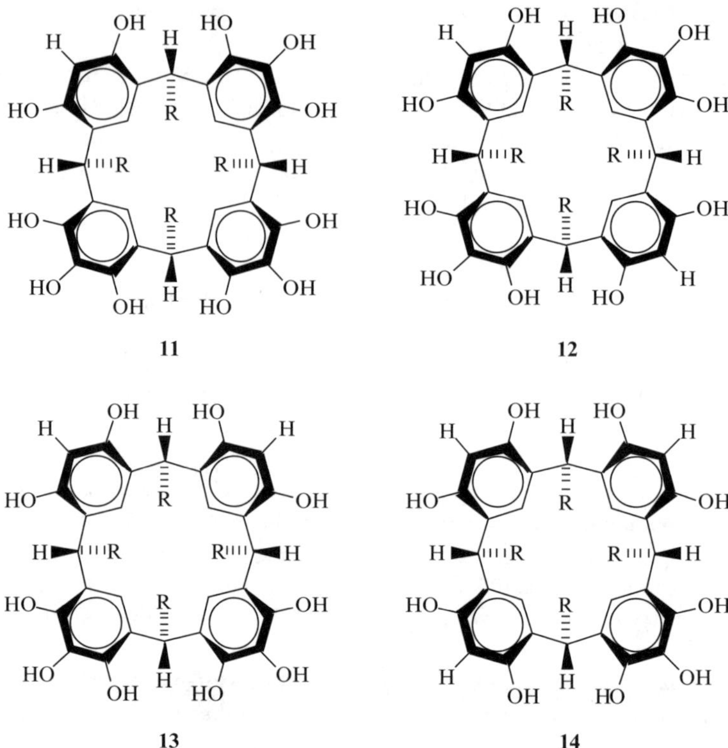

Figure 18 Macrocycles composed of resorcinol and pyrogallol units.

building block **7**. In the capsule of mixed macrocycle **11**, each mixed marcocycle is rimmed with 11 hydroxyl groups, providing a potential of 66 hydrogen bond donors in the hexamer. A detailed study of Figure 19 shows that 24 hydrogen bond donors are used in intermolecular hydrogen bonds and 24 are used in intramolecular hydrogen bonds (as in capsule **8**). Of the $66 - 48 = 18$ remaining hydrogen bond donors, six are oriented toward the inside and six toward the outside, bonding the diethyl ether molecules. The internal diethyl ether molecules are bound by a hydrogen bond to one of the resorcinol ring OH groups, while the external diethyl ether molecules are bound by a hydrogen bond to one of the 1-OH groups of a pyrogallol ring.

The accounting of the hydrogen bond donors discussed above may be completed by noting that the hexamer in Figure 19 has 24 intermolecular hydrogen bonds, 24 intramolecular hydrogen bonds, and 12 hydrogen bonds to diethyl ether molecules, for a total of 60 hydrogen bonds. The remaining six hydrogen bond donors which are not involved in the O–H···O hydrogen bonding scheme are found external to the capsule, as may be seen in Figure 19.

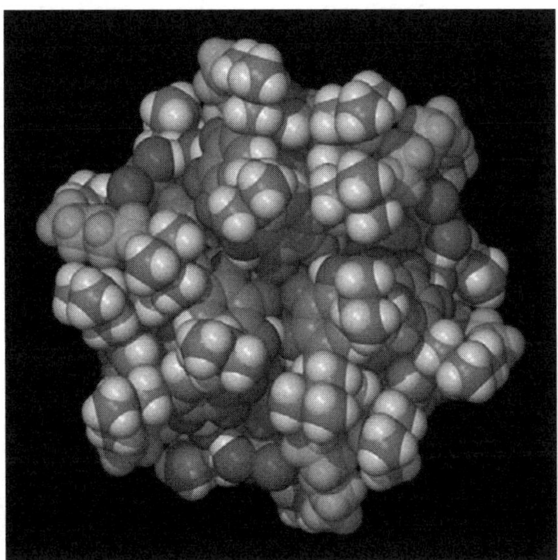

Figure 19 Space-filling representation of hexamer **15**, or mixed macrocycle **11**, viewed along the 3 bar axis of the capsule (see also **Plate 25**).

It is interesting that capsule **15** assumes the shape of a trigonal antiprism with the centers of **11** at its corners (25). Figure 20 emphasizes this geometrical similarity. Figure 20a presents a view of capsule **15** with the framework in stick bond representation and the diethyl ether molecules as space filling models. The orientation of the capsule is identical with that of Figure 18. Figure 20b displays the trigonal antiprism constructed from the centroids of the centers of the rings of mixed macrocycle **11**. The superposition of capsule **15** and the trigonal antiprism is shown in Figure 20c.

It is appropriate to draw an analogy between **10a**, the complex with eight acetone guests, four of which are hydrogen bonded to the framework of the host, per cavity, and **11**, the capsule which binds six Et_2O molecules on the interior. We have not yet discovered a capsule analog of **10b**, but it will surely be possible to obtain such intricate order with a molecular capsule, probably even with that of **11**.

The enclosure of chemical space can be effected using existing supramolecular strategies. We have now shown that order at the highest level can be enforced by hydrogen bonding from the framework to the guests. It is possible to envision molecular capsules as nanoscale reaction vessels in which extraordinary control over the reactants can be obtained. The main theme of this discussion is that understanding leads to control. This is a basic tenet of crystal engineering, and it holds equally well for the engineering of supramolecular capsules.

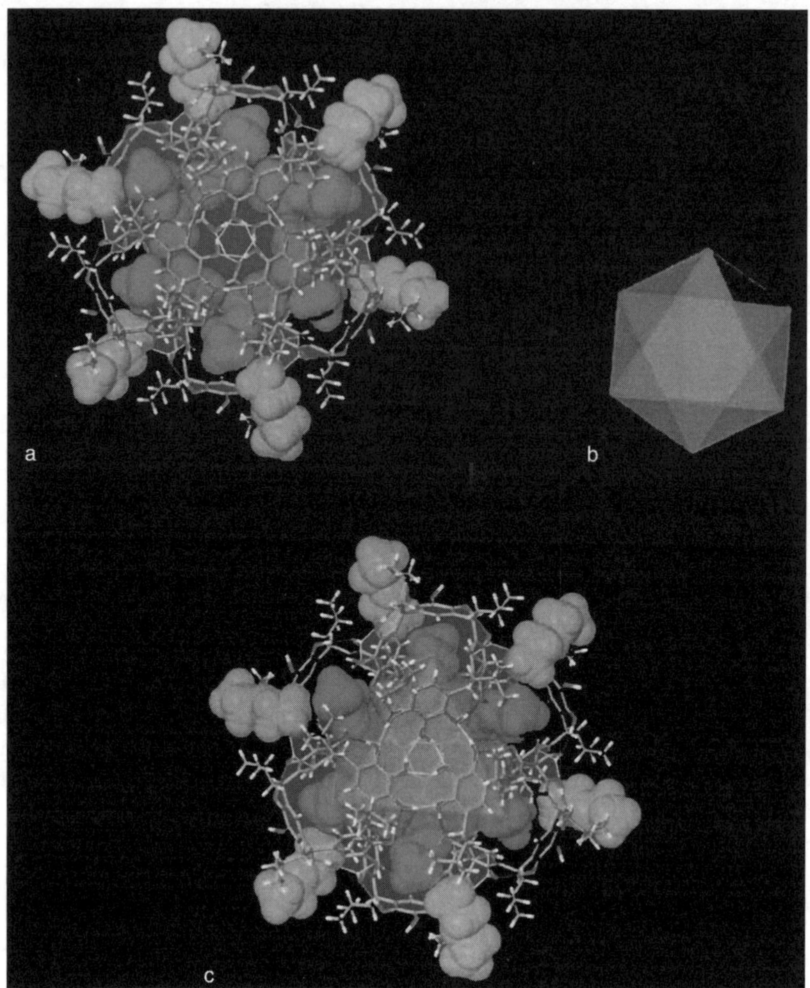

Figure 20 (a) Capsule **15** shown in stick bond representation with the diethyl ether guests given in space filling representation. The orientation of the capsule is identical with that given in Figure 18. (b) The trigonal antiprism that results from connection of the centroids of the centers of the aromatic rings of macrocycle **11**. (c) Superposition of the trigonal antiprism and capsule (see also **Plate 26**).

REFERENCES

1. Avers, C. J. *Molecular Cell Biology*. Addison-Wesley, Reading, MA, **1986**, pp. 768–770.
2. The term 'chemical space' is used to specify that area within a molecular capsule which may be used to house a guest.
3. Caspar, D.; Klug, A. *Cold Spring Harbor Symp. Quant. Biol.* **1962**, *27*, 1–24.

4. Branden, C.; Tooze, J. *Introduction to Protein Structure*. Garland, New York, **1991**, pp. 161–177.
5. Hogle, J. M., Chow, M.; Filman, D. J. *Science* **1985**, *229*, 1358–1365.
6. Douglas, T.; Young, M. *Nature* **1998**, *191*, 152–155.
7. Douglas, T. in *Biomimetic Approaches to Materials Science*, ed. Mann, S. VCH, New York, **1996**, pp. 91–115.
8. Cram, D. J.; Karbach, S.; Kim, Y. H.; Baczynskyj, L.; Kalleymeyn, G. W. *J. Am. Chem. Soc.* **1985**, *107*, 2575.
9. Gabard, J.; Collet, A. *Chem. Commun.* **1981**, 1137–1138.
10. Chapman, R. G.; Sherman, J. C. *J. Am. Chem. Soc.* **1995**, *117*, 9081–9082.
11. Chopra, N.; Sherman, J. C. *Angew. Chem., Int. Ed. Engl.* **1999**, *38*, 1955–1957.
12. Wyler, R.; de Mendoza, J.; Rebek, J., Jr. *Angew. Chem., Int. Ed. Engl.* **1993**, *32*, 1699–1701.
13. Conn, M. M.; Rebek, J., Jr. *Chem. Rev.* **1997**, *97*, 1647–1668.
14. Rebek, J., Jr. *Chem. Commun.* **2000**, 637–643.
15. Shivanyuk, A.; Rebek, J., Jr. *Chem. Commun.* **2001**, 2374–2375.
16. Fujita, M.; Kwon, Y. J.; Washirzu, S.; Ogura, K. *J. Am. Chem. Soc.* **1994**, *116*, 1151–1152.
17. Takeda, N.; Umemoto, K.; Yamaguchi, K.; Fujita, M. *Nature* **1999**, *398*, 794–796.
18. Olenyuk, B.; Whiteford, J. A.; Fechtenkotter, A.; Stang, P. J. *Nature* **1999**, *398*, 796–799.
19. Fox, O. D.; Dalley, N. K.; Harrison, R. G. *J. Am. Chem. Soc.* **1998**, *120*, 7111–7112.
20. Hardie, M. J.; Raston, C. L. *J. Chem. Soc., Dalton Trans.* **2000**, 2483–2492.
21. Atwood, J. L., Barbour, L. J., Hardie, M. J.; Raston, C. L. *Coord. Chem. Rev.* **2001**, *222*, 3–32.
22. Biradha, K., Seward, C.; Zaworotko, M. J. *Angew. Chem., Int. Ed. Engl.* **1999**, *38*, 492–495.
23. Rose, K. N.; Barbour, L. J.; Orr, G. W.; Atwood, J. L. *Chem. Commun.* **1998**, 407–408.
24. Gibb, C. L. D.; Gibb, B. C. *J. Supramol. Chem.* **2001**, *1*, 39–52.
25. MacGillivray, L. R.; Atwood, J. L. *Nature* **1997**, *389*, 469–472.
26. Wenninger, M. J. *Polyhedron Models*. Cambridge University Press, New York, **1971**.
27. MacGillivray, L. R.; Atwood, J. L. *Angew. Chem., Int. Ed. Engl.* **1999**, *38*, 1019–1033.
28. MacGillivray, L. R.; Atwood, J. L. in *Advances in Supramolecular Chemistry*, ed. Gokel, G. W. JAI, Stamford, **2000**, pp. 157–183.
29. Atwood, J. L.; Hamada, F.; Robinson, K. D.; Orr, G. W.; Vincent, R. L. *Nature* **1991**, *349*, 683–685.
30. Atwood, J. L.; Barbour, L. J.; Hardie, M. J.; Raston, C. L. *Coord. Chem. Rev.* **2001**, *222*, 3–32.
31. Orr, G. W., Barbour, L. J.; Atwood, J. L. *Science* **1999**, *285*, 1049–1052.
32. Gerkensmeier, T., Iwanek, W., Agena, C., Frohlich, R., Kotila, S., Nather, C.; Mattay, J. *Eur. J. Org. Chem.* **1999**, 2257–2262.
33. Meissner, R. S.; Rebek, J.; de Mendoza, J. *Science* **1995**, *270*, 1485.
34. Atwood, J. L., Barbour, L. J.; Jerga, A. *Chem. Commun.* **2001**, 2376–2377.
35. Atwood, J. L., Barbour, L. J.; Jerga, A. *J. Supramol. Chem.*, **2001**, *1*, 131–134.
36. Heinz, T., Rudkevich, D. M.; Rebek, J., Jr. *Nature* **1998**, *394*, 764–766.
37. MacGillivray, L. R.; Holman, K. T.; Atwood, J. L. J. *Supramol. Chem.*, **2001**, *1*, 125–130.
38. Desiraju, G. R. *Angew. Chem., Int. Ed. Engl.* **1995**, *34*, 2311.

Chapter 4

Molecular Tectonics: Molecular Networks Based on Inclusion Processes

Julien Martz, Ernest Graf, André De Cian and Mir Wais Hosseini

Université Louis Pasteur, Institut Le Bel, Strasbourg, France

1 INTRODUCTION

Molecular solids are composed of molecular entities and are defined by the chemical nature of their molecular components and by their interactions with respect to each other in the solid phase. Molecular crystals are molecular solids presenting order and periodicity. With our present level of knowledge, the complete understanding, and therefore prediction, of all intermolecular interactions in the crystalline phase seems, unreachable. Thus, it appears that crystal structures are not entirely predictable [1]. This fact is at the origin of active debates on whether the concept of crystal engineering is justifiable. However, following the statement by Jack Dunitz, 'A crystal is, in a sense, the supramolecule *par excellence*' [2], using concepts developed in the area of supramolecular chemistry [3], one may predict some of the inter-motive interactions. In other terms, as stated above, although all intermolecular forces may not be controlled currently with accuracy, nevertheless, by programming some of the interactions between the molecular components, one may design molecular networks in the crystalline phase. Thus, our aim is centred on the design and formation of molecular networks in the crystalline phase rather than crystal engineering.

Crystal Design: Structure and Function, edited by G. R. Desiraju

2 MOLECULAR NETWORKS

A molecular network in the solid state is defined as a supramolecular structure, theoretically composed of infinite number of molecules named tectons (from the Greek TEKTON, builder) [4] capable of mutual recognition. A tecton is an informed molecular component containing in its backbone recognition sites disposed in a specific manner. The specific intermolecular interaction between complementary tectons leads to a recognition pattern which may be called an assembling core. One may draw the analogy between the assembling core defined in the context of molecular networks and a supramolecular synthon elaborated in the area of crystal engineering [5]. Within the network, the assembling core becomes a structural node and by translation of the latter the network is formed (Figure 1). The dimensionality of the network is defined by the number of translations operating on the structural nodes. Whereas a 1-D network (also called α-network) is obtained upon a single translation of the node (Figure 1a), two-(2-D or β-network) and three-(3-D or γ-network) dimensional networks are obtained upon two (Figure 1b) and three (Figure 1c) translations of identical or different structural nodes in two and three independent directions of space respectively [6–8]. In dealing with molecular networks, the critical point is concerned with the definition in terms of energy of interaction of the structural nodes. Indeed, a crystal is by definition a 3-D structure. Defining a network within a crystalline arrangement requires an energy hierarchy between specific recognition processes leading to the formation of the assembling core and other interactions (van der Waals, electrostatic, etc.) responsible for the cohesion of the solid.

It is interesting that whereas molecules are described as assemblies of atoms interconnected by covalent bonds, by analogy, one may describe molecular networks as hypermolecules in which the connectivity between tectons is ensured by noncovalent inter-tecton interactions.

The construction of large molecular networks (10^{-6}–10^{-3} m scale) in the crystalline phase with predicted and programmed structure may hardly be envisaged through step by step type strategy. Indeed, taking into account the dimension of tectons, usually in the order of 1 nm, a millimetric crystal would require the controlled interconnection of ca 10^6 tectons. If the connectivity between tectons was achieved step by step using the formation of covalent bonds, the formation of the network would necessitate 1 million operations! However, the preparation of such higher order solid materials may be reached through iterative self-assembling processes [9–12] between tectons containing within their structure recognition sites and a specific assembling algorithm which will operate as soon as complementary tectons are allowed to interact. This strategy, called molecular tectonics [13], seems to be the most viable one and is of efficient and frequent use in nature. The molecular tectonics approach relies on a double analysis. One is at the molecular level, dealing with the individual tectons composing the solid, and the other is at the supramolecular level, dealing with the inter-tecton interactions [3,5].

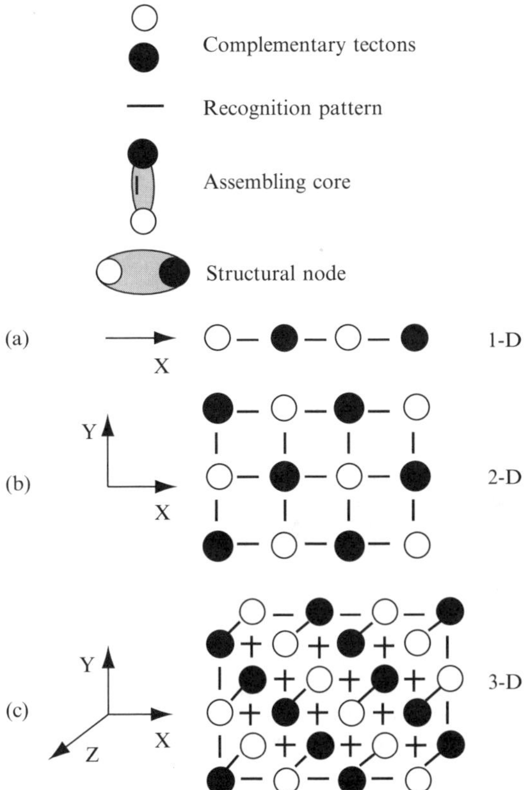

Figure 1 Schematic representation of 1-D (a), 2-D (b) and 3-D (c) molecular networks.

Obviously the latter type of analysis is more subtle. The design of specific tectons leading to predicted molecular networks in the crystalline phase still remains a challenge.

Dealing with design of the tecton, in order to achieve an iterative assembling process, the tecton must fulfil both structural and energetic criteria. In particular, as stated above, the complementary tectons must, on the one hand, recognise each other and thus generate the assembling core, and on the other hand, must allow the translation of the recognition patterns allowing the formation of the network. These two criteria require tectons possessing at least two or more recognition sites located in a divergent fashion. Indeed, whereas endo molecular receptors form discrete molecular complexes with appropriate substrates (Figure 2a), exo receptors may either form discrete species with substrates behaving as stoppers

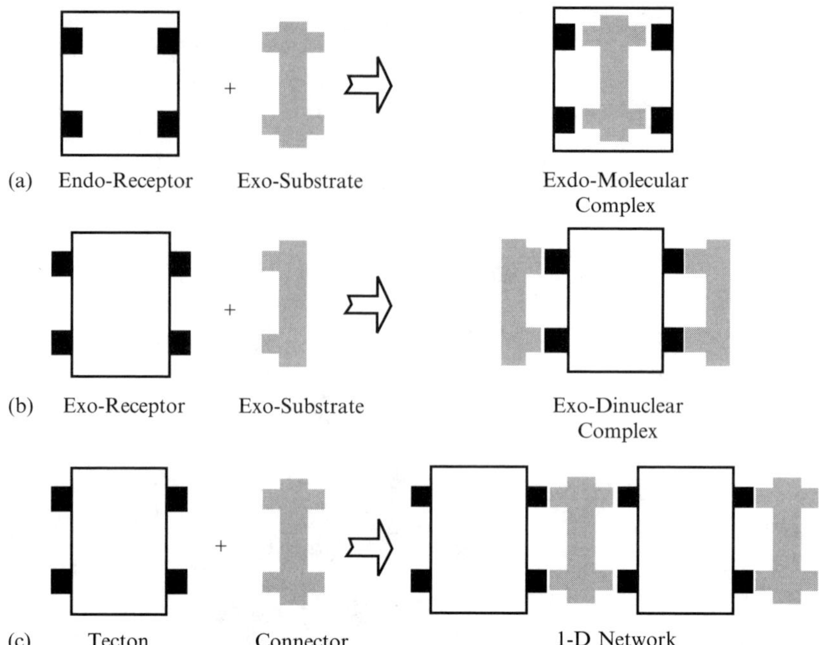

Figure 2 Schematic representations of formation of discrete molecular complexes by an endo receptor and a exo substrate (a), by an exo receptor and two exo substrate acting as stoppers (b) and of a linear molecular network by an exo receptor and a connector (c).

(Figure 2b) or, by acting as tectons, molecular networks in the presence of appropriate connectors (Figure 2c).

Since the formation of molecular networks takes place under self-assembly conditions, the recognition process between tectons must be reversible. This point is crucial because the reversibility of the assembling processes allows self-reparation by the system composed of the molecular components. In other terms, the reversibility allows the system to find a thermodynamic minimum while it explores the entire possibility space. Thus, in terms of interaction energy between tectons one may use any type of reversible bonding processes. Hydrogen [14] and coordination [15–19] bonds have been almost exclusively used for the formation of molecular networks. During our own research on molecular networks, we have also used charge-assisted H-bonding [20–32] and coordination processes [33–43]. Some time ago, we proposed to design molecular networks based on inclusion processes in the crystalline phase using rather weak van der Waals interactions [44–47]. This chapter will mainly deal with inclusion molecular networks in the solid state.

3 INCLUSION MOLECULAR NETWORKS

3.1 Design of Koilates

The chemistry of inclusion complexes based on host and guest molecules (Figure 3a), i.e. the inclusion of a substrate within the cavity of a receptor molecule, is an established area [48–50]. Indeed, many endo receptors, possessing an endomolecular cavity, have been designed to bind substrates in a inclusion fashion. In some cases, the inclusion phenomena has been established in the crystalline phase by X-ray diffraction methods revealing their discrete nature. Because the connection between the discrete complexes is not ensured by specific interaction in the solid, these molecular crystals can not be classified as molecular networks which require the repetition of recognition patterns. On the other hand, the formation of clathrates [51,52] in the solid state has been truly investigated during the past century by solid-state chemists (Figure 3b). Clathtrates are formed when molecular components with specific shapes are arranged in the crystalline phase in such a way that cavities are formed. The exo molecular cavities thus obtained are occupied by guest molecules. We have extended the concept of inclusion complexes in solution and in the solid state and the formation of clathrates exclusively in the crystalline phase to the construction of a new type of molecular networks in the solid state (Figure 3c). This strategy, based on inclusion phenomena between concave and convex molecular objects, combines endo and exo molecular cavities. The basic concept guiding the design of inclusion networks, is based on the use of koilands (from the Greek KOILOS, hollow) [44–47], multicavity receptor molecules composed of at least two cavities arranged in a divergent fashion (concave tectons), and connectors (convex tectons) capable of being included within the cavities of the koilands and thus connecting consecutive koilands by inclusion processes. Indeed, since each individual cavity offers the possibility of forming inclusion complexes with a convex molecules, connecting two or more of such cavities must lead in the presence of an appropriate connector to non-covalently assembled polymeric species called koilates or inclusion networks.

The driving forces for the formation of koilates are, on the one hand, interactions between cavities and connectors, and on the other hand, the best compaction indispensable for the formation of the crystalline phase. The attractive forces between the koiland and the connector may be of different nature such as electrostatic charge–charge, charge–dipole, dipole–dipole or van der Waals interactions. Here, we shall mainly focus on the last case. As mentioned above, recognising a network in a molecular crystal is a matter of translation of the recognition pattern. As long as the energy of interaction defining the assembling core is larger as in the case of strong H-bonds (charge assisted or not) or substantially greater as in the case of coordination bonds than other interactions responsible for the formation of the crystal (three-dimensional compaction of the individual networks), the network may be defined without ambiguity. However, for inclusion

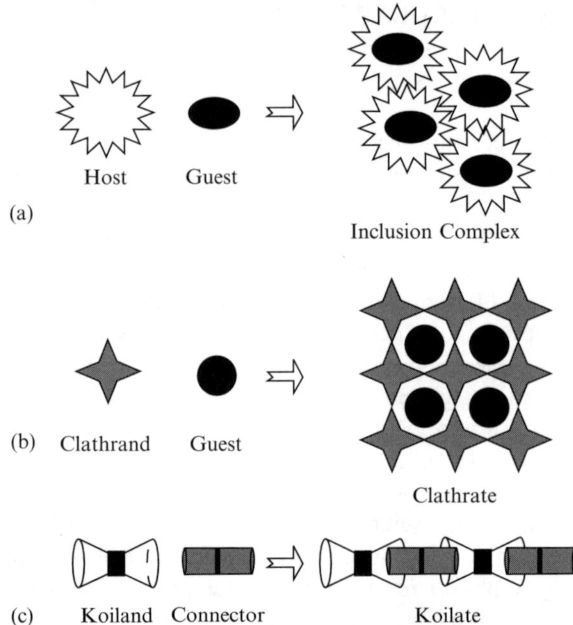

Figure 3 Schematic representation of the formation of inclusion complexes based on the recognition of a guest molecule by a host receptor (a), of a clathrate based on the inclusion of guest molecules within cavities generated upon packing of clathrands in the solid state (b), and of a 1-D inclusion molecular network named koilate formed through interconnection of hollow tectons (koilands) by connector molecules (c).

networks excessively based on van der Waals interaction or weak H-bonds (1–10 kJ/mol of interaction), the task is not obvious since the packing of the components may also be ensured by the same type of interactions. This difficulty may be overcome by considering the recognition pattern, formed by inclusion of a section of the connector by the cavity of the koiland, as a supramolecular synthon [5]. The term supramolecular synthon here means a recognition pattern which by translation would become a structural node of the network.

In terms of design, one may envisage a variety of koilates with different dimensionality. Considering a two-component system, 1-D koilates may be formed using koilands possessing two divergent cavities. When using connectors with two extremities each capable of being included within one of the two cavities and thus connecting consecutive koilands, 1-D koilates may be formed (Figure 4a). Another possibility may be based on a single self complementary tecton possessing both a concave moiety acting as receptor and a convex part acting as the connector. In that case, the 1-D inclusion network will be formed by self-inclusion processes (Figure 4b and c). Dealing with 1-D koilates, as in the case of any type of 1-D molecular networks, the design of directional koilates remains a challenge.

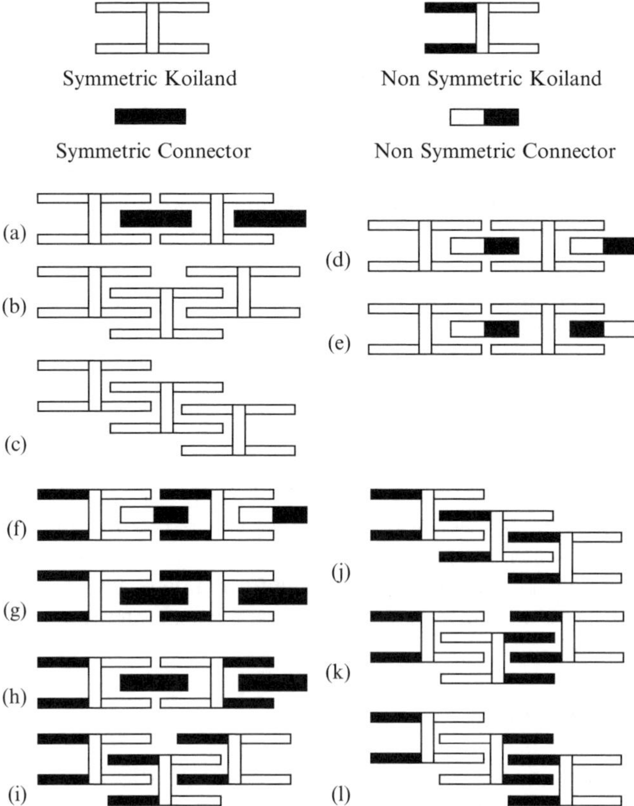

Figure 4 Schematic representations of the formation of 1-D inclusion networks based on koiland, building block possessing two divergent cavities, and connector molecules (a, d, e–h) or through self-inclusion by self complementary koilands (b, c, i–l). For definition, see text.

Directional molecular networks are of great interest with respect to vectorial physical properties such as non-linear optics, dipolar coupling etc. In principle, the formation of this type of network may be envisaged in a variety of ways. In the case mentioned above (Figure 4a), owing to the centrosymmetric nature of both the koiland and the connector, no particular directionality may be expected for the koilate since the centre of symmetry is translated. In order to impose a directionality, a combination of non-symmetric koilands and or connectors is required (Figure 4d-l). By using centrosymmetric koilands and noncentrosymmetric connectors, one may form either directional (Figure 4d) or nondirectional (Figure 4e) koilates. Similarly, one may envisage the same type of possibility [Figure 4g (directional) and h (non directional)] when using noncentrosymmetric koilands and symmetric connectors. The most promising strategy would take advantage of recognition processes between the two cavities of the koiland and the two extremities of the connector (Figure 4f). Indeed, noncentrosymmetric koilands and connectors, if the recognition process is properly programmed, must lead exclusively to the formation of directional koilates. As for symmetric koilates mentioned above, again a single nonsymmetric self-complementary tecton may be used. The koilate that can be formed by self-inclusion may either be directional (Figure 4i and j) or nondirectional (Figure 4k and l). As will be presented in detail and on real cases below, almost all possibilities mentioned above (Figure 4) have been observed.

The 2- and 3-D koilates may be designed using koilands possessing three and four cavities, respectively, oriented in a divergent fashion. Koilands based on three interconnected cavities with an angle of 120° between them were obtained. In principle, this type of koiland may be assembled into a 2-D network using linear connectors. Dealing with 3-D inclusion networks, one may envisage two possibilities. Koilands possessing four cavities located on the apices of a tetrahedron which in the presence of linear connectors would lead to 3-D koilates of the diamondoid type, or koilands with six cavities occupying the apices of a octahedron. In that case, in principle 3-D inclusion networks of the cubic type may be envisaged, again using linear connectors. The formation of both 2- and 3-D inclusion networks has not been yet achieved and remains an interesting challenge.

3.2 Design of Koilands

Calix[4]arene derivatives [53–56] are candidates of choice for the design of koilands (Scheme 1). Indeed, these compounds offer a preorganised and tuneable hydrophobic pocket surrounded by aryl moieties as well as four hydroxy groups for further functionalisation. However, because of the junctions (CH_2 or S) between the aromatic moieties, calix[4]arenes present a conformational flexibility (Figure 5). Calix[4]arenes bearing four free OH groups or *O*-alkylated derivatives with short alkyl chains present four limit conformations [cone (c), partial cone (pc), 1,2-alternate (1,2-a) and 1,3-alternate (1,3-a)]. Among all conformers, only the cone may be used for the design of koilands. Furthermore, these compounds, easily prepared in large quantities, present another advantage in terms of design related to the possibility of controlling both the entrance and the depth of the preorganised cavity by the nature of the substituents R at the *para* position (Scheme 1), i.e. H (**1**), Me (**2**), Et (**3**), *i*Pr (**4**), $C(CH_3)_3$ (**5**), allyl (**6**) or Ph (**7**). Furthermore, one may also replace the CH_2 groups connecting the aromatic rings by sulfur atoms (compound **8**) [57–60], and finally, the substitution of the methylene groups by S atoms as well as the replacement of the OH groups by SH moieties may also be performed (compound **9**)[61]. Whereas compounds **1** (Figure 6a), **3** (Figure 6b), **4** (Figure 6c), **5** (Figure 6d), **7** (Figure 6e) and **8** (Figure 6f) adopt the cone conformation in the crystalline phase, compound **9** was shown to

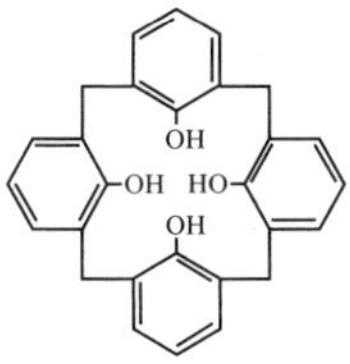

OH
OH HO
OH
1
OH
OH HO
OH
2
OH
OH HO
OH
3
OH
OH HO
OH
4
OH
OH HO
OH
5
OH
OH HO
OH
6
OH
OH HO
OH
7
S S
OH
OH HO
OH
S S
8
S S
SH
SH HS
SH
S S
9
SH
OH HO
SH
10
SH
SH HS
SH
11

Scheme 1

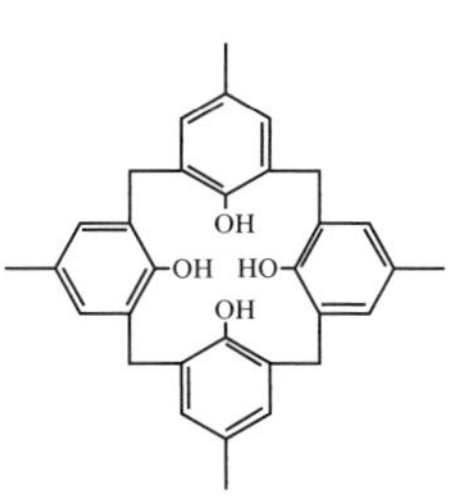

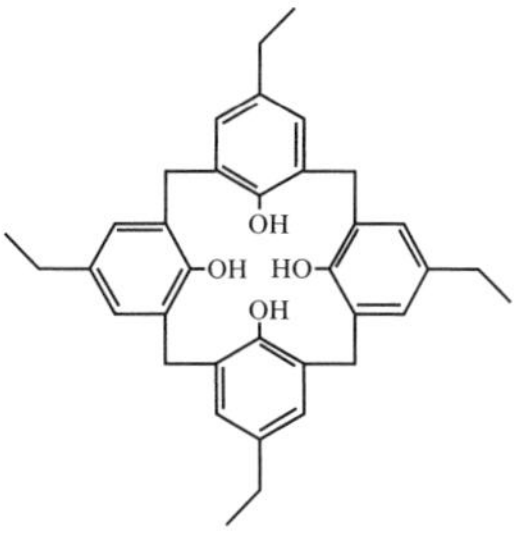

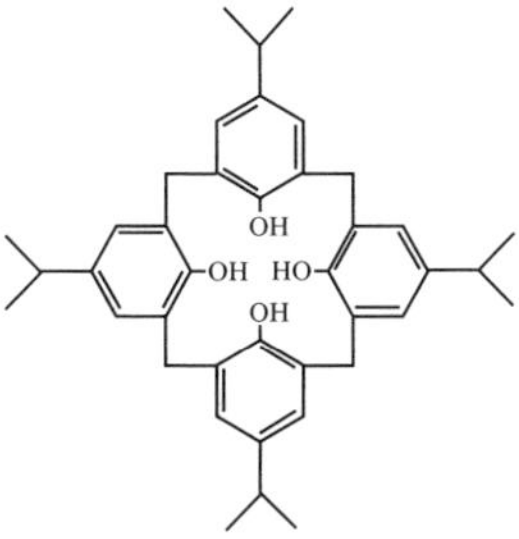

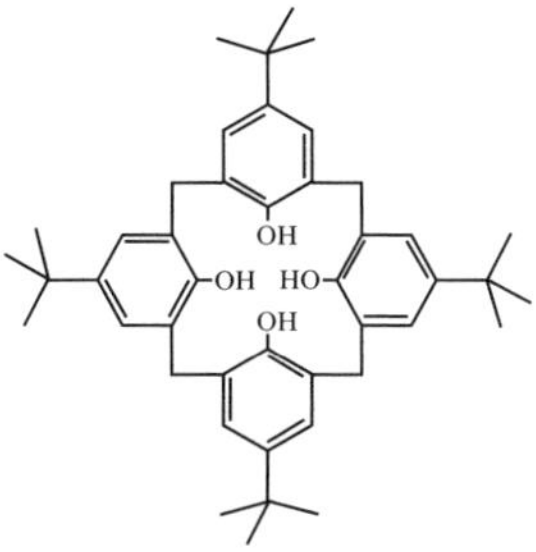

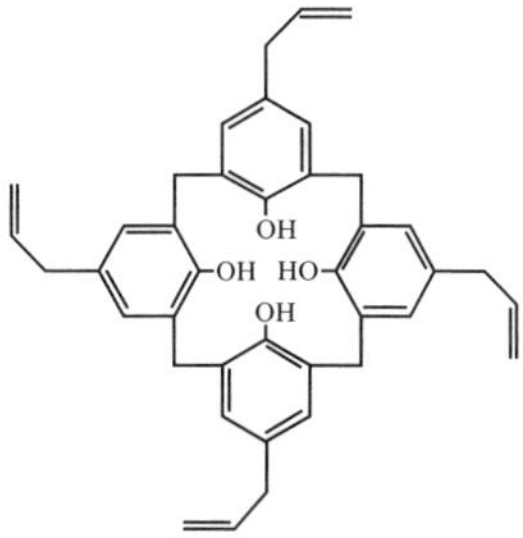

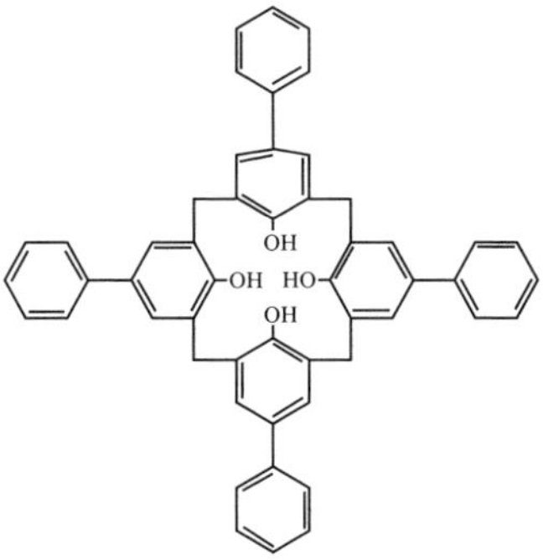

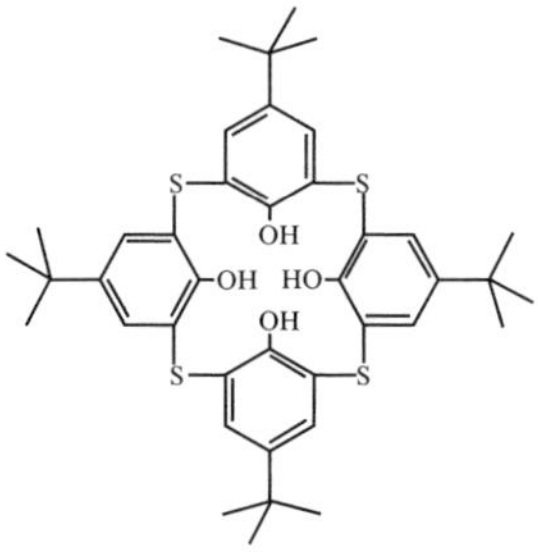

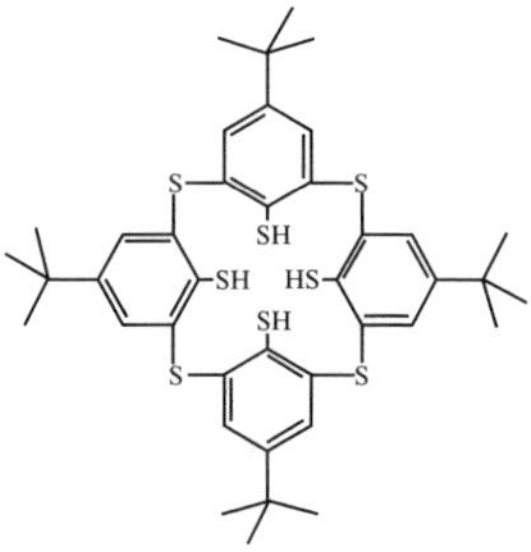

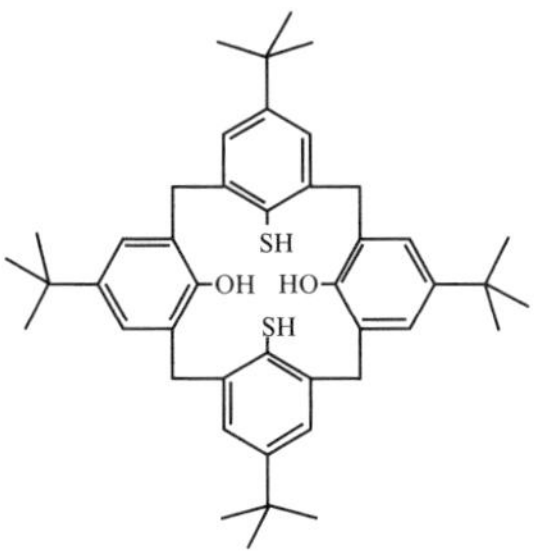

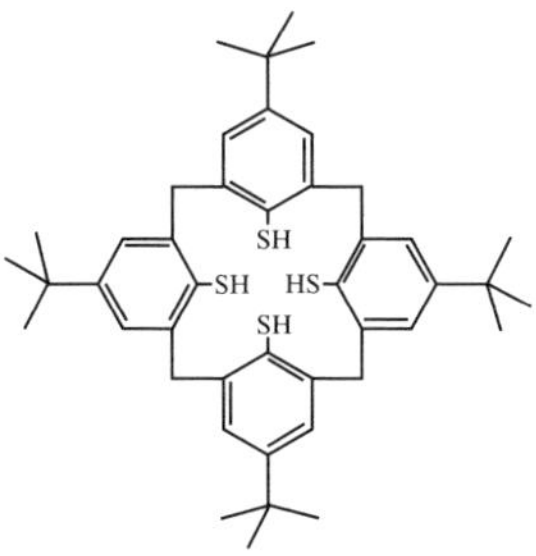

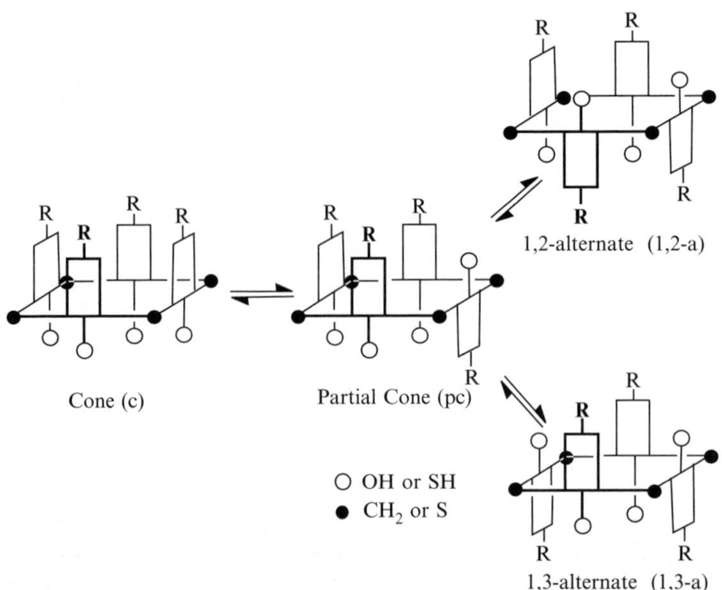

Figure 5 Schematic representations of the conformational processes allowing the interconversion between four limit conformers of calix[4]arenes.

be in a 1,3-alternate conformation (Figure 6g) [61]. The basket-type cavity of the calix[4]arene derivatives in a cone conformation has been also shown to accommodate in the solid state a variety of neutral guest molecules [62]. The same behaviour leading to the inclusion of MeOH (Figure 7a), CH_2Cl_2 (Figure 7b) and $CHCl_3$ (Figure 7c) was established in the crystalline phase for the thiacalix **8** (Figure 7) [58].

Taking into account the fact that a calix[4]arene moiety offers four OH groups, two such molecules in cone conformation and in a face to face-type arrangement would present eight convergent hydroxy groups. Thus, the design of koilands was based on the fusion of two calix[4]arene derivatives in the cone conformation by two centres adopting a tetrahedral coordination geometry. The initial study was concerned with silicon atoms as the fusing element (Figure 8a) [44,46]. Examples of fused calix[4]arene using titanium(IV), niobium(V), aluminium(IV) and zinc have been reported [63]. Two calix[4]arenes have also been interconnected by organic [64] and organometallic bridges[65].

3.2.1 Tuning the cavity

The general strategy for the synthesis of koilands was based on the treatment of the calix derivatives under basic conditions (NaH–THF) by addition of $SiCl_4$. The

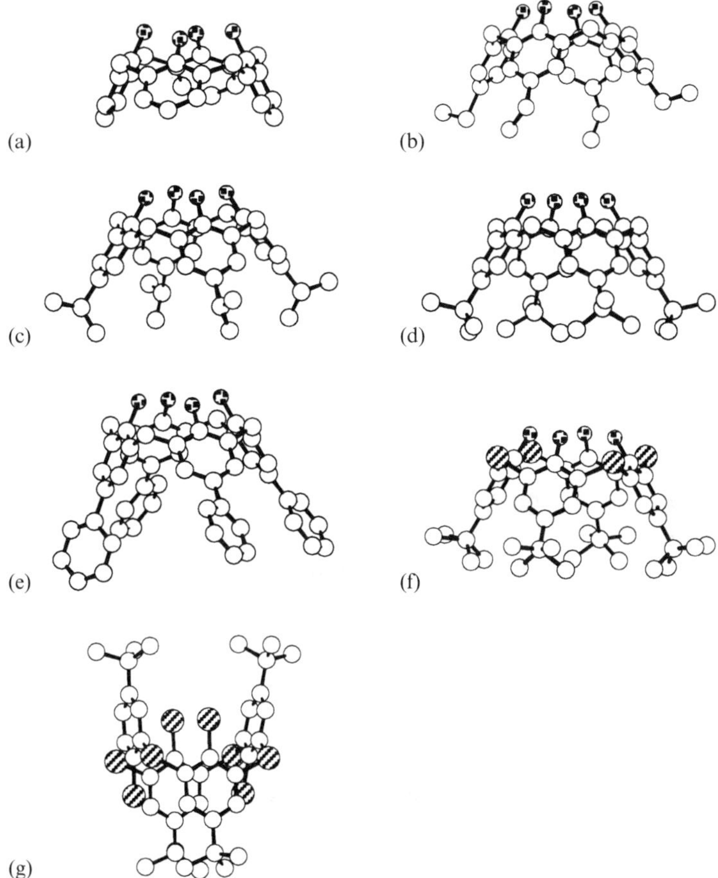

Figure 6 X-ray structures of calix[4]arene derivatives: **1** (a), **3** (b), **4** (c), **5** (d) **7** (e) and **8** (f) adopting the cone conformation and **9** (g) adopting the 1,3-alternate conformation in the crystalline phase For the sake of clarity, solvent molecules and hydrogen atoms are not represented.

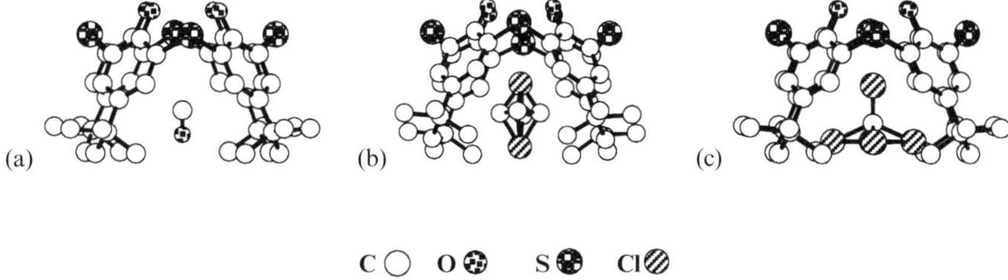

Figure 7 Structures of inclusion complexes of the thiacalix[4]arene **8** with MeOH (a), CH_2Cl_2 (b) and $CHCl_3$ (c). For the sake of clarity, hydrogen atoms are not represented.

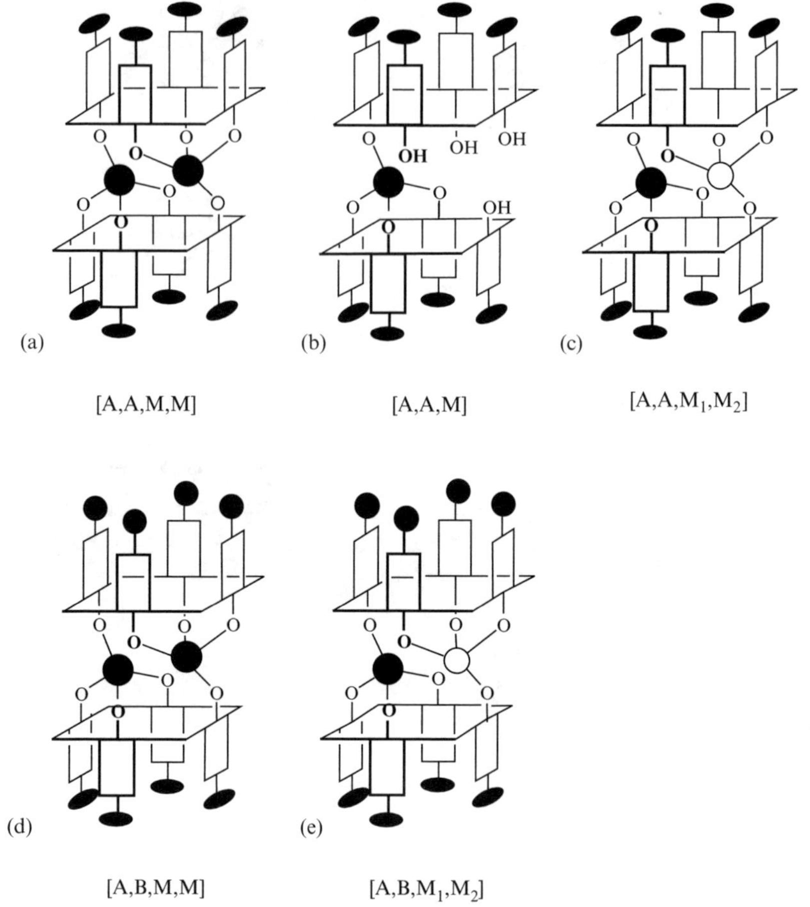

Figure 8 Schematic representation of koilands (hollow molecular building blocks bearing two preorganised cavities oriented at a 180° angle): centrosymmetric koiland (a), noncentrosymmetric koilands based on: electronic differentiation (b and c), geometrical differentiation (d) and both electronic and geometrical differentiation (e). A and B represent calix[4]arene units and M, M_1 and M_2 atoms used to fuse two calix units.

desired compounds were usually obtained after chromatography on silica and/or by crystallisation. The starting material for the synthesis of all koilands reported was the *p-tert*-butylcalix[4]arene **5** prepared according to the published procedure [66]. The calix[4]arene **1** was obtained by aluminium chloride de-*tert*-butylation of **5** [67]. The latter was the common starting material for the synthesis of *p*-methylcalix[4]arene **2**, *p*-ethylcalix[4]arene **3**, *p*-allylcalix[4]arene **6** and *p*-phenylcalix[4] arene **7**. Although the syntheses of **2** [68] and **7** [69] were first reported by Gutsche and co-workers, the procedures of Ungaro and co-workers [70] and Atwood and co-workers [71], respectively, were found to be more convenient. Compounds **3** [72] and **6** [73] were prepared according to published procedures. Dealing with the preparation of **4**, although the Ni-catalysed direct isopropylation of calix[4]arene using propene was recently reported [74], for the sake of experimental simplicity we modified the reported Zinke–Conforth procedure [75,76]. Finally, the two tetrathiacalix[4]arenes **8** [57] and **9** [61] were synthesised following reported procedures.

Based on calix[4]arenes **1–7** and two Si atoms, a library containing 22 different koilands (**n[Y,Z]**, with **n** the tecton number and **Y** and **Z** the calix derivatives **1–7** composing the koiland) were obtained (Figure 9). Whereas koilands **9[1,1]** [77], **10[2,2]** [77], **11[3,3]** [78], **12[4,4]** [76], **13[5,5]** [44], **14[6,6]** [79] and **15[7,7]** [77] are centrosymmetric tectons of the **[X,X]** type, koilands **16[1,2]** [80], **17[1,5]** [80], **18[2,5]** [80], **19[1,7]** [80], **20[2,7]** [80], **21[5,7]** [80] and **22[5,6]** [81] are noncentrosymmetric of the **[X,Y]** type.

3.2.2 Formation of discrete binuclear inclusion complexes

Since a koiland possesses two independent cavities, it forms discrete binuclear complexes with molecules such as solvents acting as stoppers (Figure 10a). This fact was demonstrated in the case of the koiland **11** (Figure 11a) based on double fusion of two ethylcalix **3** (Scheme 1) by two Si atoms (Figure 9) and *p*-xylene **31** (Scheme 2). Each cavity of the koiland **11** indeed includes a host *p*-xylene molecule **31** [78]. In the case of koiland **12** [76] (Figure 11b) and **14** [79] Figure 11c), the same behaviour was observed when CH_2Cl_2 was used in the crystallisation event. Discrete binuclear complexes were also obtained for koiland **13** with $CHCl_3$ [44] (Figure 11d), anisol **32** (Scheme 2) [82] (Figure 11f) and *p*-xylene **31** [82] (Figure 11e). For both *p*-xylene **31** and anisole **32** inclusion complexes, the crystal study showed that their CH_3 groups were deeply inserted into the cavity of the hollow module. Although in the case of *p*-xylene **31** a sandwich complex composed of the parent calix **5** and the solvent molecule with a 1:1 ratio was reported in the solid state [83], for the koiland **13**, as stated above, a discrete binuclear inclusion complex was observed. Whereas the formation of discrete binuclear complexes with $CHCl_3$ and anisole **32** may be explained by their restricted dimensions and unfavourable geometry, in the case of *p*-xylene **31**, possessing the right topology, one could, *a priori*, expect the formation of a koilate. However, because of the size of the *tert*-butyl groups, it appeared from the X-ray study that the distance between the two CH_3 groups ($d_{C\text{–}C=5.83\,\text{Å}}$) of *p*-xylene **31** is to short to allow the interconnection of two consecutive units and thus the formation of the koilate.

9 [1,1]
10 [2,2]
11 [3,3]
12 [4,4]
13 [5,5]
14 [6,6]
15 [7,7]
16 [1,2]
17 [1,5]
18 [2,5]
19 [1,7]
20 [2,7]
21 [5,7]
22 [5,6]
23 [5,5]
24 [5,5]
25 [5,5]
26 [8,8]
27 [8,8]
28 [8,8]

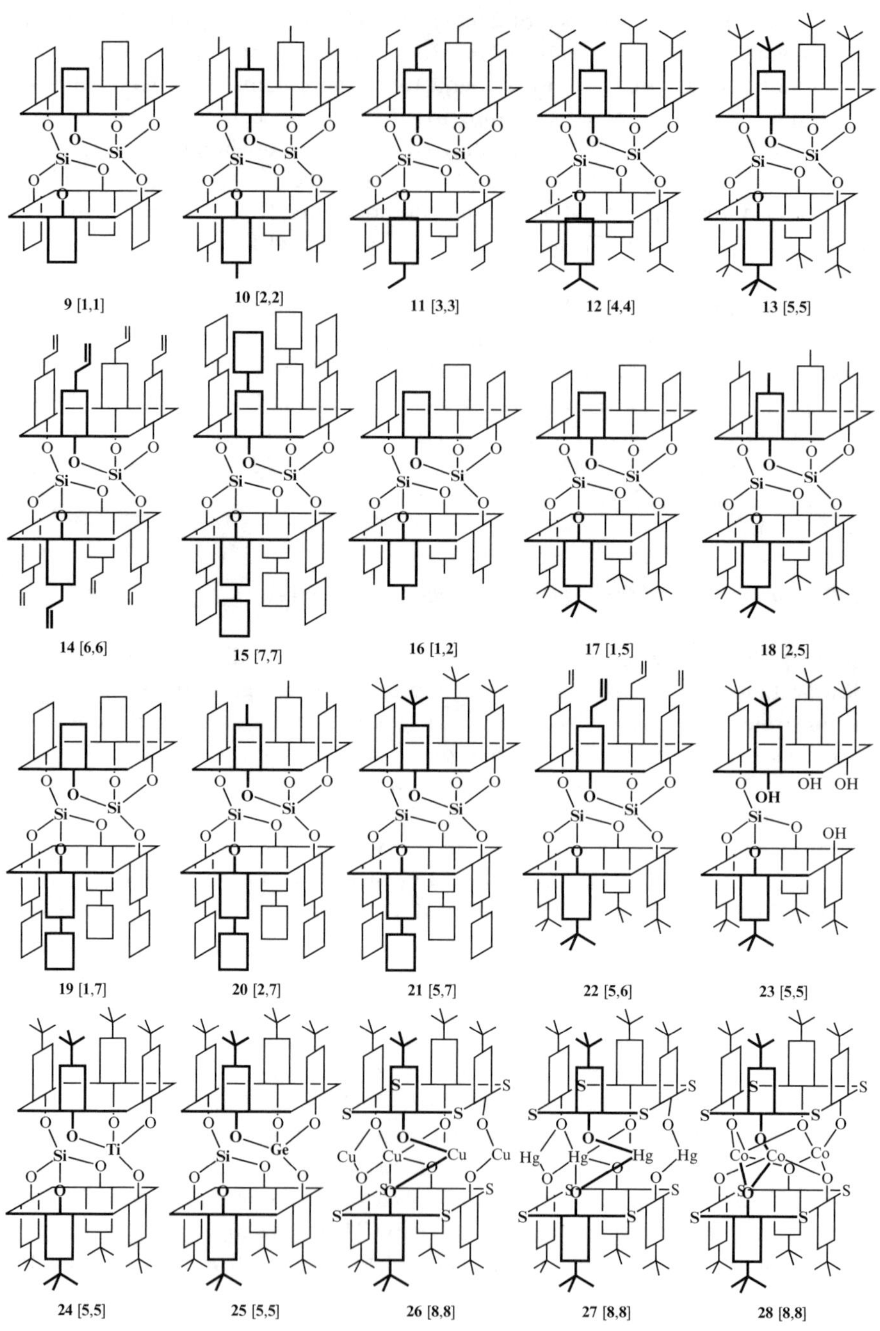

(*continues*)

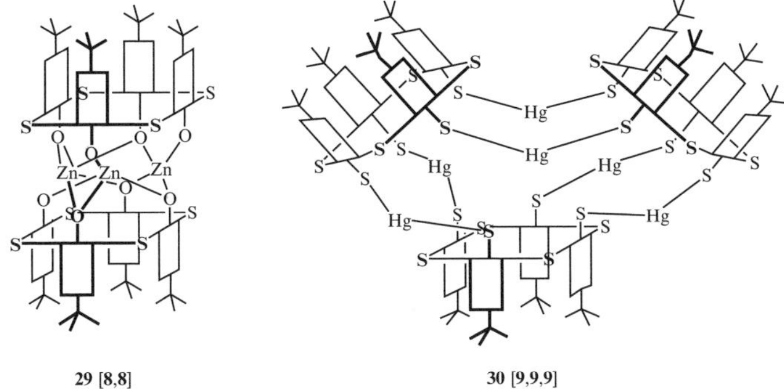

Figure 9 Schematic representation of koilands based on fusion of two or three calix units by connecting atoms such as Si (**9–23**), Si and Ti (**24**), Si and Ge (**25**), Cu (**26**), Hg (**27**, **30**), Co (**28**) and Zn (**29**).

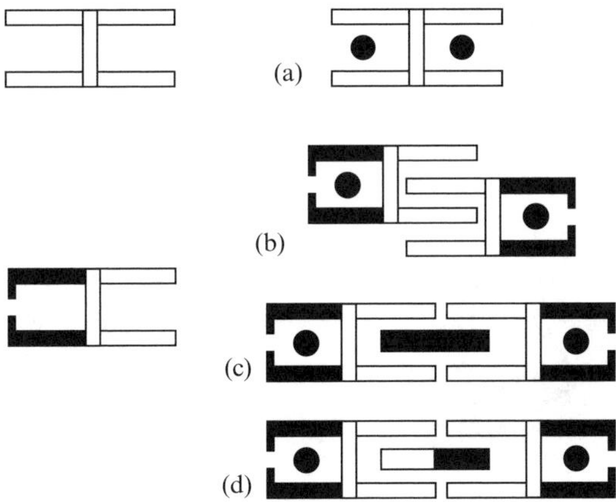

Figure 10 Schematic representations of the formation of discrete inclusion complexes based on homo- (a) and hetero-koilands (b–d). For more details, see text.

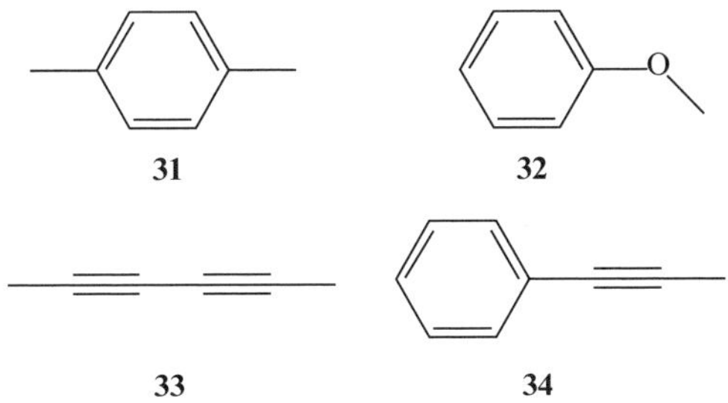

Scheme 2

In the case of unsymmetrical koilands such as **16–22**, depending on the steric bulk of the substituents located at the upper rim of the calix units, a variety of discrete complexes may be envisaged in the absence of connector molecules (Figure 10). In particular, for all koilands **13, 17, 18** and **22** bearing at least one calix unit **5**, owing to the steric bulk of the *tert*-butyl group, no self-inclusion between calix **5** units may occur. However, the other face of the molecule bearing less bulky groups may undergo a self-inclusion process, thus leading to discrete polynuclear complexes in the solid state (Figure 10b–d). This was indeed observed in the case of **17[1,5]** [Figure 12a). The crystal study revealed the formation of a discrete complex for which two koilands **17** are interconnected by self-inclusion through their cavities based on calix **1** units. The remaining two cavities based

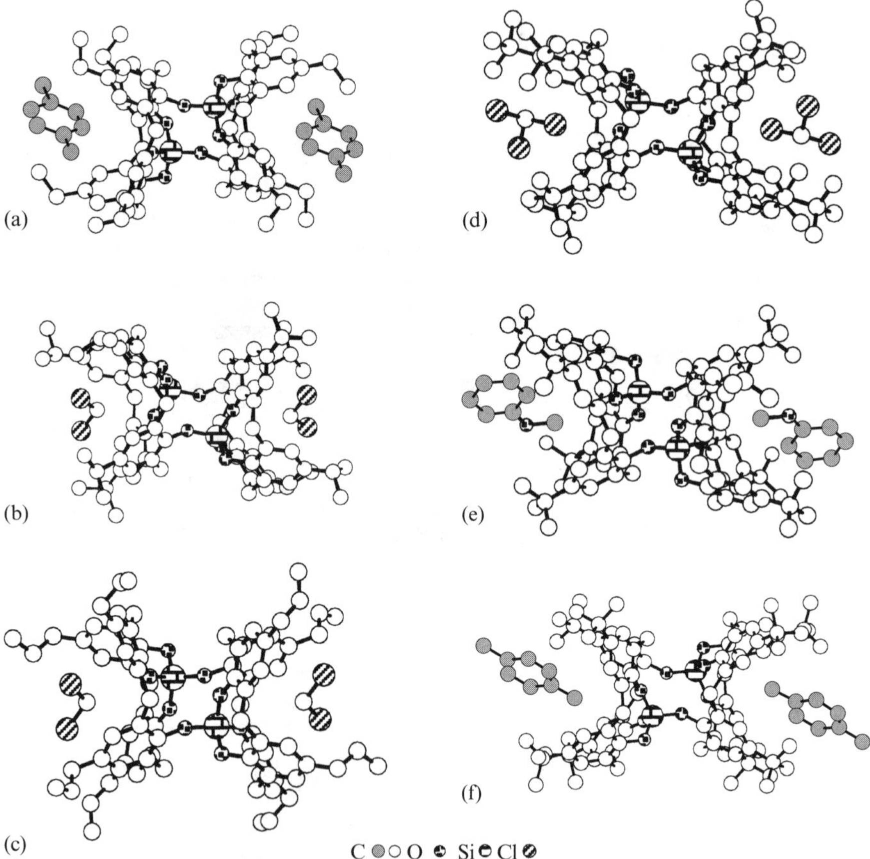

Figure 11 Structures of discrete binuclear inclusion complexes formed by koilands: **11** with *p*-xylene **31** (a), **12** with CH_2Cl_2 (b), **14** with CH_2Cl_2 (c), **13** with $CHCl_3$ (d), **13** with *p*-xylene **31** (e) and **13** with anisole **32** (f). For the sake of clarity, hydrogen atoms are not represented.

on calix **5** units form each an inclusion complex with EtOH solvent molecule, leading to a discrete tetranuclear entity (Figure 10b). The same behaviour was also demonstrated for the unsymmetrical koiland **21[5,7]** (Figure 12b). In that case the self-inclusion takes place between the calix units **7** bearing phenyl groups and the other two cavities based on calix units **5** located on the other face of the koiland are occupied by THF molecules. Although not observed to date, one may also envisage two other possibilities of discrete complexes based on the interconnection of two of such units by either symmetrical (Figure 10c) or unsymmetrical (Figure 10d) connector molecules.

3.2.3 Formation of koilates

Using hexadiyne **33** (Scheme 2), a rod-type molecule, possessing both the required linear geometry and distance between its two terminal CH_3 groups ($d_{C-C=6.65\ \text{Å}}$),

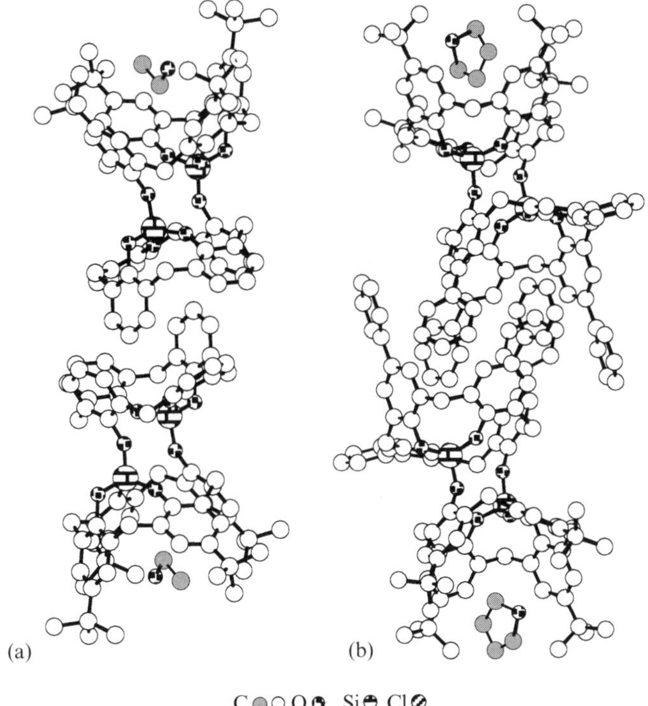

Figure 12 Structures of discrete tetranuclear inclusion self-inclusion complexes formed by noncentrosymmetric koilands: **17** with EtOH (a) and **21** with THF (b). For the sake of clarity, hydrogen atoms are not represented.

the formation of koilates was demonstrated for both koilands **12** (Figure 13a) [76] and **13** (Figure 13b) [82]. For both cases, crystalline material was obtained from either $CHCl_3$–MeOH or $CHCl_3$–hexane mixtures of koiland **12** or **13** and hexadiyne **33** in large excess. The X-ray analysis revealed in both cases the presence of $CHCl_3$ molecules in addition to **12** or **13** and hexadiyne in a 1:1 ratio. The $CHCl_3$ solvent molecules do not participate in the formation of the inclusion networks. As expected, the lattice in both cases is indeed composed of linear koilates formed between koilands **12** or **13** and hexadiyne **33** as connectors. Each connector bridges two consecutive koilands by penetrating their cavities through its terminal CH_3 groups. The methyl groups of hexadiyne **33** are deeply inserted into the cavity of the koiland [distance between the CH_3 group of the connector and the aromatic carbon atom (CO) of the koiland of ca 3.6 Å]. It is worth noting that the hexadiyne molecules **33** are extremely well encapsulated by two consecutive koilands **12** or **13**.

3.2.4 Functionalisation of koilands

The calix[4]arene backbone, in addition to the features stated above, presents further advantages by being readily modified at the *para* position by a large variety of substituents bearing functionalities. In particular, one may introduce at the *para*

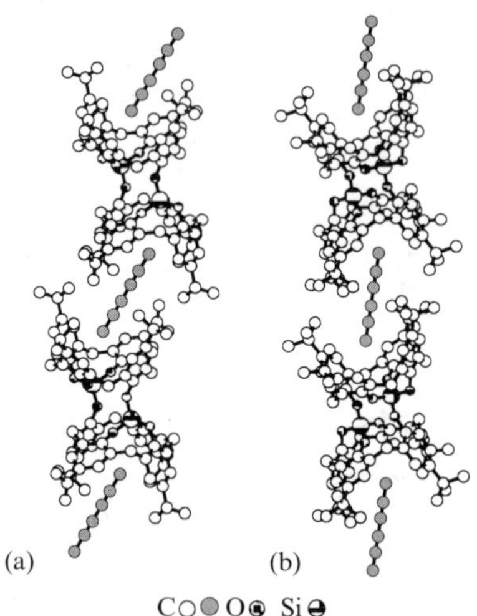

Figure 13 A portion of the X-ray structure of the koilate formed between the koilands **12** (a) and **13** (b) with hexadiyne (**33**) as connector. The 1-D networks are obtained by a single translation of the assembling core defined by the inclusion of the connector into the cavity of the receptor. For the sake of clarity, the carbon atoms of the connector are coloured in grey and the hydrogen atoms are not represented.

position functional groups such as allyl moieties [84]. It has been shown that although compound **6** shows temperature-dependent conformational mobility, below room temperature it adopts a cone conformation [84]. The centrosymmetric koiland **14** (Figure 9) possessing functional groups at both faces of the molecule was obtained upon double fusion of the parent compound **6** with two silicon atoms [79]. The design of the koiland **14** was motivated by the possibility that such a functionality may offer to promote [2 + 2] cycloaddition reactions in the solid state using irradiation. The formation of a koilate was demonstrated using *p*-xylene **31** as a rigid, compact and ditopic connector. Suitable crystals were obtained from a mixture of compound **14** and *p*-xylene in excess and *i*PrOH at room temperature. The X-ray analysis (Figure 14) revealed that the crystal was composed exclusively of **14** and *p*-xylene **31** disposed in a alternate fashion. As predicted, a linear koilate was formed through the interconnection of two cavities belonging to consecutive koiland **14** with the *p*-xylene **31**. In the crystal one could identify an assembling core which was defined as the inclusion of a CH_3 group of the connector within a cavity of **14** (Figure 14a). The shortest C–C distance of 3.63 Å between the CH_3 group of the connector and one of the carbon atoms belonging to one of the phenolic group at the bottom of the cavity indicated a high degree of inclusion.

Although the formation of the koilate was predicted, the X-ray analysis revealed another unexpected feature. Interestingly, when looking at lateral packing of consecutive linear koilates, it was observed that among the eight allyl groups present in each koiland **14**, two of them were oriented centrosymmetrically in a peculiar manner (Figure 14b). Indeed, two alkene moieties belonging to two consecutive koilands were located perfectly parallel to each other with an alkene-to-alkene distance of 4.52 Å between them, suggesting the possibility of interconnecting covalently the koilates through a [2 + 2] cycloaddition reaction in the solid state. Unfortunately, probably owing too the slightly too large distance between the allyl groups, all attempts failed [85].

It is interesting that although, using koilands **12** [76] or **13** [82] and hexadiyne **33**, the first examples of fully characterised linear koilates were obtained in the solid state, one could nevertheless argue that, owing to the presence of $CHCl_3$ molecules in the lattice, the formation of both koilates was governed by solvent molecules. The formation of the koilate obtained when using the koiland **14** and *p*-xylene **31** precludes this objection since in that case the linear koilate was formed in the solid without the presence of any solvent molecules in the lattice.

3.2.5 Formation of koilates by self-inclusion

As demonstrated above, koilands may lead in the solid state either to discrete molecular complexes in the presence of molecules acting as stoppers (Figure 10)

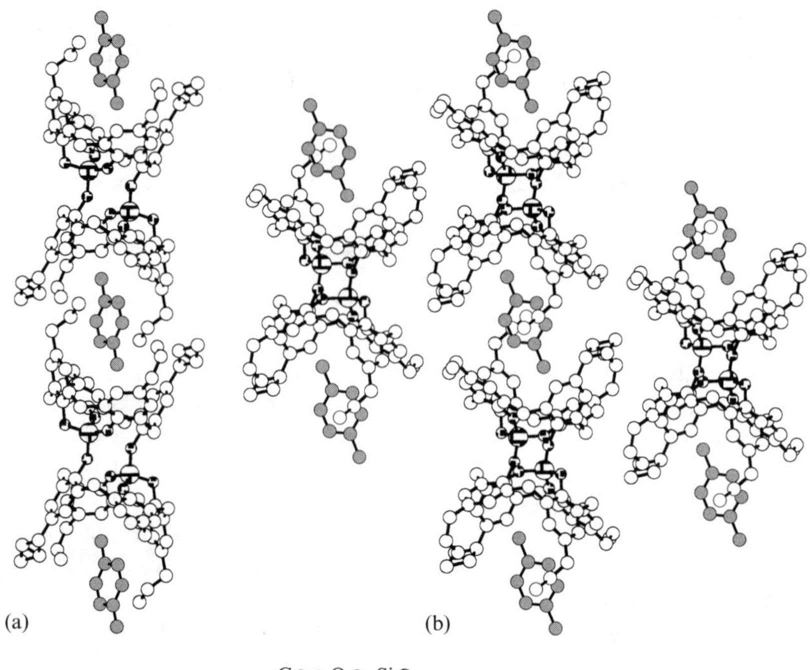

Figure 14 A portion of the X-ray structure of the koilate formed between the koiland **14** and *p*-xylene **31** as connector. The 1-D network is obtained by a single translation of the assembling core defined by the inclusion of the connector into the cavity of the receptor (a) and a portion of the structure showing the lateral packing of linear koilates leading to parallel disposition of the allyl moieties (b). For the sake of clarity, the carbon atoms of the connector are coloured in grey and the hydrogen atoms are not represented.

or to linear molecular networks in the presence of appropriate connectors (Figure 4). Another alternative route to the formation of linear koilate may be based on the use of self-complementary koilands possessing simultaneously two divergent cavities and connecting moieties (Figure 4). Compound **14** fulfils these requirements. Indeed, the latter possesses two preorganised cavities as well as allyl groups at both faces of the molecule which may act as connectors [86].

In the crystalline phase, the self-complementary koiland **14** forms a linear koilate through the interconnection of two cavities belonging to consecutive koilands **14** by van der Waals interactions (Figure 15a). The assembling core leading by a single translation to the 1-D self-inclusion network was identified as the inclusion of one of the four allyl groups located at one of the two faces of the molecule **14** by one of the two cavities of the consecutive self-complementary compound **14**. The shortest C–C distance between the terminal CH_2 of the allyl group acting as connector and carbon atoms of one of the aromatic moieties of the cavity belonging to the next unit varied from 3.63 to 3.92 Å, indicating a rather high

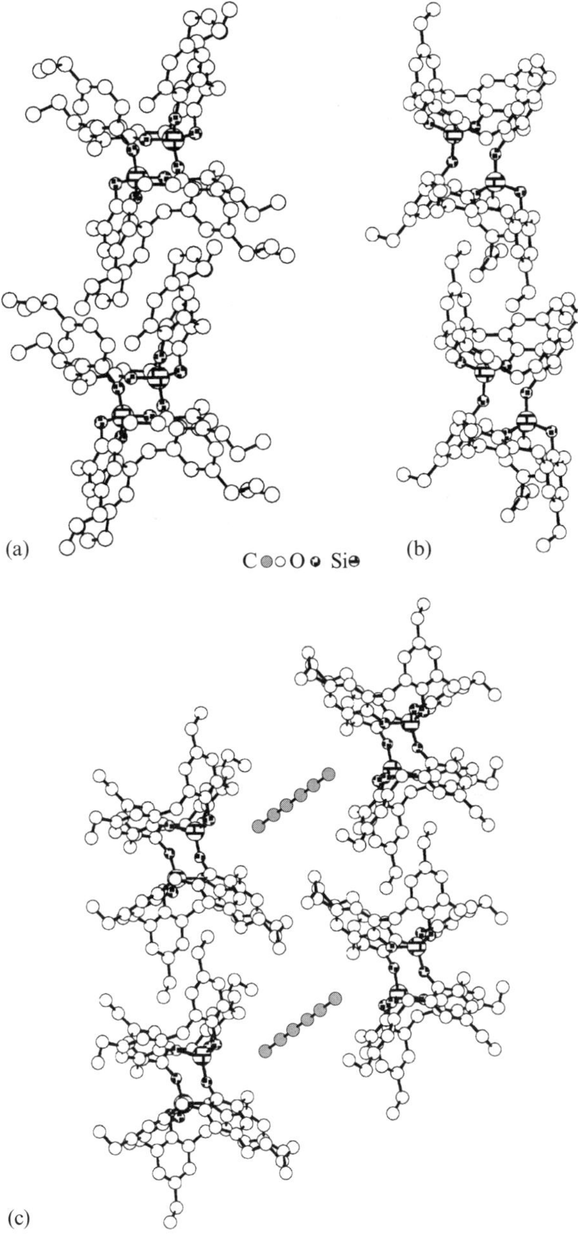

Figure 15 A portion of the structure of koilates formed through self-inclusion by the centrosymmetric self-complementary koilands **14** (a) and **11** (b) and the interconnection of consecutive koilates by hexadiyne molecules **33** (c) (see text for more details). For the sake of clarity, hydrogen atoms are not represented.

degree of inclusion. Interestingly, when crystals of compound **14** were grown from other solvent systems such as toluene–EtOH, benzene–EtOH, $CHCl_3$–*i*PrOH and toluene–*i*PrOH, in all cases crystals of the same morphology and stability were obtained and their investigation by X-ray (lattice parameters) confirmed that they were the same. However, when crystals of **14** were grown by the same technique from a CH_2Cl_2–EtOH mixture, two types of crystals, one with a rod-type morphology and the other with rhombic shape, were observed. Whereas the crystals with rhombic morphology were stable at room temperature and proved by X-ray analysis to be of the same type as those mentioned above, the rod-shape crystals were unstable outside the solution. Their analysis by X-ray diffraction revealed the presence of **14** and three molecules of CH_2Cl_2. In the lattice two types of discrete inclusion complexes were present and therefore no network based on inclusion processes could be identified.

The same type of behaviour was also demonstrated for the koiland **11** based on double fusion of two ethylcalix[4]arenes (Figure 15b). In the presence of hexadiyne **33** and $CHCl_3$, a self-inclusion 1-D koilate was obtained. Again, the assembling core leading to the 1-D self-inclusion network was identified as the inclusion of one of the four ethyl groups located on one of the two faces of the koiland **11** by one of the two cavities of consecutive self-complementary koilands. The hexadiyne molecules **33** were not participating to the formation of the koilate but were occupying the empty space between koilates (Figure 15c) [78].

Rather unpredictably, when the koiland **11** was crystallised in the absence of hexadiyne **33** but in the presence of a mixture of benzene and 2-propanol as the crystallising mixture, a 2-D self-inclusion koilate was obtained (Figure 16). The 2-D inclusion network is formed through inclusion of two ethyl groups belonging to two consecutive koiland **11** by each face of the koiland [78].

3.2.6 Directional koilates

Dealing with 1-D networks, the control of directionality is an important and challenging issue. In particular, such control is crucial for the exploitation of directional physical properties such as dipolar interactions. For all cases of koilates mentioned above, although the dimensionality of the molecular assembly could be controlled (1- or 2-D-inclusion networks) by the geometrical features of the koilands, owing to the their centrosymmetric nature (Figure 8a), no directionality may be expected (Figure 4a–c). In order to control both the dimensionality and the directionality of koilates (Figure 4f–l), one may use nonsymmetric koilands (Figure 8). The design of non-centrosymmetric koilands may be based on electronic (Figure 8b and c), geometric (Figure 8d) or both electronic and geometric (Figure 8 e) differentiation of the two cavities.

Electronic differentiation may be achieved by using a single atom to connect the two calix units. Indeed, one of the two calixes is triply coordinated to an

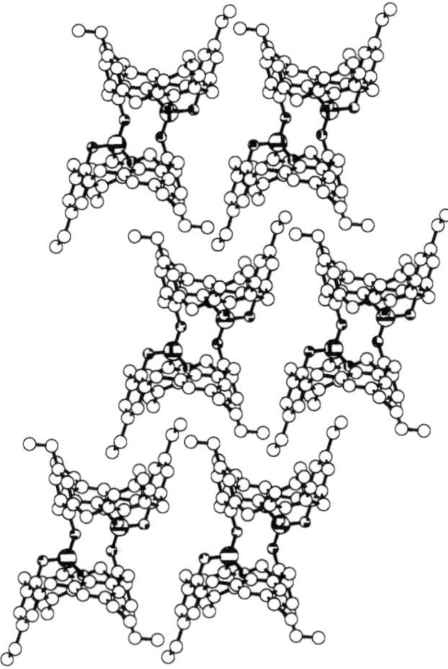

Figure 16 A portion of the structure of the 2-D koilates formed through self-inclusion by the centrosymmetric self-complementary koiland **11** (see text for more details). For the sake of clarity, hydrogen atoms are not represented.

Si atom whereas the other unit is coordinated by a single bond through one of its phenolic oxygen atoms to the Si atom (Figure 8b). Another possibility may be based on two different fusing atoms with the same oxidation state IV such as Si and Ti or Ge. The nonsymmetric nature of the Si–Ti heterobinuclear koiland is based on the induced difference between the two calix units by their coordination to two different metals with different elctronegativity. Indeed, one of the calixes is triply coordinated to an Si atom whereas the other unit is triply coordinated to a Ti atom (Figure 8c). The same holds for the heterobinuclear koiland based on Si and Ge atoms. All three cases were investigated. The koiland **23** based on the single fusion of two calix **5** units by a single Si atom was prepared [87]. The latter was the starting material for the preparation of the heterobinuclear koiland **24** based on double fusion of two calix **5** units by a Si and Ti atoms [87] as well as the koiland **25** based on interconnection of two calix **5** units by a Si and Ge atoms [78]. Indeed, a stepwise strategy based on first the preparation of the mono fused compound **23** and then introduction of the other heteroatom was achieved. In solution, koiland **23** was shown to adopt a face to face or '*syn*' conformation by NMR-ROESY experiments. The same conformation was also confirmed in the solid state by X-ray diffraction methods.

Geometric differentiation (Figure 8d) may be accomplished, while keeping the same fusing element such as Si, by connecting two different calix units. Mixed koilands **16–22**, based on the double fusion of two different calix units by two Si atoms, were prepared. The general strategy for the preparation of nonsymmetric koilands was based on a combinatorial approach. Upon treatment of a combination of two different calix[4]arenes (X,Y) (X = Y = **1**–**7**] with base, followed by reaction with $SiCl_4$, as expected, three different koilands [X–X], [Y–Y] of the homo type (centrosymmetric) and [X–Y] of the hetero type (noncentrosymmetric) were obtained [88].

In solution, as expected, ^{29}Si NMR studies revealed the presence of two signals for all noncentrosymmetric heterodimers, whereas for the centrosymmetric koilands a unique signal was observed.

The ability of noncentrosymmetric koilands **22–24** to form discrete binuclear inclusion complexes was demonstrated in the solid state by X-ray diffraction methods on single crystals [78]. Inclusion complexes for both the doubly fused koiland **22** [Figure 17a) and the mono fused koiland **23** (Figure 17b) with CH_2Cl_2 were obtained. In the presence of phenylpropyne **34**, both koilands **23** (Figure 17c) and **24** (Figure 17d) led to the formation of binuclear inclusion complexes.

The third possibility, consisting in using two different calix derivatives and two different atoms for their fusion, has not yet been exploited.

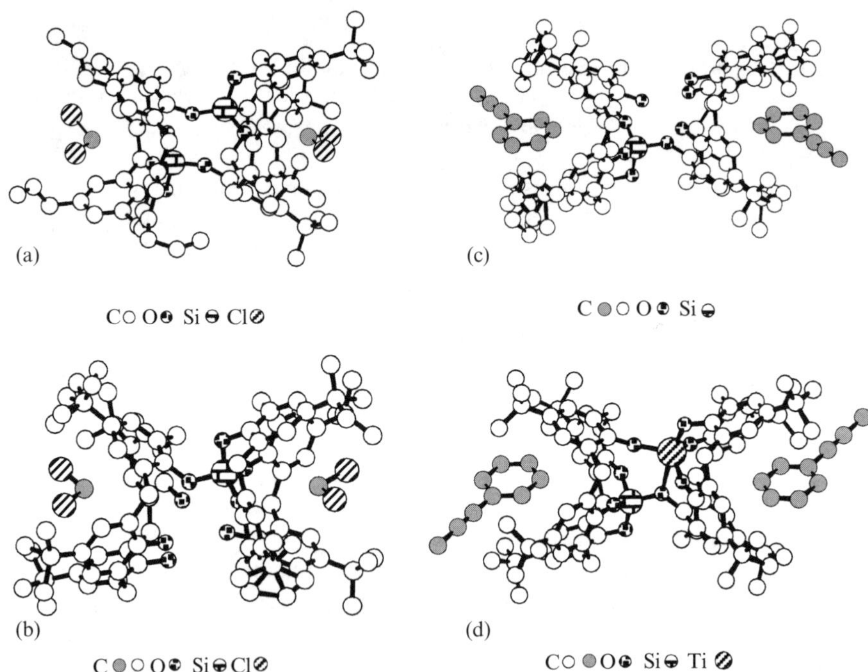

Figure 17 Structures of discrete binuclear inclusion complexes formed by noncentrosymmetric koilands: **22** with CH_2Cl_2 (a), mono fused koiland **23** with CH_2Cl_2 (b), mono fused koiland **23** with phenylpropyne **34** (c) and heterobinuclear koiland **24** with CH_2Cl_2 (d) (see text for more details). For the sake of clarity, hydrogen atoms are not represented.

Since hexadiyne **33** was shown to be the most appropriate connector for koiland **13** based on double fusion of two calix **5** units by two Si atoms, the same connector was used to generate directional koilate with the monofused koiland **23**. In the crystalline phase a directional koilate (Figure 4g) was observed [78]. The 1-D inclusion network (Figure 18a) is analogous to the one mentioned above when using the centrosymmetric koiland **13** and the connector **33**. Indeed, the connectivity between koilands by the connector leading to the koilate is the same but, owing to the noncentrosymmetric nature of the koiland **23**, the network is directional. The packing of koilates within the crystal is parallel. However, owing to the opposite orientation of linear arrays, a nondirectional arrangement was obtained (Figure 18b).

For the geometrically differentiated noncentrosymmetric koiland **22** based on the double fusion of the *p*-*tert*-butyl calix **5** and *p*-allyl calix unit **6** by two Si atoms, the formation of directional koilates was studied using *p*-xylene **31** as a centrosymmetric (Figure 4g and h) and phenylpropyne **34** as a noncentrosymmetric (Figure 4f) connectors [81].

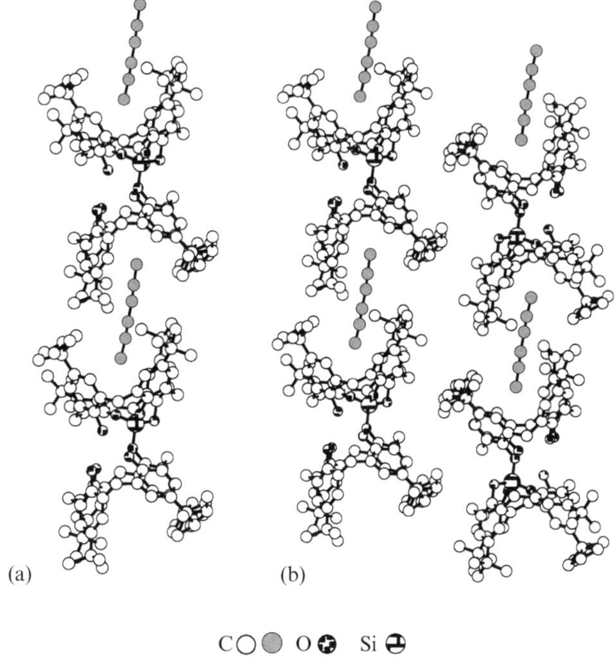

Figure 18 A portion of the X-ray structure of the koilate formed between the mono fused koiland **23** and hexadiyne **33** as connector. The directional 1-D network is obtained by a single translation in a noncentric fashion of two different assembling cores defined by the inclusion of the connector into the two differentiated cavities of the koiland (a) and a portion of the structure showing the centrosymmetric packing of consecutive directional koilates leading to a nonpolar solid (b). For the sake of clarity, the carbon atoms of the connector are coloured in grey and the hydrogen atoms are not represented.

Upon slow diffusion of 2-propanol into a solution of **22** in *p*-xylene used as solvent, colourless crystals were obtained and studied by X-ray diffraction methods. The crystal is composed of compound **22** and two *p*-xylene molecules **31**. Of the two *p*-xylene molecules present, one of them acts as a connector by bridging through inclusion processes consecutive koilands, thus leading to the formation of the koilate, whereas the other one does not participate in the formation of the koilate but rather occupies the empty space between them. Dealing with the koilate, the assembling core is defined as the encapsulation of a *p*-xylene molecule by two consecutive koilands **22** in an unsymmetrical manner (Figure 19a). In other terms, one methyl group of *p*-xylene is included within the cavity of **22** bearing *p*-allyl groups, whereas the other methyl group is penetrating the cavity of a consecutive koiland **22** bearing *tert*-butyl groups. Owing to the unsymmetrical nature of the assembling core (Figure 4g), upon translation, a 1-D directional koilate is obtained. Again, the directional linear koilates are positioned in a parallel fashion, but the packing is again symmetrical, i.e. consecutive linear koilates are oriented in opposite directions, thus generating a centre of symmetry. Consequently, the overall system is non directional (Figure 19b).

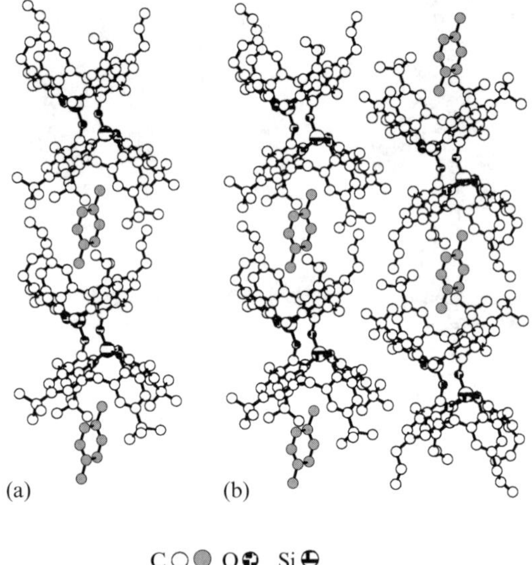

Figure 19 A portion of the X-ray structure of the koilate formed between the noncentrosymmetric koiland **22** and *p*-xylene **31** as connector. The directional 1-D network is obtained by a single translation in a noncentric fashion of two different assembling cores defined by the inclusion of the connector into the two different cavities of the koiland (a) and a portion of the structure showing the centrosymmetric packing of consecutive directional koilates leading to a nonpolar solid (b). For the sake of clarity, the carbon atoms of the connector are coloured in grey and the hydrogen atoms are not represented.

For the preparation of directional linear koilates, as mentioned above, another alternative consisting in using a nonsymmetrical koiland and connector may be exploited by using the koiland **22** and phenylpropyne **34** as connector, respectively (Figure 4f). Again, upon slow diffusion of 2-propanol into a solution of **22** in phenylpropyne used as solvent, colourless crystals were obtained. The X-ray defined as the encapsulation of a phenylpropyne molecule by two consecutive koilands **22** (Figure 20a). Whereas the propyne moiety of the connector is included within the cavity of **22** bearing *tert*-butyl groups, the phenyl moiety penetrates the cavity of a consecutive koiland **22** bearing *p*-allyl. Again, owing to the unsymmetrical nature of the assembling core, upon translation, a 1-D directional koilate is obtained (Figure 20a). However, again, the directional linear koilates are positioned in a parallel fashion but with opposite orientation, thus leading to a nondirectional packing (Figure 20b).

The formation of directional 1-D koilates based on self-inclusion was also investigated. The self-complementary koiland **16** based on the double fusion of the calix derivatives **1** and **2** by two Si atoms was found to generate a self-inclusion network (Figure 21). However, the koilate formed was not directional because the self-inclusion processes, leading to the formation of the network, take

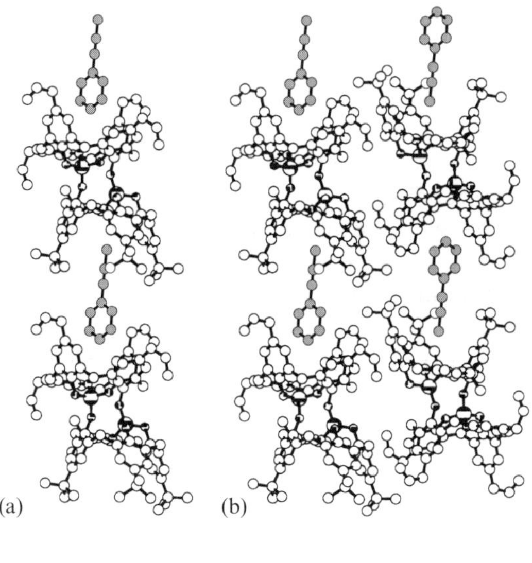

Figure 20 A portion of the X-ray structure of the koilate formed between the noncentrosymmetric koiland **22** and phenylpropyne **34** as connector. The directional 1-D network is obtained by a single translation in a noncentric fashion of two different assembling cores defined by the inclusion of the connector into the two different cavities of the koiland (a) and a portion of the structure showing the centrosymmetric packing of consecutive directional koilates leading to a nonpolar solid (b). For the sake of clarity, the carbon atoms of the connector are coloured in grey and the hydrogen atoms are not represented.

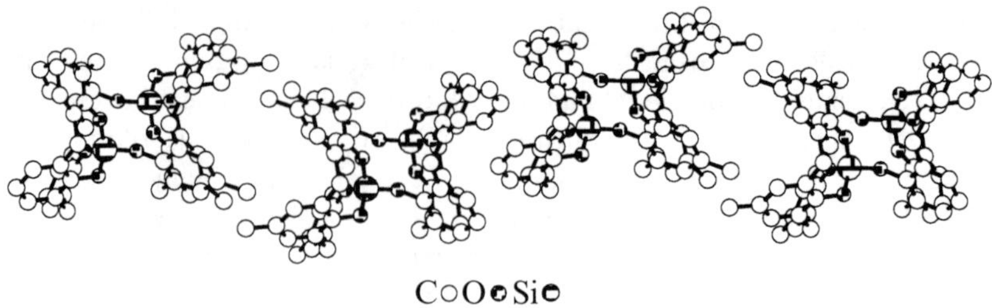

Figure 21 A portion of the structure of koilates formed through self-inclusion by the noncentrosymmetric self-complementary koiland **16** (see text for more details). For the sake of clarity, hydrogen atoms are not represented.

place through self-recognition between the two faces of the koiland (Figure 4k). Indeed, the face of **16** composed of **1** is included within the same face of the consecutive koiland whereas the other face based on **2** is included with the cavity of the next koiland of the same nature.

3.2.7 Koilands based on transition metals

The coordination ability of calix[4]arenes such as **5** towards alkaline and transition as well as main group elements has been established. Although in a few cases the contribution of the aromatic π cloud to the binding process was demonstrated, for the majority of reported examples the binding of metal cations takes place at the OH groups [53–56]. The replacement of CH_2 junctions by S atoms leading to thia calix derivatives **8** and **9** enhances the number of coordination sites which may be exploited for the design of new koilands. Using compound **8** and a paramagnetic transition metal such as Cu(II), the koiland **26** was obtained and structurally characterised [89]. The solid-state structure of **26** was elucidated by X-ray crystallography on single crystals. The Cu_4 array, close to exactly square, is sandwiched between two thiacalix entities **8** in a 'cone' conformation similar to that of the free ligand **8** [58]. Interestingly, the copper koiland obtained with the tetrathiacalix **8**, as in the case of silicon koilands **13** obtained with calix **5**, forms a binuclear inclusion complex with two molecules of CH_2Cl_2. Each of the solvent molecule penetrates deeply into the cavity of the koiland (Figure 22a).

Other structures obtained using the same strategy showed that this mode of binding is certainly not restricted to Cu(II) [90–94]. For example, solid-state structural analysis of the Co(II) (Figure 22b) and Zn(II) (Figure 22c) complexes formed with the tetrathiacalix **8** revealed the formation of koilands **28** and **29**, respectively, in which the two calix units are fused by three metal centres. In both cases, again, both cavities were occupied by solvent molecules, demonstrating the inclusion ability of the new koilands [90]. Recently, the interconnection of two thiacalix **8** by four ethylene fragments was reported [95].

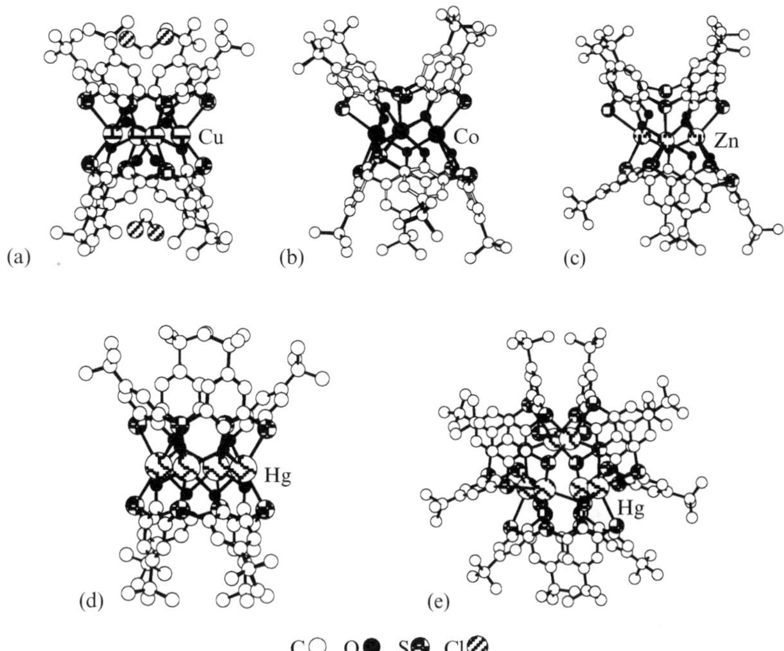

Figure 22 Structures of discrete transition metal containing koilands based on thia (**8**) and mercaptothia (**9**) calix derivatives: **26** (a), **28** (b), **29** (c), **27** (d) and **30** (e) (see text for more details). For the sake of clarity, hydrogen atoms are not represented.

A final strategy used to generate new koilands was based on the use of the calix derivative **8** and **9** and mercury cation. In marked contrast with the mono- and binuclear mercury complexes obtained when using compounds **10** [96] and **11** [97], respectively, in the case of **8** and **9** tetra- and hexanuclear mercury-based koilands **27** and **30** were obtained, respectively [98]. Both koilands were characterised in the solid state. Whereas for the koiland **27** the formation of 1-D koilates using a variety of connectors may be envisaged, for the koiland **30**, based on connection of three units **9** by six Hg(II) cations, the generation of new types of 2-D koilates appears interesting.

4 PERSPECTIVES

The viability of the strategy based on the use of concave and convex molecular tectons to generate inclusion networks in the crystalline phase was demonstrated. A further step consisting in designing directional inclusion networks was also shown. However, so far noncentrosymmetric packing of the linear directional arrays has not been achieved. This remains a challenge. In principle, if the latter point of concern could be solved, the strategy presented here may be used to generate polar solids of fundamental importance Furthermore, in terms of applications, the design of predicted polar solids is important since one may then exploit vectorial physical properties.

ACKNOWLEDGEMENTS

This work was supported by the Université Louis Pasteur, Institut Universitaire de France and Centre National de la Recherche Scientifique. Thanks are also due to Dr X. Delaigue, Dr F. Hajek and N. Kyritsakas-Gruber.

REFERENCES

1. A. Gavezzotti, *Acc. Chem. Res.*, **27**, 309 (1994).
2. J. D. Dunitz, *Pure Appl. Chem.*, **63**, 177 (1991).
3. J.-M. Lehn, *Supramolecular Chemistry, Concepts and Perspectives*, VCH, Weinheim, 1995.
4. M. Simard, D. Su, J. D. Wuest, *J. Am. Chem. Soc.*, **113**, 4696 (1991).
5. G. R. Desiraju, *Angew. Chem., Int. Ed. Engl.*, **34**, 2311 (1995).
6. M. C. Etter, *Acc. Chem. Res.*, **23**, 120 (1990).
7. G. M. Whitesides, J. P. Mathias, T. Seto, *Science*, **254**, 1312 (1991).
8. F. W. Fowler, J. W. Lauher, *J. Am. Chem. Soc.*, **115**, 5991 (1993).
9. J. S., Lindsey, *New J. Chem.*, **15**, 153 (1991).
10. D. S. Lawrence, T. Jiang, M. Levett, *Chem. Rev.*, **95**, 2229 (1995).
11. X. Delaigye, E. Graf, F. Hajek, M. W. Hosseini, J. -M. Planeix in NATO ASI Series, Eds. G. Tsoucaris, J. L. Atwood, J. Lipkowski, Series C, Vol. 480, Kluwer, Dordrecht, 1996, p. 159.
12. D. Philp, J. F. Stoddart, *Angew. Chem., Int. Ed. Engl.*, **35**, 1155 (1996).
13. S. Mann, *Nature*, **365**, 499 (1993).
14. C. B. Aakeröy, K. R. Seddon, *Chem. Soc. Rev.*, **22**, 397 (1993); S. Subramanian, M. J. Zaworotko, *Coord. Chem. Rev.*, **137**, 357 (1994); D. Braga, F. Grepioni, *Acc. Chem. Res.*, **27**, 51 (1994); V. A. Russell, M. D. Ward, *Chem. Mater.*, **8**, 1654 (1996).
15. R. Robson, in *Comprehensive Supramolecular Chemistry*, Eds J. L. Atwood, J. E. D. Davies, D. D. MacNicol, F. Vögtle, Pergamon Press, Oxford, Vol. 6 (Eds D. D. MacNicol, F. Toda, R. Bishop), 1996, p. 733.
16. A. J. Blake, N. R. Champness, P. Hubberstey, W.-S. Li, M. A. Withersby, M. Schöder, *Coord. Chem. Rev.*, **183**, 117 (1999).
17. O. M. Yaghi, H. Li, B. Chen, C. Davis, D. Richardson, T. L. Groy, *Acc. Chem. Res.*, **31**, 474 (1998).
18. M. W. Hosseini, in *NATO ASI Series*, Eds D. Braga, F. Grepiono, G. Orpen, Series C, Vol. 538, Kluwer, Dordrecht, 1999, p. 181.
19. B. Moulton, M. J. Zaworotko, *Chem. Rev.*, **101**, 1629 (2001).
20. G. Brand, M. W. Hosseini, R. Ruppert, A. De Cian, J. Fischer, N. Kyritsakas, *New J. Chem.*, **19**, 9 (1995).
21. M. W. Hosseini, R. Ruppert, P. Schaeffer, A. De Cian, N. Kyritsakas, J. Fischer, *J. Chem. Soc. Chem. Commun.*, 2135 (1994).

22. M. W. Hosseini, G. Brand, P. Schaeffer, R. Ruppert, A. De Cian, J. Fischer, *Tetrahedron Lett.*, **37**, 1405 (1996).
23. G. Brand, M. W. Hosseini, O. Félix, P. Schaeffer, R. Ruppert, in *Magnetism: a Supramolecular Function*, Ed. O. Kahn, *NATO ASI Series*, Series C, Vol. 484, Kluwer, Dordrecht, 1996, p. 129.
24. O. Félix, M. W. Hosseini, A. De Cian, J. Fischer, *Angew. Chem., Int. Ed. Engl.*, **36**, 102 (1997).
25. O. Félix, M. W. Hosseini, A. De Cian, J. Fischer, *Tetrahedron Lett.*, **38**, 175 (1997).
26. O. Félix, M. W. Hosseini, A. De Cian, J. Fischer, *Tetrahedron Lett.*, **38**, 1933 (1997).
27. O. Félix, M. W. Hosseini, A. De Cian, J. Fischer, *New J. Chem.*, **21**, 285 (1997).
28. O. Félix, M. W. Hosseini, A. De Cian, J. Fischer, *New J. Chem.*, **22**, 1389 (1998).
29. O. Félix, M. W. Hosseini, A. De Cian, J. Fischer, *Chem. Commun.*, 281 (2000).
30. D. Braga, L. Maini F. Grepioni, A. De Cian, O. Félix, J. Fischer, M. W. Hosseini, *New J. Chem.*, **24**, 547 (2000).
31. O. Félix, M. W. Hosseini, A. De Cian, *Solid State Sci.*, **3**, 789 (2001).
32. S. Ferlay, O. Félix, M. W. Hosseini, J.-M. Planeix, N. Kyritsakas, *Chem. Commun.*, in press (2002).
33. C. Kaes, M. W. Hosseini, C. E. F. Rickard, B. W. Skelton, A. White, *Angew. Chem., Int. Ed. Engl.*, **37**, 920 (1998).
34. G. Mislin, E. Graf, M. W. Hosseini, A. De Cian, N. Kyritsakas, J. Fischer, *Chem. Commun.*, 2545 (1998).
35. M. Loï, M. W. Hosseini, A. Jouaiti, A. De Cian, J. Fischer, *Eur. J. Inorg. Chem.*, 1981 (1999).
36. M. Loï, E. Graf, M. W. Hosseini, A. De Cian, J. Fischer, *Chem. Commun.*, 603 (1999).
37. C. Klein, E. Graf, M. W. Hosseini, A. De Cian, J. Fischer, *Chem. Commun.*, 239 (2000).
38. H. Akdas, E. Graf, M. W. Hosseini, A. De Cian, J. McB. Harrowfield, *Chem. Commun.*, 2219 (2000).
39. A. Jouaiti, M. W. Hosseini, A. De Cian, *Chem. Commun.*, 1863 (2000).
40. B. Schmaltz, A. Jouaiti, M. W. Hosseini, A. De Cian, *Chem. Commun.*, 1242 (2001).
41. A. Jouaiti, V. Jullien, M. W. Hosseini J.-M. Planeix, A. De Cian, *Chem. Commun.*, 1114 (2001).
42. C. Klein, E. Graf, M. W. Hosseini, A. De Cian, J. Fischer, *New J. Chem*, **25**, 207 (2001).
43. S.Ferlay, S. Koenig, M. W. Hosseini, J. Pansanel, A. De Cian N. Kyritsakas, *Chem. Commun.*, 218 (2002).
44. X. Delaigue, M. W. Hosseini, A. De Cian, J. Fischer, E. Leize, S. Kieffer, A. Van Dorsselaer, *Tetrahedron Lett.*, **34**, 3285 (1993).
45. X. Delaigue, M. W. Hosseini, R. Graff, J.-P. Kintzinger, J. Raya, *Tetrahedron Lett.*, **35**, 1711 (1994).
46. X. Delaigue, E. Graf, F. Hajek, M. W. Hosseini, J.-M. Planeix, in *Crystallography of Supramolecular Compounds*, Eds G. Tsoucaris, J. L. Atwood, J. Lipkowski, *NATO ASI Series*, Series C, Vol. 480, Kluwer, Dordrecht, 1996, p. 159.
47. M. W. Hosseini, A. De Cian, *Chem. Commun.*, 727 (1998).
48. C. J. Pedersen, *Angew. Chem., Int. Ed. Engl.* **27**, 1021 (1988).
49. D. J. Cram, *Angew. Chem., Int. Ed. Engl.*, **27**, 1009 (1988).
50. J.-M. Lehn, *Angew. Chem., Int. Ed. Engl.*, **27**, 89 (1988).
51. H. M. Powell, *J. Chem. Soc.*, 61 (1948).
52. *Molecular Inclusion and Molecular Recognition. Clathtrates I and II. Topics in Current Chemistry*, Ed., E. Weber, Springer, New York, Vols 140 and 149, 1987 and 1988.
53. C. D. Gutsche, *Calixarenes, Monographs in Supramolecular Chemistry*, Ed. J. F. Stoddart, Royal Society of Chemistry, London, 1989; C. D. Gutsche, *Calixarenes Revisited,*

Monographs in Supramolecular Chemistry, Ed. J. F. Stoddart, Royal Society of Chemistry, London, 1998.
54. *Calixarenes. A Versatile Class of Macrocyclic Compounds*, Ed. J. Vicens and V. Böhmer, Kluwer, Dordrecht, 1991.
55. V. Böhmer, *Angew. Chem., Int. Ed. Engl.*, **34**, 713 (1995).
56. *Calixarenes 2001*, Eds J. Vicens, Z. Asfari, J. M. Harrowfield, V. Böhmer, Kluwer, Dordrecht, 2001.
57. H. Kumagai, M. Hasegawa, S. Miyanari, Y. Sugawa, Y. Sato, T. Hori, S. Ueda, H. Kamiyama, S. Miyano, *Tetrahedron Lett.*, **38**, 3971 (1997), T. Sone, Y. Ohba, K. Moriya, H. Kumada, K. Ito, *Tetrahedron*, **53**, 10689 (1997).
58. H. Akdas, L. Bringel, E. Graf, M. W. Hosseini, G. Mislin, J. Pansanel, A. De Cian, J. Fischer, *Tetrahedron Lett.*, **39**, 2311 (1998).
59. M. W. Hosseini, *Calixarene Molecules for Separation*, Eds G. J. Lumetta, R. D. Rogers, A. S. Gopalan, *ACS Symposium Series*, Vol. 557, American Chemical Society, Washington, DC, 2000, p. 296.
60. M. W. Hosseini, in *Calixarenes 2001*, Eds J. Vicens, Z. Asfari, J. M. Harrowfield, V. Böhmer, Kluwer, Dordrecht, 2001, pp. 110–129.
61. P. Rao, M. W. Hosseini, A. De Cian, J. Fischer, *Chem. Commun.*, 2169 (1999).
62. C. D. Gutsche, B. Dhawan, K. H. No, R. Muthukrishnan, *J. Am. Chem. Soc.*, **103**, 3782 (1981); G. D. Andreetti, R. Ungaro, A. Pochini, *Chem. Commun.*, 1005 (1979); R. Ungaro, A. Pochini, G. D. Andreetti, P. Domiano, *J. Chem. Soc., Perkin Trans. 2*, 197 (1985); M. Coruzzi, G. D. Andreetti, V. Bocchi, A. Pochini, R. Ungaro, *J. Chem. Soc., Perkin Trans. 2*, 1133 (1982).
63. M. M. Olmstead, G. Sigel, H. Hope, X. Xu, P. Power, *J. Am. Chem. Soc.*, **107**, 8087 (1985); F. Corazza, C. Floriani, A. Chiesti-Villa, C. Guastini, *Chem. Commun.*, 1083 (1990); J. L. Atwood, S. G. Bott, C. Jones, C. L. Raston, *Chem. Commun.*, 1349 (1992); J. L. Atwood, P. C. Junk, S. M. Lawrence, C. L. Raston, *Supramol. Chem.*, **7**, 15 (1996).
64. D. Kraft, J.-D. van Loon, M. Owens, W. Verboom, W. Vogt, M. A. McKervey, V. Böhmer, D. N. Reinhoudt, *Tetrahedron Lett.*, **31**, 4941 (1990); J.-D. van Loon, D. Kraft, M. J. K. Ankoné, W. Verboom, S. Harkema, V. Böhmer, D. N. Reinhoudt, *J. Org. Chem.*, **55**, 5176 (1990); P. Schmidtt, P. D. Beer, M. G. B. Drew, P. D. Sheen, *Angew. Chem., Int. Ed. Engl.*, **36**, 1840 (1997).
65. P. D. Beer, A. D. Keefe, A. M. Z. Slawin, D. J. Williams, *J. Chem Soc., Dalton Trans.*, 3675 (1990).
66. C. D. Gutsche, M. Iqbal, *Org. Synth.*, **68**, 234 (1989).
67. C. D. Gutsche, J. A. Levine, *J. Am. Chem. Soc.*, **104**, 2652 (1982).
68. C. D. Gutsche, K. C. Nam, *J. Am. Chem. Soc*, **110**, 6153 (1988).
69. C. D. Gutsche, K. H. No, *J. Org. Chem.*, **47**, 2708 (1982); K. H. No, C. D. Gutsche, *J. Org. Chem.*, **47**, 2713 (1982).
70. M. Almi, A. Arduini, A. Casnati, A. Pochini, R. Ungaro, *Tetrahedron*, **45**, 2177 (1989).
71. R. K. Juneja, K. D. Robinson, C. P Johnson, J. L. Atwood, *J. Am. Chem. Soc*, **115**, 3818 (1993).
72. A. Arduini, A. Pochini, A. Rizzi, A. R. Sicuri, F. Ugozzoli, R. Ungaro, *Tetrahedron*, **48**, 905 (1992).
73. C. D. Gutsche, J. A. Levine, *J. Am. Chem. Soc*, **104**, 2653 (1982).
74. B. Yao, J. Bassus, R. Lamartine, *New J. Chem.*, **20**, 913 (1996).
75. B. Dhawan, S. I. Chen, C. D. Gutsche, *Makromol. Chem.*, **188**, 921 (1987).
76. J. Martz, E. Graf, M. W. Hosseini, A. De Cian, J. Fischer, *J. Mater. Chem*, **8**, 2331 (1998).
77. F. Hajek, E. Graf, M. W. Hosseini, *Tetrahedron Lett.*, **37**, 1409 (1996).

78. J. Martz, E. Graf, M. W. Hosseini, A. De Cian, to be published.
79. F. Hajek, E. Graf, M. W. Hosseini, A. De Cian, J. Fischer, *Angew. Chem., Int. Ed. Engl.*, **36**, 1760 (1997).
80. F. Hajek, E. Graf, M. W. Hosseini, A. De Cian, J. Fischer, *Tetrahedron Lett.*, **38**, 4555 (1997).
81. J. Martz, E. Graf, M. W. Hosseini, A. De Cian, J. Fischer, *J. Chem. Soc., Dalton Trans.*, 3791 (2000).
82. F. Hajek, E. Graf, M. W. Hosseini, X. Delaigue, A. De Cian, J. Fischer, *Tetrahedron Lett.*, **37**, 1401 (1996).
83. M. Perrin, F. Gharnati, D. Oehler, R. Perrin, S. Lecocq, *J. Inclus. Phenom.*, **14**, 257 (1992).
84. C. D. Gutsche, J. A. Levine, *J. Am. Chem. Soc.*, **104**, 2653 (1982).
85. F. Hajek, B. M. Foxman, M. W. Hosseini, unpublished results.
86. F. Hajek, E. Graf, M. W. Hosseini, A. De Cian, J. Fischer, *Cryst. Eng.* **1**, 79 (1998).
87. X. Delaigue, M. W. Hosseini, E. Leize, S. Kieffer, A. van Dorsselaer, *Tetrahedron Lett.*, **34**, 7561 (1993).
88. F. Hajek, E. Graf, M. W. Hosseini, A. De Cian, J. Fischer, *Tetrahedron Lett.*, **38**, 4555 (1997).
89. G. Mislin, E. Graf, M. W. Hosseini, A. Bilyk, A. K. Hall, J. McB. Harrowfield, B. W. Skelton, A. H. White, *Chem. Commun.*, 373 (1999).
90. A. Bilyk, A. K. Hall, J. M. Harrowfield, M. W. Hosseini, G. Mislin, B. W. Skelton, C. Taylor, A. White, *Eur. J.Inorg. Chem.*, 823 (2000).
91. A. Bilyk, A. K. Hall, J. M. Harrowfield, M. W. Hosseini, B. W. Skelton, A. White, *Aust. J. Chem*, **53**, 895 (2000).
92. H. Akdas, E. Graf, M. W. Hosseini, A. De Cian, J. M. Harrowfield, *Chem. Commun.*, 2219 (2000).
93. Z. Asfari, A. Bilyk, J. W. C. Dunlop, A. K. Hall J. M. Harrowfield, M. W. Hosseini, B. W. Skelton, A. H. White, *Angew. Chem., Int. Ed. Engl.*, **40**, 721 (2001).
94. A. Bilyk, A. K. Hall, J. M. Harrowfield, M. W. Hosseini, B. W. Skelton, A. H. White, *Inorg. Chem.*, **40**, 672 (2001).
95. S. E. Matthews, V. Felix, M. G. B. Drew, P. D. Beer, *New J. Chem*, **25**, 1355, (2001).
96. X. Delaigue, J. McB. Harrowfield, M. W. Hosseini, A. De Cian, J. Fischer, N. Kyritsakas, *Chem. Commun.*, 1579 (1994).
97. X. Delaigue, M. W. Hosseini, A. De Cian, N. Kyritsakas, J. Fischer, *Chem. Commun.*, 609 (1995).
98. H. Akdas, E. Graf, M. W. Hosseini, A. De Cian, A. Bilyk, B. W. Skelton, G. A. Koutsantonis, I. Murray, J. M. Harrowfield, A. H. White, *Chem Commun.*, 1042 (2002).

Chapter 5

Layered Materials by Design: 2D Coordination Polymeric Networks Containing Large Cavities/Channels

Kumar Biradha

Indian Institute of Technology Kharagpur, Kharagpur, India

Makoto Fujita

The University of Tokyo, Tokyo, Japan

1 INTRODUCTION

Crystal engineering deals with the engineering of crystalline materials with desired physical and chemical properties and is a multidisciplinary area as it has implications for materials chemistry, supramolecular chemistry, molecular recognition and biology [1]. In general, the properties of crystalline materials are the result of the arrangement or assembly of molecules in a crystal lattice. Therefore, the properties of a solid can be engineered by arranging or assembling molecules in a required fashion through intermolecular interactions such as hydrogen or coordination bonds [2]. Currently, crystal engineering with coordination bonds is an attractive area of research owing to the inherent stability of coordination bonds and also to the versatility of coordination modes of transition metal atoms [3]. Porosity is one of the most targeted properties of coordination polymers as channels/cavities exist in the coordination networks [4]. Owing to the existence of such large channels/cavities, polymeric coordination materials are expected to mimic the functions of

Crystal Design: Structure and Function, edited by G. R. Desiraju

zeolites and clays. Further, they are expected to show some special functions owing to their inherent hydrophobicity which is not present in clays and zeolites.

The predictable formation of networks or assemblies through intermolecular interactions in order to derive the entire crystal lattice is the main goal of crystal engineering. In the case of coordination polymers the predictability will be influenced by several factors such as metal to ligand ratios, metal coordination modes and the presence of anions and guest molecules. Interpenetration of the networks is an issue that will hinder the formation of networks with large cavities [3a,5]. However, to some extent, interpenetration can be avoided by carrying out the reaction in the presence of large guest molecules that favor interaction with host frameworks. For successful crystal engineering, besides the selection of right components that form the network, one should equally consider all the above-mentioned factors.

Rigid rod-like bidentate ligands are known to form several coordination networks (one-, two- and three-dimensional) upon assembling with transition metals. In this chapter, our focus is on two-dimensional or layered coordination materials that are formed purely based on coordination bonds between bis(pyridine) units (Scheme 1) and transition metal atoms. Each single layer in these solids can be considered as a giant macromolecule and these pack on each other to form a molecular crystal. The atoms within the layers are bound to each other with covalent and coordinate bonds while the adjacent layers are bound with mostly aromatic interactions. The majority of the layered structures that are explored using coordination bonds can be classified as two types based on the connectivity of metal atoms: (a) four-connected and (b) three-connected networks. Square grid networks and rectangular grid networks come under the four-connected type

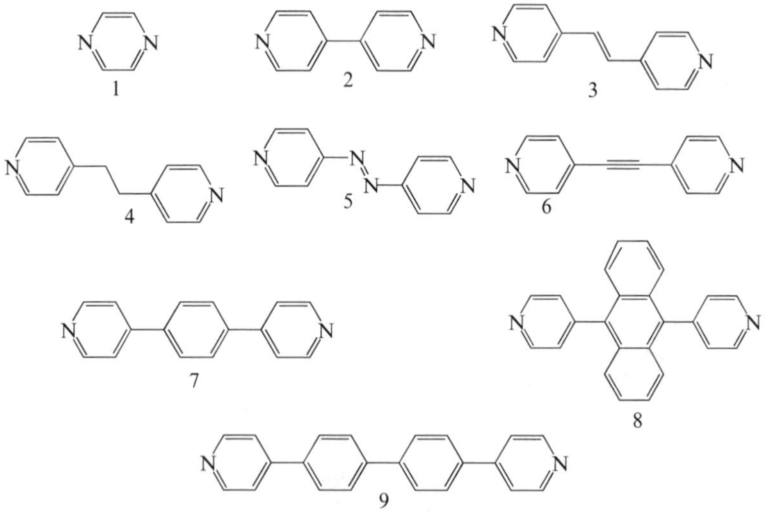

Scheme 1

Scheme 1 Library of bi-exodentate linear ligands.

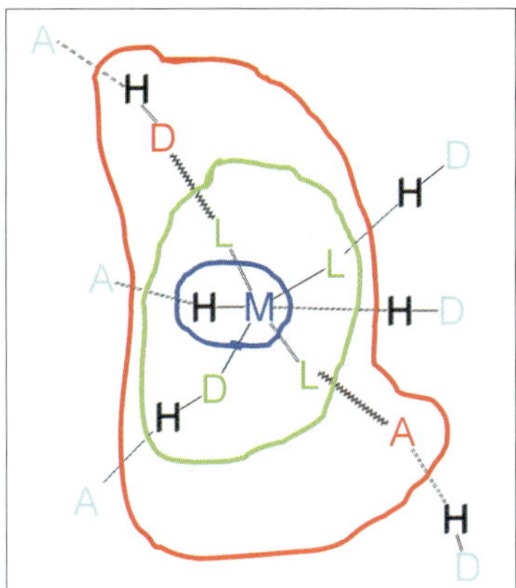

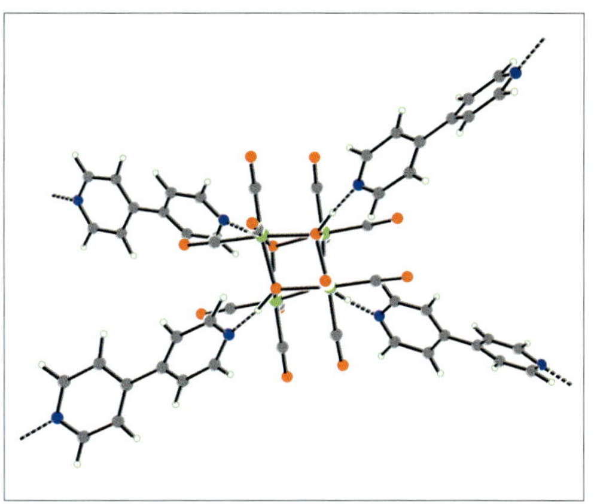

Plate 1 (Figure 1.3). (left). Domain model for hydrogen bonding involving metal complexes. Metal Domain (blue); Ligand Domain (green); Periphery Domain (red); Environment (cyan).

Plate 2 (Figure 1.4). (right). Diamondoid (M)O–H···N hydrogen-bonded network in crystalline $[Mn(\square_3\text{-OH})(CO)_3]_4\cdot 2(4,4'\text{-bipy})\cdot 2CH_3CN$. O–H groups (red); 4,4'-bipy (blue) [29].

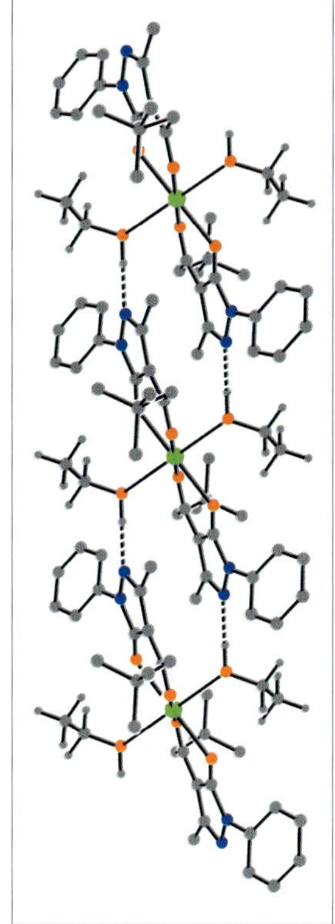

Plate 3 (Figure 1.5). (M)O–H···N hydrogen-bonded network involving coordinated ethanol. O (red); N (blue); Cd (green) [31a].

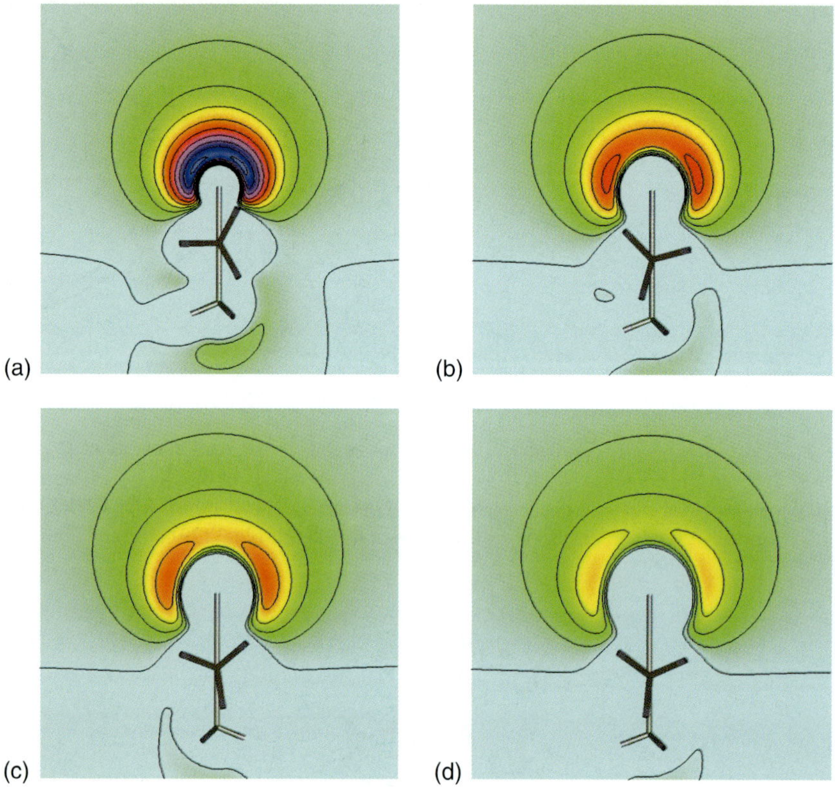

(a) (b)

(c) (d)

Plate 4 (Figure 1.6). Calculated negative electrostatic potential for *trans*-[$PdX(CH_3)(PH_3)_2$] illustrating positions of potential minima; X = F (a); Cl (b); Br (c); I (d). (Reproduced from ref. 27c with permission of the American Chemical Society).

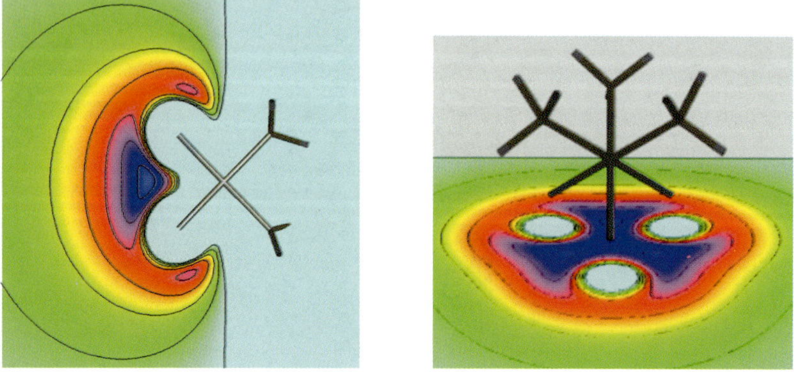

Plate 5 (Figure 1.7). Calculated negative electrostatic potential for *cis*-[$PdCl_2(PH_3)_2$] (left) and *fac*-[$RhCl_3(PH_3)_3$] (right) identifying recognition sites (potential minima, deep blue) for hydrogen bond donors. (Reproduced from ref. 39 with permission of the National Academy of Sciences, USA).

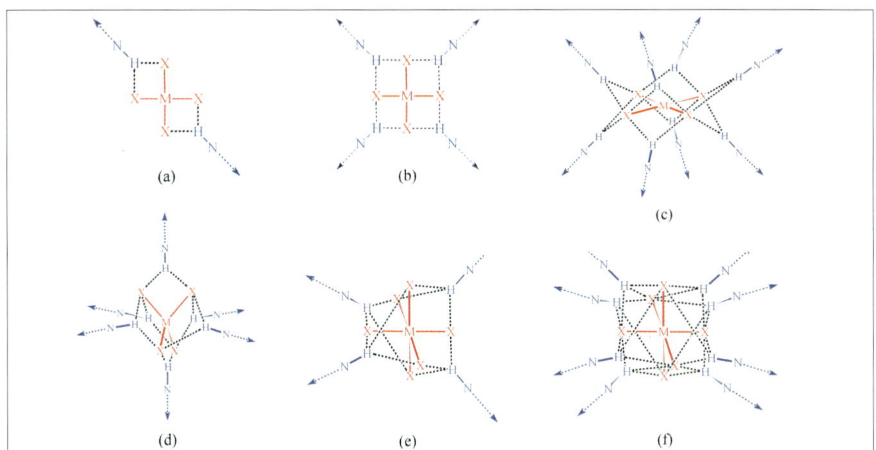

Plate 6 (Figure 1.9). Perhalometallate ions as potential hydrogen-bonded network nodes. (Reproduced from ref. 39 with permission of the National Academy of Sciences, USA).

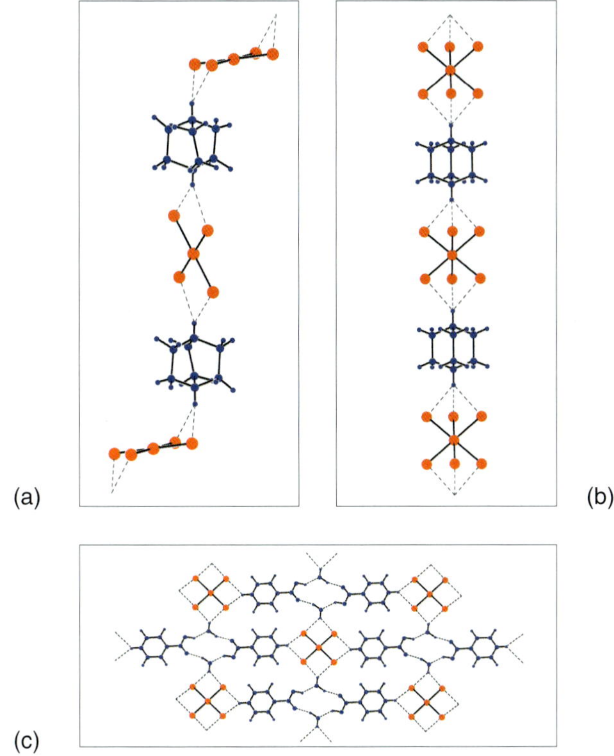

Plate 7 (Figure 1.10). One-dimensional networks in $[(DABCO)H_2][PtCl_4]$ (a) and $[H_2(DABCO)]$ $[PtCl_6]$ (b) employing synthons **I** and **II**, respectively [41a]. Two-dimensional network in $[\{(isonicotinic\ acid)H\}_2(OH_2)_2][PtCl_4]$ (c) employing synthon **I** [42b]. (Reproduced from ref. 39 with permission of the National Academy of Sciences, USA).

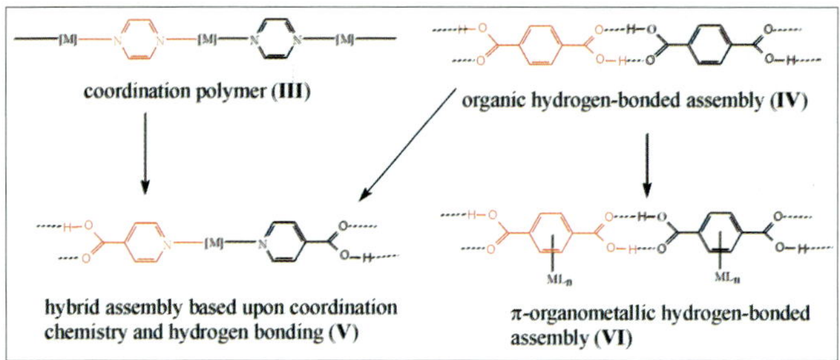

Plate 8 (Figure 1.12). Strategies for designing networks by combining hydrogen bonds with coordination chemistry or π-arene organometallic chemistry. (Adapted from ref. 27a with permission of the Royal Society of Chemistry).

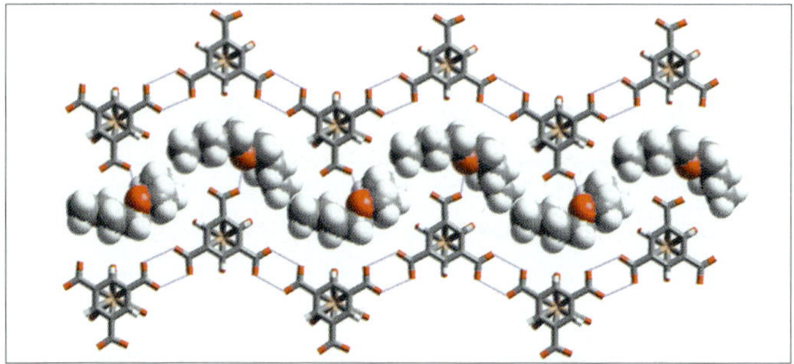

Plate 9 (Figure 1.39). Hydrogen bonded network in crystal structure of [Cr{η^6-1,3,5-$C_6H_3(CO_2H)_3$} $(CO)_3$]·$^nBu_2O$ [27a].

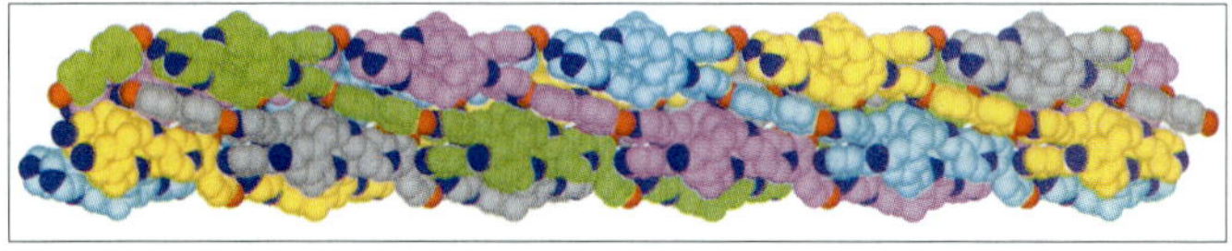

Plate 10 (Figure 2.29). Quintuple helical supramolecular assembly of different complexes **27•4**.[49]

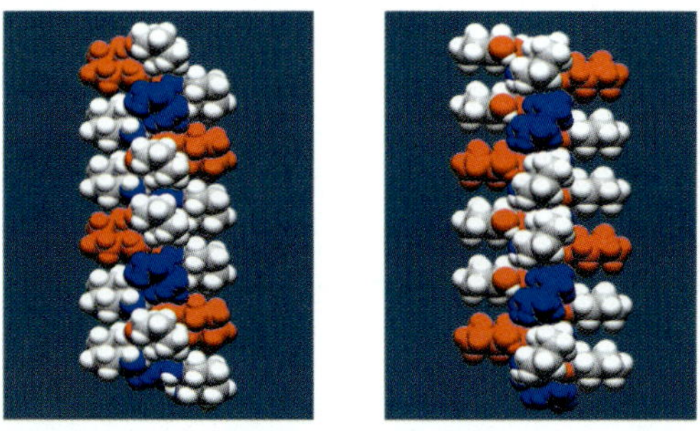

Plate 11 (Figure 2.31). CPK representation of the triple-stranded helical structures of **28•30** (left) and **29•30** (right).[50]

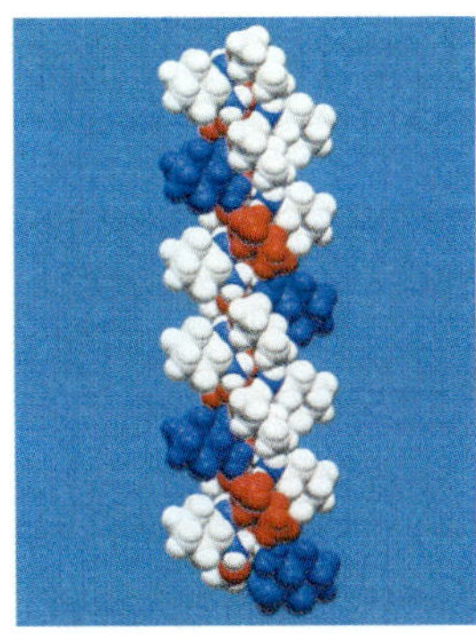

Plate 12 (Figure 2.34). CPK representation of the X-ray structure of **29•32**.[51]

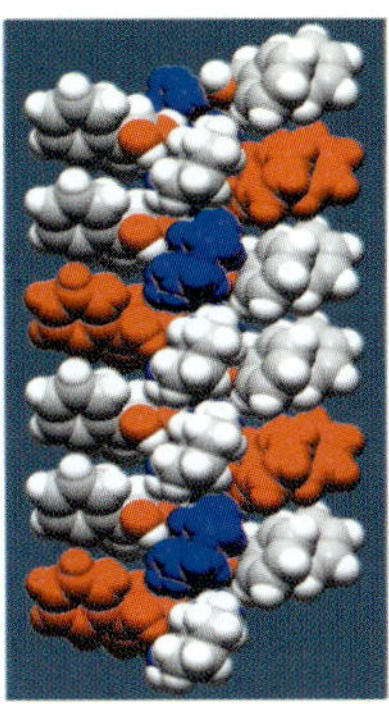

Plate 13 (Figure 2.37). CPK representation of the triple-stranded helical structures of **29•34**.[51]

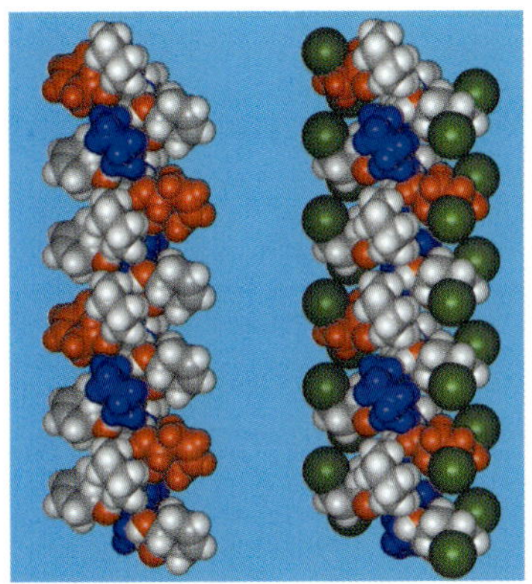

Plate 14 (Figure 2.40). CPK representations of the adducts **29•35** and **29•36**.[60]

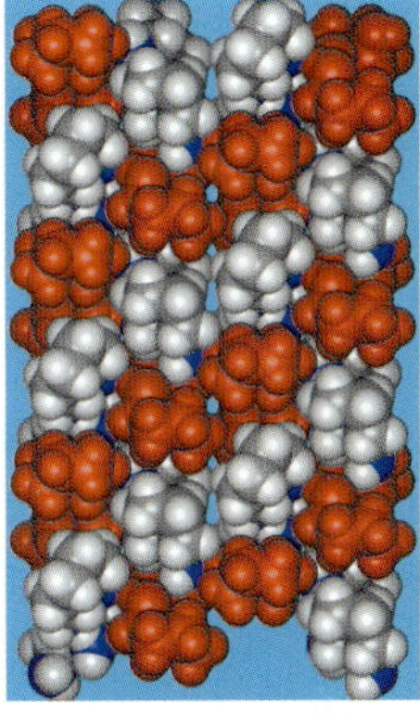

Plate 15 (Figure 2.44). CPK representation of the adduct **29•38**.[60]

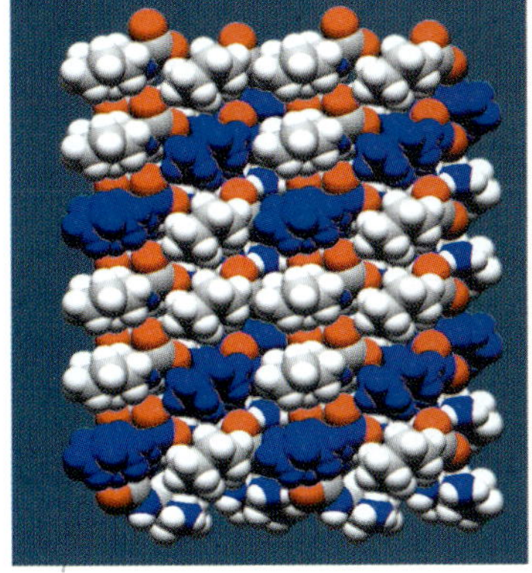

Plate 16 (Figure 2.49). CPK representation of the adduct **41** along the *a* axis.[51]

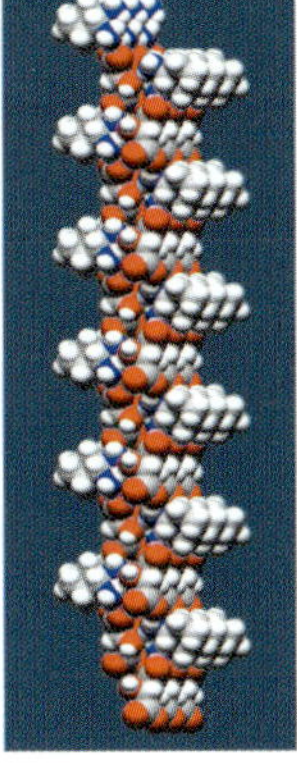

Plate 17 (Figure 2.51). CPK representation of the adduct (*R*,*R*)-**42**•(*R*,*R*)-**29**.[51]

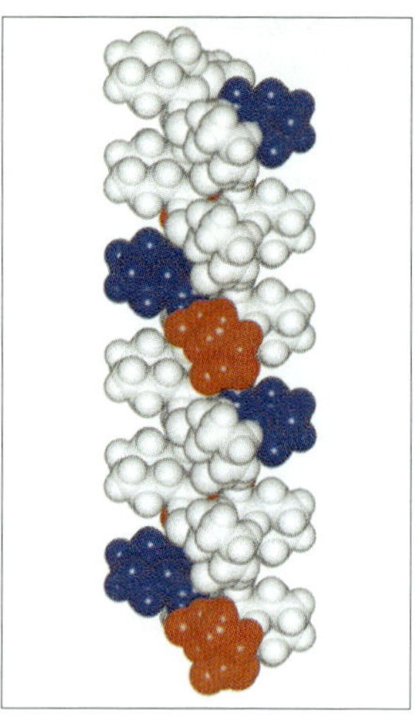

Plate 18 (Figure 2.54). CPK representation of the adduct (*R,R*)-**29•**(*R,R*)-**44**.[63]

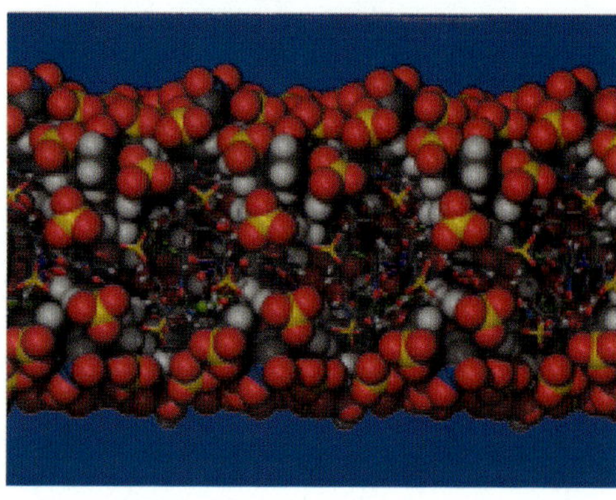

Plate 19 (Figure 3.5). Structure of the tubular assembly **6**.

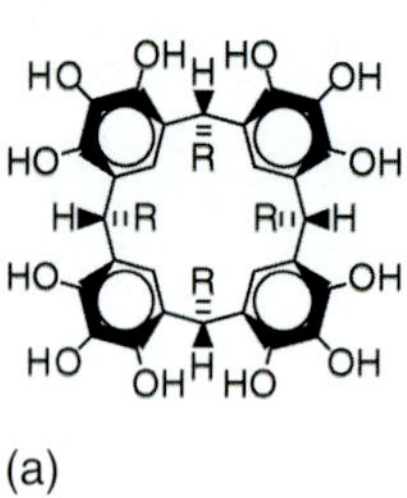

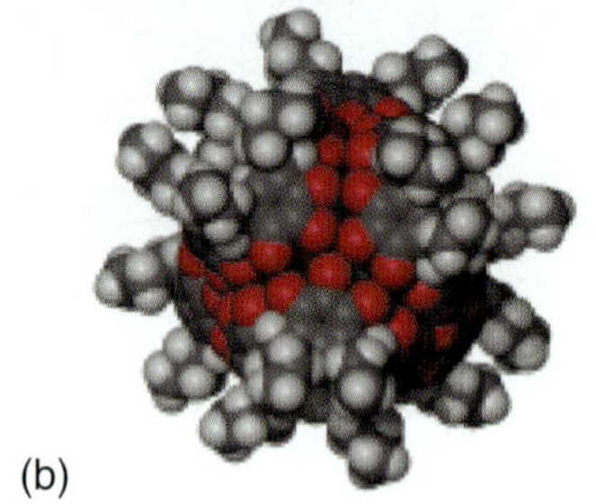

Plate 20 (Figure 3.8). (a) General forumla for pyrogallol[4]arene, **7**; (b) structure of the hexamer, **8**, *C*-propylresorcin[4]arene with the oxygen atoms shown in red.

Plate 21 (Figure 3.9). Stick-bond representation of hexameric capsule with the enclosed space represented in green.

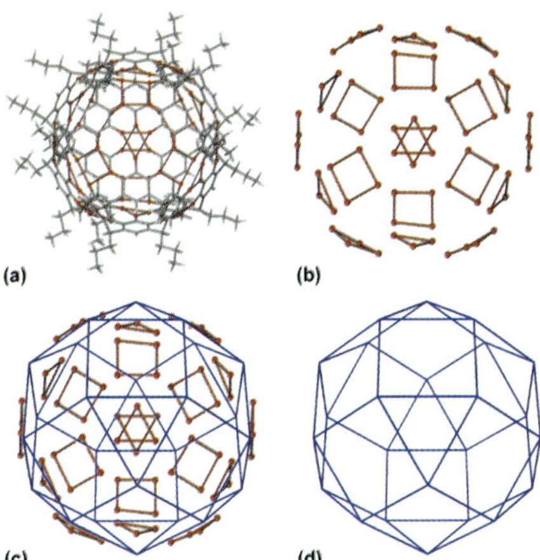

Plate 22 (Figure 3.12). **(a)** he spherical capsule consisting of six pyrogallol[4]arene molecules shown in the capped-stick metaphor, and **(b)** with the carbon and hydrogen atoms removed. Hydrogen bonds are shown as thin, solid red lines. Parts **(c)** and **(d)** show the remarkable correspondence of the hydrogen bonded pattern with the Archimediean solid, the small rhombicuboctahedron.

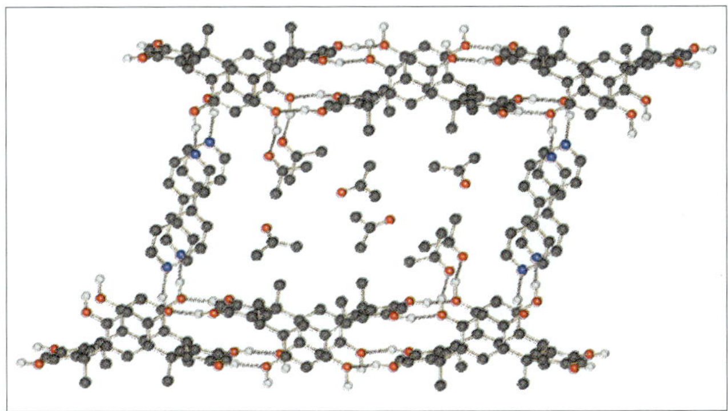

Plate 23 (Figure 3.13). Skewed molecular bricks made from *C*-ethylresorcin[4]arene and 4,4'-bipyridine.

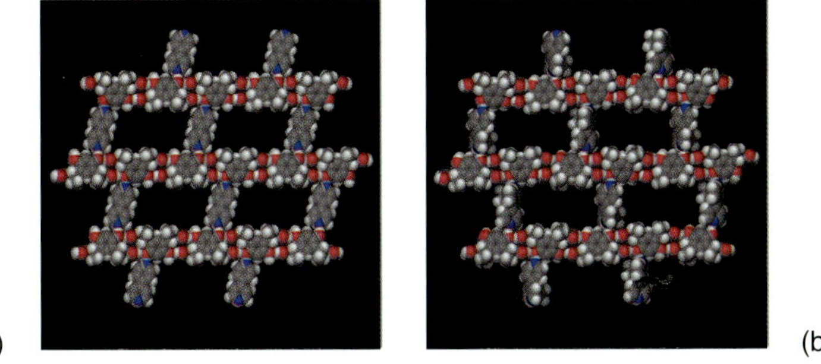

Plate 24 (Figure 3.15). Space filling view of the layer structure of **10a** and **10b**.

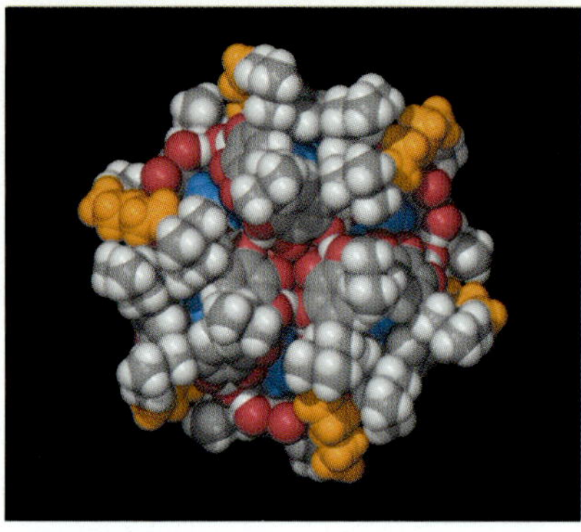

Plate 25 (Figure 3.19). Space-filling representation of hexamer **12**, or mixed macrocycle **11**, viewed along the 3 bar axis of the capsule.

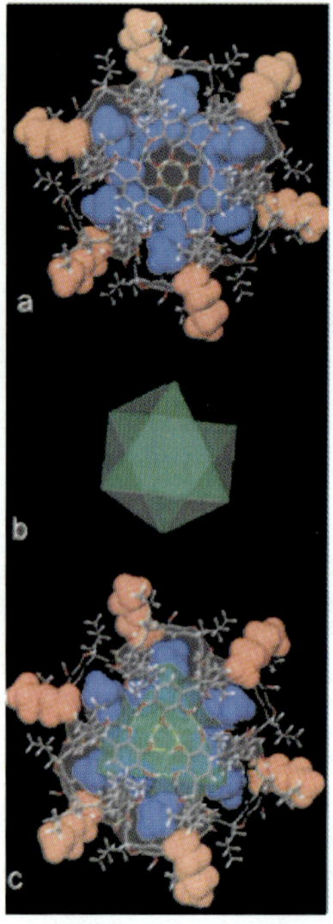

Plate 26 (Figure 3.20). **(a)** Capsule **12** shown in stick bond representation with the diethyl ether guests given in space filling representation. The orientation of the capsule is identical to that given in Figure 17a; **3.20 (b)** the trigonal antiprism that results from connection of the centroids of the centers of the aromatic rings of macrocycle **11**; **(c)** superposition of the trigonal antiprism and capsule.

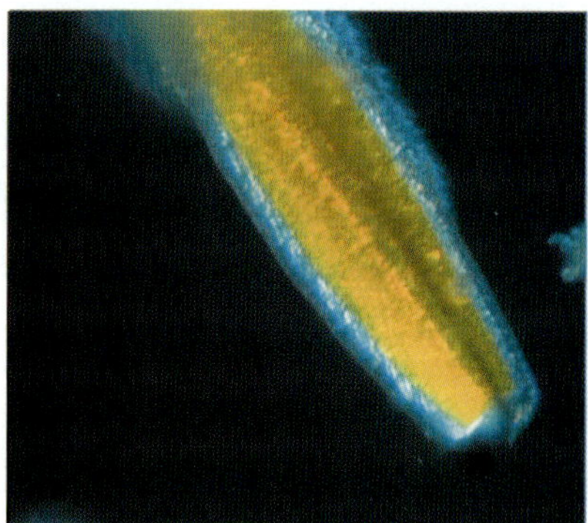

Plate 27 (Figure 9.9). Snapshot of a crystal of **27** during the release of SO_2, forming **26**. (Reproduced by permission of Nature)

while honeycomb, brick wall, bilayer and herringbone networks come under the three-connected type. Three-dimensional networks can also be generated from two-dimensional networks either by the interpenetration of the layers or by connecting the metal centers of the adjacent layers.

2 SQUARE GRIDS

All the ligands **1–9** except **6** were known to assemble into square grid networks with M(II) salts. The dimensions of the square grid cavities, depending on the length of the ligands, can vary from 7 to 20 Å. In networks the metal atom adopts octahedral geometry with four ligands at equatorial positions and two counteranions or water molecules at axial positions. In most of these grid networks, except those of **1**, the guest molecules occupy more than 40 % of the crystal volume. Especially using ligand **2** several structures containing open square grid networks have been reported since 1994. These square grid networks will be described briefly in Sections 2.1–2.5 in increasing order of the size of the grid cavity.

2.1 Square Grids of Dimensions 7 × 7 Å

The shortest ligand **1** has been shown to form a square grid network (Figure 1) upon reaction with $CoCl_2$ [6]. The grids pack in an offset fashion on both the edges such that there exist small channels of dimensions of the quarter of the square cavity and with an interlayer separation of 5.3 Å. Subsequently there have been several related reports using **1** with different metal salts such as $Cu(ClO_4)_2$, $Cu(CH_3SO_3)_2$, $Fe(NCS)_2$, $Cu(NCO)_2$ and $Co(NCS)_2$ [7–11]. However the square grid cavities in these structures are too small for the inclusion of guest molecules and mostly the counterions occupy those cavities.

2.2 Square Grids of Dimensions 11 × 11 Å

The first square grid coordination polymer using ligand **2** was reported by Robson and co-workers in the crystal structure of $\{[Zn(\mathbf{2})_2(H_2O)_2][SiF_6]\}_n$ [12]. The cavities in the layer have an approximate square shape with a Cd to Cd separation of 11.44 Å. However, given the large size of the grid cavity, the networks are doubly interpenetrated (Figure 2). Owing to the interpenetration, the crystal lattice achieved efficient packing in which the average volume occupied by nonhydrogen atoms is only 14.6 Å^3.

The noninterpenetrated or open square grid polymer with **2** was first observed in the crystal structure of $\{[Cd(\mathbf{2})_2(NO_3)_2] \cdot 2(o\text{-dibromobenzene})\}_n$ [13]. The guest molecules were included in the cavities by eschewing the interpenetration of the

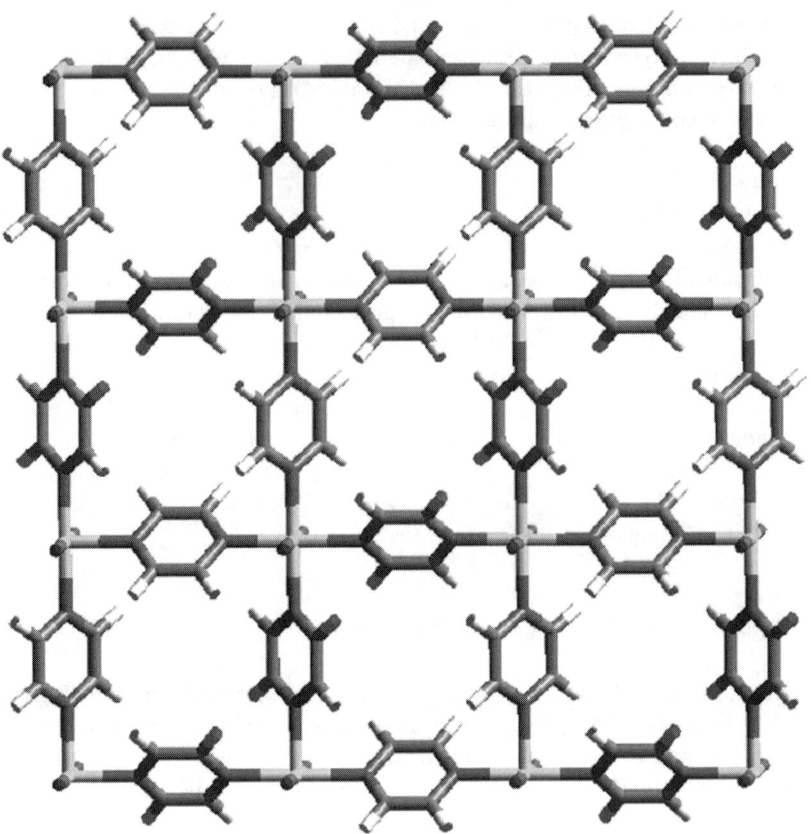

Figure 1 Square grid network formed by ligand **1** and $CoCl_2$.

networks (Figure 3). The layers in the structure consist of perfectly planar squares with a Cd to Cd separation of 11.77 Å. Each square cavity accommodates two molecules of *o*-dibromobenzene. The layers packed on each other with an interlayer separation of 6.30 Å. Further, the square cavities were found to show shape-selective inclusion: only *o*-dihaloaromatic compounds and not their *meta* and *para* analogues were included. The catalytic activity of these square cavities is described in Section 5.1. Similar grid structures were also reported using ligand **2** itself as a guest molecule [12b].

The formation of the above noninterpenetrated square grid in the EtOH–H_2O solvent system was studied using different ratios of **2** and $Cd(NO)_3$ [14]. These results indicate that at a **2**:$Cd(NO_3)_2$ ratio of ≥ 2.5, the formation of the square grid network was always favored. Notably in this structure the interlayer separation is much smaller (4.8 Å versus 6.3 Å) than that of above-described structure owing to the absence of large guest molecules. Interestingly, the hydrophobic

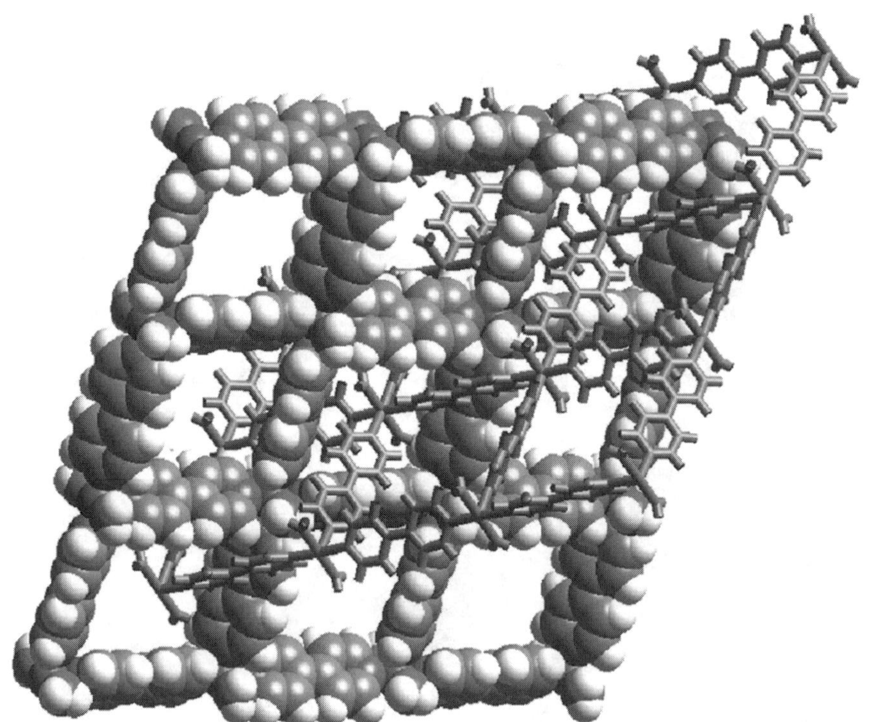

Figure 2 Interpenetrated square grid network in the crystal structure of $\{[Zn(\mathbf{2})_2(H_2O)_2][SiF_6]\}_n$.

square cavities were occupied by the hydrogen-bonded tetramers of two nitrates and two water molecules (Figure 4). At a **2**:$Cd(NO_3)_2$ ratio of ≥ 3, the square grid network was found to be in competition with the 1D chain, which is a 1:1 aggregate of **2** and $Cd(NO_3)_2$, owing to the presence of excess of **2**. At higher **2**:$Cd(NO_3)_2$ ratios it was found that the more predominant species was a bimetallic aggregate.

The square grid networks of ligand **2** and $Ni(NO_3)_2$ or $Co(NO_3)_2$ were studied by Zaworotko and co-workers [15]. In these studies, the square grid networks were prepared in the presence of several types of guest molecules. For example, hydrocarbons as diverse as benzene and pyrene and also both electron-rich (naphthalene, anisole and veratrole) and electron-poor (nitrobenzene) guests were included. From these studies it is understood that depending on the host–guest stoichiometry, the interlayer separation can vary from 6 to 8 Å and the square grid netw can pack on each other in three types of packing modes. Further, some of these grid structures reveal that the guest molecules can form networks of their own by interacting with each other via aromatic interactions. Such noncovalent networks of guest molecules were found to interpenetrate through square grid networks (Figure 5) [16]. These structures provide a unique opportunity for the

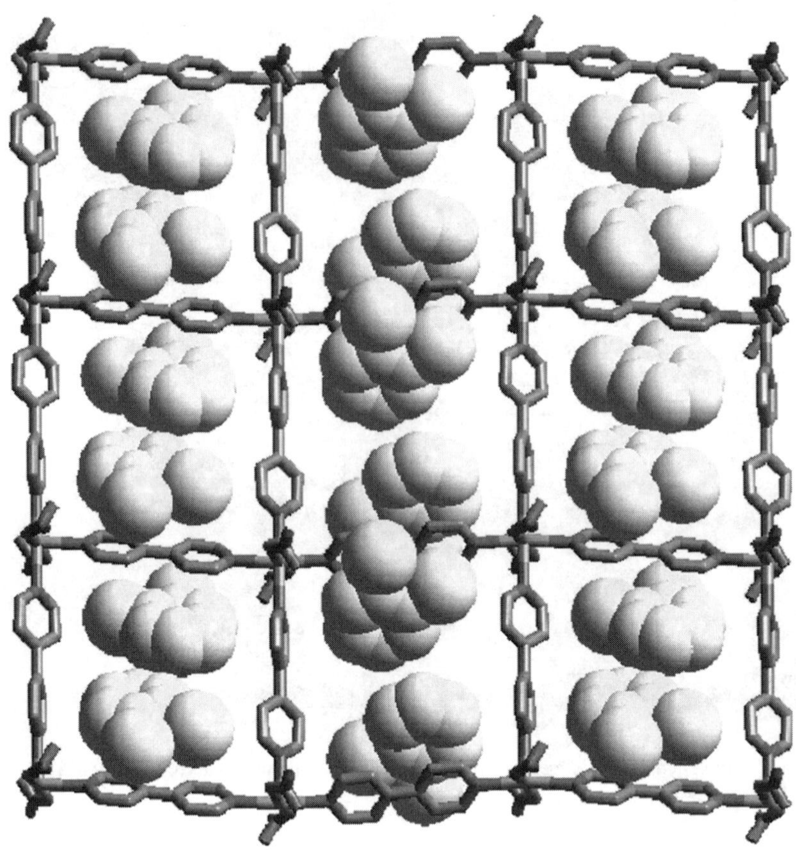

Figure 3 Inclusion of *o*-dibromobenzene (space filling mode) in square grid networks of **2** and $Cd(NO_3)_2$.

coexistence of covalent and noncovalent networks in a crystal lattice. The ligand **2** was also shown to form similar open square grid polymers with $Cd(ClO_4)_2$, $Zn(ClO_4)_2$ and $Cu(ClO_4)_2$ [17].

2.3 Square Grids of Dimensions 13 × 13 Å

A larger square grid network than that of **2** was first reported in the product formed between ligand **3** and Fe(II)–NCS^- [18]. In the network the Fe(II) ions are separated by **3** with a distance of 13.66 Å. However, the networks are doubly interpenetrated in diagonal-to-diagonal fashion, similar to that of **2** (Figure 2), with an interplanar separation of 11.3 Å between the parallel sheets. The solvent (MeOH) molecules occupy the channels formed between the interpenetrated

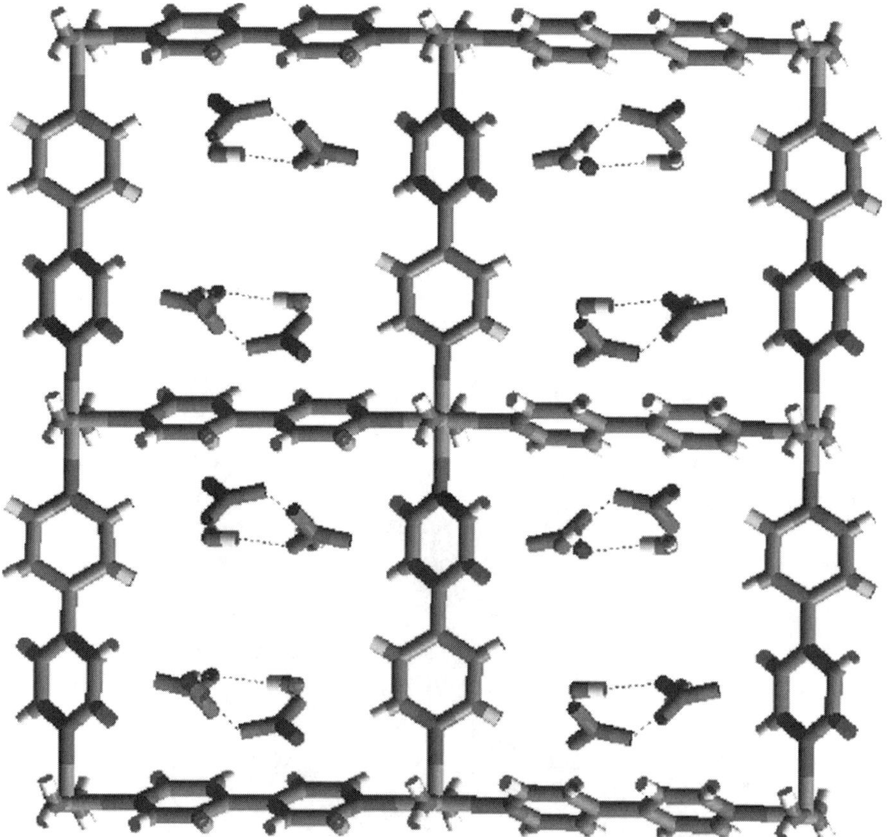

Figure 4 Inclusion of nitrates and water molecules in the square grid networks of **2** and $Cd(NO_3)_2$.

layers. This structure was found to show an interesting property of spin crossover that will be discussed in the Section 5.2. Similar types of interpenetrated networks were also observed between **3** and $Ni(NO_3)_2/Co(NO_3)_2$ [15] and also between **4** and $Co(SCN)_2$ [19]. The relatively new ligand 4,4′-azobipyridine, **5**, was also shown to act as ligand **3** to form interpenetrated and noninterpenetrated square grid architectures with $Co(NCS)_2$ or $ZnSiF_6$ and $Cd(NO_3)_2$, respectively [20].

2.4 Square Grids of Dimensions 15 × 15 Å

The ligand **7** with $Ni(NO_3)_2$ in the presence of benzene and MeOH was shown to form a coordination polymer $\{[Ni(\mathbf{7})_2(H_2O)_2]\cdot\text{benzene}\cdot 5(\text{MeOH})\cdot 2(NO_3)\}_n$ [21]. The crystal structure of this complex exhibited open square grid cavities with

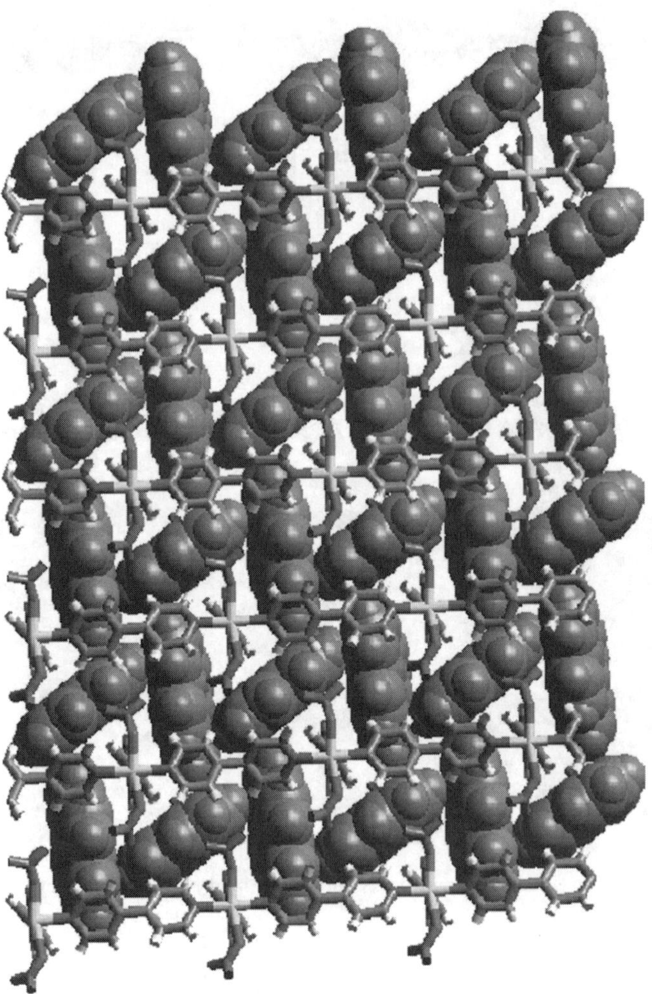

Figure 5 Interpenetration of pyrene network (space filling mode) through the square grid networks of **2** and $Ni(NO_3)_2$. The coordination and pyrene networks are represented perpendicular and parallel, respectively, to the plane of the paper.

slightly distorted square dimensions of 15.63 × 15.56 Å (Figure 6a). Interestingly, the benzene molecules are enclathrated not in the cavity but between the layers that have an interlayer separation of 4.7 Å. The square cavities were occupied by two O–H···O dimers of MeOH and NO_3 ion (O···O 2.766 and 2.883 Å). These dimers link their neighboring layers with the apical coordinated water molecules via a hydrogen-bonded motif (Figure 6b). In a sense, the whole structure can be described as a doubly interpenetrated 3D hydrogen bonded network. The hydrogen-bonded motif consists of a total of 12 atoms that are assembled from two

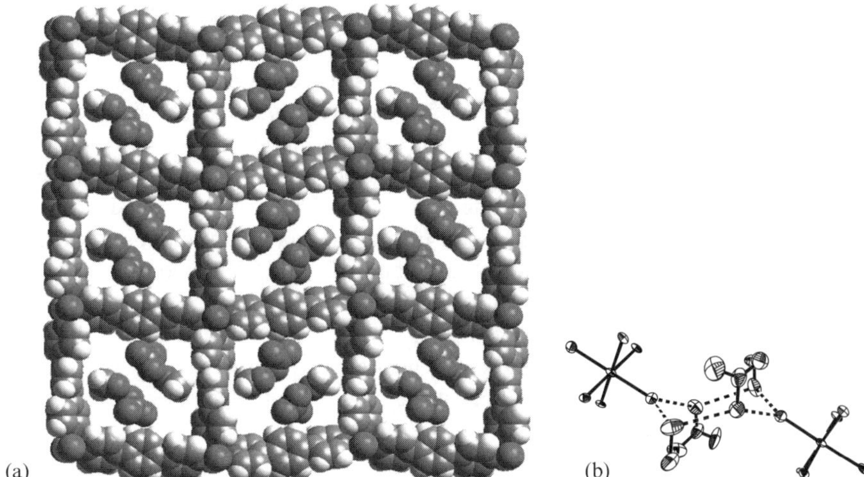

Figure 6 (a) Inclusion of nitrate and methanol molecules in square grids of ligand **7** and $Ni(NO_3)_2$; (b) hydrogen-bonded synthon that joins the square grid layers of **7** and $Ni(NO_3)_2$.

nitrates, two MeOH and two H_2O, and resembles the geometry of chair conformation of cyclohexane (Figure 6b). As a result of linking the alternate layers through hydrogen-bonded motif, the 2D layers are superimposed on each other in an offset fashion on both edges of the grid such that there are continuous channels of dimensions of one quarter of the square cavity.

The electron-rich ligand **8** was shown to form open square grid networks with nitrate salts of Ni(II), Cd(II) and Cu(II) and iodide salts of Cd(II) and Co(II) only in the presence of electron-deficient guest molecules such as nitrobenzene or cyanobenzene [22]. The general formula of these complexes is $\{[M(\mathbf{8})_2(\text{anion})_2] \cdot 2\,(\text{guest})\}_n$. In these structures the metal atoms adopt the usual octahedral geometry as the equatorial positions are occupied by N atoms of **8** while the axial positions are occupied by O atoms of NO_3 or I anions. The dimensions of the square grids vary from 15.6 to 16.0 Å depending on the M–N bond lengths. The cavities of the grid are occupied by three guest molecules: one of them fully immersed in the grids, while the other two exist at the portals of the grids, and only half of these molecules are within the grid (Figure 7). The one that is fully immersed in the grid is in fact sandwiched at the corners of the grid by two anthracene moieties. As a result, the grids are slightly distorted from a square nature.

Interestingly, the grid layers in all these structures superimpose on each other in a slightly slipped manner (the interlayer separation is ca 5.4 Å) such that there exist continuous channels of dimensions slightly less than the dimensions of the grid. It is worth noting that the anions have almost no role in packing of the layers as they face in the channels that are formed across the layers. These structures have a guest-available volume of ca 45 % of their crystal volume. A column of guest molecules occupies the channels that are formed across the packing of the grid layers (Figure 7b).

The ligand **8** with CuI in the presence of nitrobenzene was also shown to form an open square grid network, further proving the point that the robustness of host–guest interactions can lead to a predictable formation of network structures [22]. It is interesting that the M(I) serves the purpose of M(II) by forming a dimer (Cu_2I_2 unit) to act as a four-connected node [23]. This four-connected node can be regarded as a secondary building unit in the construction of coordination nets. Here each Cu_2I_2 unit coordinates to four ligands [Cu–N 2.062(5) and Cu–I 2.6408 (8) Å] such that the Cu_2I_2 plane is perpendicular to the plane of the grids layer (Figure 8). This type of arrangement forces the planes of pyridine and anthracene moieties to lie parallel and perpendicular, respectively, to the plane of grids layer. The packing of the grid layers is almost identical with that of the grid layers of the above structures of **8**.

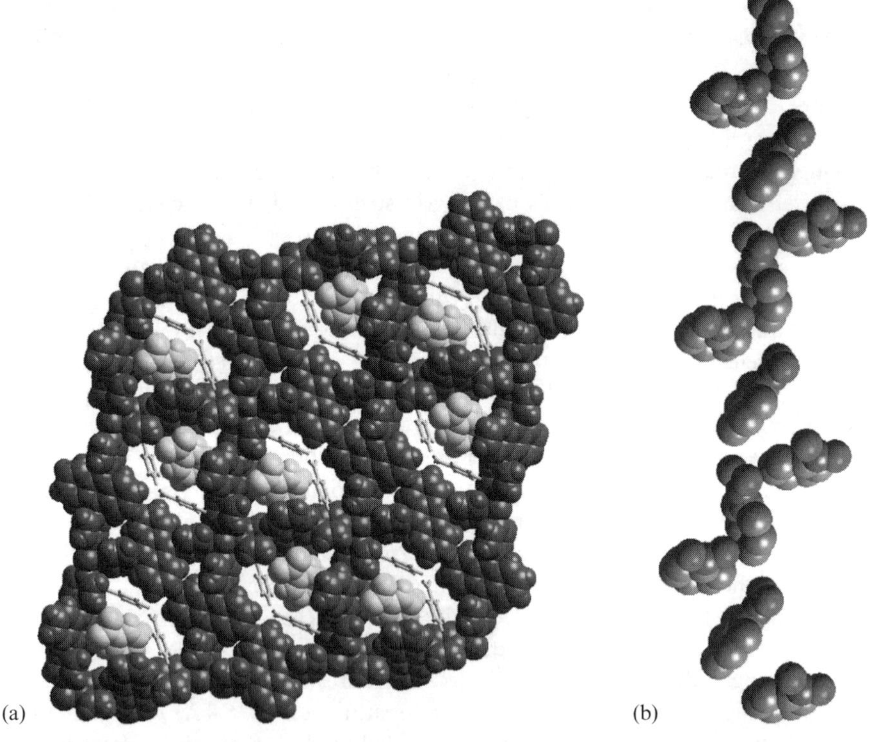

Figure 7 (a) Inclusion of nitrobenzene in the square grids of ligand **8** and $Ni(NO_3)_2$. The nitrobenzene that is represented in space filling mode was sandwiched in the corner of the grids. (b) Column of the guest molecules that form across the packing of square grids of **8** and $Ni(NO_3)_2$.

Figure 8 Square grid network exhibited in the complex of **8** and CuI.

The importance of host–guest interactions in the predictable formation of open square grid networks was further exemplified with the reaction of $Ni(NO_3)_2$ or $Cu(NO_3)_2$ and ligand **8** in the presence of benzene [21]. With $Ni(NO_3)_2$ it formed a doubly interpenetrated network as the anthracene moiety prefers to interact with itself rather than with benzene (Figure 9). With $Cu(NO_3)_2$ it failed to form a 2D network but formed two types of one-dimensional chains.

2.5 Square Grids of Dimensions 20 × 20 Å

In a similar fashion to the above ligands, the ligand **9** was also shown to form an open square grid network containing the largest square cavities reported so far [24]. A noninterpenetrated 2D network containing square grids of dimensions 20 × 20 Å was obtained when **9** was treated with $Ni(NO_3)_2$ in the presence of either benzene

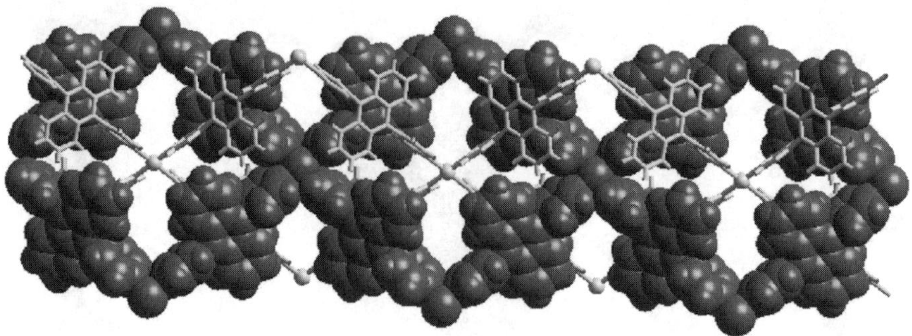

Figure 9 Doubly interpenetrated square grids of ligand **8** and $Ni(NO_3)_2$.

or *o*-xylene. In the case of benzene, five molecules of benzene and two molecules of MeOH were included per metal atom (Figure 10a), whereas in the case of *o*-xylene, four *o*-xylene molecules were included per metal atom (Figure 10b). The common feature in both the structures is the formation of large rectangular channels (ca 10×20 Å) across the packing of the grid networks. These channels were occupied by columns of guest molecules formed through aromatic interactions (Figure 10c). The 2D layers in these structures, unlike those in square grids of **2** which have a large interlayer separation (6–8 Å), have a very short interlayer separation of 4.1 Å and the ligands of adjacent grids interact with each other via edge-to-face aromatic interactions. Thermogravimetric analysis, powder X-ray and single-crystal X-ray diffraction studies of the complex reveal that the guest molecules can be removed without destroying the network or crystal packing. More surprisingly, this material exhibited an unprecedented property of sliding of the layers in crystal-to-crystal fashion when guest *o*-xylene molecules exchanged with mesitylene (see Section 5.5).

3 RECTANGULAR GRIDS

The synthesis of rectangular grids at will allows the modulation of the size and function of the cavity. Only a few rectangular grid polymers have been designed either using two charged ligands or using charged and neutral ligands [25–28]. However, rectangular grids that are formed from two neutral ligands are relatively rare as they have to be formed selectively from two different ligands and a metal atom (Scheme 2).

3.1 Rectangular Grids of Dimensions 7×11 Å

Mak and co-workers have shown how the ligands **1** and **2** react with Cu(II) to form a rectangular grid network of dimensions 6.83×11.15 Å (Figure 11) [29]. The Cu atom adopts an elongated octahedral geometry with two moieties of **1** and two

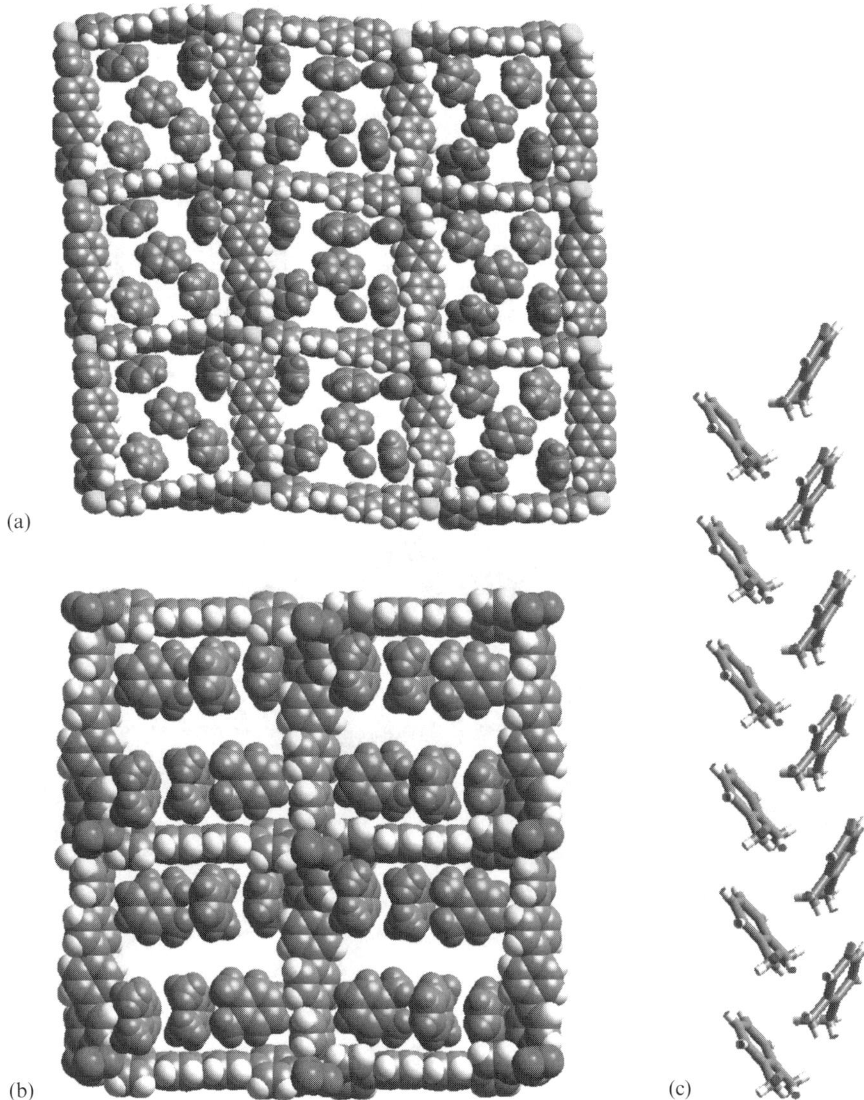

Figure 10 Inclusion of (a) benzene and MeOH and (b) *o*-xylene in the 20×20 Å square grids of **9** and $Ni(NO_3)_2$; (c) column of *o*-xylene molecules observed in the crystal structure of $\{[Ni(\mathbf{9})_2(NO_3)_2] \cdot 4(o\text{-xylene})\}_n$.

moieties of **2** at the equatorial positions and two water molecules at the axial positions. The offset superposition of each pair of adjacent layers by half of the longer edges resulted in smaller rectangular channels (ca 5.6×6.8 Å) that are occupied by PF_6^- ions.

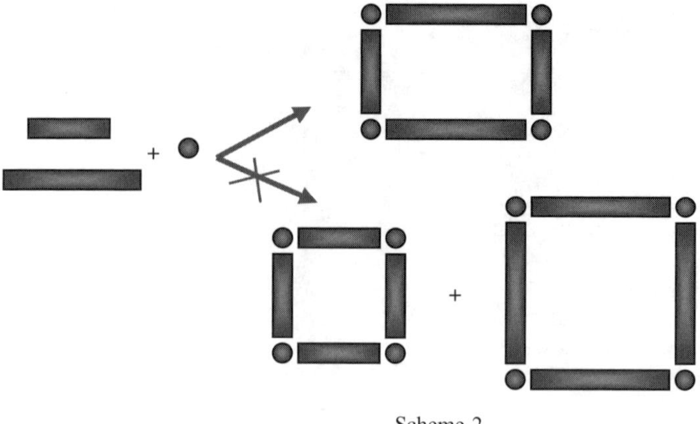

Scheme 2 Selective formation of rectangular grids from two different ligands and transition metal atoms.

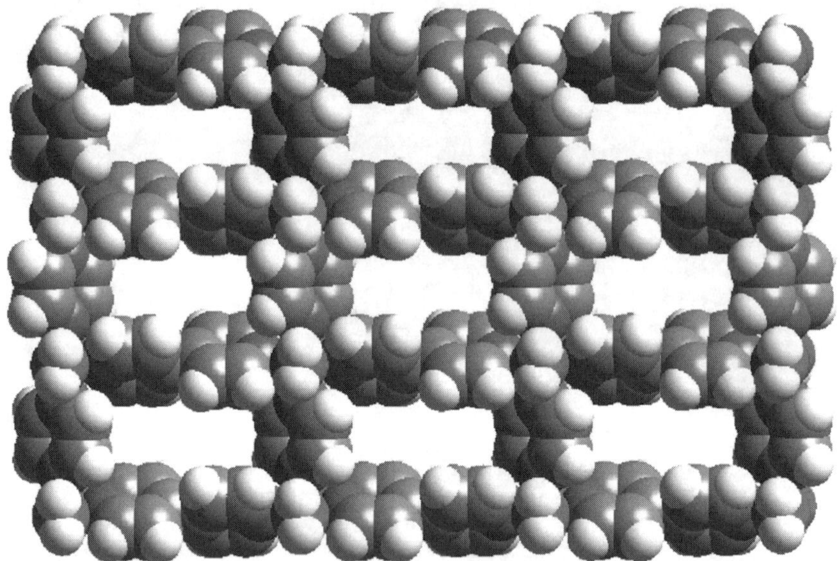

Figure 11 Rectangular grids of dimensions 7 × 11 Å formed by **1**, **2** and Cu(II).

3.2 Rectangular Grids of Dimensions 11 × 15 Å

The ligands **2** and **7** upon reaction with $Ni(NO_3)_2$ in the presence of benzene were found to form rectangular grids of dimensions 11.3 × 15.6 Å with the inclusion of six benzene molecules per metal atom, two of them are embodied in the rectangular cavities and the other four between the layers (Figure 12a) [30]. Further benzene molecules form a two-dimensional layer (Figure 12b) which is similar to that of naphthalene molecules in the crystal structure of $\{M(4,4'\text{-bipyridine})_2(NO_3)_2 \cdot 3\text{naphthalene}\}_n$ (M = Co or Ni) [16]. The network of benzene molecules can be described as one of the semi-regular planar networks, which were described by Wells [31], when benzene molecules are depicted as nodes and aromatic interactions as node connections. The packing of the grids is as follows: the moieties of **2** deposit on each other such that 2,6 and 2′ ,6′-C–H groups of **2** form C–H ⋯ O hydrogen bonds (C ⋯ O 3.126, 3.568 Å; C–H ⋯ O 151°, 148°) with the uncoordinated O-atoms of the $Ni(NO_3)_2$ moiety. As a result, it forms a C–H ⋯ O hydrogen-bonded layer that divides the whole structure into an infinite number of two-dimensional compartments with the width of the longer ligand (Figure 12c). Each of these compartments accommodates two benzene layers which interpenetrate through the moieties of **7** such that each moiety of **7** interacts with 12 benzene molecules via edge-to-face aromatic interactions.

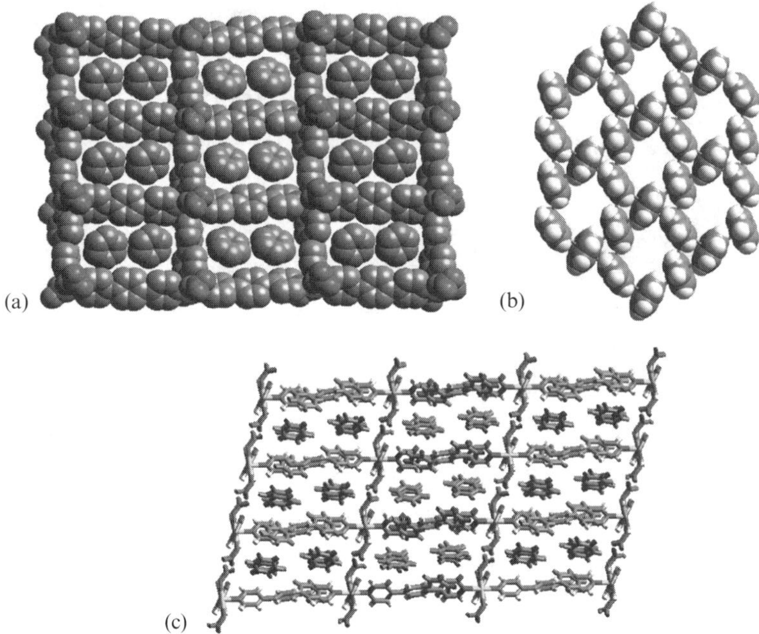

Figure 12 (a) Inclusion of benzene in the rectangular grids of dimensions 11 × 15 Å formed by **2, 7** and $Ni(NO_3)_2$; (b) network of benzene molecules observed in the same structure; (c) side view of packing of the grids and benzene layers. Note that the C–H ⋯ O bonded and benzene layers are parallel to each other but perpendicular to grid planes.

3.3 Rectangular Grids of Dimensions 11 × 20 Å

The 2D network containing rectangular grids of dimensions 11.3 × 19.9 Å was obtained when the ligands **2** and **9** were treated with $Ni(NO_3)_2$ in the presence of benzene and the complex included eight benzene molecules per metal atom [30]. Similarly to the above structure, only two benzene molecules were encapsulated in the cavities with the remaining six between the layers (Figure 13). The packing of the grids is similar to that in the above-described structure but now there are three layers of benzene molecules accommodated in each compartment as the width of the compartment increased from 15.6 to 19.9 Å (Figure 13b). In this triple layer the outer layers have the same topology as shown in Figure 12b and are linked together by a middle layer, (6,3) net, that is generated by two disordered benzene

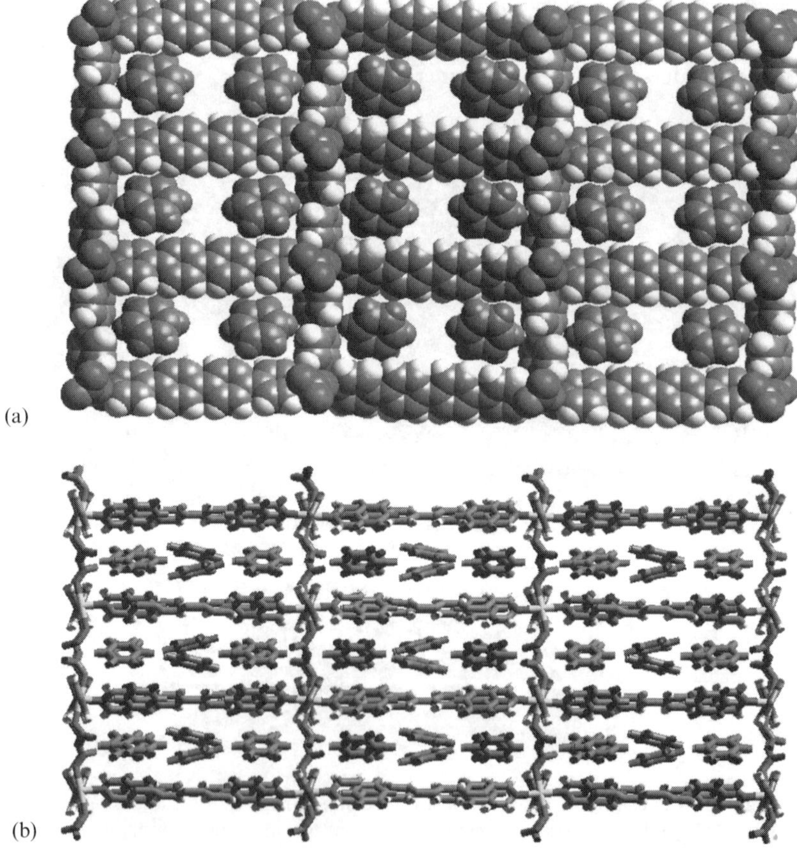

Figure 13 (a) Inclusion of benzene in the rectangular grids of dimensions 11 × 20 Å formed by **2, 9** and $Ni(NO_3)_2$; (b) side view of the packing of the grids. Note that each compartment accommodates three layers of benzene.

molecules. In effect each moiety of **9** is surrounded by 18 benzene molecules via edge-to-face aromatic interactions.

3.4 Rectangular Grids of Dimensions 15 × 20 Å

The reaction of the ligands **8** and **9** with $Ni(NO_3)_2$ in the presence of benzene was shown to form a 2D network containing a 15 × 20 Å rectangular grid (Figure 14) [30]. Each grid accommodates four benzene molecules (two are only partly in the cavity), disordered nitrate and H_2O. Unlike the above-mentioned rectangular grid structures it has an interlayer separation of only 4 Å as no guests exist between the layers. The packing of the grids is different from that in the above structures as the layers here deposit on each other in an offset fashion over the two ligands.

4 OTHER LAYERED STRUCTURES

All these networks contain three connected metal centers that can exist in trigonal or T-shaped geometries. The T-shape geometry can be adopted by M(II) metal atoms when the metal has a coordination number of seven: four occupied by the chelation of two nitrates and three occupied by the ligands. The honeycomb or hexagonal networks were shown to form when the metal atom adopts a trigonal

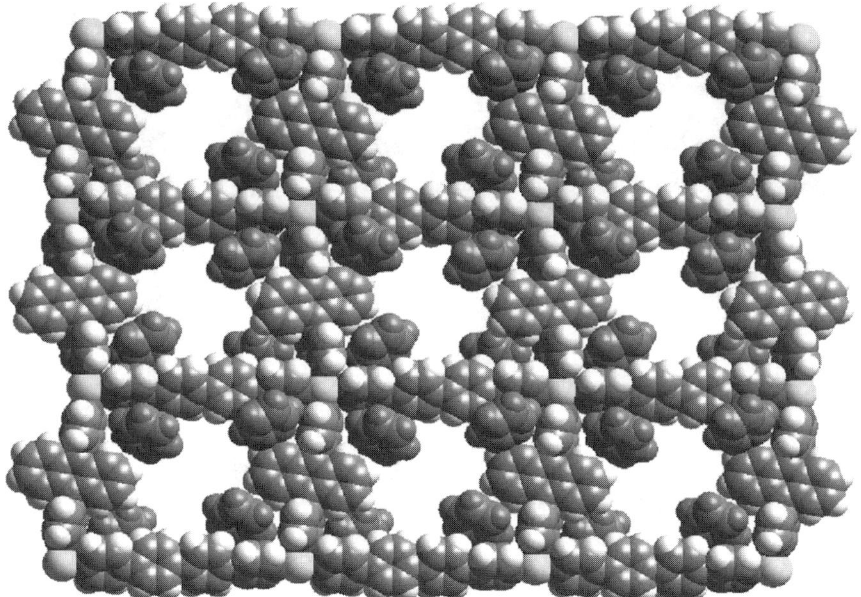

Figure 14 Inclusion of benzene molecules in the rectangular grids of **8, 9** and $Ni(NO_3)_2$.

coordination geometry whereas the brick wall, bilayer and herringbone networks were shown to form when the metal atom adopts a T-shaped coordination geometry.

4.1 Honeycomb or Hexagonal Networks

The 2,5- and 2,6-dimethyl-substituted analogues of **1** were shown to form a honeycomb network upon assembling with Cu(I) (Figure 15) [32,33]. The layers in these structures can be compared with those of graphite. Further, the ligand **1** or 2,3-dimethyl-substituted analogue of **1** were also shown to form a similar type of network with Cu(I) but in these complexes the Cu atoms form a chair-type conformation of cyclohexane instead of a planar hexagon [34,35]. Subsequently the doubly interpenetrated honeycomb network of **1** was observed when the ligand **1** was treated with $[Cu(CH_3CN)_4]SiF_6$ [36]. The SiF_6 ions are located in the channels that are formed between two enmeshed honeycomb networks.

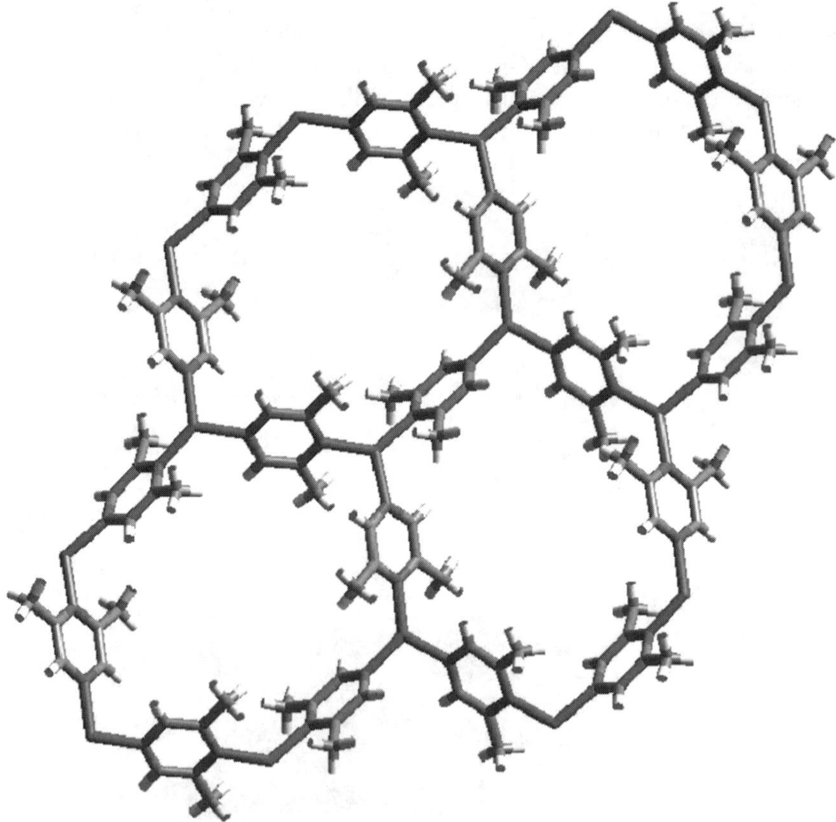

Figure 15 Honeycomb network formed by the 2,6-dimethyl analogue of **1** and Cu(I).

The ligand **2** was also shown to form a honeycomb network upon reaction with CuCl in the presence of ethylene glycol (Figure 16) [37]. In this structure the Cl atoms bridge the two CuCl units to form a dimer that acts as a four-connected node to propagate a honeycomb network in an interesting fashion. In each hexagon two of the edges are doubly bridged through the two moieties of **2**. Two of these layers interpenetrate through each other to form a three-dimensional network with small channels of dimensions 2×4 Å.

4.2 Brick Wall Network

The brick wall network is a distorted version of the above honeycomb network as the metal centers act as a T-shape node. This network was first observed to form between a more flexible ligand such as 1,4-bis[(4-pyridyl)methyl]-2,3,5,6-tetrafluorophenylene and $Cd(NO_3)_2$ (Figure 17) [38]. However, given the large nature of the

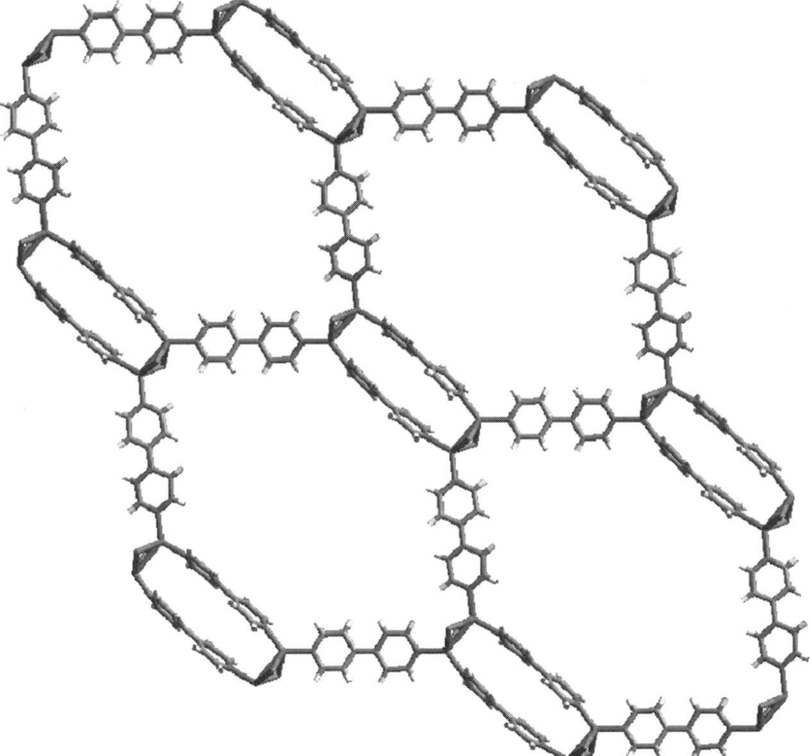

Figure 16 Honeycomb network formed by ligand **2** and CuCl.

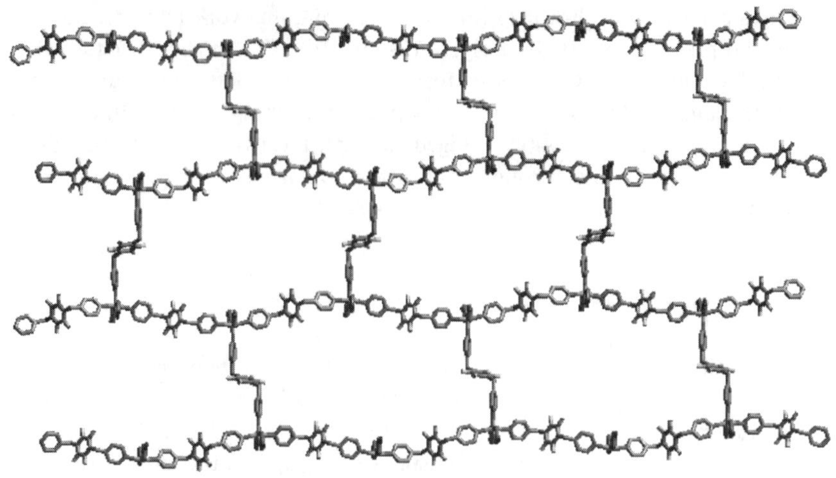

Figure 17 Brick wall network formed by 1,4-bis[(4-pyridyl)methyl]-2,3,5,6-tetraflurophenylene and $Cd(NO_3)_2$.

cavities and undulating nature of the sheets, three of these sheets interpenetrated in a parallel fashion to form a triply interpenetrated 2D layer with an approximate thickness of 14 Å. Similar brick wall-like layers that are quadruply interpnetrated were reported to be obtained by using **5** and $Co(NO_3)_2$ [20a]. Interestingly, the reaction of the same ligand **5** with $Ni(NO_3)_2$ produced a structure that has two types of interpenetrated layers (square grid and brick wall) in one crystal lattice [20b]. Different ligands such as 1,3,5-trimesic acid or 1,3-pyrazine were also shown to form brick wall-like architectures upon assembling with transition metals [39].

4.3 Herringbone Network

Recently, it was reported that the T-shape coordination geometry of metal atoms can propagate a new type of network that has a herringbone or parquet type of geometry [40]. The ligand 1,2-bis(4-pyridyl)ethyne, **6**, upon reaction with $Co(NO_3)_2$ in a solution of MeOH–MeCN afforded this herringbone-type network. Similarly, **5** with $Co(NO_3)_2$ or $Cd(NO_3)_2$ was also shown to form a herringbone network (Figure 18) but these sheets are triply interpenetrated [20a,41].

4.4 Bilayer Network

This bilayer is different from that of the bilayers of the lyotropic liquid crystals in which the hydrophobic and hydrophilic layers arranged in an alternate fashion.

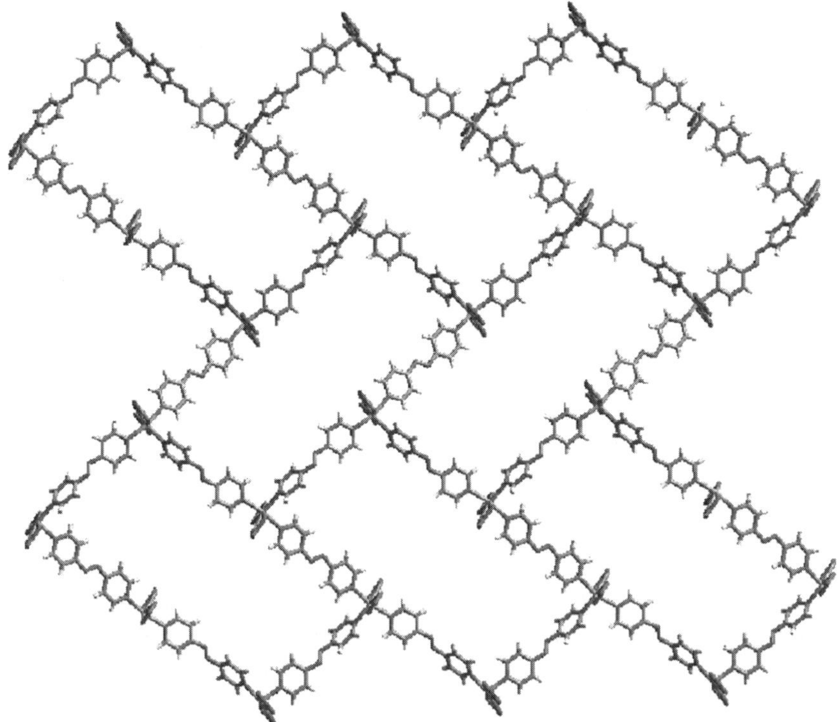

Figure 18 Herring bone network formed by ligand **5** and $Cd(NO_3)_2$.

However, in these structures the two layers have the same aromatic nature but are separated by a spacer ligand. Both the layers of a bilayer contain a set of one-dimensional chains arranged in parallel fashion. Further, the chains in one layer make an approximate angle of 90° with those of the other layer. The first bilayer structure was observed between the somewhat flexible ligand 1,2-bis(4-pyridyl) ethane, **4**, and $Co(NO_3)_2$ [42]. In this structure for two metal atoms there are three ligands: two of them exist in an *anti* conformation and propagate the chains of bilayers; the remaining ligand exists in a *syn* conformation and connects the layers (Figure 19). Later the bilayer structures were also observed to form in the reactions of **2** with $Co(NO_3)_2$ or $Ni(NO_3)_2$ in the presence of CS_2 or H_2O [43,44].

5 CRYSTALS TO PROPERTIES

Several functional properties can be expected from the materials of coordination polymers owing to their extended nature and high thermal stability and the existence of cavities and channels. Here we describe some of the properties that have been explored using above-described layered structures.

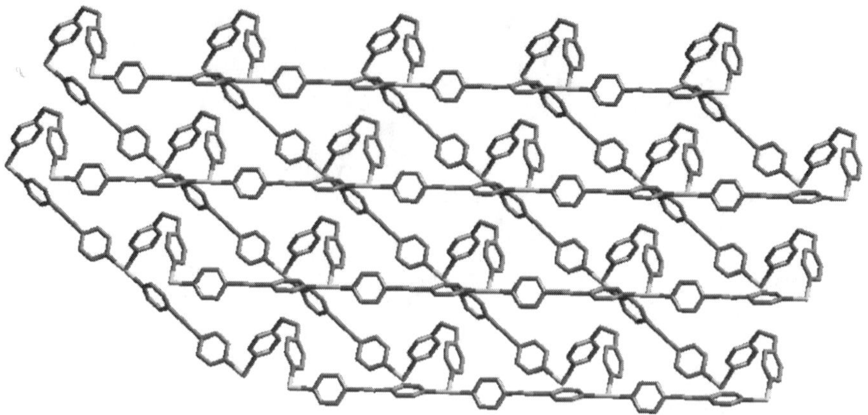

Figure 19 Bilayer network formed by ligand **4** and $Co(NO_3)_2$.

5.1 Catalysis with 11 × 11 Å Square Grid Network

The square grid network formed between $Cd(NO_3)_2$ and **2** was shown to catalyze the cyanosilylation of aldehydes [13]. The treatment of benzaldehyde with cyanotrimethylsliane in a CH_2Cl_2 suspension of powdered $\{Cd(\mathbf{2})_2(NO_3)_2\}_n$ at 40 °C for 24 h resulted in 2-(trimethylsiloxy)phenylacetonitrile in 77% yield. Further, no reaction was observed with powdered $Cd(NO_3)_2$ or **2** or the supernatant liquid of a CH_2Cl_2 suspension of $\{Cd(\mathbf{2})_2(NO_3)_2\}_n$ alone, and also the shape selectivity was observed in catalysis, 2- and 3-tolualdehyde being cyanosilylated in 40 and 19% yields, respectively. These observations indicate that the reaction was promoted heterogeneously by square grid cavities of $\{Cd(\mathbf{2})_2(NO_3)_2\}_n$.

5.2 Magnetism in the 13 × 13 Å Interpenetrated Square Grid Network

The bridging of metal centers in a controlled fashion using suitable ligands enhances the communication or cooperativity between metal centers, which is essential for the electronic properties. For example, the crystal of doubly interpenetrated network formed by **3** and Fe(II) was found to show spin crossover behavior in the temperature range 100–250 K [18]. The $\chi_m T$ product remains practically constant from room temperature to 250 K and then decreases to reach a plateau in which the residual paramagnetism is present. In Mössbauer spectra, a dominant doublet was observed at room temperature corresponding to the $S = 2$ high-spin ground state of Fe(II) and at 5 K it shows the characteristics of $S = 0$ which corresponds to the low-spin ground state of Fe(II). These kinds of systems are important in the development of electronic devices such as molecular switches.

5.3 Porosity by Linking Square Nets with Anions

Subramanian and Zaworotko found that the square grid networks formed by ligand **2** with Zn(II) can be linked with SiF_6 counterions in order to form a three-dimensional network with large channels (Figure 20) [45]. Subsequently, Kitagawa and co-workers studied the porosity of these materials by preparing a similar kind of structure but by using Cu(II) salt instead of Zn(II) [46]. In this structure the bridging of metal centers with SiF_6 resulted in channels with effective dimensions of about 8×8 Å which are filled with solvent water molecules. Powder X-ray diffraction studies have shown that the network is stable even after removal of the encapsulated water molecules. The guest removed material was analyzed for methane adsorption–readsorption in the pressure range 0–36 atm at 298 K. The adsorption–readsorption followed the same isotherm indicating that the channel structure was retained throughout this process. At 36 atm the density of adsorbed methane (0.21 g ml^{-1}) was found to be almost same as that of compressed methane (0.16 g mL^{-1}) at 300 K and 280 atm and indicative of effective filling of the channels of the 3D network. Further, the quantity of methane adsorbed is much higher than that of conventional zeolite 5A, which is known to absorb the highest quantity of methane [47].

5.4 Porosity of Bilayer Architecture

Bilayer architectures formed in $\{M_2(\mathbf{2})_3(NO_3)_4\}_n$ (where M = Co, Ni and Zn) were one of the first systems of coordination polymers to be shown as porous materials [43]. The bilayer architectures interdigitate with each other leaving small channels in the crystal lattice which were occupied by solvated water molecules. Powder X-ray studies indicate that the water molecules can be removed from the network without causing any distortion or decomposition of the network. The adsorption studies of water removed and dried sample indicated that the material is capable of adsorbing CH_4, N_2 and O_2. About 2.3 mmol of CH_4 and 0.80 mmol of N_2 or O_2 are adsorbed per gram of anhydrous material. The adsorption–readsorption followed the same isotherm, indicating the stability of the network throughout the process. Further, the isotherms for the adsorption–readsorption can be classified as type I in the IUPAC classification [48].

5.5 Dynamics in 20×20 Å Square Grid Network

The 2D network materials provide a unique opportunity to adjust the dimensions of the channels, which are formed across the layers, according to the demands of the substrate which is targeted to be incorporated [49]. Here we discuss one such

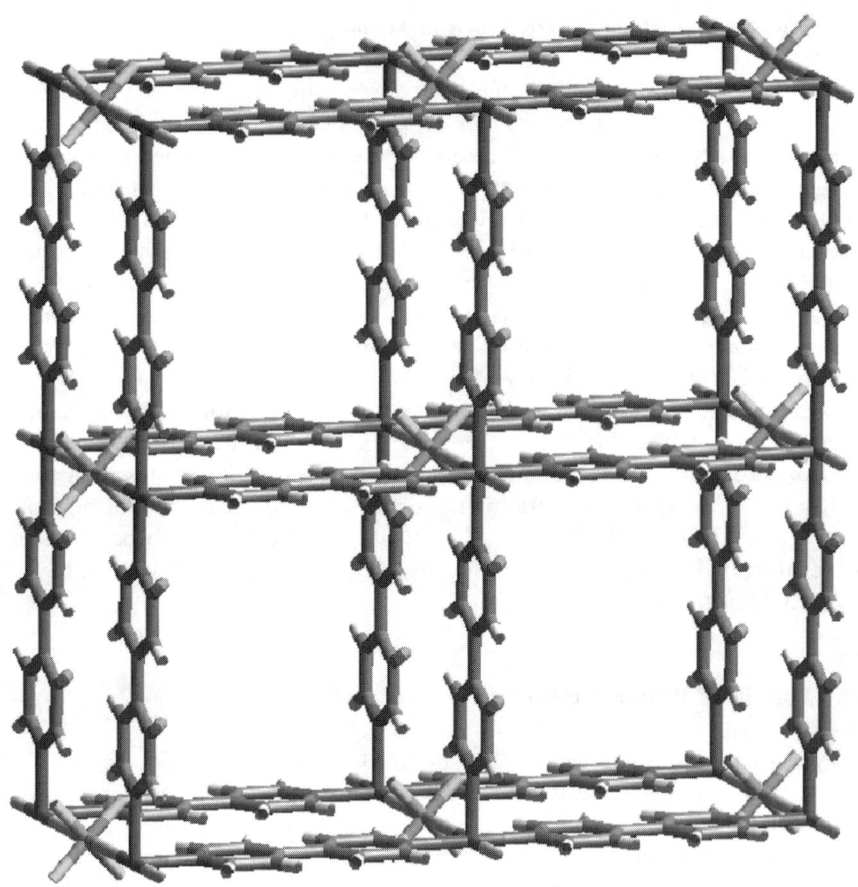

Figure 20 Linking of 11 × 11 Å square grid network with PF_6 anions.

example in crystals of a 2D network containing square grids of dimensions 20 × 20 Å, $\{Ni(\mathbf{9})_2(NO_3)_2{\cdot}4(o\text{-xylene})\}_n$. In this structure the layers have a short interlayer separation of 4.1 Å and also have a flexibility to slide on each other as the binding forces between the layers are only weak aromatic and C–H···O interactions which can be re-formed after the sliding is completed. Accordingly, the crystal-to-crystal sliding of the 2D networks between two packing modes A and B (Scheme 3) was observed when the *o*-xylene molecules in the above complex were exchanged with mesitylene molecules.

Single-crystal X-ray analysis of the crystals after exchanging with mesitylene (MEC) revealed that the considerable sliding of the layers on each other occurred such that the dimensions of the channels were larger (Figure 21) than those of the parents crystal (PC). For example, the closest distance between Ni atoms of the

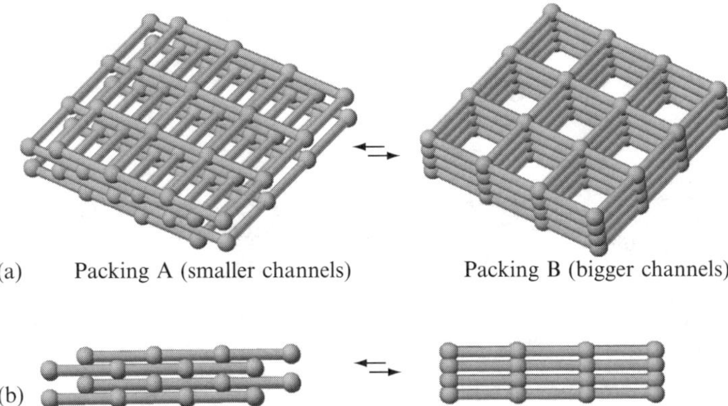

Scheme 3 Illustration of changes in the packing of the grids: (a) top view; (b) side view.

adjacent layers in PC is 11.46 Å whereas in MEC it is 8.06 Å. Assuming that one of the layers is fixed, the approximate sliding distance of its neighboring layer can be calculated as 4.2 Å based on the trigonometry of Ni atoms. In PC the layers pack on each other such that they are slipped in one direction and are offset in the other direction, which resulted in the approximate dimension of channels as 10×20 Å (packing A). In contrast, the offset packing was not observed in MEC as the layers slide on each other such that the channel dimensions are approximately 15×20 Å (packing B).

The mechanism for guest exchange and sliding of the layers has been established by monitoring the reaction using single-crystal diffraction at various time intervals. The unit cell parameters were determined at various time intervals by fixing PC in a glass capillary which was filled with mesitylene. Interestingly, the results indicate no change in cell parameters (similar to PC) up to 6 h, which is the time required for the exchange of guest molecules as indicated by GC analysis. Between 6 and 9 h the crystal stopped indexing but showed diffraction. At 10 h the crystal started indexing to give cell parameters that are similar to those of MEC. These observations clearly indicate that the transformation occurred in a crystal-to-crystal manner in which the first step is guest exchange and the second step is the sliding of the layers.

Unlike PC, which exhibits selectivity in guest exchange, MEC exhibits a greater ability to exchange mesitylene with any liquid aromatic guest molecules without destroying or altering the crystal nature. The unit cell parameters after the exchange of mesitylene with guests such as nitrobenzene and *o*-xylene were found to be the same as those of MEC. Further, the guest absorption reactions were carried out for MEC after removing the mesitylene by heating the material at 100 °C at 5 mmHg for 6 h. Interestingly, mestylene-removed crystals exhibited liquid or gaseous absorption of aromatic guest molecules.

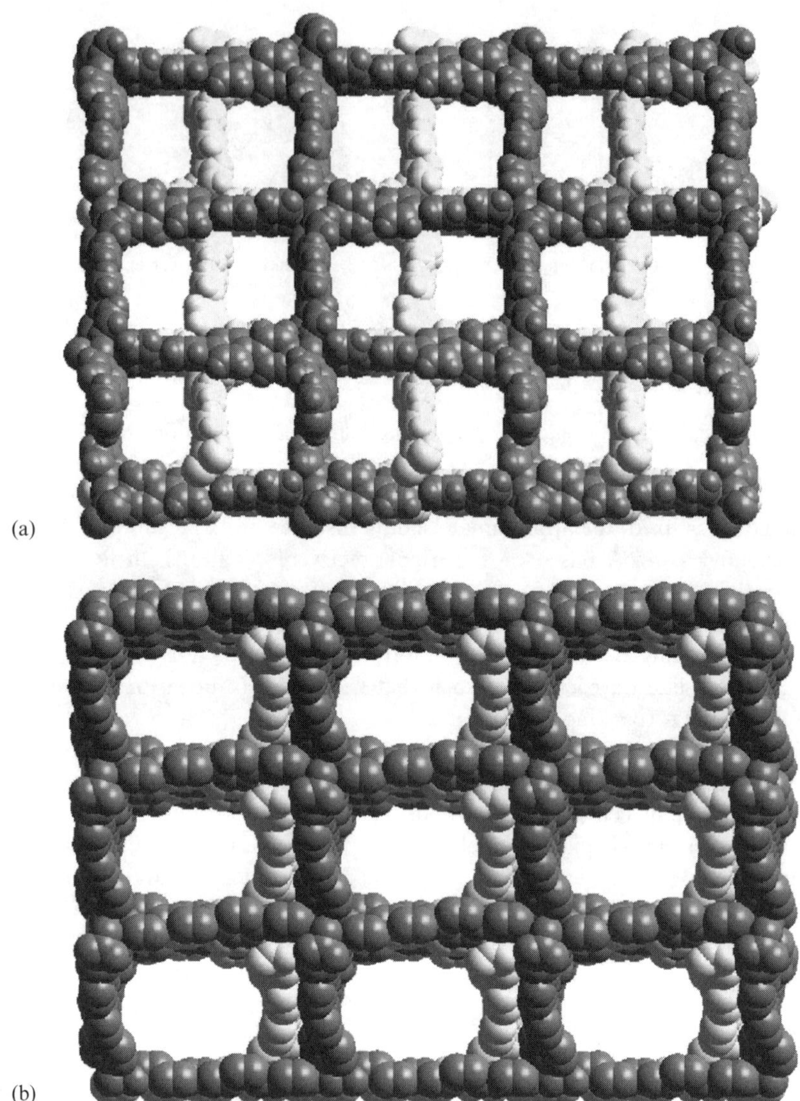

Figure 21 Top view of the space filling representation of packing of the layers in the crystal structure (a) $\{[Ni(\mathbf{9})_2(NO_3)_2] \cdot 4(o\text{-xylene})\}_n$ and (b) after exchange of *o*-xylene with mesitylene. Alternate layers are shown in the same shade.

6 SUMMARY

A family of simple rod-like exo-bidentate ligands were discussed to form several types of 2D networks upon assembling with transition metal atoms. The layered networks containing square and rectangular grid cavities exemplified the importance of a modular approach in fine tuning the shape and size of the cavities. The T-shaped nodes exhibited three types of layered structures so far: (1) brick wall; (2) bilayer; and (3) herringbone. The exhibition of a variety of networks by the same structural unit has been described as supramolecular isomerism by Zaworotko and co-workers. Concerning the porosity, the emphasis so far has been on designing coordination networks that are stable for reversible adsorption and readsorption of guest molecules without undergoing phase or morphology changes. Accordingly, in recent reports several coordination network materials were shown to exhibit such porous properties. However, the example presented in Section 5.5 comes under a new class of compound in which the solid responds to the demands of the external stimuli, exchange of guest molecules in this particular case. Such solids were termed third-generation solids by Kitagawa and Kondo [3c] and may appear frequently in the future.

REFERENCES

1. (a) G. M. J. Schmidt, *Pure Appl. Chem.*, **27**, 647 (1971); (b) G. R. Desiraju, *Crystal Engineering: the Design of Organic Solids*, Materials Science Monographs 54, Elsevier, Amsterdam (1989).
2. (a) G. R. Desiraju, *Angew. Chem., Int. Ed. Engl.*, **34**, 2311 (1995); (b) Y. Aoyama, *Topp. Curr. Chem.*, **198**, 131 (1998); (c) M. J. Zaworotko, *Chem. Commun.*, 1 (2001).
3. (a) R. Robson, *J. Chem. Soc., Dalton Trans.*, 3735 (2000); (b) D. Braga, *J. Chem. Soc., Dalton Trans.*, 3705 (2000); (b) O. M. Yaghi, H. Li, C. Davis, D. Richardson and T. L. Groy, *Acc. Chem. Res.*, **31**, 474 (1998); (c) S. Kitagawa and M. Kondo, *Bull. Chem. Soc. Jpn.*, **71**, 1739 (1998); d) D.-L. Long, A. J. Blake, N. R. Champness, C. Wilson and M. Schröder, *Angew. Chem.*, **113**, 2510 (2001); (e) S. A. Bourne, J. Lu, A. Mondal, B. Moulton and M. J. Zaworotko, *Angew. Chem.*, **113**, 2169 (2001); (f) H. Li, M. Eddaoudi, M. O'Keeffe and O. M. Yaghi, *Nature*, **402**, 276 (1999); (g) J. S. Seo, D. Whang, H. Lee, S. I. Jun, J. Oh, Y. J. Jeon and K. Kim, *Nature*, **404**, 982 (2000).
4. (a) M. Eddaoudi, J. Kim, N. Rosi, D. Vodak, J. Wachter, M. O'Keeffe and O. M. Yaghi, *Science*, **295**, 469 (2002); (b) J.-S. Seo, D. Whang, H. Lee, S. I. Jun, J. Oh, Y. J. Jeon and K. Kim, *Nature*, **404**, 982 (2000); (c) R. Kitaura, K. Fujimoto, S.-I. Noro, M. Kondo and S. Kitagawa, *Angew. Chem., Int. Ed. Engl.*, **41**, 133 (2002); (d) K. S. Min, M. P. Suh, *J. Am. Chem. Soc.*, **122**, 6834 (2000).
5. (a) S. R. Batten and R. Robson, *Angew. Chem., Int. Ed. Engl.*, **37**, 1460 (1998); (b) S. R. Batten, *Cryst. Eng. Commun.*, **18**, 1 (2001).
6. (a) P. W. Carreck, M. Goldstein, E. M. McPartlin and W. D. Unsworth, *J. Chem. Soc. D*, 1634 (1971); (b) C. Gairing, A. Lentz, E. Grosse, M. Haseidl and L. Walz, *Z. Kristallogr.*, **211**, 804 (1996).
7. J. Darriet, M. S. Haddad, E. N. Duesler and D. N. Hendrickson, *Inorg. Chem.*, **18**, 2679 (1979).
8. J. S. Haynes, S. J. Rettig, J. R. Sams, R. C. Thompson and J. Trotter, *Can. J. Chem.*, **65**, 420 (1987).
9. J. A. Real, G. De Munno, M. C. Munoz and M. Julve, *Inorg. Chem.*, **30**, 2701 (1991).
10. T. Oteino, S. J. Rettig, R. C. Thompson and J. Trotter, *Inorg. Chem.*, **32**, 1607 (1993).

11. J. Lu, T. Paliwala, S. C. Lim, C. Yu, T. Niu and A. J. Jacobson, *Inorg. Chem.*, **36**, 923 (1997).
12. (a) R. W. Gable, B. F. Hoskins, and R. Robson, *J. Chem. Soc., Chem. Commun.*, 1677 (1990); (b) R. Robson, B. F. Abrahams, S. R. Batten, R. W. Gable, B. F. Hoskins and J. Liu, in *Supramolecular Architecture*, ed. T. Bein, ACS Symposium Series, Vol. 499, American Chemical Society, Washington, DC, chapt. 19 (1992).
13. M. Fujita, Y. J. Kwon, S. Washizu and K. Ogura, *J. Am. Chem. Soc.*, **116**, 1151 (1994).
14. M. Aoyagi, K. Biradha and M. Fujita, *Bull. Chem. Soc. Jpn.*, 1369 (2000).
15. K. Biradha, D. Dennis, V. A. MacKinnon, C. Seward and M. J. Zaworotko, in *Current Challenges on Large Supramolecular Assemblies*, ed. G. Tsoucaris, Kluwer, Dordrecht, pp. 115–132 (1999).
16. a) K. Biradha, K. V. Domasevitch, B. Moulton, C. Seward and M. J. Zaworotko, *Chem. Commun.*, 1327 (1999); (b) K. Biradha, K. V. Domasasevitch, C. Hogg, B. Moulton, K. N. Power and M. J. Zaworotko, *Cryst. Eng.*, **2**, 37 (1999).
17. M.-L. Tong, B.-H. Ye, J.-W. Cai, X.-M. Chen, and S. W. Ng, *Inorg. Chem.*, **37**, 2645 (1998).
18. J. A. Real, E. Andrés, M. C. Muñoz, M. Julve, T. Granier, A. Bousseksou and F. Varret, *Science*, **268**, 265 (1995).
19. S. H. Park, K. M. Kim, S.-G. Lee and O.-S. Jung, *Bull. Korean Chem. Soc.*, **19**, 79 (1998).
20. (a) M. Kondo, M. Shimamura, S.-I. Noro, S. Minakoshi, A. Asami, K. Seki and S. Kitagawa, *Chem. Mater.*, **12**, 1288 (2000); (b) L. Carlucci, G. Ciani and D. M. Proserpio, *New J. Chem.*, 1319 (1998).
21. K. Biradha and M. Fujita, *J. Chem. Soc., Dalton. Trans.*, 3805 (2000).
22. K. Biradha and M. Fujita, *J. Inclus. Phenom.*, **41**, 201–208 (2001).
23. K. Biradha, M. Aoyagi and M. Fujita, *J. Am. Chem. Soc.*, **122**, 2397 (2000).
24. K. Biradha, Y. Hongo and M. Fujita, *Angew. Chem., Int. Ed. Engl.*, **39**, 3843 (2000)
25. S. Kawata, S. Kitagawa, M. Konda, I. Furuchi and M. Munakata, *Angew. Chem., Int. Ed. Engl.*, **33**, 1665 (1994).
26. L. R. MacGillivray, R. H. Groeneman and J. L. Atwood, *J. Am. Chem. Soc.*, **120**, 2677 (1998).
27. R. H. Groeneman, L. R. MacGillivray and J. L. Atwood, *Chem. Commun.*, 2735 (1998).
28. L.-M. Zheng, X. Fang, K.-H. Lii, H.-H. Song, X.-Q. Xin, H.-K. Fun, K. Chinnakali and I. A. Razak, *J. Chem. Soc., Dalton Trans.*, 2311 (1999).
29. M.-L. Tong, X.-M. Chen, X.-L. Yu and T. C. W. Mak, *J. Chem. Soc., Dalton Trans.*, 5 (1998).
30. K. Biradha and M. Fujita, *Chem. Commun.* 15 (2001).
31. A. F. Wells, *Structural Inorganic Chemistry*, 5th edn, Clarendon Press, Oxford p. 82 [also see p. 613 (M_xO_3), p. 1006 (-quartz), p. 1302 ($CaCu_5$)].
32. T. Otieno, S. J. Rettig, R. C. Thompson, and J. Trotter, *Inorg. Chem.*, **32**, 1607 (1993).
33. S. Kitagawa, S. Kawata, M. Kondo, Y. Nozaka and M. Munakata, *Bull. Chem. Soc. Jpn.*, **66**, 3387, (1993).
34. S. Kitagawa, M. Munakata and T. Tanimura, *Inorg. Chem.*, **31**, 1714 (1992).
35. M. M. Turnbull, G. Pon and R. D. Willett, *Polyhedron*, **10**, 1835 (1991).
36. L. R. MacGillivray, S. Subramanian and M. J. Zaworotko, *J. Chem. Soc., Chem. Commun.*, 1325 (1994).
37. O. M. Yaghi and G. Li, *Angew. Chem., Int. Ed. Engl.*, **34**, 207 (1995).
38. M. Fujita, Y. J. Kwon, O. Sasaki, K. Yamaguchi and K. Ogura, *J. Am. Chem. Soc.*, **117**, 7287 (1995).

39. (a) H. J. Choi and M. P. Suh, *J. Am. Chem. Soc.*, **120**, 10622 (1998); (b) C. V. K. Sharma and R. D. Rogers, *Cryst. Eng.*, **1**, 19 (1998).
40. Y.-B. Dong, R. C. Layland, N. G. Pschirer, M. D. Smith, U. H. F. Bunz and, H.-C. zur Loye, *Chem. Mater.*, **11**, 1413 (1999).
41. M. A. Withersby, A. J. Blake, N. R. Champness, P. A. Cooke, P. Hubberstey and M. Schröder, *New J. Chem.*, **23**, 573 (1999).
42. (a) T. L. Hennigar, D. C. MacQuarrie, P. Losier, R. D. Rogers and M. J. Zaworotko, *Angew. Chem., Int. Ed. Engl.*, **36**, 972 (1997); (b) R. Atencio, K. Biradha, T. L. Hennigar, K. M. Poirier, K. N. Power, C. M. Seward, N. S. White and M. J. Zaworotko, *Cryst. Eng.*, **1**, 203 (1998).
43. M. Kondo, T. Yoshitomi, K. Seki, H. Matsuzaka and S. Kitagawa, *Angew. Chem., Int. Ed. Engl.*, **36**, 1725 (1997).
44. K. N. Power, T. L. Hennigar and M. J. Zaworotko, *New J. Chem.* **22**, 177 (1998)
45. S. Subramanian and M. J. Zaworotko, *Angew. Chem., Int. Ed. Engl.*, **34**, 2127 (1995).
46. S.-I. Noro, S. Kitagawa, M. Kondo and K. Seki, *Angew. Chem., Int. Ed. Engl.*, **39**, 2082 (2000).
47. (a) S.-Y. Zhang, O. Talu, D. T. Hayhurst, *J. Phys. Chem.* **95**, 1722 (1991); (b) J. L. Zuech, A. L. Hines, E. D. Solan, *Ind. Eng. Chem. Process Des. Dev.*, **22**, 172 (1983); (c) L. Mentasty, A. M. Woestyn and G. Zgrablich, *Adsorpt. Sci. Technol.*, **11**, 123 (1994).
48. IUPAC, *Pure Appl. Chem.*, **57**, 603 (1985).
49. K. Biradha, Y. Hongo and M. Fujita, *Angew. Chem., Int. Ed.,* **41**, 3395 (2002).

Chapter 6

The Construction of One-, Two- and Three-Dimensional Organic–Inorganic Hybrid Materials from Molecular Building Blocks

Robert C. Finn, Eric Burkholder, and Jon Zubieta
Syracuse University, Syracuse, NY, USA

1 INTRODUCTION

The contemporary interest in inorganic oxide materials reflects their impressive diversity of compositions [1], structures [2], physical properties [3] and applications [4–27]. However, whereas naturally occurring oxides, such as silicates, ores and gems, may possess complex crystal structures, the majority are of simple composition and have highly symmetrical structures with rather small unit cells. Although such 'simple' oxides can have unique and specific properties such as piezoelectricity, ferromagnetism or catalytic activity, as a general rule there is a correlation between the complexity of the structure of a material and its functionality [28,29]. In the representative case of biomineralization [30], the inorganic oxide contributes to the increased functionality via assimilation as one component in a hierarchical assembly in which there exists a synergistic interaction between an organic component and the inorganic oxide. Consequently, the exploitation of organic compounds in the construction of solid-state inorganic oxides has received considerable contemporary attention [31–33].

Crystal Design: Structure and Function, edited by G. R. Desiraju

The interface between organic and inorganic substructures has been manipulated in the preparation of hybrid materials which exhibit composite or even new properties, representative examples of which include zeolites [34–36], mesoporous oxides of the MCM-41 class [37] and metal phosphates with entrained organic cations [38–55] (Refs 39–55 are representative examples of the expanding family of oxovanadium–phosphate–organic cation materials).

While the organic components in these examples serve as charge-compensating cations and space-filling structural subunits, these species may also be introduced as ligands, tethered directly to the metal oxide substructure or to a secondary metal site. Thus, the structural influences of organonitrogen ligands in vanadium and molybdenum oxides are apparent in materials such as $[MoO_3(4, 4'\text{-bipyridine})_{0.5}]$, $[MoO_3(\text{triazole})_{0.5}]$ and $[V_9O_{22}(\text{terpyridine})_3]$ [56,57]. Principles of fundamental coordination chemistry may also be applied to the modification of oxide microstructure in hybrid materials. In this case, the organic component acts as a ligand to a secondary metal site, which is in turn directly coordinated through bridging oxo groups to the oxide substructure. Consequently, the overall structure reflects both the geometric constraints of the ligand, as manifested in the size, shape, relative dispositions of the donor groups and denticity, as well as the coordination preferences of the secondary metal site, as reflected in the coordination number and geometry, degree of aggregation into oligomeric units and mode of attachment to the primary metal oxide scaffolding. This general strategy has been exploited by us in the development of the structural chemistries of two families of materials: the vanadium oxides of the type $V/O/M'$ ligand [58–64] and the molybdenum oxides of the $Mo/O/M'$ ligand family [65–82].

The molybdenum oxide hybrid materials in many cases are constructed from molecular cluster building blocks, derived from the vast family of polyoxomolybdate anions [83]. Consequently, these materials are related to general efforts to construct specific architectures from molecular building blocks, exemplified by coordination polymers [84] and cluster-based materials [85] from the expansive literature of 'crystal engineering' [86,87]. The use of well-defined molecular oxide clusters for the construction of solids with more or less predictable connectivity in the crystalline state is attractive, since secondary metal–ligand bridges should provide linkages sufficiently strong to connect the clusters into kinetically stable, crystalline architectures. In this respect, the polyoxomolybdate-organophosphonates [88] of the type $[Mo_5O_{15}(O_3PR)_2]^{4-}$, shown in Figure 1, should provide building blocks for the construction of novel molybdenum oxide materials [89].

2 HYDROTHERMAL SYNTHESIS

Conventional solid-state syntheses are carried out at high temperatures which preclude the retention of the structural characteristics of organic starting materials and which produce thermodynamic products. However, numerous reactions relevant

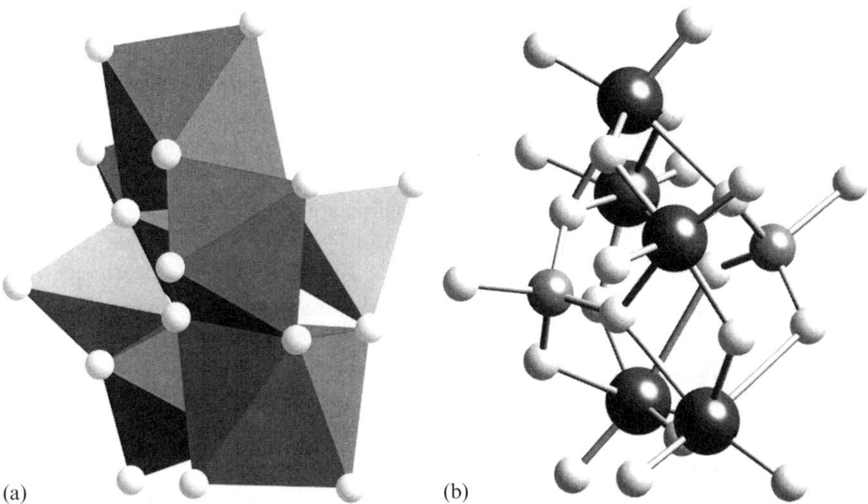

Figure 1 (a) Polyhedral and (b) ball and stick representations of the structure of $[Mo_5O_{15}(O_3PR)_2]^{4-}$. Molybdate polyhedra are dark; phosphorus are lighter and oxygen atoms are lightly shaded spheres. This convention is used throughout.

to solid-state synthesis such as cation exchange, dehydration, intercalates, redox, sol–gel and hydrothermal, occur at ambient or relatively mild temperatures [90]. These routes to the preparation of new materials are collectively referred to as 'chimie douce', a concept first introduced by Livage [91a] and Rouxel [91b]. Although well established for the preparation of aluminosilicates, hydrothermal techniques have only recently been adopted for the preparation of a wide variety for metastable materials, including transition metal phosphates, metal organophosphonates and complex polyoxoalkoxometalates [92].

Hydrothermal reactions are typically carried out in the temperature range 120–260 °C under autogenous pressure, so as to exploit the self-assembly of the product from soluble precursors. The reduced viscosity of water under these conditions enhances diffusion processes so that solvent extraction of solids and crystal growth from solution are favored. Since differential solubility problems are minimized, a variety of simple precursors may be introduced, as well as a number of organic and/or inorganic structure-directing agents from which those of appropriate size and shape may be selected for efficient crystal packing during the crystallization process. Under such nonequilibrium crystallization conditions, metastable kinetic phases rather than the thermodynamic phases are most likely isolated [93–96]. While several pathways, including that resulting in the most stable phase, are available in such nonequilibrium mixtures, the kinetically favored structural evolution results from the smallest perturbations of atomic positions. Consequently, nucleation for a metastable phase may be favored.

3 STRUCTURAL COMPONENTS

The syntheses proceed from the self-assembly of three component building blocks: a di- or multi-topic organoamine ligand, a first-row transition metal cation and an oxomolybdenum phosphate or organophosphonate cluster. While organoamine constituents have been conventionally introduced as charge-balancing counterions in zeolite synthesis, in our application the organic component serves as a ligand to the secondary metal site, the first-row transition or post-transition metal cation. Consequently, a coordination complex cation is assembled which serves to provide charge-compensation, space-filling and structure-directing roles. The structure of the organoamine–secondary metal complex cation is derived, of course, from the geometrical requirements of the ligand as well as the coordination preferences of the metal. As illustrated in Figure 2, the ligand set may include chelating agents which coordinate to a single metal center or bridging ligands of various extensions which generate binuclear secondary metal/organopolyimine coordination complex cations.

The properties of this cationic component may be tuned by exploiting the preferred coordination modes of various transition and post-transition metal cations. For example, while a Cu(II)–organoamine fragment will likely exhibit $4+2$ or $4+1$ coordination geometries, the Cu(I) counterpart will result in low coordination numbers with tetrahedral or trigonal planar geometries preferred. Similarly, an Ni(II)-based cation will adopt more regular octahedral coordination, while Zn(II) species may adopt various coordination modes. In this case, the discussion will be limited to Cu(II) sites as the secondary metal components.

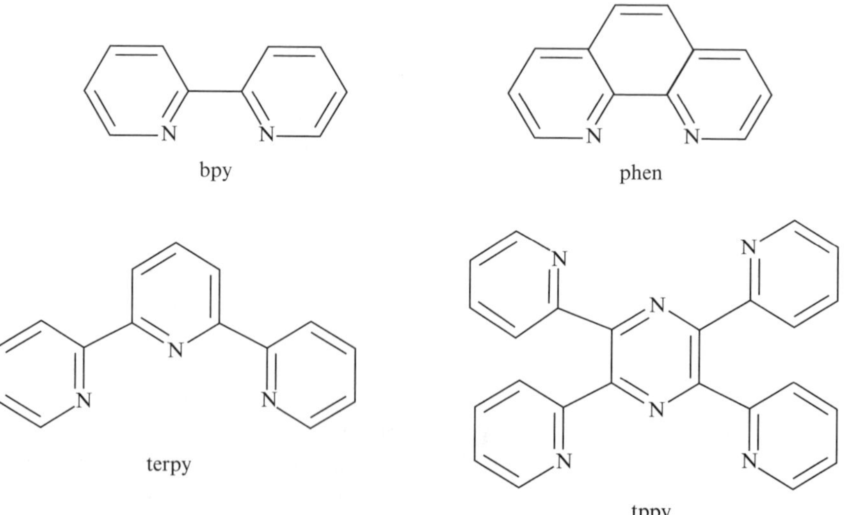

Figure 2 The ligands of this study.

The nature of the molybdenum oxide component can be directed to some extent by the choice of reactions conditions and the presence of other structure-directing agents, in this instance phosphate and organophosphonates. Thus, whereas low pH (3–5), short reaction times (10–48 h), and lower temperature ranges (110–160 °C) favor the formation of molybdate clusters as structural building blocks, conditions of higher pH (5–8) and more extreme temperatures and exposure times generally result in architectures in which the secondary metal is incorporated into a bimetallic oxide network and/or in which the molybdate cluster identity is lost upon fusing into one- or two-dimensional molybdenum oxide substructures.

Similarly, the introduction of phosphate or organophosphonate components, RPO_3^{2-}($R = OH$, *alkyl*, *aryl*), favors the assembly of molybdophosphate clusters of the general type $\{Mo_5O_{15}(O_5PR)_2\}^{4-}$, which may serve as a structural building blocks in the construction of complex bimetallic phosphate oxides incorporating organic components.

The structural versatility of the system is apparent. Modifications include: (i) the identity of the organoimine ligand to the secondary metal site, in this case Cu(II); (ii) the nature of the phosphate component $HOPO_3^{2-}$, RPO_3^{2-} or $\{O_3P(CH_2)_nPO_3\}^{4-}$; (iii) the tether length of the organic bridge in the diphosphonate materials; (iv) the nuclearity of the oxomolybdate cluster; (v) the oxoanion component, that is, $RAsO_3^{2-}$ rather than RPO_3^{2-}; (vi) in principle, the identity of the secondary metal; and (vii) the overall cluster oxidation state, that is, the introduction of reduced molybdenum centers. The last two possibilities will not be discussed in this chapter.

For the molybdophosphate building block $\{Mo_5O_{15}(O_3PR)_2\}^{4-}$, a conceptually attractive, albeit naive, scheme for the construction of one- and two-dimensional structures may be proposed. Thus, building blocks of the type $\{Mo_5O_{15}(O_3POH)_2\}^{4-}$ or $\{Mo_5O_{15}(O_3PR)_2\}^{4}$ may be linked through binuclear subunits such as $\{Cu_2(tppz\}^{4+}$ (tppz-tetrapyridylpyrazine) into one-dimensional chains, as shown in Figure 3a. Similarly, the $\{Mo_5O_{15}(O_3PR)_2\}^{4+}$ clusters may be linked through the alkyl tethers of diphosphonate components $\{O_3P(CH_2)_nPO_3\}^{4-}$ into chain structures, schematically illustrated in Figure 3b. Combining these two approaches allows extension into two dimensions, as shown in Figure 3c, an example of a three component system which inherently contains the structural information for the self-assembly of a hybrid network.

While the scheme is attractive and in practice of some predictive value, the structural chemistry of the oxomolybdate–phosphonate/organoimine–copper system is significantly more complex than the naive model suggests. Consequently, a variety of molybdophosphate subunits are encountered, depending on crystallization conditions and the nature of the phosphate components, as well as diverse linkage modes of the Cu(II) sites to the molybdophosphate subunits.

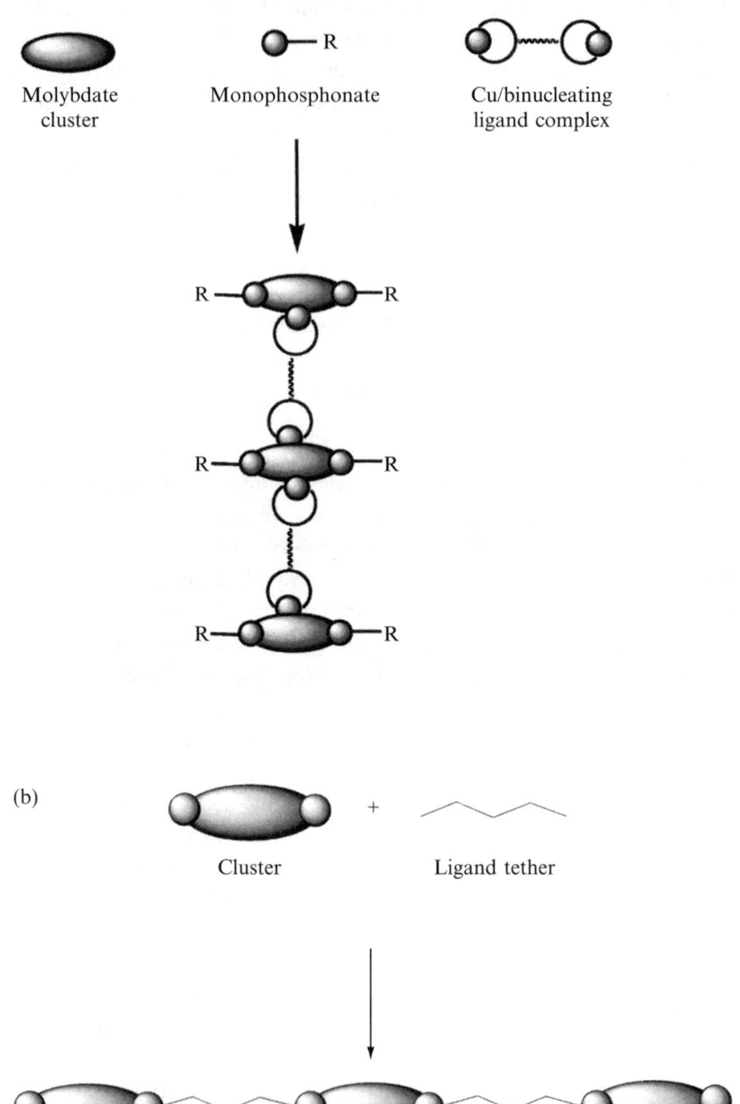

Figure 3 (a) The linking of phosphomolybdate clusters through binuclear metal organic complex subunits. (b) The linking of clusters through alkyl tethers of the diphosphonate groups. (c) The linking of clusters by both binuclear metal organic complex bridges and organic tethers of the diphosphonate ligands to produce a network structure.

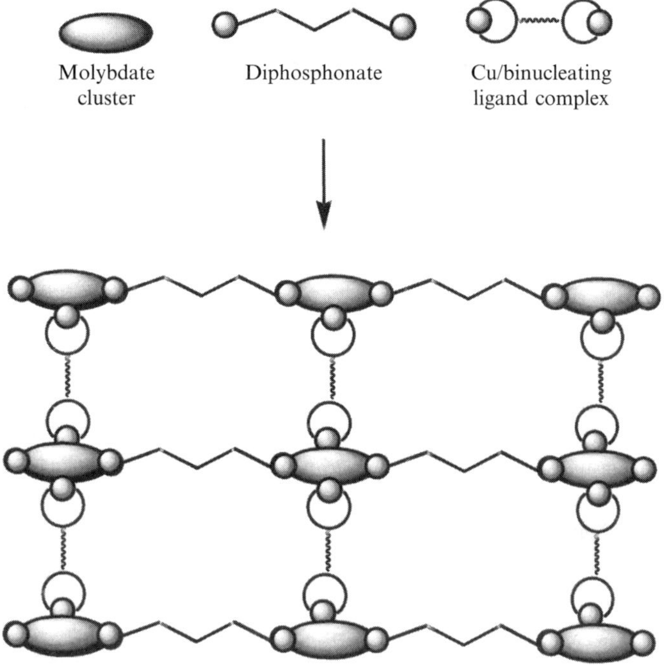

Figure 3 (*continued*)

3.1 Structures Containing the $\{Mo_5O_{15}(O_3PR)_2\}^{4-}$ Cluster as a Building Block

The synthetic strategy of linking clusters through appropriate building blocks is realized in a straightforward fashion in the prototypical structure of $[\{Cu_2(tppz)(H_2O)\}(Mo_5O_{15})(O_3PC_6H_5)_2] \cdot 2.5H_2O$(**1**·$2.5H_2O$) shown in Figure 4. The one-dimensional chain is constructed from $\{(Mo_5O_{15})(O_3PC_6H_5)_2\}^{4-}$ clusters, linked by $\{Cu_2(tppy)(H_2O)\}^{4+}$ tethers. The phosphomolybdate cluster consists of a ring of edge-and-and corner-sharing $\{MoO_6\}$ octchedra capped on one pole by a $\{O_3PC_6H_5\}$ tetrahedron which shares three oxygen vertices with the molybdenum sites and on the other pole by a phenylphosphonate tetrahedron sharing two vertices at unexceptional Mo–O distances and a third at 2.63 Å. This polyhedral connectivity is distinct from that observed in structures of the isolated polyanions $\{R_2P_2Mo_5O_{23}\}^4$($R = H$, alkyl, aryl) [97,98], where both phosphorus tetrahedra fuse from three vertices to the oxomolybdate ring to produce overall C_2 symmetry. It is noteworthy that both the isolated clusters and compound **1** exhibit three ring molybdenum sites which share two *cis* edges with the neighboring molybdate octahedra, while two share an edge and a vertex.

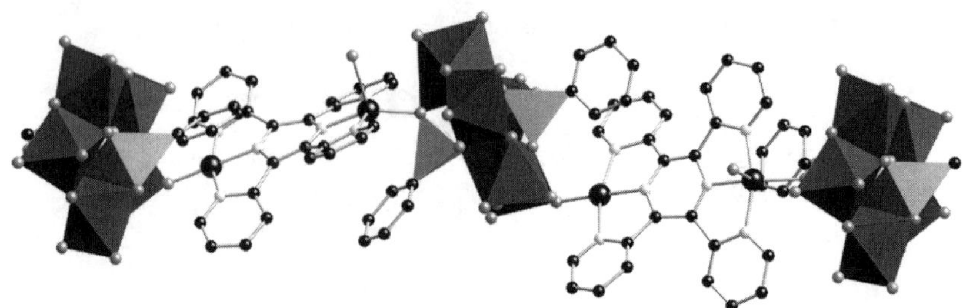

Figure 4 A polyhedral view of the structure of $[\{Cu_2(tppz)(H_2O)\}(Mo_5O_{15})(O_3PC_6H_5)_2]\cdot 2.5H_2O$(**1** $\cdot$ $2.5H_2O$). Only the phosphorus-bound carbon atoms of the phenyl groups are shown. The copper sites are shown as dark spheres.

The clusters are linked by binuclear Cu(II)–ligand tethers, $\{Cu_2(tppz)(H_2O)\}^{4+}$. Curiously, the two copper sites are structurally distinct. One is square pyramidal, with the basal plane defined by the three nitrogen donors of one terminus of the tppz ligand and an oxygen donor from a phenylphosphonate of a cluster and the apical site occupied by an aqua ligand. The second site is square planar, exhibiting ligation by three nitrogen donors of the bridging ligand and an oxo group from a molybdate of the cluster. The variability in the coordination number of the Jahn–Teller distorted Cu(II) centers of these structures which maybe 4, '4 + 1' or '4 + 2' coordinated, results in a range of tethering modes and consequently in the diverse structural chemistry described in subsequent discussions.

The significant structural consequences of introducing a secondary metal–ligand substructure are manifest in contrasting the structure of **1** to those of other oxomolybdenum phosphonate solid-state materials. Although numerous molecular clusters of the Mo–O–RPO_3^{2-} family have been described [98–101], extended structures are not only relatively rare but structurally unrelated to **1**. As shown in Figure 5a, the structure of $[MoO_2(O_3PC_6H_5)(H_2O)]$ [102] is a one-dimensional chain with isolated $\{MoO_6\}$ octahedra, while $Cs_2[(MoO_3)_3(O_3PCH_3)]$ [103] exhibits hexagonal tungsten oxide layers capped by phosphonate tetrahedra.

In an effort to introduce progressively more radical structural modifications, the substitiuent on the phosphorus subunit was changed from $-C_6H_5$ to $-OH$ in the preparations of $[\{Cu_2(tppz)(H_2O)_2\}(Mo_5O_{15})(O_3POH)_2]\cdot 2H_2O$(**2** $\cdot$ $2H_2O$) and $[\{Cu_2(tppz)(H_2O)_2\}(Mo_5O_{15})(O_3POH)_2]\cdot 3H_2O$(**3** $\cdot$ $3H_2O$). In contrast to the one-dimensional structure adopted by **1**, the structure of **2**, shown in Figure 6, consists of two-dimensional sheets, constructed from $\{Mo_5O_{15}(HOPO_3)_2\}^{4-}$ clusters linked through $\{Cu_2(tppz)(H_2O)_2\}^{4+}$ subunits. Each binuclear copper component links three phosphomolybdate clusters, while each cluster is in turn bonded to three binuclear copper subunits, which bridge to six neighboring clusters of the network. The periphery of a phosphomolybdate cluster is decorated by four corner-sharing copper(II) octahedra.

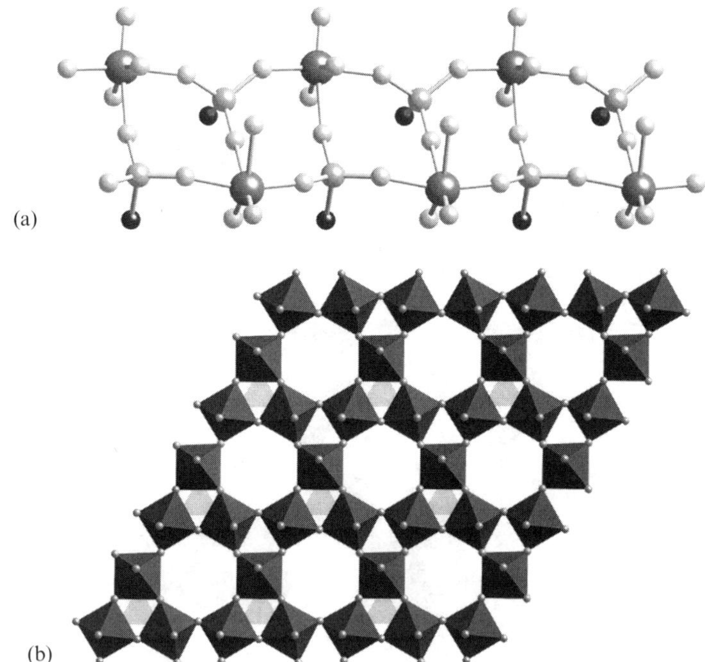

Figure 5 (a) The structure of $[MoO_2(O_3PC_6H_5)(H_2O)]$. (b) The structure of the $\{(MoO_3)_3(O_3PCH_3)\}^{2-}$ anionic network of $Cs_2[(MoO_3)_3(O_3PCH_3)]$.

In this case, the phosphomolybdate cluster is capped on both poles of the moybdate ring by $\{O_3POH\}^{2-}$ units sharing three vertices to give the common C_2 cluster symmetry [97]. The pendant oxygen atom of each phosphorus tetrahedron is protonated, as indicated by charge balance considerations and confirmed by valence sum calculations. The copper coordination of the $\{Cu_2(tppz)(H_2O)_2\}^{4+}$ subunit is defined by two bridging oxo groups from the phosphomolybdate clusters, three nitrogen donors from the chelating and bridging tppz ligands and an aqua ligand. The geometry is the common '4 + 2' variant with normal distances in the equatorial plane and long axial distances. The metal–oxide substructure $\{Cu_2Mo_5O_{15}(HOPO_3)_2\}$ is two-dimensional with intralamellar cavities occupied by the organic component and the water molecules of crystallization. It is evident that by adopting a different linkage mode from that observed for **1** and through expansion of the coordination number, the copper sites of **2** provide the loci for structural expansion in two dimensions.

Minor synthetic variations result in the isolation of **3**, a material constructed from the same building blocks as **2**, in the same ratios, but with a dramatically different structure. As shown in Figure 7(a), the structure of **3** is a three-dimensional framework constructed from phosphomolybdate clusters and Cu(II)–tppz binuclear subunits. As illustrated in Figure 7(b), the building blocks connect to form one-dimensional $\{Cu_2Mo_5O_{15}(HOPO_3)_2\}$ chains, in contrast to the $\{Cu_2Mo_5O_{15}(HOPO_3)_2\}$ sheets found in **2**. The Cu sites of a given chain link through the tppz bridges to Cu sites on four adjacent chains to provide the three-dimensional connectivity.

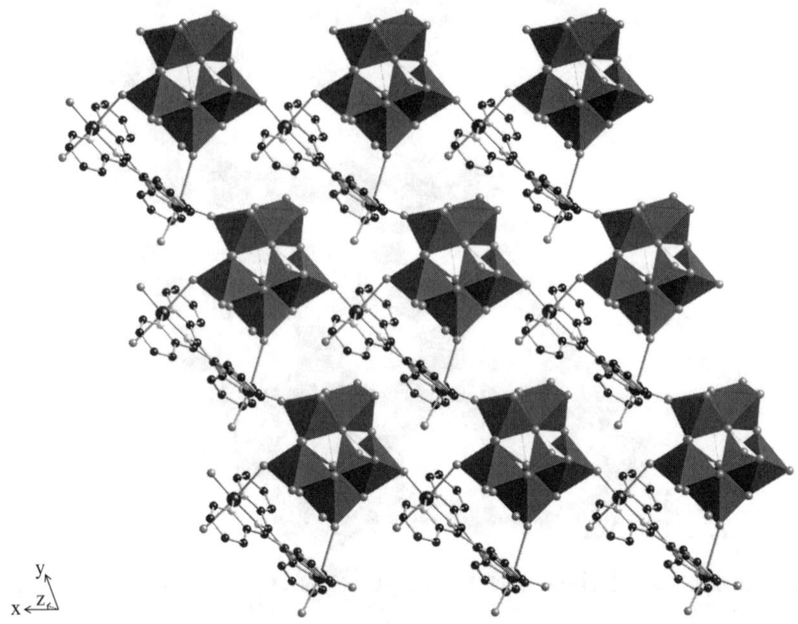

Figure 6 The structure of $[\{Cu_2(tppz)(H_2O)_2\}(Mo_5O_{15})(O_3POH)_2] \cdot 2H_2O$ (**2** · $2H_2O$).

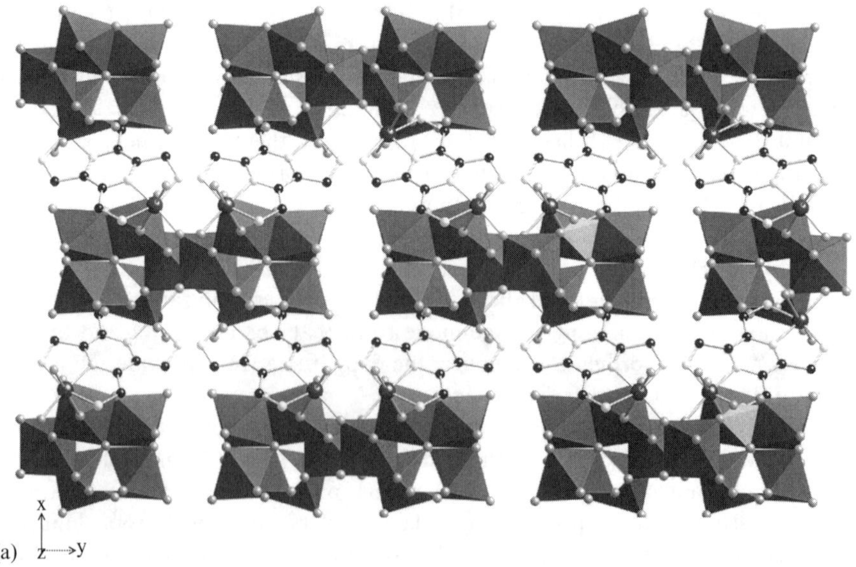

(a)

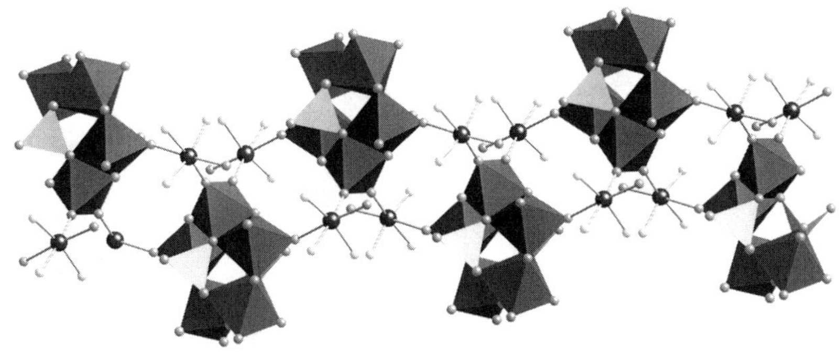

(b) 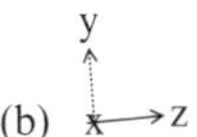

Figure 7 (a) The three-dimensional structure of $[\{Cu_2(tppz)(H_2O)_2\}(Mo_5O_{15})(O_3POH)_2]\cdot 3H_2O$(**3** · $3H_2O$). (b) The one-dimensional $\{Cu_2Mo_5O_{15}(O_3POH)_2\}$ substructure of **3**.

It is noteworthy that the structures of **2** and **3** are distinct from those of previously reported extended materials of the $Mo/O/H_nPO_4^{3-n}$ family [104–106]. While these Mo(V) and Mo(V)–Mo(VI) mixed valence materials exhibit a variety of embedded molybdate clusters, the pentanuclear ring of **1–3** is not among these.

Rather than tethering the phosphomolybdate clusters through binuclear Cu(II)–organoimine coordination complex bridges, organodiphosphonate bridges may be exploited, as shown schematically in Figure 3. Curiously, the successful exploitation of this strategy required the presence of secondary metal–ligand complex cation subunits to provide charge compensation and to participate in bonding to the periphery of the $\{(Mo_5O_{15})(O_3P(CH_2)_nPO_3)\}^{4-}$ clusters.

The prototypical structure of this type is provided by $[\{Cu(bpy)_2\}\{Cu(bpy)(H_2O)\}(Mo_5O_{15})(O_3\ PCH_2CH_2CH_2CH_2PO_3)]\bullet H_2O$(**4** · H_2O), shown in Figure 8. The one-dimensional structure of **4** is constructed from the phosphomolybdate clusters linked through butylene bridges and decorated with corner-sharing Cu(II) square pyramids. Two different Cu(II) moieties decorate the exterior of the molybdate rings. The Cu(1) square pyramid is defined by four nitrogen donors from two chelating bpy ligands in the basal plane and an apical oxo-group bridging to an $\{MoO_6\}$ octahedron. The Cu(2) site exhibits bonding to two nitrogen donors from a bpy ligand, an oxygen donor from a doubly bridging phosphonate ligand and an aqua ligand in the basal plane, with an oxo group bridging to a molybdenum site in the apical position.

As shown in Figure 9, the structure of the phenanthroline derivative $[\{Cu(phen)(H_2O)_2\}\{Cu(phen)_2\}(Mo_5O_{15})(O_3PCH_2CH_2CH_2PO_3)]\cdot 2.5H_2O$(**5** · $2.5H_2O$) is one-dimensional, in common with that of **4**. However, the details of the connectivities of the copper subunits to the clusters are not identical. Thus, the copper site featuring a single bpy chelate in **4** is bonded to a single aqua ligand and shares an edge

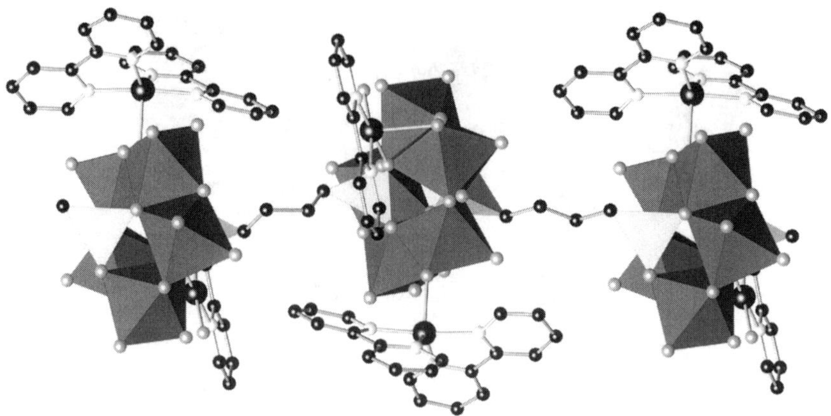

Figure 8 A view of the structure of $[\{Cu(bpy)_2\}\{Cu(bpy)(H_2O)\}(Mo_5O_{15})(O_5PCH_2CH_2CH_2CH_2PO_3)] \cdot H_2O$($\mathbf{4} \cdot H_2O$).

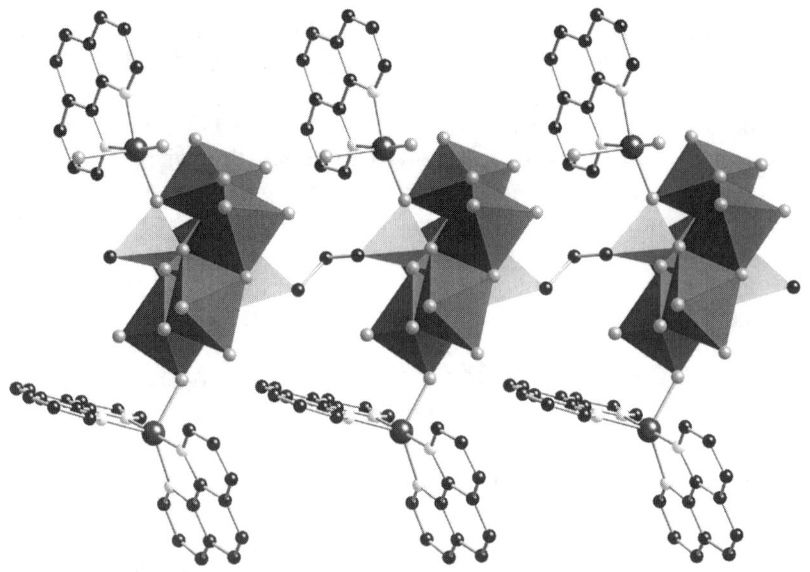

Figure 9 A polyhedral representation of the structure of $[\{Cu(phen)(H_2O)_2\}\{Cu(phen)_2\}(Mo_5O_{15})(O_3PCH_2CH_2CH_2PO_3)] \cdot 2.5H_2O$($\mathbf{5} \cdot 2.5H_2O$).

with a molybdenum site of the ring, while the similar copper site of the phenanthroline derivative **5** is bonded to two aqua ligands and has a single corner-sharing interaction with the cluster.

The minor structural differences between the Cu(II)–bpy–and Cu(II)–phen-containing materials **4** and **5** encouraged us to explore the influences of terpyridine as the ligand in the secondary metal–organoimine subunit. It was anticipated that the increased denticity of the ligand would constrain the geometry at the copper center and limit the connectivity and bridging possibilities between the cluster sites. As discussed below, this rationale proved somewhat naive.

The one-dimensional structure of $[\{Cu(terpy)(H_2O)\}_2(Mo_5O_{15})(O_3PCH_2CH_2PO_3)]\cdot 3H_2O$($\mathbf{6}\cdot 3H_2O$), shown in Figure 10, is not unexpected. While similar to the chain structures of **4** and **5**, the details of attachment of the $\{Cu(terpy)\}^{2+}$ subunits in **6** are distinct from those observed for the latter two structures. The Cu(II) square-pyramidal geometry of **6** is defined by three nitrogen donors of the terpy ligand and a bridging oxo group in the basal plane and by an aqua ligand in the apical position. Thus, each Cu(II) polyhedron shares a single vertex with the molybdate ring.

Expansion of the tether length by a single methylene group as in compound **6** to $[\{Cu(terpy)\}_2(Mo_5O_{15})(O_3PCH_2CH_2CH_2PO_3)]$ (**7**) has unexpected structural consequences. As shown in Figure 11, the structure of **7** consists of organodiphosphonate–pentamolybdate chains $\{Mo_5O_{15}(O_3PCH_2CH_2CH_2PO_3)\}_n{}^{4n-}$ linked through $\{Cu(terpy)\}^{2+}$ square pyramids into a two-dimensional network. The phosphomolybdate substructure is analogous to those of **2** and **6**. The Cu(II) square-pyramidal subunits exhibit a basal plane defined by three nitrogen donors of the terpy ligand and a bridging oxo group from a molybdate site and an apical site occupied by a second bridging oxo group. Each $\{Cu(terpy)\}^{2+}$ bridges phosphomolybdate clusters

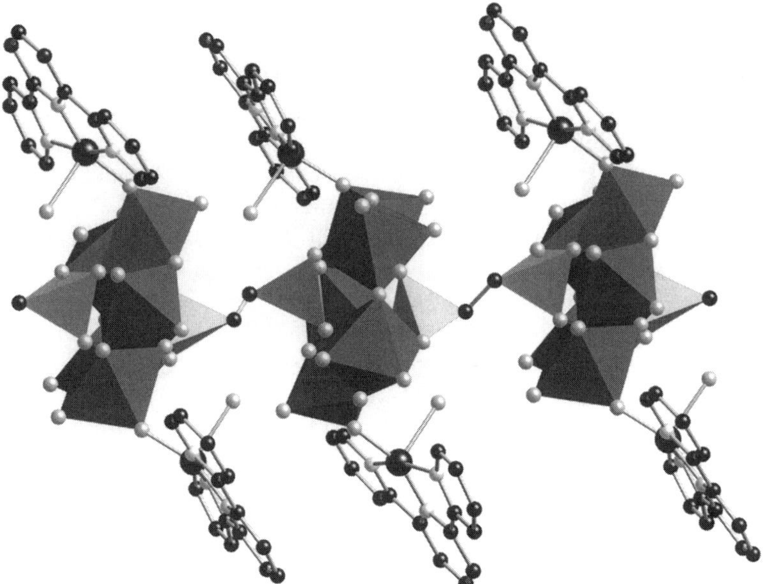

Figure 10 A view of the chain structure of $[\{Cu(terpy)(H_2O)\}_2(Mo_5O_{15})(O_3PCH_2CH_2PO_3)]\cdot 3H_2O$($\mathbf{6}\cdot 3H_2O$).

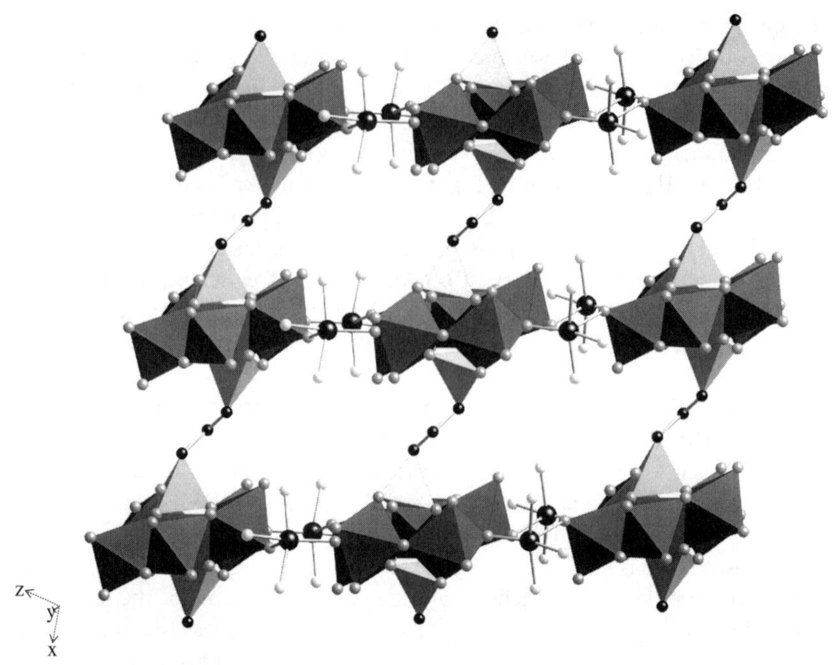

Figure 11 The structure of $[\{Cu(terpy)\}_2(Mo_5O_{15})(O_3PCH_2CH_2CH_2PO_3)]$(**7**). The terpy carbon atoms have been omitted.

from two adjacent $\{(Mo_5O_{15})(O_3PCH_2CH_2CH_2PO_3)\}_n{}^{4n-}$ chains. Consequently, the structure may be alternatively described as copper molybdate chains $\{Cu_2Mo_5O_{15}\}_n{}^{4n+}$ linked through propylenediphosphonate ligands into a two-dimensional network. The cross linking $\{Cu_2Mo_5O_{15}\}_n{}^{4n+}$ and $\{(Mo_5O_{15})(O_3PCH_2CH_2CH_2PO_3)\}_n{}^{4n-}$ chains intersect at an angle of ca 78°.

Based on the synthetic strategies schematically illustrated in Figure 3a and b, the naive expectation was the isolation of one-dimensional chains of tethered clusters. While this was realized in structures **1, 4, 5** and **6**, compounds **2** and **7** are two-dimensional whereas **3** is three-dimensional, reflecting the versatility of the linkage modes adopted by the secondary metal components. However, it was anticipated that combining a diphosphonate linker with a binuclear Cu(II)–organoimine subunit would result in a two-dimensional material, with clusters linked in one direction by the diphosphonate and cross-tethered by the binuclear binder. The implementation of this strategy is reflected in $[\{Cu_2(tppz)(H_2O)_2\}(Mo_5O_{15})(O_3PCH_2CH_2PO_3)]\cdot 5.5H_2O$(**8** $\cdot$ $5.5H_2O$), shown in Figure 12. The structure consists of one-dimensional $\{(Mo_5O_{15})(O_3PCH_2CH_2PO_3)\}_n{}^{4n-}$ chains linked through $\{Cu_2(tppz)(H_2O)_2\}^{4+}$ subunits into a two-dimensional network. While the molybdophosphonate chain of **8** is constructed from $\{(Mo_5O_{15})(O_3P\text{-})_2\}$ clusters identical with those described for **2** and **7**, the dispositions of the secondary metal–ligand subunits about the clusters are quite different. Thus, each molybdate cluster of **8** associates with three Cu(II) sites, one octahedral and two square pyramidal. The square-pyramidal $\{CuN_3O_2\}$ centers bridge Mo sites on two adjacent rings in a chain, while the octahedral $\{CuN_3O_3\}$ sites shares a single corner with one cluster. The remaining coordination sites are occupied by the three nitrogen donors of one terminus of the tppz ligand and two aqua ligands. The coordinated water molecules project into a cavity in the network, the perimeter of which is defined by terminal oxo groups from the Mo sites of adjacent chains. The water of crystallization occupies interlamellar sites disposed above and below these cavities.

3.2 Structural Consequences of Diphosphonate Tether Length

3.2.1 Methylenediphosphonate structures

The characteristic structural feature of the methylenediphosphonate ligand is the formation of {M–O–P–C–P–O–} chelate ring, rather than the tethering of distinct metal or cluster sites. Consequently, the reduced tether length of this ligand precludes its use as an extender in the construction of chain motifs. Furthermore, it is unsuitable as a cap for pentanuclear molybdate rings of the type observed in **1–8**, thus preventing the incorporation of $\{(Mo_5O_{15})(O_3PR)_2\}^{4-}$ subunits into the resultant materials.

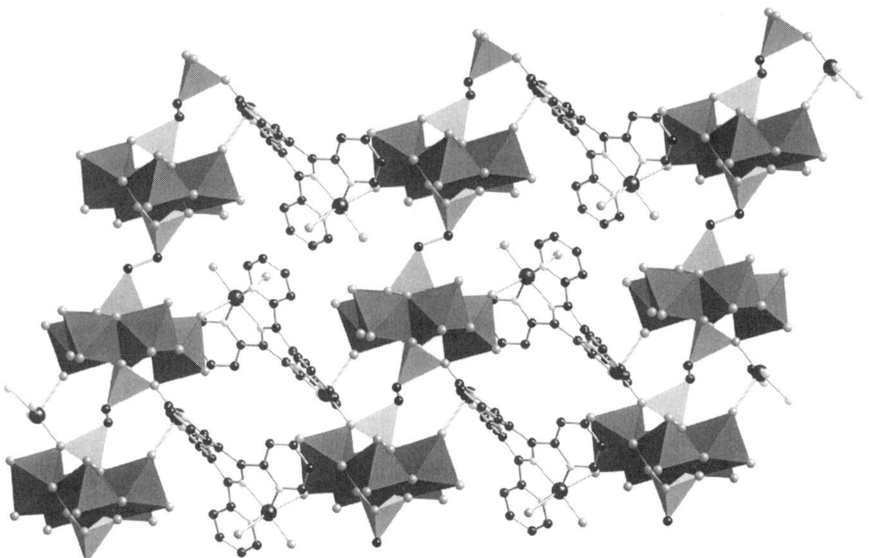

Figure 12 A polyhedral view of the structure of $[\{Cu_2(tppz)(H_2O)_2\}(Mo_5O_{15})(O_3PCH_2CH_2PO_3)]\cdot 5.5H_2O$(**8** $\cdot$ $5.5H_2O$).

These characteristics are illustrated in the representative structure [Cu(bpy)(MoO_2)(H_2O)($O_3PCH_2PO_3$)] (**9**), shown in Figure 13, which consists of one-dimensional chains, constructed from an {MoO_6} octahedral, {CuN_2O_3} square pyramids and phosphorus tetrahedra. The coordination sphere of the molybdenum is defined by three phosphonate oxygen donors, a terminal oxo group, an oxo group bridging to the copper and an aqua ligand. Each Mo(VI) site is linked to three methylenediphosphonate ligands. The Cu(II) site adopts '4 + 1' square-pyramidal geometry, with the basal plane occupied by two oxygen atoms from methylenediphosphonate ligands and two nitrogen donors from the bipyridine groups while the axial position is defined by a bridging oxo group. Each methylenediphosphonate ligand chelates to a Cu(II) site through oxygen atoms on each phosphorus terminus to form a six-membered {Cu–O–P–C–P–O} ring and corner-shares with three {MoO_6} octahedra. Consequently, one oxygen is pendant.

The overall structure of **9** may be described as $\{(MoO_2)(H_2O)(O_3PCH_2PO_3)\}_n^{2n-}$ ribbons decorated with $\{Cu(bpy)\}^{2+}$ subunits. The aqua ligands project above and below the 12-membered ring of the ribbon while the bpy ligands are directed into the intrachain regions where they interdigitate with bpy groups from neighboring chains.

Substitution of phenanthroline for bipyridine in the Cu(II)–ligand subunit results in a material with a binuclear molybdate building block and diphosphonate chelation to the molybdenum site rather than the copper site, [Cu(phen)(Mo_2O_5)($O_3PCH_2PO_3$)] (**10**), shown in Figure 14. The structure consists of one-dimensional chains constructed from corner- and edge-sharing Mo(VI) octahedra. The coordination sphere of the Cu(II) site is defined by the nitrogen donors of the chelating phenanthroline ligand, oxygen donors from each {PO_3} group of the chelating methylenediphosphonate ligand and an oxo group bridging to a molybdenum site. Binuclear Mo sites constructed from edge-sharing octahedra are embedded in

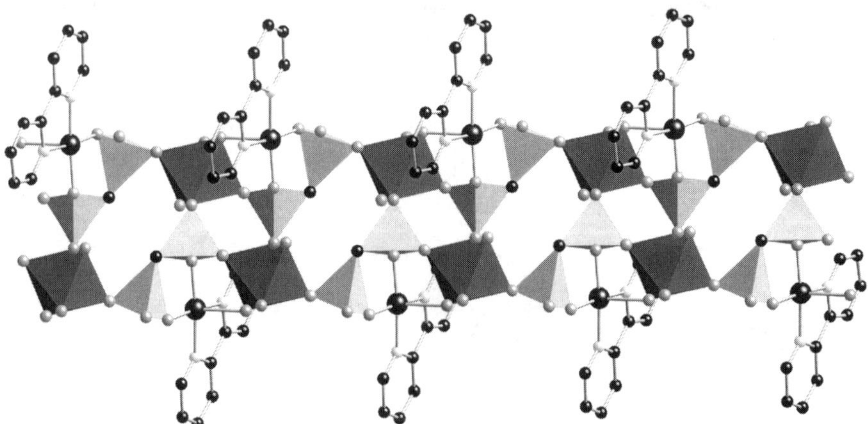

Figure 13 The structure of [Cu(bpy)(MoO_2)(H_2O)($O_3PCH_2PO_3$)](**9**).

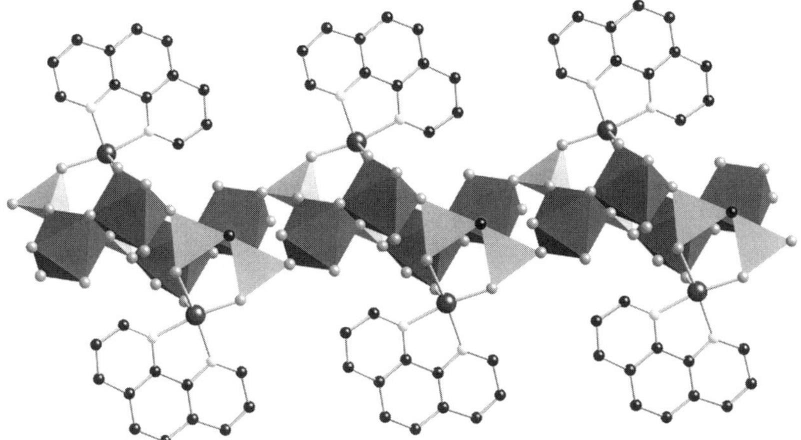

Figure 14 A view of the structure of $[Cu(phen)(Mo_2O_5)(O_3PCH_2PO_3)]$(**10**).

the chain, each bonding to three diphosphonate ligands. The diphosphonate units, in turn, chelate a Cu site and employ the remaining four oxygen donors to link three adjacent binuclear molybdenum units.

The unanticipated structural consequences of replacing the bidentate bpy and phen by terpyridine are demonstrated by the structure of $[Cu(terpy)(Mo_2O_5)(O_3PCH_2PO_3)]$ (**11**), shown in Figure 15. The two-dimensional structure of **11** is constructed from binuclear units edge-sharing $\{MoO_6\}$ octahedra linked into a one-dimensional phosphomolybdate chain by the methylenediphosphonate ligand which is connected into a two-dimensional network by binuclear subunits of edge-sharing Cu(II) square pyramids. The six-coordinate Mo(VI) sites exhibit two *cis*-terminal oxo groups, a bridging oxo group between Mo sites and three oxygen donors from two methylenediphosphonate groups. Each methylenediphosphonate ligand spans a binuclear molybdate subunit, sharing two vertices from one phosphorus terminus and one from the second, so as to chelate and bridge both Mo sites of the subunit. Two of the remaining oxygen donors bridge to adjacent molybdate binuclear units, while the third oxygen bridges the copper sites of a binuclear copper subunit. Consequently, each methylenediphosphonate ligand links three molybdate centers and one copper subunit.

The Cu(II) site exhibits '4 + 1' square-pyramidal geometry with three nitrogen donors of a terpy ligand and an oxygen donor of a methylenediphosphonate in the basal plane and the apical position occupied by the oxygen from a second diphosphonate ligand. The copper sites of the binuclear unit share an axial/basal edge defined by the oxygen donors at the vertices. In contrast to all the previously described structures of this type, the Cu polyhedra do not share corners or edges with the Mo polyhedra. The structure can thus be described as phosphomolybdate ribbons linked by binuclear Cu subunits into a two-dimensional network.

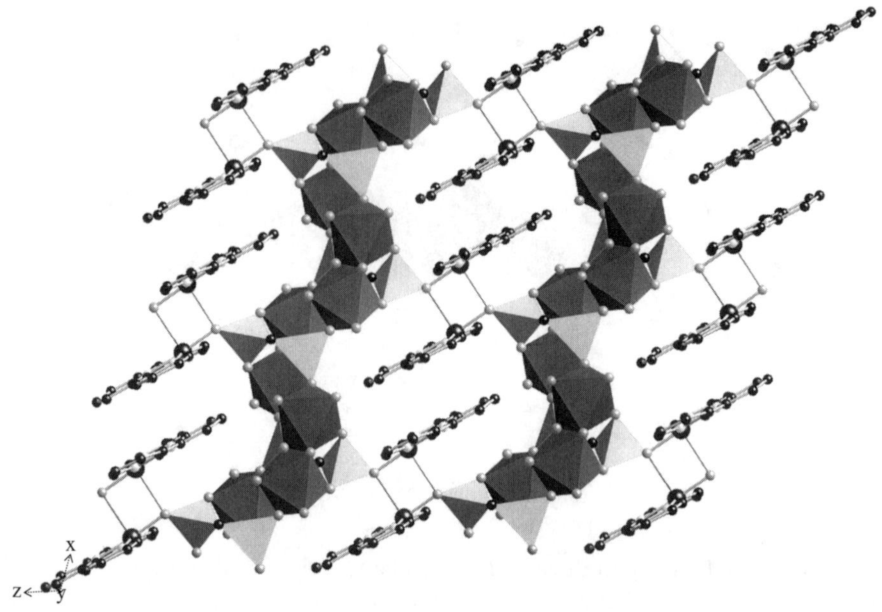

Figure 15 A polyhedral and ball and stick representation of the two-dimensional structure of [Cu(terpy)(Mo_2O_5)($O_3PCH_2PO_3$)](**11**).

The connectivity pattern produces 12 polyhedral connects or 24-membered $\{Mo_4Cu_2P_6O_{12}\}$ rings. The terpyridyl ligands project into these intralamellar cavities. Consequently, there is no water of crystallization associated with the layer cavities as in **12** · H_2O, where the reduced steric requirements leave the cavities unencumbered.

The influence of reaction conditions is often dramatic. Thus, reducing the reaction time for the preparation of **9** produces a two-dimensional phase [Cu(bpy) (Mo_2O_5)(H_2O) ($O_3PCH_2PO_3$)] · H_2O(**12** · H_2O), shown in Figure 16.

The network is constructed from binuclear units of edge-sharing $\{MoO_6\}$ octahedra, Cu(II) square pyramids and phosphorus tetrahedra. The two molybdenum environments of the binuclear subunit are quite distinct. One site is defined by two oxygen donors from the methylenediphosphonate ligands, a bridging oxo group to a second molybdenum site, two *cis*-oriented terminal oxo groups and an aqua ligand. The second site exhibits three oxygen donors from methylenediphosphonate ligands, a bridging oxo group to the first site, a bridging oxo group to the Cu(II) site and a terminal oxo group. The first site shares two vertices with two methylenediphosphonate ligands and does not link to the Cu(II) center, whereas the second site shares three vertices with two methylenediphosphonate ligands and one vertex with the copper. The Cu(II) site exhibits the common '4 + 1' square pyramidal geometry, with the basal plane defined by two oxygen atoms of a chelating methylenediphosphonate ligand and the nitrogen donors of the bpy

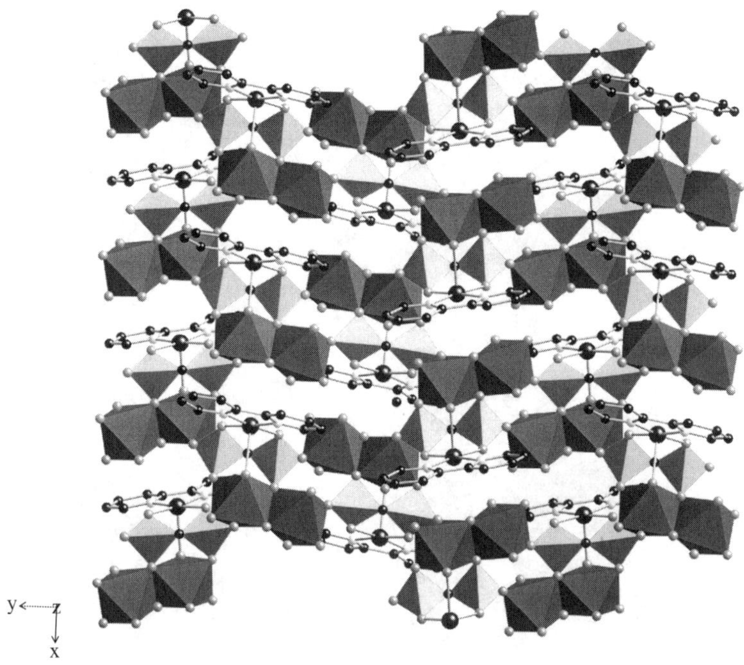

Figure 16 The structure of $[Cu(bpy)(Mo_2O_5)(H_2O)(O_3PCH_2PO_3)] \cdot H_2O$(**12** $\cdot$ H_2O).

ligand and the apical position occupied by the oxo group bridging to a molybdenum center.

Each methylenediphosphonate chelates not only to the Cu(II) site but also to one molybdenum center to produce a $\{MoCu(O_3PCH_2PO_3)\}$ subunit. The remaining two oxygen atoms link the diphosphonate to two molybdenum sites. Consequently, each diphosphonate ligand bonds to three binuclear Mo sites. In contrast to the structure of **9**, all of the oxygen atoms of the methylenediphosphonate ligand of **12** are involved in bonding.

The molybdophosphonate $\{(Mo_2O_5)(H_2O)(O_3PCH_2PO_3)\}_n^{2n-}$ substructure of **12** is two-dimensional, in contrast to the $\{(MoO_2)(H_2O)(O_3PCH_2PO_3)\}_n^{2-}$ ribbon of **9**. The polyhedral connectivity generates 24-membered rings of six $\{MoO_6\}$ octahedra and six $\{O_3PR\}$ tetrahedra. The water molecules of crystallization occupy the internal cavities of these rings. There are also eight-membered {Mo–O–P–O–} and six-membered {Mo–O–P–C–P–O–} rings in the molybdophosphonate network. The $\{Cu(bpy)\}^{2+}$ moieties occupy positions associated with the {Mo–O–P–C–P–O–} rings above and below the molybdophosphonate network.

The binucleating ligand tppz may also be exploited in this series of solids, with a rather unusual result. As shown in Figure 17, the structure of $[\{Cu_2(tppz)(H_2O)\}(Mo_3O_8)(O_3PCH_2PO_3H)_2] \cdot xH_2O$(**13** $\cdot$ xH_2O) is a one-dimensional spiral, constructed from $\{(Mo_3O_8)(O_3PCH_2PO_3H)_2\}^{4-}$ clusters linked by $\{Cu_2(tppz)$

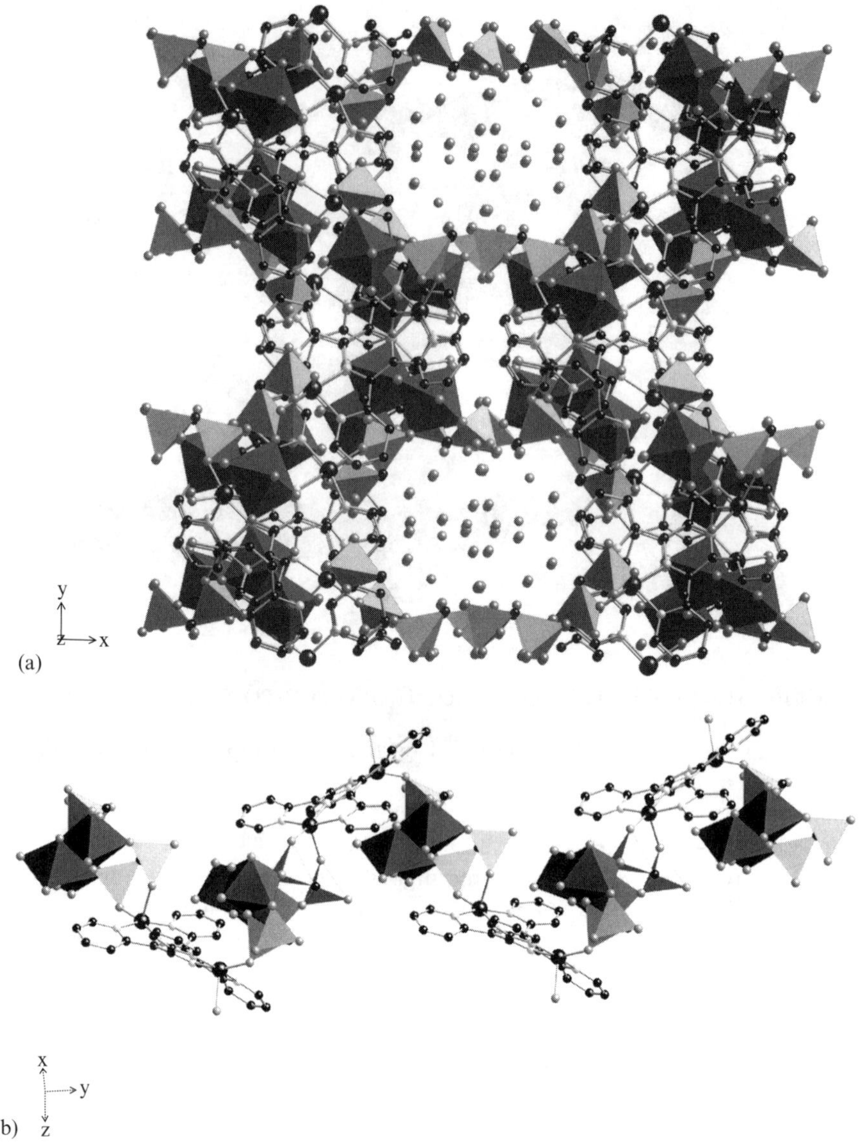

Figure 17 (a) A view of the structure of $[\{Cu_2(tppz)(H_2O)\}(Mo_3O_8)(O_3PCH_2PO_3)_2] \cdot xH_2O$(**13** · xH_2O). (b) A view of the nesting of the spiral chains of **13** to produce large cavities occupied by water of crystallization.

$(H_2O)\}^{4+}$ tethers. The unusual cluster is constructed of a trinuclear unit of edge-sharing $\{MoO_6\}$ octahedra, with two diphosphonate ligands decorating the periphery. Both diphosphonate moieties are monoprotonated at the pendant oxygen atoms. One diphosphonate additionally chelates to a Cu(II) site of a tethering binuclear unit, while the second shares a single vertex with the second copper of this subunit. Furthermore, this diphosphonate also exhibits an unusual pendant–P=O group. A most unusual feature of the structure is the nesting of the spirals to produce a structure with large cavities occupied by water molecules. The pendant OH groups of the diphosphonate ligands project into these cavities and engage in hydrogen bonding with the water of crystallization. The less polar–P=O groups project into a smaller cavity which is free of associated water. Consequently, the structure exhibits amphiphilic properties with crystal packing dependent on hydrophilic–hydrophobic interactions.

An obvious feature of the structures of the methylenediphosphonate derivatives is the absence of phosphomolybdate cluster building blocks. This is a consequence of the limited spatial extension between the $\{PO_3\}$ termini of the diphosphonate ligand provided by a single methylene tether. Consequently, rather than spanning component building blocks, the methylenediphosphonate ligand is geometrically suited to forming six-membered chelate rings to a single metal site, {M–O–P–C–P–O–}. This coordination preference is apparent in the previously reported oxovanadium phosphonate structures $Cs[(VO)(O_3PCH_2PO_3)]$[107], $[H_2N(CH_2)_4NH_2][(VO)(O_3PCH_2PO_3)]$[108], $Cs[(VO)_2V(O_3PCH_2PO_3)_2(H_2O)_2]$[107], $[(VO)_2(O_3PCH_2PO_3)(H_2O)_4]$ [109] and $(NH_4)_2[(VO)(O_3PCH_2PO_3)]$[110]. The consequences of tether expansion are dramatically illustrated in the series of oxovanadate materials: the 1-D $[H_2N(CH_2)_4NH_2][(VO)(O_3PCH_2PO_3)]$, the 2-D $[H_3N(CH_2)_2NH_3][(VO)(O_3PCH_2CH_2PO_3)]$ and the 3-D $[H_3N(CH_2)_2NH_3][(VO)_4(OH)_2(H_2O)_2(O_3PCH_2CH_2CH_2PO_3)_2]$[108]. The dimensional expansion reflects the tether length of the diphosphonate ligands, which in the ethylene and propylenediphosphonate materials link spatially distinct building blocks rather than chelating to a single vanadate polyhedron. The naive expectation was that the oxomolybdate–diphosphonate–Cu(II)–organoimine system would exhibit a similar trend and that under appropriate reaction conditions phosphomolybdate building blocks would be incorporated into the overall architectures.

3.2.2 *Other diphosphonate-linked structures*

The $\{(Mo_5O_{15})(O_3P(CH_2)_nPO_3\}^{4-}$ cluster is not a ubiquitous building block of the materials of this family, even in cases where the diphosphonate ligand can in principle link cluster sites. For example, as shown in Figure 18, the structure of $[\{Cu(bpy)\}_2(Mo_4O_{12})(H_2O)_2(O_3PCH_2CH_2PO_3)]\cdot 2H_2O$(**14** $\cdot$ $2H_2O$) is a two-dimensional network constructed from tetranuclear oxomolybdate clusters linked through corner-sharing copper octahedra and diphosphonate ligands. The tetranuclear clusters consist of an 'S'-shaped array of edge-sharing Mo(VI) octahedra and square pyramids. An interior pair of base edge-sharing square pyramids are each involved in *cis*-base edge-sharing with the exterior Mo(VI) octahedra. The square-pyramidal site is defined in the basal plane by a doubly bridging oxo group to the octahedral Mo center, two triply bridging oxo groups to the neighboring square-pyramidal center and to the octahedral Mo site, and an oxo group bridging to the Cu site, while a terminal oxo group occupies the apical position. The octahedral molybdate exhibits bonding to a terminal oxo group, a doubly bridging oxo group to the square-pyramidal site, a triply bridging oxo group to this site, the oxygen donor of a phosphonate ligand, an oxo group bridging to the copper center and an aqua ligand. The Cu(II) sites exhibit '4 + 2' distorted octahedral geometry. The four equatorial bonds are to two oxygen donors of diphosphonate ligands and to the nitrogen donors of the chelating bpy ligand, while long axial interactions are formed to molybdate polyhedra of neighboring clusters. Each terminus of the diphosphonate ligand bridges two Cu sites and one Mo center.

The structure may be described as molybdate clusters linked through copper octahedra into one-dimensional chains which are in turn cross-linked by ethylenediphosphonate subunits into a two-dimensional network. The $\{Cu(bpy)(Mo_4O_{12})(H_2O)_2\}^{2+}$ substructure is shown in Figure 18b.

The polyhedral connectivity generates large $\{Mo_8Cu_2P_4O_{12}C_4\}$ rings of 14 polyhedra connects within the layers. The water molecules of crystallization occupy these intralamellar cavities. The bpy ligands project from the surfaces of the

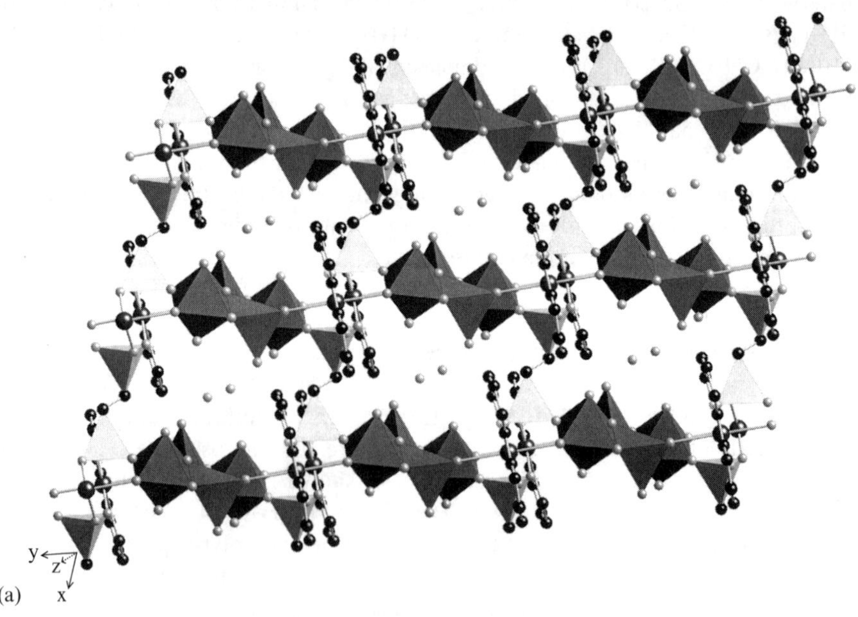

Figure 18 (a) The structure of $[\{Cu(bpy)\}_2(Mo_4O_{12})(H_2O)_2(O_3PCH_2CH_2PO_3)]\cdot 2H_2O$ (**14** · $2H_2O$). (b) The $\{Cu(bpy)(Mo_4O_{12})(H_2O)_2\}^{2+}$ substructure of **14**.

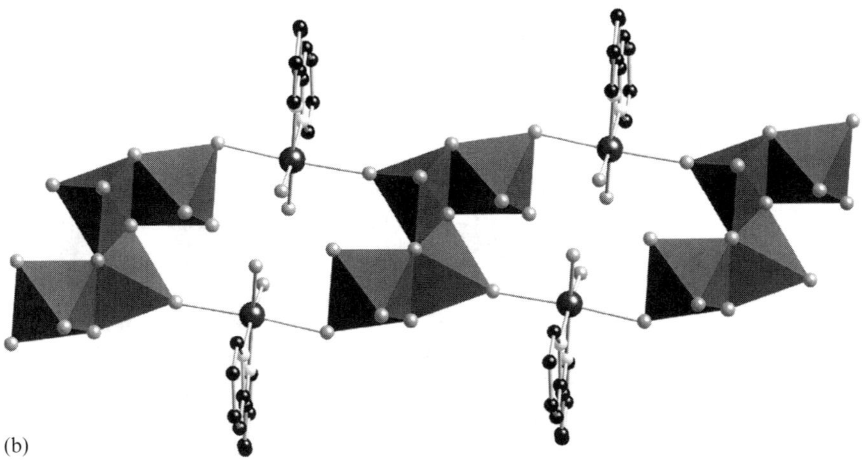

Figure 18 (*continued*)

layers into the interlamellar spaces and interdigitate with rings from adjacent layers. The structure of **14**•$2H_2O$ is isomorphous with that of the phenanthroline derivative $[\{Cu(phen)\}_2(Mo_4O_{12})(H_2O)_2(O_3PCH_2CH_2PO_3)] \cdot 2H_2O$(**15** $\cdot 2H_2O$).

Expansion of the tether length to three methylene groups in the propylenediphosphonate derivative $[Cu(bpy)(Mo_2O_5)(O_3PCH_2CH_2CH_2PO_3)]$ (**16**) has unexpected structural consequences. As shown in Figure 19a, the structure of **16** is layered and constructed of binuclear units of face-sharing molybdate octahedra linked through copper octahedra and phosphorus tetrahedra. The Mo geometries are defined by three oxygen donors from diphosphonate ligands, a doubly bridging oxo group to the second Mo site of the binuclear unit, an oxo group bridging to a Cu(II) site and a terminal oxo group. The Cu(II) site exhibits '4 + 2' six coordination with an equatorial plane defined by the nitrogen donors of the chelating bpy ligand and two oxygen donors from diphosphonate groups; the long axial interactions are with oxo groups from two neighboring molybdate binuclear units.

The most unusual feature of the structure of **16** is the disposition of the propylenediphosphonate ligand. Two oxygen donors of each phosphorus terminus link to molybdate groups of two neighboring binuclear subunits; consequently, each diphosphonate group bridges four binuclear subunits. Curiously, the remaining oxygen donors are employed in chelation to a single Cu(II) center, such that the diphosphonate folds to form a most unusual eight-membered {Cu–O–P–C–C–C–P–O–} chelate ring. One consequence of this bonding pattern is to produce a much less open network than that associated with **14**•$2H_2O$, a feature consistent with the absence of water of crystallization within layers of **16**.

The copper molybdate substructure of **16**, $\{Cu(bpy)(Mo_2O_5)\}^{4+}$, is a one-dimensional chain of binuclear molybdate subunits bridged by $\{Cu(bpy)\}^{2+}$ moieties. These chains are connected in turn by the propylenediphosphonate ligands. As shown in Figure 19b, the structure may be alternatively described as a two-dimensional organophosphomolybdate network with $\{Cu(bpy)\}^{2+}$ moieties embedded in the cavities produced by the long propylene tethers of the disphosphonate ligands.

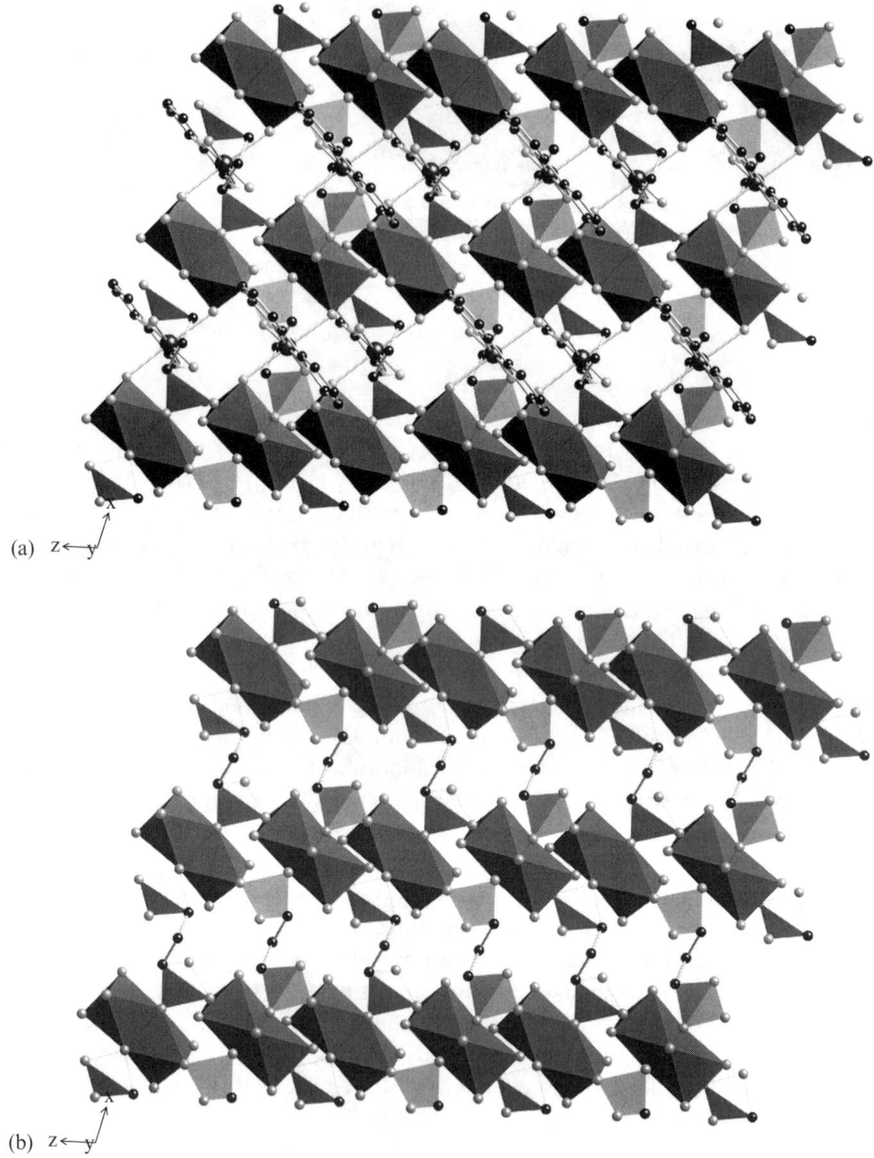

Figure 19 (a) A view of the layer structure of $[Cu(bpy)(Mo_2O_5)(O_3PCH_2CH_2CH_2PO_3)\}$(**16**). (b) The organophosphomolybdate substructure of **16**.

3.3 Variation of the Tetrahedral Subunit: Organoarsonate Materials

Substitution of As for P in the tetrahedral substructure $\{RAsO_3\}^{2-}$ has predictable consequences as shown in Figure 20 for the structure of $[\{Cu_2(tppz)\}(Mo_6O_{18})(O_3AsC_6H_5)_2]$ (**17**). The two-dimensional structure of **17** is constructed from $\{(Mo_6O_{18})(O_3AsC_6H_5)_2\}^{4-}$ clusters linked by $\{Cu_2(tppz)\}^{4+}$ binuclear subunits. The clusters consist of a ring of six edge-sharing $\{MoO_6\}$ octahedra, capped on both poles by $\{O_3AsR\}^{2-}$ groups each sharing three oxygen vertices with the ring. Each cluster is associated with four copper sites, each from a different $\{Cu_2(tppz)\}^{4+}$ subunits. Each binuclear Cu(II) component in turn links four clusters to provide the two-dimensional connectivity. Thus, the bimetallic oxide substructure $\{(Cu_2Mo_6O_{18})(O_3AsR)_2\}_n$ is itself two-dimensional, with the tppz ligand occupying both intralamellar and interlamellar vacancies. In contrast to other structures described in this chapter, both Cu(II) sites of the $\{Cu_2(tppz)\}^{4+}$ subunit exhibit '4 + 2' geometry, a necessary feature for the network construction.

The embedded $\{(Mo_6O_{18})(O_3AsC_6H_5)_2\}^{4-}$ cluster exhibits the D_3 symmetry characteristic of previously reported isolated clusters of the type $\{(Mo_6O_{18})(O_3AsR)_2\}^{4-}$ {101}. The cluster expansion observed upon substituting As for P in these materials reflects in part the larger covalent radius of As compared with P.

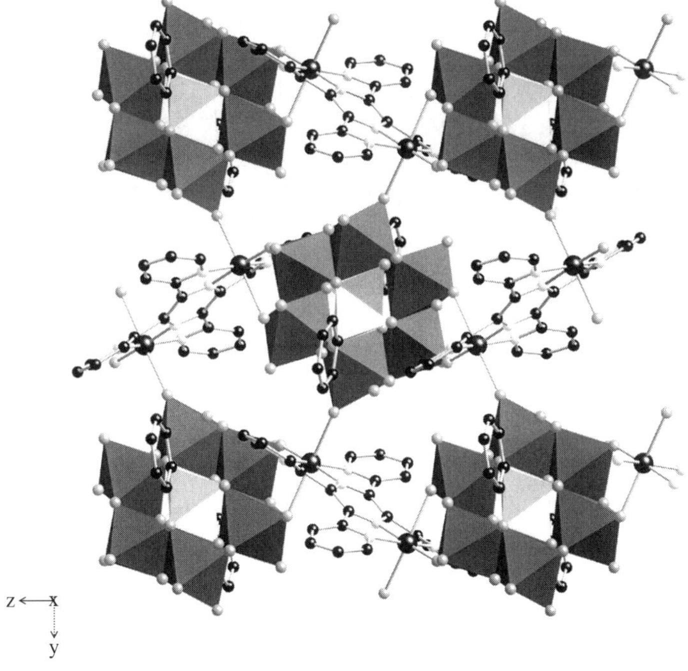

Figure 20 The structure of $[\{Cu_2(tppz)\}(Mo_6O_{18})(O_3AsC_6H_5)_2]$(**17**)

This in turn results in an expansion of the O_3 face of the tetrahedral $\{O_3ER\}$ subunit, which fuses to the molybdate ring. The O···O distances for this triangular face are ca 2.53 Å for phosphorus and 2.80 Å for As. This increase in the dimensions of this structural building block allows a concomitant expansion of the molybdate ring nuclearity.

4 TRENDS AND CLASSIFICATIONS

The structures of this study exhibit a remarkable range of component substructures and polyhedral connectivities. To date, the family of materials of the type oxomolybdate–$REO_3{}^{2-}$–Cu(II)–organoimine is represented by E = As and, P and by the organoimine ligands 2, 2′-bipyridine, *o*-phenanthroline, 2, 2′:6′, 2″-terpyridine and tetra(4-pyridyl)pyrazine. While the $\{(Mo_5O_{15})(O_3P\text{-})_2\}^{4-}$ cluster is a recurring motif, it is not ubiquitous, as manifested in the structural summary of Table 1. Despite the overall complexity of the structural characteristics of this family of materials, a relatively simple method of classification based on the nuclearity of the molybdenum subunits embedded within the solid has been adopted. Thus, six subclasses of materials have been identified, incorporating $\{MoO_6\}_x$ subunits with $x = 1, 2, 3, 4, 5$ and 6.

The structural summary evokes a number of observations. While a significant database has been compiled for this unusually fruitful family of materials, it is clear that the set of compounds is not exhaustive in view of the structural variability associated with the constituent building blocks. This latter point is reflected in the different coordination geometries accessible for Mo(VI), including tetrahedral, square-pyramidal and octahedral, in addition to the aggregation for molybdenum polyhedra into oligomers of various nuclearities. Furthermore, the diphosphonate ligand may adopt a variety of coordination modes with respect to the molybdate subunit and the secondary metal component, which also enjoys considerable latitude in coordination geometry and polyhedral connectivity to the molybdophosphonate substructure. Finally, coordinated water molecules are often constituents of both the molybdenum and the copper coordination sphere in phases which have been prepared hydrothermally at moderate temperature. Thus, it may be anticipated that for a given combination of organoimine coligand to the Cu(II) subunit and organodiphosphonate linker more than one phase may be accessible by appropriate manipulation of the hydrothermal conditions. Hydrothermal parameter space, which includes stoichiometries, pH, temperature and fill volume, is vast and even minor variations can result in the isolation of new metastable phases. This point is illustrated by the isolation of two phases **9** and **12** · H_2O for the methylenediphosphonate and 2, 2′-bipyridine combination of ligands. It is likely that other combinations of organodiphosphonates and

Table 1 Summary of structural characteristics for materials of the oxomolybdate–phosphonate–Cu(II)–organoimine family.

Class and compound	Overall dimensionality	Cu(II) component structure	Copper molybdate substructure	Phosphomolybdate substructure
Type I: isolated $\{MoO_6\}$ *octahedra*				
$[Cu(bpy)(MoO_2)(H_2O)(O_3PCH_2PO_3)]$ (**9**)	1-D	$\{CuN_2O_3\}$ square pyramid	Binuclear unit	$\{MoO_2(H_2O)(O_3PCH_2PO_3)\}_n^{2n-}$ chain
Type II: bioctahedral molybdate unit				
$[Cu(phen)(Mo_2O_5)(H_2O)(O_3PCH_2PO_3)]$ (**10**)	1-D	$\{CuN_2O_3\}$ square pyramid	Trinuclear unit	$\{(Mo_2O_5)(O_3PCH_2PO_3)\}_n^{2n-}$ chain
$[Cu(terpy)(Mo_2O_5)(O_3PCH_2PO_3)]$ (**11**)	2-D	Binuclear unit of edge-sharing square pyramids	No connectivity between Cu and Mo polyhedra	$\{(Mo_2O_5)(O_3PCH_2PO_3)\}_n^{2n-}$ chain
$[Cu(bpy)(Mo_2O_5)(O_3PCH_2CH_2CH_2PO_3)]$ (**16**)	2-D	$\{CuN_2O_4\}$, '4 + 2' octahedron	$\{Cu(bpy)(Mo_2O_5)\}_n^{4n+}$ chain	$\{(Mo_2O_5)(O_3PCH_2CH_2CH_2PO_3)\}_n^{2n-}$ network
$[Cu(bpy)(Mo_2O_5)(H_2O)(O_3PCH_2PO_3)]\cdot H_2O$(**12** $\cdot H_2O$)	2-D	$\{CuN_2O_3\}$ square pyramid	Trinuclear unit	$\{(Mo_2O_{15})(H_2O)(O_3PCH_2PO_3)\}_n^{2n-}$ network
Type III: trioctahedral molydate				
$[\{Cu_2(tppz)(H_2O)(Mo_3O_8)(O_3PCH_2PO_3H)_2]\cdot xH_2O$(**13** $\cdot H_2O$)	1-D	$\{CuN_3O_2\}$ square pyramid	No connectivity between Cu and Mo polyhedra	$\{(Mo_2O_8)(O_3PCH_2PO_3H)_2\}^{4-}$ cluster
Type IV: tetraoctahedral molydate unit				
$[\{Cu(bpy)\}_2(Mo_4O_{12})(H_2O)_2(O_2PCH_2CH_2PO_3)]\cdot 2H_2O$ (**14**)	2-D	$\{CuN_2O_4\}$ '4 + 2' octahedron	$\{Cu(bpy)(Mo_4O_{12})(H_2O)_2\}_n^{4n+}$ chain	$\{(Mo_4O_{12})(H_2O)_2(O_3PCH_2CH_2PO_3)\}_n^{4n-}$ Network
$[\{Cu(phen)\}_2(Mo_4O_{12})(H_2O)_2(O_3PCH_2CH_2PO_3)]\cdot 2H_2O$ (**15**)	2-D	$\{CuN_2O_3\}$ square pyramid	$\{Cu_2(phen)_2(Mo_4O_{12})\}_n^{4n+}$ chain	$\{(Mo_4O_{12})(O_3PCH_2PO)\}_n^{4n-}$ chain
Type V: pentaoctahedral molybdate unit				
$[\{Cu_2(tppz)(H_2O)\}(Mo_5O_{15})(O_3PC_6H_5)_2]\cdot 2.5H_2O$(**1** $\cdot 2.5H_2O$)	1-D	$\{CuN_3O\}$ square plane and $\{CuN_3O_2\}$ square pyramid	$\{Cu(tppz)_{0.5}Mo_5O_{15}\}^{2+}$ cluster	$\{(Mo_5O_{15})(O_3PC_6H_5)_2\}^{4-}$ cluster
$[\{Cu_2(tppz)(H_2O)\}(Mo_5O_{15})(O_3POH)_2]\cdot 2H_2O$(**2** $\cdot 2H_2O$)	2-D	$\{CuN_3O_3\}$'4 + 2' octahedra	$\{Cu_2(Mo_5O_{15})\}^{4+}$ network	$\{(Mo_5O_{15})(O_3POH)_2\}^{4-}$ cluster

(continues)

Table 1 Summary of structural characteristics for materials of the oxomolybdate–phosphonate–Cu(II)–organoimine family. (*continued*)

Class and compound	Overall dimensionality	Cu(II) component structure	Copper molybdate substructure	Phosphomolybdate substructure
$[\{Cu_2(tppz)(H_2O)_2\}(Mo_5O_{15})(O_3POH)_2]\cdot 3H_2O$(**3** $\cdot$ $3H_2O$)	3-D	$\{CuN_3O_3\}$'4 + 2' octahedra	$\{Cu_2(Mo_5O_{15})\}^{4+}$ chain	$\{(Mo_5O_{15})(O_3POH)_2\}^{4-}$ cluster
$[\{Cu(bpy)_2\}\{Cu(bpy)(H_2O)\}(Mo_5O_{15})(O_3PCH_2CH_2CH_2CH_2PO_3)]\cdot H_2O$(**4** $\cdot$ H_2O)	1-D	$\{CuN_4O\}$ and $\{Cu_2O_3\}$ square pyramids	$\{Cu_2(bpy)_3(H_2O)(Mo_5O_{15}\}^{4+}$ cluster	$\{(Mo_5O_{15})(O_3PCH_2CH_2CH_2CH_2PO_3)\}_n^{4n-}$ chain
$[\{Cu(phen)_2\{Cu(phen)(H_2O)_2\}(Mo_5O_{15})(O_3PCH_2CH_2CH_2PO_3)]\cdot 2\cdot 5H_2O$ (**5** $\cdot$ $2.5H_2O$)	1-D	$\{CuN_4O\}$ and $\{CuN_2O_3\}$ square pyramids	$\{Cu_2(phen)_3(H_2O)_2(Mo_5O_{15})\}^{4+}$ cluster	$\{(Mo_5O_{15})(O_3PCH_2CH_2CH_2PO_3)\}_n^{4n-}$ chain
$[\{Cu(terpy)(H_2O)\}_2(Mo_5O_{15})(O_3PCH_2CH_2PO_3)]\cdot 3H_2O$(**6** $\cdot$ $3H_2O$)	1-D	$\{CuN_3O_2\}$ square pyramid	$\{Cu_2(terpy)_2(Mo_2O_{15})\}^{4+}$ cluster	$\{(Mo_5O_{15})(O_3PCH_2CH_2PO_3)\}_n^{4n-}$ chain
$[\{Cu(terpy)\}_2(Mo_5O_{15})(O_3PCH_2CH_2CH_2PO_3)]$ (**7**)	2-D	$\{CuN_3O_2\}$ square pyramid	$\{Cu_2(terpy)_2(Mo_5O_{15})\}_n^{4+}$ chain	$\{(Mo_5O_{15})(O_3PCH_2CH_2CH_2PO_3)\}_n^{4n-}$ chain
$[\{Cu_2(tppz)(H_2O)_2\}\{Mo_5O_{15})(O_3PCH_2CH_2PO_3)]\cdot 5.5H_2O$(**8** $\cdot$ $5.5H_2O$)	2-D	$\{CuN_3O_2\}$ square pyramid and $\{CuN_3O_3\}$ octahedron	$\{Cu_2(terpy)(H_2O)_2(Mo_5O_{15})\}_n^{4n+}$ chains	$\{(Mo_5O_{15})(O_3PCH_2CH_2PO_3\}_n^{4n-}$ chain
Type VI: hexaoctahedral molybdate unit				
$[\{Cu_2(tppz)\}(Mo_6O_{18})(O_3AsC_6H_5)_2]$ (**17**)	2-D	$\{CuN_3O_3\}$'4 + 2' octahedron	$\{Cu_2(tppz)(Mo_6O_{18})\}_n^{4n+}$ network	$\{(Mo_6O_{18})(O_3AsC_6H_5)_2\}^{4-}$ cluster

2, 2′-bipyridine or terpyridine will also yield more than one phase as the hydrothermal conditions are more fully explored.

The most apparent structural trend is observed for the methylenediphosphonate derivatives **9, 10, 12** and **13**. The methylene spacer does not allow extension of the diphosphonate ligand to link two discrete substructural motifs, but rather constrains the ligand to act as a chelate forming six-membered {M–O–P–C–P–O–} rings. However, within this limitation, there is considerable flexibility as the methylenediphosphonate may chelate either the Mo or the Cu sites and may establish a variety of connectivity patterns with the remaining oxygen donors, and observation manifest in the distinct structures of **9, 10, 12** and **13**.

The copper coordination geometry is also variable in this structural set. While mononuclear square-pyramidal sites are most common, mononuclear six-coordinate and even binuclear sites are also incorporated into the overall architectures. Water coordination is also evident, often giving rise to more than one coordination geometry in a material, as illustrated by **1, 4, 5** and **8**.

Although the syntheses were carried out under conditions favoring the formation of $\{(Mo_5O_{15})(O_3PR)_2\}^{4-}$ clusters, a variety of molybdenum substructures are observed, including discrete $\{MoO_6\}$ octahedra, binuclear, trinuclear, tetranuclear, pentanuclear and hexanuclear (in the arsonate case only) units. While the $\{(Mo_5O_{15})(O_3P\text{–})\}_2{}^{4-}$ motif appears in eight of 17 structures, the overall architecture may be one-dimensional (**1, 4–6, 9, 10** and **13**), two-dimensional (**2, 7, 8, 11, 12** and **14–16**) or three dimensional (**3** and **17**). The $\{(Mo_5O_{15})(O_3P\text{–})_2\}^{4-}$ cluster core is associated exclusively with phosphomolybdate chain substructures when the phosphonate component is a diphosphonate tethering ligand, as in structures **4–8**. In the monophosphonate or hydrogen phosphate structures of Type v (**1–3**), the phosphomolybdate substructure is an embedded cluster. While the cluster motif spans diphosphonates $\{O_3P(CH_2)_nPO_3\}^{4-}$ for $n = 2$, 3 and 4, its presence is also contingent on the organoimine coligand. For example, the cluster substructure is observed for the propylenediphosphonate derivatives **5** and **7**, but is absent in the structure of **16**. Similarly, the cluster substructure is present for the ethylenediphosphonate series in the case of **6** and **8** but absent for **14** and **15**.

While four phases exhibit binuclear molybdate building blocks, three involve edge-sharing octahedra, but the fourth is an unusual example of a face-sharing binuclear unit. A curious feature of these structures is that two exhibit two-dimensional phosphomolybdate substructures, **12** and **16**, while two contain one-dimensional phosphomolybdate components, **10** and **11**.

The two phases incorporating the tetranuclear molybdate core, **14** and **15**, are the only isomorphous pair to be revealed to date. This may suggest that under appropriate conditions, isomorphous phenanthroline derivatives of **4, 9, 12** and **16** may be accessible.

5 CONCLUSIONS

A series of novel and complex structure types of the oxomolybdate–phosphonate–Cu(II)–organoimine family have been isolated from hydrothermal media and structurally characterized. The results reinforce the observation that hydrothermal chemistry offers an effective synthetic tool for the isolation of composite organic–inorganic materials. It is also clear that organic ligands with specific geometric requirements may be introduced as structural components of materials. Bridging ligands may be used to propagate the architecture about a metal site, while the steric influences of coligands to a secondary metal site may be exploited to dictate the dimensionality of the product. Manipulation of the microstructure of the solid is thus achieved by tuning the coordination influences of the metal to the geometric requirements of the ligand.

It is noteworthy that the materials of this study require the presence of the secondary metal–ligand coordination complex cation for isolation. This structural component serves not only a space-filling and charge-compensating role but is also intimately involved in structural propagation in one or two dimensions.

While it is now evident that organic components may be introduced into the synthesis solid-state inorganic materials in order to manipulate the coordination chemistry of the metal and consequently the structure of the material, designed extended structures remain elusive in the sense of predictability of the final structure. However, this observation reflects the compositional and structural versatility of inorganic materials and should be considered an opportunity for development rather than an occasion for lamentation. The subtle interplay of metal oxidation states, coordination preferences, polyhedral variability, ligand donor group types and orientations, spatial extension and steric constraints provides a limitless set of construction components for solid-state materials. As the products of empirical observations are elucidated, the structure–function relationships of such components will begin to emerge and to provide further guidelines for synthetic methodologies.

ACKNOWLEDGMENT

This work was supported by a grant from the National Science Foundation (CHE 9987471).

REFERENCES

1. Greenwood, N. N.; Earnshaw, A., *Chemistry of the Elements*. Pergamon Press, New York, 1984.
2. Wells, A. F., *Structural Inorganic Chemistry*, 4th edn. Oxford University Press, Oxford, 1978.

3. Cheetham, A. J., *Science* **1994**, *264*, 794.
4. Reynolds, T. G.; Buchanan, R. C., *Ceramic Materials for Electronics*, 2nd edn. Marcel Dekker, New York, 1991, p. 207.
5. Büchner, W., Schliebs, R.; Winter, G.; Büchel, K. H., *Industrial Inorganic Chemistry*. VCH, New York, 1989.
6. McCarroll, W. H., *Encyclopedia of Inorganic Chemistry*, Vol. 6. Wiley, New York, 1994, p. 2903.
7. Haertling, G. H., *Ceramic Materials for Electronics*, 2nd edn. Marcel Dekker, New York, 1991, p. 129.
8. Leverenz, H. W., *Luminescence of Solids*. Wiley, New York, 1980.
9. Einzinger, R., *Annu. Rev. Mater. Sci.* 1987, *17*, 299.
10. Tarascon, J.-M.; Barboux, P.; Miceli, R. F.; Greene, L. H.; Hull, G. W.; Elbschutz, M.; Sunshine, S. A., *Phys. Rev. B* **1988**, *37*, 7458.
11. Matkin, D. I., *Modern Oxide Materials*. Academic Press, New York, 1972, p. 235.
12. Rao, C. N. R.; Rao, K. J., in *Solid State Chemistry: Compounds*. Eds. A. K. Cheetham and P. Day Clarendon Press, Oxford, 1992, p. 281.
13. Bierlein, J. D.; Arweiler, C. B., *Appl. Phys. Lett.* **1987**, *49*, 917.
14. Centi, G.; Trifuro, Ebner, J. R.; Franchett, V. M., *Chem. Rev.* **1988**, *88*, 55.
15. Grasselli, R. K., *Appl. Catal.* **1985**, *15*, 127.
16. Gasior, M.; Gasior, I.; Grzybowska, B., *Appl. Catal.* **1984**, *15*, 87.
17. Okuhara, T.; Misono, M., *Encyclopedia of Inorganic Chemistry*, Vol. 6. Wiley, New York, 1994, p. 2889.
18. Niiyama, H.; Echigoya, E., *Bull. Chem. Soc. Jpn.* **1972**, *45*, 83.
19. Yamaguchi, T., *Appl. Catal.* **1990**, *61*, 1.
20. Clearfield, A., *Chem. Rev.* **1988**, *88*, 125.
21. Newsam, J. M., in *Solid State Chemistry: Compounds*. Eds. A. K. Cheetham and P. Day Clarendon Press, Oxford, 1992, p. 234.
22. Ruthven, D. M., *Principles of Absorption and Absorption Processes*. Wiley-Interscience, New York, 1984.
23. Szostak, R., *Molecules Sieves. Principles of Synthesis and Identification*. Van Nostrand Rheinhold, New York, 1988.
24. Raleo, J. A., *Zeolite Chemistry and Catalysis*, ACS Monograph No. 7. American Chemical Society, Washington, DC, 1976.
25. Murakami, Y.; Iijima, A.; Ward, J. W. (Eds), *New Developments in Zeolite Science*. Elsevier, Amsterdam, 1986.
26. (a) Vaughan, D. E. W., *Properties and Applications of Zeolites*, Chemical Society Special Publication No. 33. Chemical Society, London, 1979, p. 294; (b) Cheetham, A. K., *Science* **1994**, *264*, 794.
27. Venuto, P. B., *Microporous Mater.* **1994**, *2*, 297.
28. Stupp, S. I.; Braun, P. V. *Science* **1997**, *277*, 1242.
29. Whitesides, G. M.; Ismagilov, R. F. *Science* **1999**, *284*, 89.
30. Hench, L. L., Inorganic Biomaterials, in *Materials Chemistry, an Emerging Discipline*, ACS Symposium Series, Vol. 245, Interrante L. V.; Casper L. A.; Ellis, A.-B. (Eds). American Chemical Society, Washington, DC, 1995, p. 523; Mamm, S. *J. Chem. Soc., Dalton Trans.* 3953 (1997).
31. Hagrman, P. J.; Hagrman, D.; Zubieta, J. *Angew Chem., Int. Ed. Engl.* **1999**, *38*, 2638.
32. Yaghi, O. M.; Li, Hi., Davis, C., Richardson, D.; Gray, T. L. *Acct. Chem. Res.* **1998**, *31*, 474.
33. Cheetham, A. K.; Fuay, G.; Loisean, T. *Angew. Chem., Int. Ed. Engl.* **1999**, *38*, 3268.
34. (a) Raeo, J. A., *Zeolite Chemistry and Catalysis*, ACS Monograph No. 7. American Chemical Society, Washington, DC, 1976; (b) Murakami, Y.; Iijima, A.; Ward, J. W.

(Eds), *New Development in Zeolite Science*. Elsevier, Amsterdam, 1986; (c) Vaughan, D. E. W., *Properties and Applications of Zeolites*, Chemical Society Special Publication No. 33, Chemical Society, London, 1979, p. 294; (d) Venuto, P. B., *Microporous Mater.* **1994**, 2, 297; (e) Szostak, R., *Molecular Sieve – Principles of Synthesis and Identification*, 2nd edn. Van Nostrand Rheinhold, New York, 1997.

35. (a) Smith, J. V., *Chem. Rev.* **1988**, *88*, 149; (b) Occelli, M. L.; Robson, H. C., *Zeolite Synthesis*. American Chemical Society, Washington, DC, 1989.
36. Barrer, R. M., *Hydrothermal Chemistry of Zeolites*. Academic Press, New York, 1982.
37. Kresge, C. T., Leonowicz, M. E.; Roth, W. J.; Vartuli, J. C.; Beck, J. S., *Nature* **1992**, *359*, 710.
38. The oxomolybdenum phosphates with entrained organic captions have been reviewed: Haushalter, R. C.; Mundi, L. A., *Chem. Mater.* **1992**, *4*, 31.
39. Khan, M. L.; Meyer, L. M.; Haushalter, R. C.; Schweitzer, C. L.; Zubieta, J.; Dye, J. L., *Chem. Mater.* **1996**, *8*, 43.
40. Khan, M. I.; Haushalter, R. C.; O'Connor, C. J.; Tao, C.; Zubieta, J., *Chem. Mater.* **1995**, *7*, 593.
41. Bircsak, Z.; Hall, A. K.; Harrison, W. T. A., *J. Solid State Chem.* **1999**, *142*, 168.
42. Chippindale, A. M., *Chem. Mater.* **2000**, *12*, 818.
43. Soghomonian, V.; Haushalter, R. C.; Chen, Q.; Zubieta, J., *Inorg. Chem.* **1994**, *31*, 25.
44. Lu Y.; Haushalter, R. C.; Zubieta, J., *Inorg. Chim. Acta* **1997**, *257*, 268.
45. Bircsak, Z.; Harrison, W. T. A., *Inorg. Chem.* **1998**, *37*, 3204.
46. Finn, R. C.; Zubieta, J., unpublished results.
47. Soghomonian, V.; Chen, Q.; Zhang, Y.; Haushalter, R. C.; O'Connor, C. J.; Tao, C.; Zubieta, J., *Inorg. Chem.* **1995**, *34*, 3509.
48. Soghomonian, V.; Chen, Q.; Haushalter, R. C.; Zubieta, J.; O'Connor, C. J.; Lee, Y.-S., *Chem. Mater.* **1993**, *5*, 1690.
49. Loiseau, T.; Fèrey, G., *J. Solid State Chem.* **1994**, *111*, 416.
50. Soghomonian, V.; Chen, Q.; Haushalter, R. C.; Zubieta, J., *Chem. Mater.* **1996**, *35*, 2826.
51. Soghomonian, V.; Chen, Q.; Haushalter, R. C.; Zubieta, J., O'Connor, C. J., *Inorg. Chem.* **1996**, *35*, 2826.
52. Bu, X.; Feng, P.; Stucky, G. D., *Chem. Commun.* **1995**, 1337.
53. Soghomonian, V.; Chen, Q.; Haushalter, R. C.; Zubieta, J., *Angew. Chem., Int. Ed. Engl.* **1993**, *32*, 610.
54. Harrison, W. T. A.; Hsu, K.; Jacobson, A. J., *Chem. Mater.* **1995**, *7*, 2004.
55. Zhang, Y.; Clearfield, A.; Haushalter, R. C., *Chem. Mater.* **1995**, *7*, 1221.
56. Hagrman, P. J.; LaDuca, R. L. Jr; Koo, H.-J.; Rarig, R. S. Jr; Haushalter, R. C.; Whangbo, M.-H.; Zubieta, J., *Inorg. Chem.* **2000**, *39*, 4311.
57. Hagrman, P. J.; Zubieta, J., *Inorg. Chem.* **2000**, *39*, 3252.
58. Zhang, Y.; DeBord, J. R. D.; O'Connor, C. J.; Haushalter, R. C.; Clearfield, A.; Zubieta, J., *Angew. Chem., Int. Ed. Engl.* **1996**, *35*, 989.
59. Debord, J. R. D.; Zhang, Y.; Haushalter, R. C.; Zubieta, J.; O'Connor, C. J., *Solid State Chem.* **1996**, *122*, 251.
60. LaDuca, Jr., R. L.; Finn, R. C.; Zubieta, J., *Chem. Commun.* **1999**, 1669.
61. LaDuca, R. L. Jr; Rarig, R. S. Jr; Zubieta, J., *Inorg. Chem.* **2001**, **40**, 40.
62. Ollivier, P. J.; DeBord, J. R. D.; Zapf, P. J.; Zubieta, J,; Meyer, L. M.; Wang, C.-C.; Mallouk, T. E.; Haushalter, R. C., *J. Mol. Struct.* **1998**, *470*, 49.
63. Hagrman, P. J.; Bridges, C.; Greedan, J. E.; Zubieta, J., *J. Chem. Soc., Dalton Trans.* **1999**, 2901.
64. LaDuca, Jr. R. C.; Brodkin, C.; Finn, R. C.; Zubieta, J., *Inorg. Chem. Commun.* **2000**, *3*, 248.

65. Hagrman, D.; Zubieta, C.; Haushalter, R. C.; Zubieta, J., *Angew. Chem., Int. Ed. Engl.* **1999**, *38*, 2638.
66. Hagrman, D.; Zubieta, C.; Haushalter, R. C.; Zubieta, J., *Angew. Chem., Int. Ed. Engl.* **1997**, *36*, 873.
67. Hagrman, D.; Zubieta, C.; Haushalter, R. C.; Zubieta, J., *J. Chem. Soc., Dalton Trans.* **1999**, 3707.
68. Hagrman, D.; Hagrman, P.; Zubieta, J., *Inorg. Chim. Acta* **2000**, *300–302*, 212.
69. DeBord, J. R. D.; Haushalter, R. C.; Meyer, L. M.; Rose, D. J.; Zapf, P. J.; Zubieta, J., *Inorg. Chim. Acta* **1997**, *256*, 165.
70. Hagrman, D.; Zapf, P. J.; Zubieta, J., *Chem. Commun.* **1998**, 1283.
71. Hagrman, D.; Hagrman, P. J.; Zubieta, J., *Angew. Chem., Int. Ed. Engl.* **1999**, *38*, 3165.
72. Hagrman, D.; Zubieta, J., *C. R. Acad. Sci., Ser. IIc* **2000**, *3*, 231.
73. Hagrman, D.; Hagrman, P. J. Zubieta, J., *Comments Inorg. Chem.* **1999**, *21*, 225, and references therein.
74. Chesnut, D. J.; Hagrman, D.; Zapf, P. J.; Hammond, R. P.; LaDuca, Jr; Haushalter, R. C.; Zubieta, J., *Coord. Chem. Rev.* **1999**, *190–192*, 737.
75. Zapf, P. J.; Hammond, R. P.; Haushalter, R. C.; Zubieta, J., *Chem. Mater.* **1998**, *10*, 1366.
76. Hagrman, D.; Warren, C. J.; Haushalter, R. C.; Seip, C.; O'Connor, C. J.; Rarig, R. S. Jr.; Johnson, K. M. III; LaDuca, R. L. Jr; Zubieta, J., *Chem. Mater.* **1998**, *10*, 3294.
77. Hagrman, D.; Haushalter, R. C.; Zubieta, J., *Chem. Mater.* **1998**, *10*, 361.
78. Laskoski, M. C.; LaDuca, R. L. Jr; Rarig, R. S. Jr; Zubieta, J., *J. Chem. Soc., Dalton, Trans.* **1999**, 3467.
79. Hagrman, D. E.; Zubieta, J., *J. Solid State Chem.* **2000**, *152*, 141.
80. LaDuca, R. L. Jr.,; Desciak, M.; Laskoski, M.; Rarig, R. S.; Zubieta, J., *J. Chem. Soc., Dalton Trans.* **2000**, 2255.
81. Zapf, P. J.; Warren, C. J.; Haushalter, R. C.; Zubieta, J., *Chem. Commun.* **1997**, 1543.
82. Aschwanden, S.; Schmalle, H. W.; Reller, A.; Oswald, H.R., *Mater. Res. Bull.* **1993**, *28*, 45. For other examples of the V/O/M′ ligand family see also: (a) Lin, B.-Z .; Liu, S.-X., *Polyhedron* **2000**, *19*, 2521; (b) Zheng, L.-M.; Zhao, J.-S.; Li, K.-H.; Zhang, L.-Y.; Liu, Y.; Xin, X.-Q., *J. Chem. Soc., Dalton Trans.* **1999**, 939; (c) Chen, R.; Zavalij, P. Y.; Whittingham, M. S.; Greedan, J. E.; Raju, N. P.; Bieringer, M., *J. Mater. Chem.* **1999**, 93; (d) Shi, Z.; Zhang, L.; Zhu, G.; Yang, G.; Hua, J.; Ding, H.; Feng, S., *Chem. Mater.* **1999**, *11*, 3565; (e) Law, T. S.-C.; Williams, I. D., *Chem. Mater.* **2000**, *12*, 2070; (f) Zheng. L.-M.; Whitfield, T.; Wang, X.; Jacobson, A. J., *Angew. Chem., Int. Ed. Engl.* **2000**, *39*, 4528.
83. Pope, M. T., *Heteropoly and Isopoly Oxometalates*. Springer, New York, **1983**.
84. See, for example: (a) Hoskins, B. F.; Robson, R., *J. Am. Chem. Soc.* **1990**, *112*, 1546; (b) Yaghi, O. M.; Li, H., *J. Am. Chem. Soc.* **1995**, *117*, 10401; (c) Carlucci, L.; Gianfranco, C.; Prosperio, D. M.; Sironi, A., *J. Am. Chem. Soc.* **1995**, *117*, 12861; (d) MacGillivray, L. R.; Subramanian, L. R.; Zaworotko, M. J., *J. Chem. Soc. Chem. Commun.* **1994**, 1325; (e) Blake, A. J.; Champness, N. R.; Chung, S. S. M.; Li, W.-S.; Schröder, M., *Chem. Commun.* **1997**, 1005; (f) Noro, S.; Kitagawa, S.; Kondo, M.; Seki, K., *Angew. Chem. Int. Ed. Engl.* **2000**, 39, 2082; (g) Carlucci, L.; Ciani, G.; Proserpio, D. M.; Rizzata, S., *Chem. Commun.* **2000**, 1319; (h) Reineke, T. M.; Eddaoudi, M.; Moler, O'Keefe, M.; Yaghi, O. M., *J. Am. Chem. Soc.* **2000**, *122*, 4843.
85. See, for example: (a) Naumov, N. G.; Virovets, A. V.; Sokolov, M. N.; Artemkina, S. B.; Fedorov, V. E., *Angew. Chem., Int. Ed. Engl.* **1998**, *37*, 1943; (b) Beauvais, L. G.; Shores, M. P.; Long, J. R., *J. Am. Chem. Soc.* **1999**, *121*, 775; (d) Müller, A.; Das, S. K.; Kögerler, P.; Bögge, H.; Schmidtmann, M.; Trautwein, A. X.; Schünemann, V.;

Krickmeyer, E.; Preetz, W. *Angew. Chem., Int. Ed. Engl.* **2000**, *39*, 3414; (e) Khan, M. I.; Yohannes, E.; Powell, D., *Inorg. Chem.* **1999**, *38*, 212; (f) Ouahab, L., *C. R. Acad. Sci., Ser. IIc* **1998**, 369; (g) Galán-Mascarós, J.; Giménez-Saiz, C.; Triki, S.; Gómez-Garcia, C. J.; Coronado, E.; Quahab, L., *Angew. Chem., Int. Ed. Engl.* **1995**, *34*, 1460; (h) Sadakane, M.; Dickman, M. H.; Pope, M. T., *Angew. Chem., Int. Ed. Engl.* **2000**, *39*, 2914; (i) Krebs, B.; Loose, I.; Bösing, M.; Nöh, A.; Droste, E., *C. R. Acad. Sci., Ser. IIc* **1998**, 351; (j) Stein, A.; Fendorf, M.; Jarvie, T. P.; Mueller, K. t.; Bensei, A. J.; Mallouk, T. E., *Chem. Mater.* **1995**, *7*, 304; (k) Holland, B. T.; Isbester, P. K.; Munson, E. J.; Stein, A., *Mater. Res. Bull.* **1999**, *34*, 471.

86. Férey, G., *J. Solid State Chem.* **2000**, *152*, 37.
87. Zaworotko, M. J., *Angew. Chem., Int. Ed. Engl.* **2000**, *39*, 3052.
88. Kwak, W.; Pope, M. T.; Scully, T. F., *J. Am. Chem. Soc.* **1975**, *97*, 5735.
89. Finn, R. C.; Zubieta, J., *Inorg. Chem.* **2000**, *40*, 2466.
90. Figlarz, M., *Chem. Scr.* **1988**, *28*, 3.
91. (a) Livage, J., *Chem. Scr.* **1988**, *28*, 9; (b) Rouxel, J. *Chem. Scr.* **1988**, *28*, 33.
92. Khan, M.I.; Zubieta, J., *Prog. Inorg. Chem.* **1995**, *43*, 1.
93. Gopalakrishnan, J., *Chem. Mater.* **1995**, *7*, 1265
94. Lobachev, A. N., *Crystallization Processes Under Hydrothermal Conditions.* Consultants Bureau, New York, 1973.
95. Rabenau, A., *Angew. Chem., Int. Ed. Engl.* **1985**, *24*, 1026.
96. Feng, S.; Xu, R., *Acc. Chem. Res.* **2001**, *34*, 239.
97. Hedman, B., *Acta Chem. Scand.* **1977**, *27*, 3335.
98. Kwak, W.; Pope, M. T.; Scully, T. F., *J. Am. Chem. Soc.* **1975**, *97*, 5735.
99. Chang, Y.-D., Zubieta, J., *Inorg. Chim. Acta* **1996**, *248*, 177.
100. Cao, G.; Haushalter, R. C.; Strohmaier, K. G., *Inorg. Chem.* **1993**, *32*, 127.
101. Khan, M. I.; Chen, Q.; Zubeita, J., *Inorg. Chim. Acta* **1993**, *206*, 131.
102. Poojary, D. M.; Zhang, Y.; Clearfield, A., *Chem. Mater.* **1995**, *7*, 822.
103. Harrison, W. T. A.; Dussack, L. L.; Jacobson, A., J. *Inorg. Chem.* **1995**, *34*, 4774.
104. Lu, M.-Y.; Wang, S.-L., *Chem. Mater.* **1999**, *11*, 3588.
105. Canadell, E.; Provost, J.; Guesder, A.; Borel, M. M.; Leclair, A., *Chem. Mater.* **1997**, *9*, 68.
106. Kierkegaard, P., *Acta Chem. Scand.* **1958**, *12*, 1701; Weller, M. T.; Bell, R. G., *Acta Crystallogr.* **1988**, *C44*, 1516.
107. Bonavia, G.; Haushalter, R. C.; O'Connor, C. J.; Zubieta, J., *Inorg. Chem.* **1996**, *35*, 5603.
108. Soghomonian, V.; Chen, O.; Haushalter, R. C.; Zubieta, J., *Angew. Chem., Int. Ed. Engl.* **1995**, *34*, 223.
109. Huan, G.; Johnson, J. W.; Jacobson, A. J.; Merola, J. S., *J. Solid State Chem.* **1990**, *89*, 220.
110. Hinclaus, C.; Serre, C.; Riou, D.; Férey, G., *C. R. Acad. Sci., Ser. IIc* **1998**, *1*, 551.

Chapter 7

A Rational Approach for the Self-Assembly of Molecular Building Blocks in the Field of Molecule-Based Magnetism

MELANIE PILKINGTON AND SILVIO DECURTINS

University of Berne, Berne, Switzerland

1 INTRODUCTION

Although, historically, crystal engineering has its roots in the field of solid-state organic chemistry, in recent years it has developed into a hybrid discipline spanning many areas of chemistry with relevant interdisciplinary interactions with biology, informatics and physics [1]. This expansion can largely be attributed to the 'marriage' of supramolecular chemistry with materials science. To date, the field of crystal engineering draws its strength from the synergistic interaction between the design and synthesis of supermolecules on the one hand, and the design and synthesis of crystalline materials with desired solid-state properties on the other. It has been recently pointed out [2] that the definition of supramolecular chemistry put forward by Jean-Marie Lehn [3], namely 'chemistry beyond the molecule bearing on the organised entities of higher complexity that result in the association of two or more chemical species held together by intermolecular forces', also encompasses crystal engineering, since one can describe a molecular crystal as being an 'organised entity of higher complexity held together by intermolecular forces'.

Crystal Design: Structure and Function, edited by G. R. Desiraju

During the last two decades, the field of supramolecular chemistry has altered the way in which chemists think about crystals. Instead of focusing on molecular geometry, with a tendency perhaps to overlook the way in which the molecules are arranged in three-dimensional space, the approach is now to think 'supramolecularly' with a higher hierarchy of a crystalline arrangement in mind. The need for this new awareness 'making crystals with a purpose' has perhaps been most visible within the inorganic chemistry community, since it is in the area of coordination chemistry that the directed synthesis of molecules for their packing properties has been redefined within the field of 'supramolecular chemistry' (Figure 1).

During recent years, a plethora of eloquent research has been conducted in this discipline, resulting in an extensive array of helices, one-dimensional coordination polymers, two-dimensional grids and three-dimensional lattices [4,5]. The most common strategy is that of rational design, which utilises molecular building blocks as nucleophilic spacers that have been encoded with the information necessary to pre-determine the overall structure of the resulting product. This strategy relies on the proper programming of angular and dimensional information into both the electrophilic metals and the nucleophilic spacers [6]. In doing so, the topology of the ensuing product can be controlled and/or directed. Crystal-oriented strategies still closely resemble classical chemistry methods in the sense that molecules are modelled, synthetic routes are devised and the products are characterised and their physical properties investigated. Essentially, these processes are repeated twice, in the first instance to prepare the building blocks, and in the second to arrange them in the desired way to obtain the desired physical properties [2]. It is this second self-assembly process which has altered the way in which chemists are thinking about chemistry. One advantage in favour of this approach is that it is very often reversible, a factor that has been suggested to be crucial for the assembly of ordered networks amenable to single-crystal diffraction study. Furthermore, learning how to create large supramolecular units and the elucidation of rules mediating their macroscopic organisation into multifunctional materials offers fantastic prospects for the manipulation of not only the features of specific molecules but also the bulk, intermolecular characteristics and properties of the entire crystalline aggregate.

Driving this field forward are the potential technological applications associated with molecule-based materials. As the limits of silicon are approached, the possibility of 'engineering up' from a molecule to a functioning electronic device offers an alternative route to the design of materials for use on the nanoscale and beyond [7]. The aim is to design and assemble molecular devices that will reduce dimensions and increase speed by orders of magnitude. Chemists are therefore targeting molecular systems that are easily switchable between two available states. In this respect, molecular magnetism is a rapidly developing field since nanomagnetic materials in general easily reverse their magnetic moments in a magnetic field and are already widely used in information storage. Efforts in this area are strongly driven by the microelectronics industry, since single-molecular magnets represent the ultimate in miniaturisation. Furthermore, as we enter the twenty-first century, magnets are often the invisible, but essential, components of a large number of widely used devices which include medical implants, loudspeakers, microphones, headphones, switches, sensors, fax machines, magnetic disks, electrical motors and generators, and also magnetic shielding for high-voltage electrical lines and communication equipment [8].

Molecular magnetism can be considered to be supramolecular in nature, since it results from the collective features of components bearing free spins and on their arrangement in organised assemblies [9]. The engineering of molecular magnetic systems thus requires a search for paramagnetic species linked via organic ligands, and their arrangement in suitable supramolecular architectures so as to induce spin coupling and alignment. In comparison with the more conventional metal/intermetallic and metal oxide magnets, molecule-based magnets offer the additional advantages of low density, transparency, electrical insulation and low-temperature fabrication.

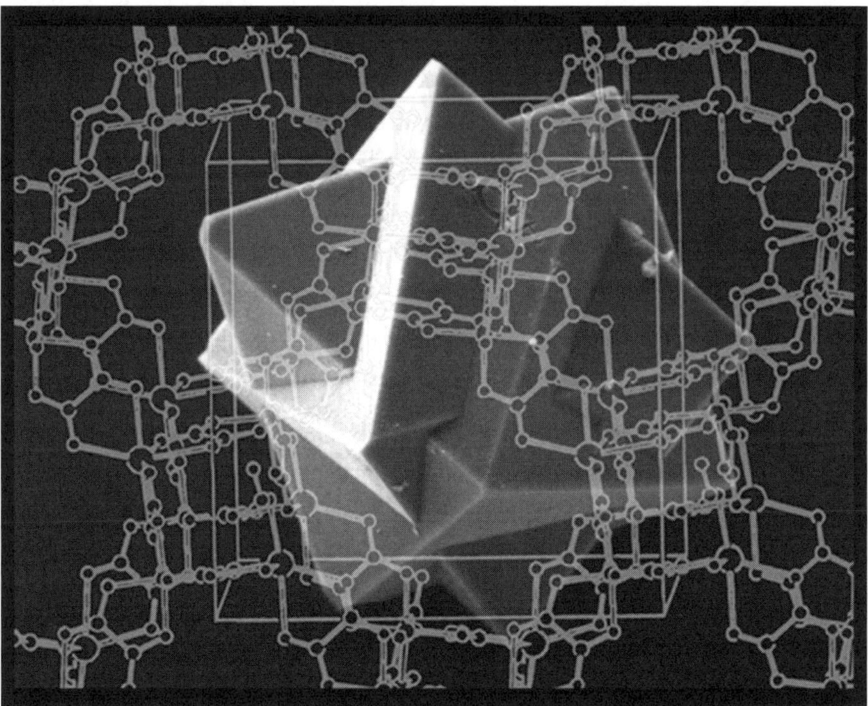

Figure 1 Single crystal of a chiral 3D molecular network compound with transition metals as spin carriers.

Over the last decade, the intense efforts of several research groups worldwide has led to the discovery of several classes of molecule-based magnets [1]. The first reported magnet reported composed of spins in isolated d-orbitals was an Fe^{III} complex of stoichiometry $Fe^{III}(S_2CNEt_2)_2Cl$ with an intermediate spin $S = \frac{3}{2}$ [10–13]. This compound orders ferromagnetically at 2.46 K, but does not exhibit the phenomenon of hysteresis. The next class of magnets prepared were the ferromagnetically ordered compounds $[Cr^{III}(NH_3)_6]^{3+}[Fe^{III}Cl_6]^{3-}(T_c = 0.66\,K)$ [14] and $[Cr^{III}(NH_3)_6]^{3+}[Cr^{III}(CN)_6]^{3-}(T_c = 2.85\,K)$ [15]. These early studies led to the development of materials with d-orbital spin sites connected via coordinate bonds using rational synthetic organic methodology. Like conventional magnets, such as Fe_3O_4, Co_5Sm and CrO_2, they have an extended network linking the d-orbital sites.

Following these discoveries, the detailed study of network-structured solids with free spins residing on adjacent sites being in orthogonal orbitals led to a ferromagnetic coupling arrangement, and ferromagnetic ordering evolved [16,17]. Molecule-based magnets currently exhibit a wide range of bonding and structural motifs. These include isolated molecules (zero-dimensional), and also those with extended bonding within chains (1-D), within layers (2-D) and within (3-D) network structures. A favourable approach for the design of crystalline solids is to use pre-organised molecules such as transition metal complexes, since a high structural organisation can be ensured through the multiple binding of transition metal ions, giving rise to a variety of extended inorganic networks in one, two or three dimensions [18]. The aim of this chapter is to highlight how a rational design strategy has been applied to cyanometalate and tris oxalato transition metal building blocks for the preparation of two classes of novel molecule-based materials displaying a range of interesting magnetic properties.

2 CYANOMETALATE BUILDING BLOCKS

The field of transition metal cyanide chemistry has a remarkable history [19], but is currently experiencing a revival, offering exciting prospects in the field of materials chemistry. In this respect, one class of interesting compounds are the novel architectures self-assembled from cyanide ligands and metallic centres that carry a magnetic moment. Since μ-cyanide linkages permit an interaction between paramagnetic metal ions, cyanometalate building blocks have currently found useful applications in the field of molecule-based magnetism [20–22].

2.1 Prussian Blue

The use of octahedral building blocks is perhaps one of the most obvious and simplest strategies for the assembly of molecule-based three-dimensional solids

[20,21]. In this context, an examination of suitable octahedral $[ML_6]$ precursors reveals that some of the most inert are the hexacyanometalate anions $[M'(III)(CN)_6]^{3-}$(M' = Cr, Mn, Fe, Co), which can be combined with paramagnetic divalent Lewis acids M(II)(M = Cu, Ni, Co, Fe, Mn) to give face-centred cubic compounds closely related to Prussian blue. Prussian blue is one of the first synthetic coordination compounds reported in the chemical literature [23], and owes its name to its colour, since mixing together an aqueous solution of $K_4[Fe(CN)_6]$ and Fe(III) chloride affords a beautiful deep-blue colour. The formula of Prussian blue itself is $Fe_4^{III}[Fe^{II}(CN)_6]_3 \cdot 14H_2O$ and the reaction leading to the neutral solid is a simple Lewis acid–base interaction:

$$q[Fe^{II}(CN)_6]^{p-}_{(aq)}(\text{Lewis base}) + pM^{IIIq+}{}_{(aq)}(\text{Lewis acid}) \longrightarrow \{Fe^{III}{}_p[M^{II}(CN)_6]_q\}^0 \cdot xH_2O_{(solid)}$$

The structure of Prussian blue was first proposed by Keggin and Miles [24], in 1936 and then reformulated by Ludi and Güdel [25] in 1973. Essentially, the face-centred cubic structure of the earlier model postulating the occurrence of interstitial Fe(III) ions was modified to a cubic lattice with a random to more ordered arrangement of a stoichiometrically determined fraction of vacant lattice sites. Applying the revised structural model to Prussian blue, it becomes apparent that due to coordinated water molecules, the Fe(III) ions are in an average $N_{4.5}O_{1.5}$ coordination environment. A recent search of the Cambridge Crystallographic Structural Database revealed that to date there is only one single-crystal structure of a 'perfectly ordered' cubic Prussian blue-like phase (Figure 2). This compound is a Cd/Pd analogue of stoichiometry $[Cd(II)Pd(IV)(CN)_6]$ [26] and does not contain any vacant metal sites or coordinated water molecules. It should be stressed, however, that these crystals are colourless and diamagnetic and so are not useful for magnetostructural correlations.

The appeal of Prussian Blue itself lies not only in its optical properties, but also in its magnetic properties, since it displays long-range ferromagnetic ordering at $T_c = 5.6\,K$. This critical temperature is low since only the high-spin d^5 Fe(III) ($S = 5/2$) sites carry a spin. As for the Fe(II) centres, they are low spin and diamagnetic, hence the magnetic interactions occur between next-nearest metal ions through 10.6 Å long Fe(III)–NC–Fe(II)–CN–Fe(III) linkages [7]. This propagation of magnetic interactions through extended bridging networks is due to the strong spin delocalisation from the metal ion towards its nearest neighbours, and seems to be specific to molecular compounds. In recent years, the Prussian blue system has been re-investigated with a view to engineering molecular solids with orbital interactions in three dimensions, in an attempt to overcome the low Curie temperatures which plagued the low-dimensional materials around at the time [27]. As a consequence, substitution of Fe^{II} and Fe^{III} by other paramagnetic metal ions M and M′ has afforded a series of Prussian blue analogues with face-centred cubic structural topologies and interesting magnetic properties [1,20,21].

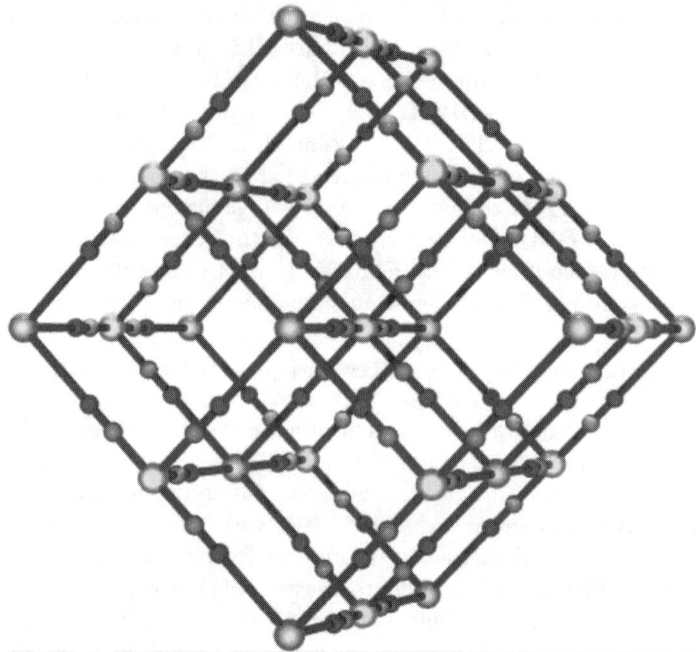

Figure 2 Three-dimensional cubic structure of an idealised Prussian blue analogue.

2.2 Prussian Blue Analogues

The appealing character of the Prussian blue-like phases is related to (i) their possibilities for developing orbital interactions in three dimensions, (ii) their tolerance to host at will, in the cubic lattice, very different paramagnetic ions, (iii) the high symmetry of the cubic system, where the linear M–CN–M′ allows the nature and amplification of the exchange interaction between the two paramagnetic metal centres to be tuned, and (iv) the possibilities for varying the number of magnetic neighbours (z) around the metal centres M and M′ with the stoichiometry, which in turn enables the Curie temperature to be varied.

The need for higher ordering temperatures has provided the driving force for several research teams to probe in detail the nature of the M′–NC–M interactions in this family of compounds. Among the pioneering studies by Anderson and co-workers [28] and Bozworth *et al.* [29] in the 1950s and Klenze *et al.* [30] in the 1980s, the results of Babel [31,32] led to the ferromagnetic compound $Cs^{I}Mn^{II}[Cr^{III}(CN)_6]$ [31], with a Curie temperature above liquid nitrogen temperature, $T_c = 90\,K$. Kahn's orbital model [9] was then applied to these systems in order to address the challenge and obtain materials that could be easily incorporated into devices. In 1995, this systematic approach paid off and Verdaguer and co-workers [33] were able to increase the Curie temperature and overcome the room-temperature barrier with the mixed-valent Prussian blue $V^{II}_{0.42}V^{III}_{0.58}[Cr(CN)_6]_{0.86} \cdot 2.8H_2O$, which orders at $T_c = 315\,K$. This compound represents a milestone in the field of molecule-based magnetism, since it is the first rationally synthesised molecule-based magnet whose T_c is above room temperature. The ferrimagnetic properties displayed by this compound directly arise from the mixed-valent V_3Cr_2 stoichiometry. Following this discovery, it was suggested that the Curie temperature could be further increased by improving the crystallinity of the sample, or by changing the stoichiometry to increase the number of magnetic neighbours. Adopting this strategy, several more complex, air-stable related materials with T_cs up to 372 K were also prepared [34]. Holmes and Girolami [35] were able to reach a T_c of 376 K, which is above the boiling point of water and currently holds the record in this series of Prussian blue analogues.

One of the appealing aspects of magnetic studies dealing with Prussian blue phases clearly resides in the fact that it is now possible to predict the nature and estimate the value of the critical temperature using simple theoretical models based on the symmetry of the singly occupied orbitals. The cubic symmetry of the Prussian blue phases does, however, come at a price. The most common problem encountered during the course of these research efforts is the continuous struggle with amorphous or poorly crystalline compounds with peculiar stoichiometries. As far as we are aware, no group to date has been successful in growing single crystals of a Prussian blue phase suitable for detailed magnetic measurements. To the contrary, in very recent work, Decurtins and co-workers [36] have succeeded in growing air-stable single crystals of a transparently coloured, magnetically ordered, mixed-valence manganese Prussian blue analogue. This compound is particularly exciting since it crystallises into a perfectly ordered face-centred cubic lattice. Mixing together aqueous solutions of $Mn(NO_3)_2$ and KCN (in an agar gel) yielded the crystalline Prussian blue phase $\{Mn[Mn(CN)_6]\}_n \cdot 2.5nH_2O$. The crystal structure consists of an uninterrupted cubic Mn–CN–Mn framework (Figure 2). The molecule crystallises in the space group $Fm3m$, so both the N-bonded and the C-bonded manganese ions sit in a perfect octahedral coordination environment. From a consideration of the electroneutrality and stoichiometry of this compound, valencies of either +4 and +2, or +3 and +3 could be predicted for the manganese ions. Magnetic studies on this compound reveal that a long-range magnetic ordering transition occurs at $T_c = 30\,K$, proposing the onset of cooperative magnetic interactions. Further work to characterize fully the structural and magnetic properties of this compound is in progress.

2.3 Devices from Room Temperature Magnets

Although the magnetisation of the Prussian blue analogues is very small, since room temperature has been reached it is now possible to find applications for these materials in molecule-based devices. Verdaguer and co-workers have developed a device using the properties of a room temperature magnet to transform light into mechanical energy [20,21], (Figure 3). The room temperature (RT) magnet is sealed in a glass vessel under argon and suspended at the bottom of a pendulum [equilibrium position (2) in the absence of a permanent magnet]. When the temperature of the RT magnet is below T_c, it is attracted by a strong permanent magnet to position (1); at this point it is heated by a light source (lamp or solar energy). When the T_c is reached, the RT magnet it is no longer attracted to the permanent magnet, and moves away under the influence of its own weight (3). In doing so, it cools and, once its temperature has dropped to below T_c, it is once again attracted to the permanent magnet and the whole cycle is repeated. This device is one of the first examples of a molecule-based thermodynamic machine working between two energy reservoirs with similar temperatures (light and shadow), allowing the conversion of light into mechanical energy. Other applications for these compounds, e.g. as thermal probes or magnetic switches, are currently being investigated.

2.4 Phototomagnetism

As the limits of silicon are approached, molecular electronics offers an alternative route to the design of materials for use not only at the macroscopic scale but also

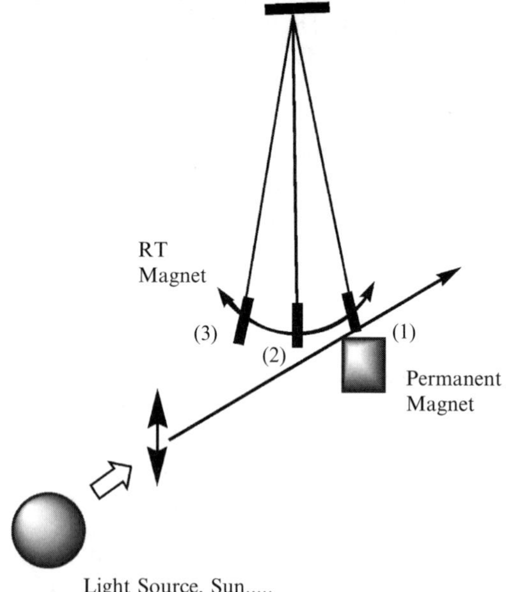

Figure 3 A molecule-based device using the properties of an RT magnet to transform light into chemical energy [20,21].

at the molecular scale. Data could be stored, transmitted and retrieved by means of molecules as in complex biological systems. Chemists are therefore looking for suitable molecular systems and for simple means to switch between them between two available states. Concerning the means, light is a very convenient and powerful way to induce molecular change. A new area in the field of molecule-based magnetism was opened up in 1996, when Hashimoto *et al.* [37] reported that they had found a way to switch the long-range magnetic properties of a simple molecule-based system derived from Prussian blue with light of different wavelengths. They were the first to observe an exciting photomagnetic effect in a Prussian blue analogue with stoichiometry close to $Co_3^{II}[Fe^{III}(CN)_6]_2 \cdot xH_2O$. The Curie temperature of the initial system is low, $T_c = 15\,K$, but a photoexcitation at the molecular level gives rise to a modification of the macroscopic properties of this material. Combining aqueous solutions of Co(II) and hexacyanoferrate(III), Hashimoto *et al.* obtained a powder which exhibited a photo-induced enhancement of the magnetisation at low temperature together with an increase in the Curie temperature. The explanation for this effect was related to the presence of isolated Co(II)–Fe(III) diamagnetic pairs in the compound, otherwise built from –Co(III)–NC–Fe(II)– units, and a photo-induced electron transfer from Fe(II) to Co(III) through the cyanide bridge. Before the excitation, the iron and cobalt ions are low spin and diamagnetic and there is no interaction between them. The red-light excitation induces spin on the ions [with CN surroundings, Fe(III) is low-spin ($S = \frac{1}{2}$), whereas Co(II) becomes high spin ($S = \frac{3}{2}$)] and a new interacting Fe(III)–Co(II) pair. The important point is that the local electron transfer at the molecular level switches on the interaction and allows the extension of the cooperative phenomenon throughout the network. It enhances the mean number of magnetic neighbours z. The ordering temperature T_c is therefore enhanced. This increase is weak (4 K), but significant. An even more exciting observation of Hashimoto *et al.* [37] is that the process can be partially reversed. Changing the wavelength of the light is enough to go back to a state that looks like the initial one and to switch off some Fe(III)–CN–Co(II) interactions (Figure 4).

In subsequent years, the preparation and study of related compounds have led to a better understanding of this phenomenon. It has now been established that the condition required to observe this effect is not only the presence of diamagnetic excitable pairs (LS)Co(III)–(LS)Fe(II), but also the presence of a certain amount of defects in the cubic structure [38]. This is due to the important expansion of the network that accompanies the photo-induced electron transfer. In a 'perfect structure', without vacant lattice sites, strain in the bulk would be so strong that only diamagnetic surface pairs would be transformed. In a structure with vacant lattice sites, however, the steric constraints are weaker, the network is more flexible and the photo-induced metastable state is able to propagate in the bulk [20]. Although the initial phenomenon observed by Hashimoto *et al.* occurs in too low a temperature range, and is too slow for practical applications, it nevertheless demonstrates that the tuning of long-range magnetic ordering is possible through

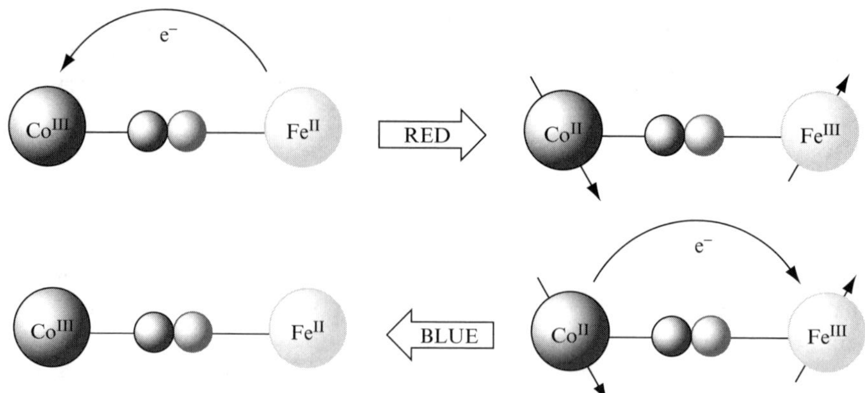

Figure 4 Back and forth electron transfer induced by photons of different wavelengths through a molecular CN bridge, and the related magnetic changes [37].

a molecular excitation induced by photons, and is clearly one of the necessary steps towards the design of molecule-based magneto-optical devices. The study of this phenomenon, with the aim of incorporating photophysical properties into magnetic materials, is currently ongoing. These observations, namely that magnetic ordering can be changed by photoinduced electron transfer through a molecular bridge, are a significant contribution towards applying molecular magnetism to molecular electronics.

The Prussian blues are not truly molecular but molecular-based, at the border between molecular and solid-state chemistry. They present many vacancies and defects and are often mixtures. Nevertheless, it is inspiring that one of the oldest coordination compounds is providing such a 'stir' in solid-state chemistry. Since the main features of the Prussian blue analogues are becoming increasingly understood, chemists have now begun to explore other directions.

2.5 1-D, 2-D and 3-D Compounds from $[M(CN)_6]^{n-}$ and Coordinatively Unsaturated $[M^{II}(L)_x]^{m+}$ Precursors

Extending this approach, using hexacyanometalate building blocks together with divalent transition metal complexes containing labile positions has afforded a variety of one-, two- and three-dimensional compounds with very different physical properties from those of the face-centred Prussian blues [20–22].

2.5.1 One-dimensional compounds

Reaction of a 3:2 molar ratio of $[Ni(en)_3]Cl_2$ and $K_3[M^{III}(CN)_6](M^{III} = Fe, Mn, Cr)$ in aqueous solution afforded large single crystals which could be characterized by X-ray crystallography [38]. The asymmetric unit consists of one $[M(CN)_6]^{3-}$ anion, one *cis*-$[Ni(en)_2]^{2+}$ cation, one half of a *trans*-$[Ni(en)_2]^{2+}$ cation and a water molecule. A polymeric zigzag chain is formed via the alternate array of $[M(CN)_6]^{3-}$ and *cis*-$[Ni(en)_2]^{2+}$ units, and two zigzag chains are combined by *trans*-$[Ni(en)_2]^{2+}$, providing a one-dimensional rope ladder type of chains running along the *c* axis of the unit cell (Figure 5). In the lattice, the chains align along the diagonal line of the *ab* plane to form a pseudo-two-dimensional sheet.

For the Fe^{III} analogue, the nearest Ni···Ni separations are 9.709, 7.713 Å and the nearest Ni···M separation is 6.494 Å. This series of compounds are classed as bimetallic one-dimensional assemblies and all display intramolecular ferromagnetic interactions as a consequence of the strict orthogonality of the $t_{2g}(M^{III})$ and $e_g(Ni^{II})$ orbitals. The exchange parameter J ($16.8\ cm^{-1}$) has been estimated by Mallah *et al.* [39] for the heptanuclear complex $[Ni(tetren)_6][Cr(CN)_6](ClO_4)_9$. No net magnetic moment is exhibited by these compounds because of an antiferromagnetic intermolecular interaction between the pseudo-2-D sheets. A polycrystalline form of the Fe(III) analogue can be prepared via a rapid precipitation method [40]. This compound displays metamagnetic behaviour, which is thought to arise from a structural disorder in the network, which provides quasi-2-D and-3-D domains. Different magnetic properties between a crystalline and polycrystalline sample have also been shown by Murrey *et al.* for $[Ni(bpm)_2]_2[Fe(CN)_6]_2 \cdot 7H_2O$ [41]. Another type of 1-D assembly has been prepared from the reaction of a 1:1:1 molar aqueous solution of $[Ni(pn)_3]Cl_2$, $K_3[M(CN)_6]$ and PPh_4Cl [42,43]. The compounds of stoichiometry $PPh_4[Ni(pn)_2][M^{III}(CN)_6] \cdot H_2O$ (M^{III} = Fe, Cr and Co) have a 1-D zigzag chain structure. The iron and chromium analogues show a ferromagnetic interaction as a consequence of the strict orthogonality of the magnetic orbitals.

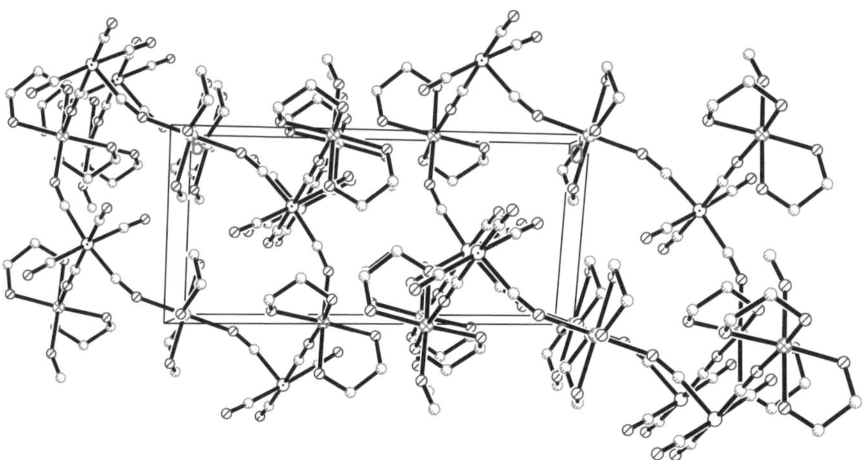

Figure 5 Projection of the 1-D structure of $[Ni(en)_2]_3[Co(CN)_6]_2 \cdot 2H_2O$ on to the *bc* plane [38].

The chromium analogue shows a tendency towards magnetic ordering, but this is overcome by an antiferromagnetic interchain interaction. The cobalt analogue is essentially paramagnetic within the 1-D chain, exhibiting only a weak antiferromagnetic interchain interaction. The use of [Mn(acacen)Cl] as the cationic constituent forms a 1-D linear chain of stoichiometry $[NEt_4]_2[Mn(acacen)][Fe(CN)_6]$, that orders ferromagnetically owing to the ferromagnetic coupling between Mn(III) and Fe(III) ions [44]. The Fe(II)Fe(III) assembly, $[Fe(cyclam)][Fe(CN)_6] \cdot 6H_2O$ was prepared via the reaction of $[Ni(cyclam)](ClO_4)_2$ with an excess of $K_3[Fe(CN)_6]$ [45]. This compound also has a 1-D chain structure and shows a ferromagnetic intrachain interaction between nearest low-spin Fe(III) ions, but no magnetic ordering in the bulk. In contrast, the Ni(II) analogue of stoichiometry $[Ni(cyclam)]_3[Fe(CN)_6]_2 \cdot 6H_2O$ prepared by Colacio *et al.* [46] crystallises in a two-dimensional honeycomb-like layered iron(III)-nickel(II)cyanide bridged complex and exhibits ferromagnetic intralayer and antiferromagnetic interlayer interactions. Above 3 K the magnetic properties of this compound are typical of a metamagnet with $H_c = 5000\,G$, whereas below 3 K a canted structure is formed, leading to long-range ferromagnetic ordering. The isostructural substitution of $[Fe(CN)_6]^{3-}$ by $[Mn(CN)_6]^{3-}$ in the above compound afforded a second two-dimensional compound, namely $[Ni(cyclam)]_3[Mn(CN)_6]_2 \cdot 16H_2O$, which exhibits metamagnetic behaviour [46].

2.5.2 *Two-dimensional compounds*

Reaction of 3:2 molar aqueous solutions of $[Ni(N\text{-men})_3]Cl_2$ and $K_3[M^{III}(CN)_6]$ (M^{III} = Fe, Co) gives a 2-D bimetallic assembly of stoichiometry $[Ni(N\text{-men})_3]\,[M(CN)_6]_2 \cdot H_2O$ [22]. X-ray crystallography reveals that this molecule crystallises in a 2-D honeycomb sheet structure. Hence the introduction of a methyl substituent into the ethylenediamine nitrogen gives rise to a drastic change in the network structure of this compound, from the 1-D zigzag chain when the ligand coordinated to the divalent metal is en, to the honeycomb sheet when the ligand is men. Three cyano nitrogens of $[M(CN)_6]^{n-}$ in the facial mode coordinate with the adjacent *trans*-$[Ni(N\text{-men})_2]^{2+}$, forming a hexagonal unit having M(III) ions at each corner and Ni(II) ions at the centre of each edge (Figure 6).

The iron analogue exhibits ferromagnetic ordering below 10.8 K [22]. A flexion point is observed at 5 kG in the magnetization curve, which has been attributed to a disorder in the network that is due to a partial dehydration. The anhydrous sample was prepared by heating the above compound at 100 °C *in vacuo*. It shows a ferromagnetic interaction, but no spontaneous magnetisation. Both the cobalt analogue and its anhydrous compound show a weak antiferromagnetic intersheet interaction. A related two-dimensional compound of stoichiometry $[Ni(cyclam)]_3[Cr(CN)_6]_2 \cdot 20H_2O$ has also been reported by Verdaguer and co-workers [47] It has a two-dimensional honeycomb structure with the nearest intersheet separations $Cr \cdots Cr = 9.06$, $Ni \cdots Ni = 9.06$ and $Ni \cdots Cr$ 8.06 Å. Magnetic studies on a

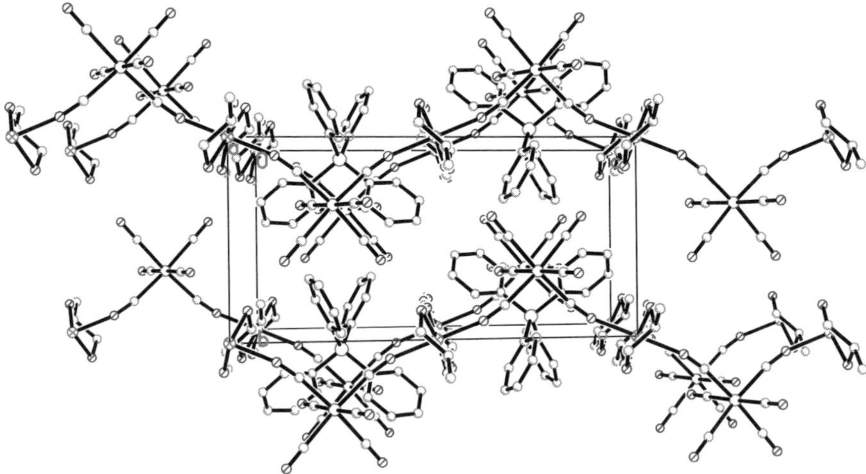

Figure 6 Projection of the structure of $PPh_4[Ni(pn)_2][Fe(CN)_6]_2 \cdot H_2O$ on to the *bc* plane [22].

powdered sample show a short-range ferromagnetic interaction between neighbouring Cr(III) and Ni(II) ions but no magnetic phase transition. Bimetallic assemblies of general formula $[Ni(L)_2]_2[Fe(CN)_6]X \cdot nH_2O$ where L = en, pn or 1,1-dmen, have been obtained with counter ions, e.g. X = ClO_4^-, BF_4^- and PF_6^- [48–50]. X-ray crystallographic studies for selected compounds have revealed a 2-D square sheet structure extended by Fe^{III}–CN–Ni^{II}–NC linkages. Four cyano nitrogens in a plane of $[Fe(CN)_6]^{3-}$ coordinate to adjacent *trans*-$[Ni(L)_2]^{2+}$ units to form an octanuclear square unit with an Fe(III) ion at each corner and an Ni(II) ion at the centre of each edge (Figure 7). The counter ion resides in the square cavity. The 2-D sheet structures in the pn and 1,1-dmen compounds are essentially similar to each other; however, the intersheet separation is generally larger in the 1,1-dmen compounds because of the bulkiness of the diamine ligand.

These compounds in general exhibit ferromagnetic intrasheet interactions and either long-range magnetic ordering or metamagnetism over the lattice, depending on the intersheet magnetic interaction [50]. Metamagnetism appears when the intersheet separation is small enough (< 10 Å) to cause an antiferromagnetic interaction between the sheets. Ferromagnetism occurs when the intersheet interaction is large (> 10 Å) so that the intersheet magnetic interaction is negligible. Dehydration of the ferromagnetic compounds results in shortening the intersheet separation and hence metamagnetism. Analogous bimetallic assemblies have been reported by Liao and co-workers [51] These compounds of stoichiometry $[Ni(L)_2]_2[Fe(CN)_6]$ $NO_3 \cdot nH_2O$, where L = en or tn, have similar two-dimensional honeycomb network topologies with a short intersheet separation of 8.9 Å. These compounds show magnetic behaviour very similar to those discussed above. Bimetallic assemblies of the general formula $A[Mn(L)]_2[M(CN)_6] \cdot nH_2O$ (A = univalent cation:

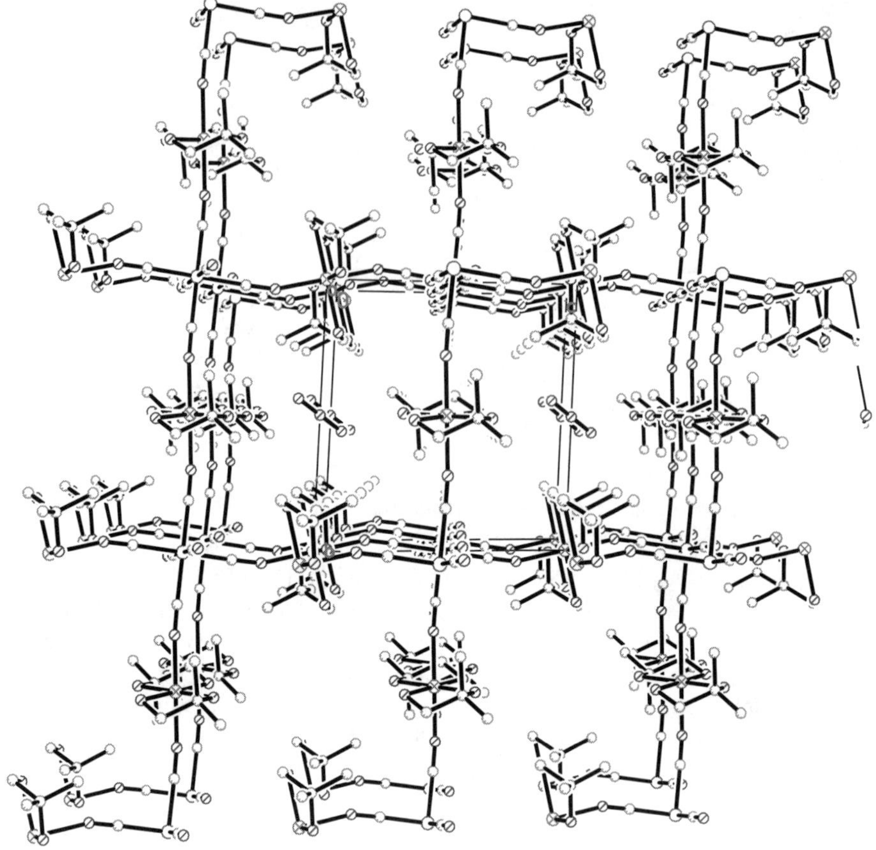

Figure 7 2-D structure of $[Ni(1,1\text{-dmen})_2]_2[Fe(CN)_6]N_3 \cdot 4H_2O$.

M^{III} = Fe, Mn, Cr, Co; L^{2-} = quadridentate Schiff bases) have been reported by Miyasaka and co-workers [52–55]. These assemblies have a 2-D structure consisting of an octanuclear repeating unit $(Mn\text{–}NC\text{–}M\text{–}CN)_4$ having M(III) ions at the corners and Mn(III) ions on the edges of a deformed square. The magnetic properties of these compounds are ferromagnetic, antiferromagnetic or metamagnetic, depending on the Mn(III)/M(III) combination and bulk structure.

2.5.3 *Three-dimensional compounds*

In order to construct a three-dimensional bimetallic network based on an $[M(CN)_6]^{3-}$ building block, the second metal complex as the connector must have at least three vacant sites for accepting cyanide nitrogens from adjacent $[M(CN)_6]^{3-4-}$ ions. These structural requirements are satisfied for the following compounds:

3-D bimetallic assemblies with the general formula $[Ni(L)_2]_3[Fe^{II}(CN)_6]X_2$(L = en, tn; X = PF_6^-, ClO_4^- were obtained from the reaction of a 3:1 molar ratio of $[Ni(L)_3]X_2$ and $K_4[Fe(CN)_6]$ [56]. X-ray crystallography reveals a 3-D network structure extended by the Fe–CN–Ni–NC– linkages, based on an Fe_8Ni_{12} cubic unit with Fe(II) ions at the corners and Ni(II) ions at the middle edges (Figure 8). Two counter ions reside in the cubic cavity and align along the diagonal axis.

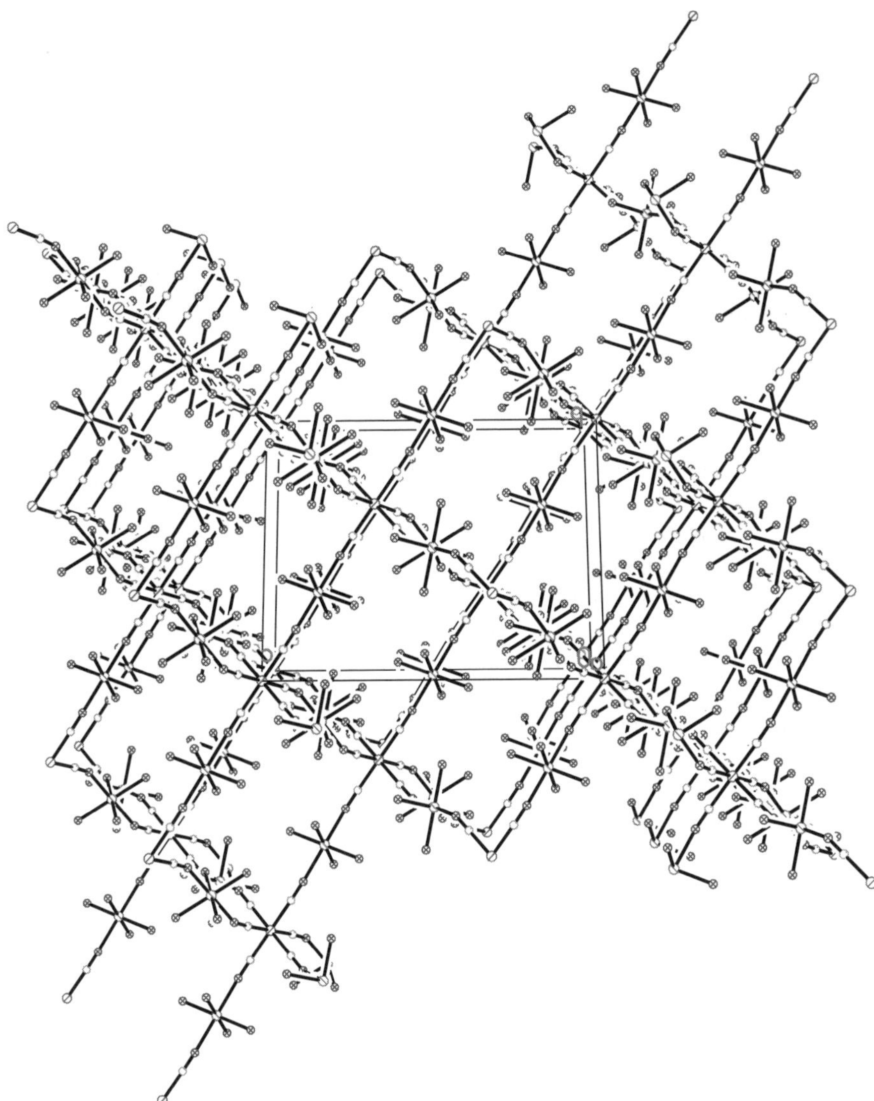

Figure 8 3-D structure of $[Ni(en)_2]_3[Fe(CN)_6](PF_6)_2$ [56].

Magnetic studies reveal a ferromagnetic exchange interaction between nearest Ni(II) ions through the diamagnetic Fe(II) ion. No magnetic ordering occurs over the lattice down to 2 K owing to the diamagnetic nature of $[Fe^{II}(CN)_6]^{4-}$. In 1996, Gatteschi and co-workers [57] reported a 3-D bimetallic assembly of formula $[Ni(tren)]_3[Fe(CN)_6]\cdot 6H_2O$ that crystallises in a complicated 3-D network and orders ferrimagnetically at 8 K. Structural strain causes a weak antiferromagntic interaction between the Fe(III) and Ni(II) ions on intersections of 1-D chains. A 3-D magnetic lattice then arises from the antiferromagnetically coupled 1-D - ferromagnetic chains. A 3-D assembly of stoichiometry $[Mn(en)]_3[Cr(CN)_6]_2\cdot 4H_2O$ was obtained by Ohba *et al.* [58]. Characterization by X-ray crystallography reveals that a 3-D network is formed by an alternate array of $[Cr(CN)_6]^{3-}$ and $[Mn(en)]^{2+}$ ions. The structure is based on a defective cubane with three Cr atoms and four Mn atoms at the corners of the cube and eight Cr–CN–Mn edges. This compound exhibits a ferromagnetic ordering at 69 K based on a short-range antiferromagnetic interaction between Mn(II) and Cr(III) ions. The magnetic hysteresis loop at 2 K is typical of soft magnets showing a weak coercive field $H_c = 28$ G. This compound can be compared with a Prussian blue-like phase where the $M(H_2O)_2]^{2+}$ is simply replaced by $[M(en)]^{2+}$. This strategy has recently been extended by Okawa and co-workers [59] to prepare two cyanide-bridged 3-D bimetallic ferrimagnets of the type $[Mn(L)]_3[Cr(CN)_6]_2\cdot nH_2O$, where L = ethylenediamine, $n = 4$, or L = glycine amide, $n = 2.5$. Once again X-ray crystallographic studies show that these compounds have a 3-D network structure based on a defective cubane unit. Magnetic studies demonstrate that both compounds are ferrimagnets with $T_c = 69$ and 71 K, respectively. These assemblies closely resemble the Prussian blue analogue $Mn_3[Cr(CN)_6]\cdot 12H_2O$, which is a ferrimagnet and orders at $T_c = 63$ K.

Many groups are now designing convergent precursors in order to lower the dimensionality of the Prussian blues into the molecular regime. Dunbar and co-workers [60] have adopted this strategy for the self-assembly of molecular squares or cubes. With the aim of obtaining square units with octahedral metal ions at the vertices, they succeeded in capping four of the coordination sites with two bidentate ligands such as 2, 2′-bipyridine or 1,10-phenanthroline [60]. This results in a 90° angle between the two unblocked sites. Along these lines, the air-sensitive complex $Mn(bpy)(CN)_2$ has been prepared and its structure has been verified by X-ray crystallography. Reaction of this compound with solvated precursors is currently under investigation. This strategy has been extended to use $[M(CN)_6]^{n-}$ anions to connect *cis*-$[M(L\text{–}L)_2]^{n+}$ units [60]. Hexacyanometalate precursors can either produce polymers via *trans* linkages, but there is also the possibility for the self-assembly of molecular squares through *cis*-cyano interactions. In order to investigate this, a series of transition metal complexes were reacted with $[Fe(CN)_6]^{3-}$. Reaction of $[Mn(2,2'\text{-bpym})_2(H_2O)_2](BF_4)_2$ with $K_3[Fe(CN)_6]$ afforded the 2-D polymer $[Mn(H_2O)_2][Mn(2,2'\text{-bpym})(H_2O)]_2[Fe(CN)_6]_2$ [60]. This compound is obtained as a consequence of the loss of a 2,2′-bpym ligand and suggests that precursors with two bidentate ‘protecting’ ligands are not always stable in the presence of cyanometalate anions. The 2,2′-bipyridine analogue of this compound was prepared in the same way and this compound is isostructural with the bypm compound. Both compounds are ferrimagnets with an ordering temperature of 11 K. In an attempt to favour the formation of discrete clusters over polymers, complexes of the type $[M(bpy)_3]^{2+}$ were used in reactions where the bpy ligands are lost slowly in favour of the nitrogen end of a cyanide ligand. Along these lines, reaction of $[Co(bpy)_3](ClO_4)_2$ with $K_3[Fe(CN)_6]$ afforded a trigonal bipyramidal (tbp) cluster of stoichiometry $\{[Co(bpy)_2]_3[Fe(CN)_6]_2\}^+$ [60]. This cluster crystallises in the chiral space group $P6_322$ and is formed as the result of a spontaneous redox reaction. The compound is diamagnetic which supports the presence of low-spin Fe(II) and Co(III) ions. The formation of molecular squares was realized from the reaction of $[Zn(phen)_2(H_2O)_2](NO_3)_2$ with $[Fe(CN)_6]^{3-}$ to afford the compound $\{[Zn(phen)_2\ (H_2O)_2][Fe(CN)_6]_2\}^{2-}$, the structure of which was confirmed by X-ray crystallography. The molecular dianion is composed of two 1,10-phenanthroline-capped Zn(II) ions linked by two *cis*-hexacyanoferrate units. The anion packing involves intermolecular interaction between 1,10-phenanthroline ligands, which leads to a buckling of the CN ligands from an ideal 180° bonding angle. The magnetic properties of the cluster indicate the presence of two isolated noninteracting $S = \frac{1}{2}$ Fe(III) atoms. A pentamer of stoichiometry $\{[Ni(bpy)_2(H_2O)][Ni(bpy)_2]_2[Fe(CN)_6]_2\}$ was prepared from the reaction of $\{[Ni(bpy)_2(H_2O)](OTf)_2$ with $K_3[Fe(CN)_6]$ in water [60]. The formation of the pentamer can be viewed as essentially the addition of an extra $[Ni(bpy)_2(H_2O)]^{2-}$ unit to the square $\{[Ni(bpy)_2]_2\ [Fe(CN)_6]_2\}^{2-}$. The extra unit is stabilized by hydrogen bonding between the remaining water molecule and the nitrogen end of a nearby terminal cyanide corner $[Fe(CN)_6]^{3-}$ unit. The pentamer is neutral which accounts for its ease of isolation. This concept of adding additional building blocks to the central square can be extended to the formation of even larger molecules, e.g. an unusual decameric cluster of stoichiometry $\{[Zn(phen)_2][Fe(CN)_6]\}_2\{Zn(phen)_2][Zn(phen)_2\text{–}(H_2O)]\text{–}[Fe\text{–}(CN)_6]\}_2$ has been isolated in this way [60]. Preliminary data on the Ni_3Fe_2 pentamer indicate that the ground state for this molecule is ferromagnetic $S = 4$. More importantly, the cluster appears to behave as a single molecule magnet (see Section 2.5.4). A.C. magnetic susceptibility measurement of a batch of single crystals revealed a strong frequency dependence both in the χ' (in-phase) and χ'' (out-of-phase) signals with a blocking temperature of $\sim$2 K. The fact that such a small cluster with only eight unpaired electrons can exhibit slow relaxation of the magnetization reversal hints at an appreciable anisotropy due, in part, to the presence of single-ion anisotropy (ZFS) in addition to shape anisotropy. Following this strategy, a variety of cluster geometries have been encountered, including trigonal bipyramids, squares, pentamers and decamers. Molecular squares based on octahedral metal ions can in turn be used in a rational approach as building blocks for larger and highly unsymmetrical molecules. This has important implications for the future design of cyanide-based single molecule magnets.

2.5.4 *Single-molecule magnets*

A few years ago, another branch in cyanide chemistry was opened up when it became apparent that under the right experimental conditions it is possible to crystallise high-nuclearity clusters from hexacyanometalates when the Lewis acid is a complex specifically designed to present one or two labile positions, e.g $[M'(II)L_5(H_2O)_3]$, where L_5 is a pentadentate ligand. Under these conditions, the labile water can be substituted by one cyanide ion from $M(CN)_6$. When the reaction is repeated n times, a polynuclear cluster is obtained, since the pentadentate ligand impedes the growing of the three-dimensional Prussian blue phase and the clusters ground state can present a large spin [20]. A cluster with an $S = {}^{27}\!/_2$ has been obtained by Verdaguer and co-workers applying this strategy [61,62]. This approach is related to the so-called bottom-up approach of nanomagnetism. Since the properties of magnetic particles scale exponentially, it must be possible to address the individual particles, or ensembles of absolutely identical particles must be available. This has presented chemists with a formidable challenge, but an attractive solution has been found via the realisation that molecules containing several transition metal ions can exhibit properties similar to nanoscale magnetic particles (nanomagnets). For this reason, polynuclear transition metal complexes exhibiting superparamagnetic-like properties are now referred to as single-molecule magnets (SMMs) [8]. In order to tune the properties of the magnets and work towards incorporating them into useful devices, research groups are working in two directions: (i) enhancement of the molecular spin, to make large clusters containing many transition metal ions with unpaired electrons, and (ii) improvement of the anisotropy of the system, since most of the interesting properties are related to the anisotropy barrier which is known to be $\sim DS_z^2$, where D is the zero-field splitting parameter of the ground state. These two requirements are difficult to satisfy at the same time, since high spin and anisotropy are often contradictory. Verdaguer and co-workers have been successful in characterising trinuclear, tetranuclear and heptanuclear complexes [20]. Initial work was carried out using tetra-and pentaamine ligands and preparing their corresponding mononuclear complexes where the divalent metal = Cu, Ni, Co, Mn. Reaction of these complexes with $[M(III)CN_6]^{3-}$ precursors afforded species with a large variety of symmetries induced by the flexibility of the cyanometalate coordination geometry. The ground-state spin of the molecules can be predicted and tuned following Kahn's model of orbital interactions. These compounds display a range of ground-state spins and diverse magnetic interactions [20].

2.5 Heptacyanometalate Building Blocks

Although the Prussian blue phases exhibit remarkable magnetic properties, their face-centred cubic structures mean that no magnetic anisotropy can be expected.

To address this problem, Kahn and co-workers have turned their attention to investigate the versatility of the $[M(CN)_7]^{n-}$ precursor. As for the Prussian blue phases, the presence of cyano ligands can lead to extended lattices, but in contrast, the heptacoordination of the precursor is not compatible with cubic symmetry. For compounds containing a heptacoordinated $[Mo(CN)_7]^{4-}$ precursor, the Mo(III) ion is generally in a low-spin pentagonal bipyramid environment. This means that the value of magnetization when applying a magnetic field depends strongly on the orientation of this field with respect to the fivefold axis of the pentagonal bipyramid.

Slow diffusion of two aqueous solutions containing $K_4[Mo^{III}(CN)_7] \cdot 2H_2O$ and an Mn^{II} salt afforded two cyano-bridged bimetallic phases with formulae $Mn_2(H_2O)_5[Mo(CN)_7] \cdot 4H_2O$(α-phase) and $Mn_2(H_2O)_5[Mo(CN)_7] \cdot 4.7H_2O$ (β-phase) [63–65]. Single-crystal X-ray diffraction studies of these two phases revealed that both compounds crystallise in the monoclinic space group $P2_1/c$. The local environments of the metal sites are similar for both phases, but the three-dimensional organization is different. For both compounds, the unit cell contains one molybdenum site and two independent manganese sites. The molybdenum ion is surrounded by seven C–N–Mn linkages, and all cyano groups are bridging. The coordination polyhedron maybe viewed as a slightly distorted pentagonal bipyramid. The two manganese ions are in distorted octahedral surroundings. The structures of the α-phase is shown in Figure 9.

The three-dimensional structural arrangement for the α-phase may be described as follows: edge-sharing lozenge motifs $(MoCNMn1NC)_2$ form bent ladders

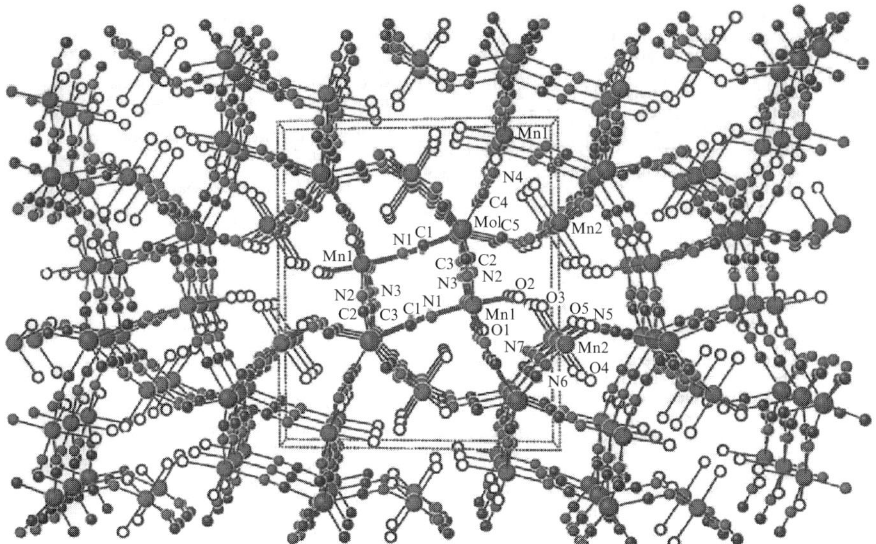

Figure 9 Crystal structure of $Mn(H_2O)_5Mo(CN)_7 \cdot 4H_2O$ (α-phase) [63].

running along the *a* direction. Each ladder is linked to four other ladders of the same type along the [011] and [01–1] directions through cyano bridges. These ladders are further connected by $Mn_2(CN)_3(H_2O)_3$ groups. An Mn2 is linked to an Mo site of one of the ladders and to two Mo sites of an adjacent ladder. In the case of the β-phase, each ladder made of edge-sharing lozenge motifs is surrounded by two instead of four identical ladders. Magnetic measurements reveal that that both phases exhibit a long-range magnetic ordering at $T_c = 51$ K, with a pronounced magnetic anisotropy in the magnetically ordered phase.

When the slow diffusion between aqueous solutions containing $K_4[Mo^{III}(CN)_7]\cdot 2H_2O$ and $[Mn(H_2O)_6](NO_3)_2$ is carried out in the presence of an excess of K^+ ions, a new two-dimensional compound of formula $K_2Mn_3(H_2O)_6[Mo(CN)_7]\cdot 2.6H_2O$ was obtained [66]. The two-dimensional structure has been characterized by single-crystal X-ray crystallography. Once again this compound contains a unique molybdendum site along with two manganese sites. Each molybdenum ion is surrounded by six –C–N–Mn linkages and a terminal –CN ligand. The geometry may be described as a strongly distorted pentagonal bipyramid, and both the Mo–CN and CN–Mn bridging angles deviate significantly from 180°. In the case of the Mn ions, they are surrounded by four NC–Mo linkages, and two water molecules in a *trans* conformation. The two-dimensional structure is made of anionic double-sheet layers parallel to the *bc* plane. K^+ and non-coordinated waters are found located between the layers. Each sheet is a type of grid in the *bc* plane made of edge-sharing lozenges of the type $[MoCNMn2NC]_2$. Two parallel sheets of a layer are further connected by $Mn1(CN)_4(H_2O)_2$ units situated between the sheets. Magnetic studies reveal that this compound exhibits ferromagnetic ordering at $T_c = 39$ K, and that below T_c a field-induced spin reorientation occurs along the c^* axis. No hysteresis was observed along *a* and *b*, while a narrow hysteresis of ca. 125 Oe was observed along c^* at 5 K. Kahn *et al.* [66] have demonstrated that the magnetic properties of this compound can also be dramatically modified through partial dehydration. After dehydration, the magnetic measurements strongly suggest that the crystallographic directions are retained. This compound once again exhibits ferromagnetic ordering, but the critical temperature has been shifted to a higher temperature, $T_c = 72$ K, compared with 39 K in the original hydrated sample. This work has succeeded in introducing magnetic anisotropy into the field of molecule-based magnets, and new structural topologies have been characterised. Furthermore, detailed magnetic measurements have revealed the existence of several magnetically ordered phases together with spin reorientations for the three compounds.

In order to lower the symmetry of the lattice further, the macrocyle 2,13-dimethyl-3,6,9,12,18-pentaazabicyclo[12.3.1]octadeca-1(18),2,12,14,16-pentaene L has been utilised to impose heptacoordination on the Mn(II) sites [67]. This macrocycle is known to adopt a planar conformation and can thus stabilize a heptacoordinated Mn(II) species of the type $[MnL(H_2O)_2]^{2+}$ with pentagonal-bipyramidyl geometry.

Reaction of $[Mo(CN)_7]^{4-}$ with $[MnL(H_2O)_2]^{2+}$ did not give the expected compound but afforded the compound with stoichiometry $[Mn^{II}L]_6[Mo^{III}(CN)_7][Mo^{IV}(CN)_8]_2 \cdot 19.5H_2O$, in which two-thirds of the $[Mo^{III}(CN)_7]^{4-}$ groups have been oxidized to $[Mo^{IV}(CN)_8]^{4-}$ [67]. The compound crystallises in a two-dimensional arrangement of edge-sharing 48-membered rings (Figures 10 and 11). Each ring has a heart shape, and contains 16 metal sites: two Mo1 sites located on a twofold axis together with six Mo2, four Mn1 and four Mn3 sites occupying general positions.

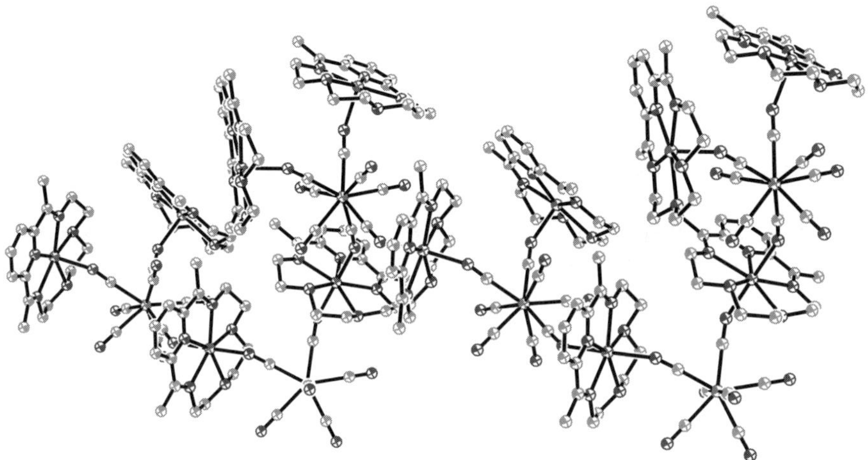

Figure 10 Structure of a chain of $[Mn^{II}L]_6[Mo^{III}(CN)_7][Mo^{IV}(CN)_8]_2$ [67].

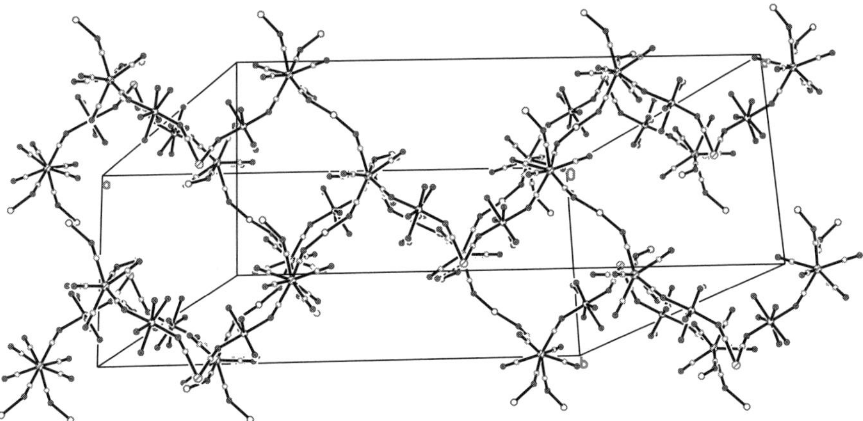

Figure 11 Projection on to the *ab* plane for $[Mn^{II}L]_6[Mo^{III}(CN)_7][Mo^{IV}(CN)_8]_2$. The macrocycles are omitted for clarity.

All of these sites are linked by cyano bridges, the carbon atoms being bound to the Mo and the nitrogen atoms to the manganese. The Mn2 site is linked to the Mo2 site through a cyano bridge, but is not directly involved in the ring. Non-coordinated water molecules occupy voids between the layers. The coordination sphere around the first molybdenum, Mo1 can be described as a distorted pentagonal bipyramid whereas the second molybdenum Mo2 has a lower symmetry closer to a square antiprism. All three manganese sites are surrounded by the five nitrogen atoms of the macrocycle in the equatorial plane. Two additional axial coordinating ligands complete a pentagonal bipyramidal geometry. Detailed magnetic studies indicate that this compound is a fully localised mixed-valence, mixed-spin species. The $(MnL)^{2+}$ groups are in three different crystallographic environments. In two cases, two axial N-coordinated cyano groups confer a low-spin state on the Mn1 and Mn3 ions. In the third case, one axial N-coordinated cyano group and an axial water molecule confer a high-spin state on the Mn2 ion. Furthermore, the Mn2 ion is pulled out of the equatorial plane of the pentadentate ligand that diminishes the ligand field exerted by the macrocycle. It follows then that the local spins of the metal sites are $S_{Mo1} = \frac{1}{2}$, $S_{Mo2} = 0$, $S_{Mn1} = \frac{1}{2}$ and $S_{Mn} = \frac{5}{2}$. Long-range ferromagnetic ordering is observed for this compound at 3 K. This low temperature is due to the fact that long-range ordering involves interactions not only through the cyano bridges, but also through the N–C–Mo^{IV}–C–N bridging linkages that are very weak.

2.6 Octacyanometalate Building Blocks

Since Prussian blue chemistry is currently enjoying an active revival, several groups are exploring the closely related octacyanometalate building block $[M(CN)_8]^{n-}$, where M is a transition metal ion, for the self-assembly of novel supramolecular coordination compounds [20,68].

2.6.1 One-Dimensional compounds

A one-dimensional chain $[Mn^{II}_2(L)_2(H_2O)][Mo^{IV}(CN)_8] \cdot 5H_2O$, where L is the macrocyclic ligand 2,13-dimethyl-3,6,9,12,18-pentaazabicyclo[12.3.1] octadeca-1(18),2,12,14,16-pentane, has been prepared from the reaction of $K_4[Mo^{IV}(CN)_8]\cdot 2H_2O$ with the complex $[Mn^{II}L(H_2O)_2]Cl_2 \cdot 4H_2O$ [69]. This compound has been structurally characterized by X-ray crystallography and its photomagnetic properties have been studied. The compound crystallizes as a polymeric arrangement of zigzag chains of alternating Mn(L) and $Mo(CN)_8$ units with an additional M(L) unit linked to the Mo as a pendant arm. This compound is paramagnetic, but shows important modifications of its magnetic properties under UV irradiation leading to the formation of ferrimagnetic chains. This non-reversible photomagnetic effect is thought to be caused by an internal photo-oxidation of Mo(IV) (diamagnetic) to Mo(V) (paramagnetic). The crystallographic data indicate that the formation of hydrogen bonds between water molecules and nitrogen atoms of non-bridging cyano groups play a key role in enabling the molybdenum photo-oxidation to be observed.

2.6.2 *High-spin clusters*

One of the most challenging issues in the field of molecule-based magnetism is the preparation of magnetic clusters with designed topology and predictable properties. Along these lines, Decurtins and co-workers [70] have recently structurally and magnetically characterized the novel cyanide-bridged high-spin molecular cluster of stoichiometry $[Mn^{II}\{Mn^{II}(MeOH)_3\}_8(\mu\text{-}(CN)_{30}\{Mo^{V}(CN)_3\}_6] \cdot 5MeOH \cdot 2H_2O$ (**1**), prepared by the self-assembly of $[Mo^{V}(CN)_8]^{3-}$ building blocks together with divalent metal ions (Figure 12). The cluster is comprised of 15 cyano-bridged metal ions, namely nine Mn(II) ions ($S = 5/2$) and six Mo(V) ions ($S = 1/2$), giving a total of 51 unpaired electrons within the cluster.

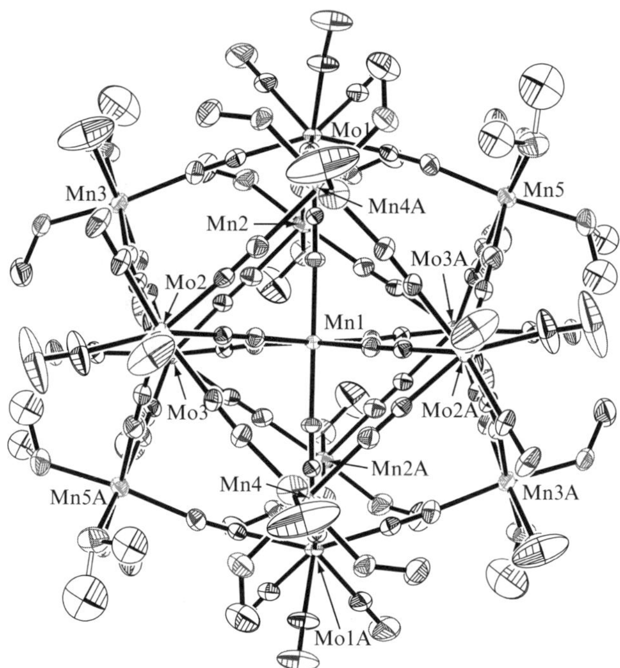

Figure 12 ORTEP representation of the molecular structure of $[Mn^{II}\{Mn^{II}(MeOH)_3\}_8(\mu\text{-}CN)_{30}\{Mo^{V}(CN)_3\}_6]$ (**1**); for clarity only the Mn(II) and Mo(V) atoms are labeled and the H atoms are omitted [70].

The Mo^{V}–CN–Mn^{II} geometry is such that the atoms are all linked to form an aesthetically pleasing topological pattern in which the polyhedron spanned by the peripheral metal ions is closest in geometry to a rhombic dodecahedron, (Figure 13). Furthermore, in parallel studies Hashimoto and co-workers have also reported the synthesis of an Mn_9W_6 cluster with an analogous structural topology [71].

The magnetic properties of the Mn_9Mo_6 cluster above 44 K are characterised by ferromagnetic intracluster coupling which finally leads to an $S = {}^{51}\!/_2$ ground-state spin. In contrast to other high spin clusters, this compound does not exhibit the typical phenomena of molecular hysteresis and slow quantum tunneling at low temperatures. In this case, a competitive interplay of intra- and intercluster interactions leads to a very interesting magnetic regime. Unfortunately, a modeling of the magnetic data below 44 K, where bulk magnetic ordering seems to arrive, is not possible since this new situation has not been dealt with theoretically so far. Further experiments are needed before we can move forward in our understanding and interpretation of the magnetic data in the low-temperature regime. It seems reasonable to suggest, however, that dipole–dipole interactions could contribute to the intercluster coupling since these interactions will become increasingly important with increasing S state values of high spin clusters. Interestingly, in contrast to these observations, antiferromagnetic intracluster interactions are reported for the Mn_9W_6 cluster, which give rise to a smaller ground-state spin value of $S = {}^{39}\!/_2$ [71].

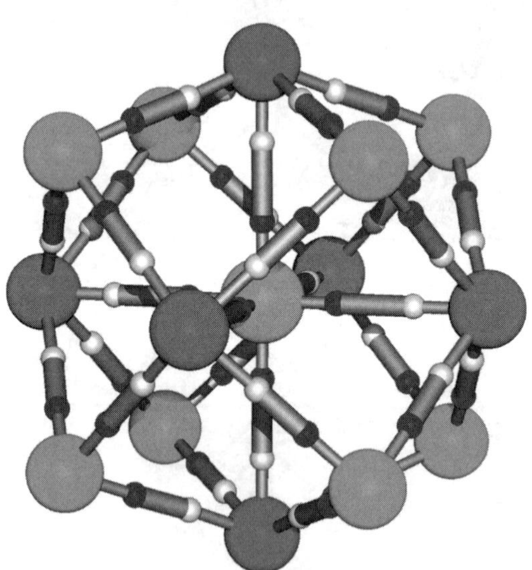

Figure 13 Representation of an idealised $[Mn^{II}{}_9(\mu\text{-CN})_{30}\text{–}Mo^{V}{}_6]$ cluster core. The light grey spheres represent Mn(II) ions and the darker spheres represent Mo(V) ions and the bonds between them represent μ-cyano ligands [70].

Although in both of the above cases the clusters exhibit a very large-spin ground state, the magnetic anisotropy of these compounds is clearly too low. This has been assigned as being mainly due to the small anisotropy of the Mn(II) ions, together with the high symmetry of the cluster topology. As a consequence, these compounds do not behave as SMMs at temperatures down to 1.8 K. In order to address these problems, this strategy has been extended, to replace Mn with Ni and prepare two new clusters of stoichiometry $[Ni^{II}\{Ni^{II}(MeOH)_3\}_8(\mu\text{-}CN)_{30}\{M^V(CN)_3\}_6]$, where M^V = Mo(V) or W(V) [72]. These compounds are isostructural with the previously mentioned $[Mn^{II}{}_9Mo^V{}_6]$ cluster. Each cluster comprises of 15 metal ions, nine Ni(II) ions and six M(V) (M = Mo or W) ions, all linked by μ-cyano ligands. Once again the compounds crystallise in the monoclinic *C*2/*c* space group, which means that the actual cluster has an overall *C*2 symmetry with the crystallographic twofold axis running through the central Ni^{II} ion. An extended intermolecular H-bonded network connects each cluster to eight nearest neighbours. All the nearest neighbour contacts between metal ions of different clusters are > 7 Å. All terminal cyano and methanol ligands not involved in intercluster H-bonding interactions take part in H-bonding interactions to solvent molecules.

Magnetic susceptibilities and magnetisation measurements both on single crystals and on the dissolved compounds indicate that the two clusters have an $S = 12$ ground state, originating from intracluster ferromagnetic exchange interactions between the μ-cyano-bridged Ni(II) and M(V) metal ions [72]. In the a.c. magnetic susceptibility, no out-of-phase (χ''_M) signal in zero field down to 1.8 K was observed, indicating that the overall magnetic anisotropy in these compounds is still too low to class them as single-molecule magnets. These findings are also in accordance with the small uniaxial magnetic anisotropy value of $D = 0.015\,cm^{-1}$, estimated from high-field, high-frequency EPR measurements. From these observations, we conclude that the small single ion magnetic anisotropy together with the highly symmetric arrangement of the metal ions in the cluster topology is still not favourable for the onset of SMM behaviour.

In an attempt to change the solid-state structure of the cluster and perhaps reduce the crystal symmetry, Decurtins and co-workers have recently prepared the ethanol analogue of compound **1** [73] . This new cluster has the stoichiometry $[Mn^{II}\{Mn^{II}(EtOH)_3\}_8(\mu\text{-}CN)_{30}\{Mo^V(CN)_3\}_6]\cdot 6EtOH\cdot MeOH$, where larger ethanol molecules now complete the coordination sphere of the peripheral Mn(II) ions. The molecular geometry of this compound has been determined by low-temperature single-crystal X-ray crystallography. The compound once again consists of cyanide bridged $Mn_9^{II}Mo_6^V$ units, with nine Mn(II) ions defining a body-centred cube and six Mo(V) ions constituting an octahedron. In contrast to the first cluster **1**, this compound crystallises in the triclinic space group (*P*1), with a central Mn^{II} ion occupying each corner of the unit cell. Structurally, a slight change in the packing motif is observed which serves to accommodate better the more bulky ethanol molecules. In this case, each cluster is surrounded by six instead of eight nearest neighbours. This change in packing motif has had a subtle effect on the closest intercluster metal–metal contacts that are in the range 6.82–7.18 Å for $Mo^{V} \cdots Mn^{II}$ and 7.36–8.23 Å for $Mn^{II} \cdots Mn^{II}$. Compared with cluster **1**, the $Mo^{V} \cdots Mn^{II}$ contacts are on average 0.4 Å shorter, whereas the $Mn^{II} \cdots Mn^{II}$ interactions are 0.4 Å longer. The cluster-to-cluster distance is found to be shorter for this compound, namely 17.2 Å compared with 17.5 Å for cluster **1**. The intermolecular network of H-bonding interactions has remained essentially intact; the ethanol ligands on neighbouring clusters are simply not bulky enough to disturb these inter-cluster interactions. They are, in fact, able to accommodate these interactions well by twisting away from each other, thus minimising steric interactions and enabling neighbouring clusters to pack close together, (Figure 14).

The shortest O–H $\cdots$ N contacts between adjacent clusters (2.74 Å) are exactly within the range of H-bonding interactions previously reported for cluster **1**. Once again, each peripheral Mn(II) ion in a cluster has one ethanol ligand that is involved in an intermolecular bond to the nitrogen of a cyano ligand in a neighbouring cluster. All remaining terminal cyano and ethanol ligands are hydrogen bonded to solvent molecules. A detailed investigation of the magnetic properties of this cluster together with the preparation of new analogues is currently ongoing.

2.6.3 Three-dimensional networks

One of the most interesting features of molecular-based materials is the way in which the magnetic properties may be transformed by small and subtle variations in molecular chemistry. Working along these lines, Decurtins and co-workers have recently investigated the effects of changing the oxidation state of the metal in the octacyanometalate building block moving from M(V) to M(IV) (M = Nb, Mo, W) [74]. Hence, reaction of the appropriate octacyanometalate building block with an M^{II} salt afforded in all cases crystalline compounds which have been shown by single-crystal X-ray analysis to be cyano-bridged extended three-dimensional networks with general stoichiometry $[M^{IV}\{(\mu\text{-}CN)_4M^{II}(H_2O)_2\}_2]$. All compounds crystallise in the tetragonal space group *I*4/*m*, with one M(IV) ion sitting on the crystallographic fourfold axis and an M(II) ion, together with two coordinating water molecules occupying a crystallographic mirror plane. Figure 15 illustrates the molecular structure of the three-dimensional compound of stoichiometry $[W^{IV}\{(\mu\text{-}CN)_4Fe^{II}(H_2O)_2\}_2]$ [75].

The network comprises W(IV) ions, which are connected to nearest neighbour Fe(II) ions through cyanide bridges in a three-dimensional grid-like arrangement. The Fe(II) ions are in an octahedral environment, bonded to four nearest neighbour W(IV) ions; two axial water molecules complete the sixfold coordination.

Despite the structural similarities, the change in electronic ground state on moving along the Periodic Table from Nb(IV) to Mo(IV) and W(IV) confers two

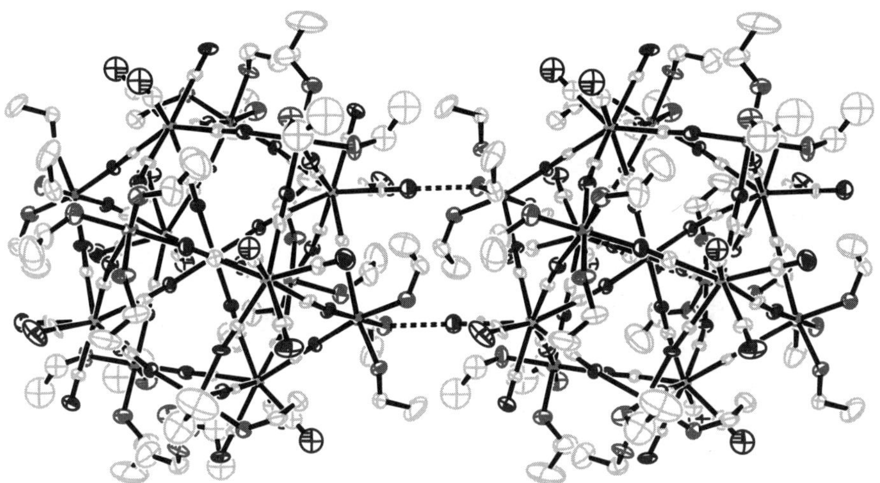

Figure 14 ORTEP representation of two nearest neighbour clusters with ethanol ligands on the outer Mn(II) ions. H atoms are omitted for clarity. Dotted lines represent the shortest O–H···N contacts (2.74 Å) between adjacent clusters [73].

different classes of magnetic properties on these compounds. The networks incorporating diamagnetic Mo(IV) or W(IV) metal ions are paramagnetic. However, Nb(IV) represents a d^1 electronic configuration, hence a paramagnetic metal centre and, accordingly, the three-dimensional networks which incorporate $[Nb^{IV}(CN)_8]^{4-}$ building blocks all show bulk magnetic behaviour. Ongoing studies are focused on the elucidation of the magnetic structures of this new class of three-dimensional materials, together with detailed structural studies to develop a straightforward concept for the synthesis of new materials from cyanometalate building blocks, having a predictable structural order and a useful set of solid-state properties.

Hashimoto and co-workers have also turned their attention to structurally and magnetically characterise the three-dimensional metal assemblies derived from $[M(CN)_8]^{3-}$ units (M = Mo, W) and simple metal ions. As a consequence, the compound of stoichiometry $[Mn^{II}_6(H_2O)_9\{W^V(CN)_8\}_4 \cdot 13H_2O]_n$ was synthesised from a hot aqueous solution containing octacyanotungstate, $Na_3[W(CN)_8] \cdot 3H_2O$ and $Mn(ClO_4)_2 \cdot 6H_2O$ [76]. The compound crystallised in a monoclinic system with the space group *P*21/*c*. The compound consists of a W^V–CN–Mn^{II} linked three-dimensional network and water molecules as crystal solvates. There are two types of W sites: one is close to dodecahedron geometry with six bridging and two terminal CN ligands and the other is closer to a bicapped, trigonal prism with seven bridging and one terminal CN ligand. Field-cooled magnetisation measurement showed that this compound exhibits a spontaneous magnetisation below

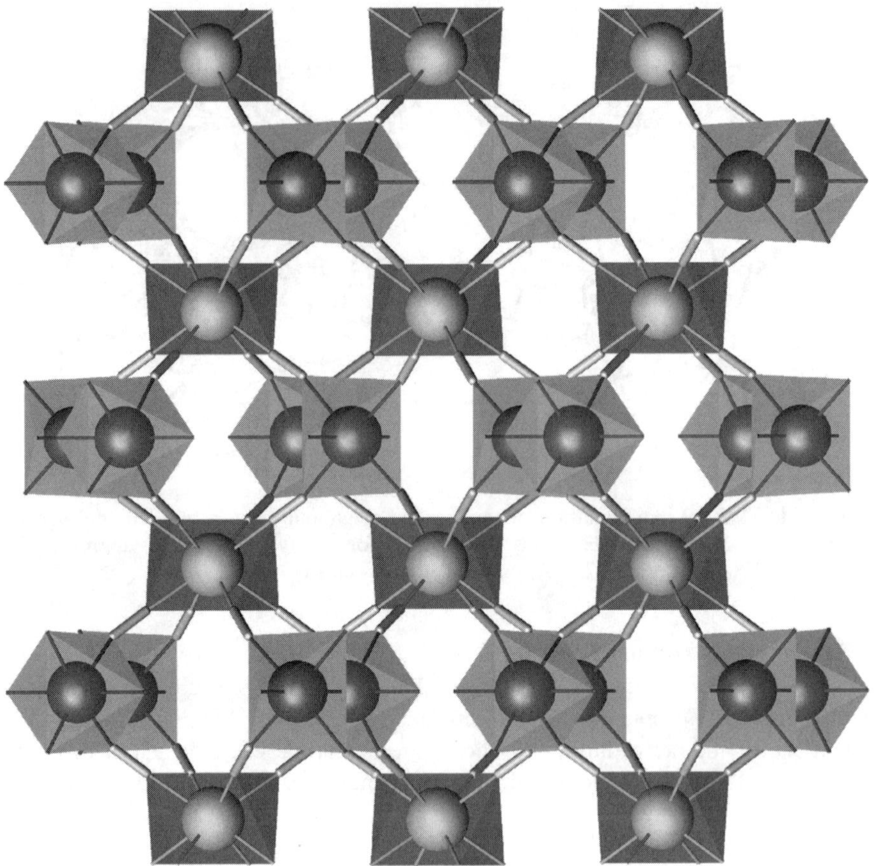

Figure 15 View of the 3D $[W^{IV}\{(\mu\text{-}CN)_4Fe^{II}(H_2O)_2\}_2]$ network [75].

$T_c = 54\,K$. Further magnetization measurements on the field dependence revealed it to be a ferrimagnet in which all of the Mn(II) ions are antiparallel to all the W(V) ions.

In parallel studies, reaction of Fe(II) with the molybdenum octacyanide afforded the Fe^{II}-based polycyanide $[Fe_2(H_2O)_4][Mo^{IV}(CN)_8] \cdot 4H_2O$ [77]. This compound, although already known, has recently been structurally and magnetically characterised by Mathionière and co-workers The compound crystallises in the tetragonal space group *I*422. There are two molybdenum sites in a distorted square antiprismatic arrangement, each site being surrounded by eight CN–Fe linkages. Four NC–Mo linkages and two axial water molecules give rise to a distorted octahedral coordination geometry around the Fe(II) ions. The structure is three-dimensional and highly symmetrical. This compound shows no evidence of long-range magnetic order, owing to the diamagnetic Mo(IV) ion, which suppresses the propagation of magnetic interactions between adjacent Fe(II) ions through the CN bridges. These properties are in contrast to the $Fe_2[Mo^{III}(CN)_7] \cdot 8H_2O$ analogue which orders below 65 K as a ferrimagnet [77]. In more recent work, Mathonière and co-workers have reported the structural and physical characterization of the compounds $[Cu(bipy)_2]_2\,[Mo(CN)_8] \cdot 5H_2O \cdot CH_3OH$, with bipy $= 2,2'$-bipyridine, and $Cu^{II}_2[Mo^{IV}(CN)_8] \cdot 7.5H_2O$ [78]. The first compound crystallises in the triclinic space group *P*-1. The structure consists of neutral trinuclear molecules in which a central $[Mo(CN)_8]^{4-}$ anion is linked to two $[Cu(bipy)_2]^{2+}$ cations through two cyanide bridges. Unfortunately, for the second compound the copper analogue crystallizes poorly, so structural information was obtained via the wide-angle X-ray scattering (WAXS) technique. The compound crystallises in an extended three-dimensional network for which the cyano groups act as bridges. An investigation of the magnetic properties revealed that both compounds behave as paramagnets. Under irradiation with light, they exhibit important modifications of their magnetic properties, with the appearance of magnetic interactions at low temperature. In the case of the first compound, the modifications are irreversible, whereas in the case of the second compound they are reversible with cycling in temperature. These photomagnetic effects are thought to be caused by the conversion of Mo(IV) (diamagnetic) to Mo(V) (paramagnetic) through a photo-oxidation mechanism for the first compound and a photoinduced electron transfer in the second. These results show that it is possible to apply a slightly different chemical strategy for the preparation of new photomagnetic extended systems, namely using light-sensitive coordination compounds as molecular building blocks. To summarise, the choice of precursors must be guided by two criteria: first, their photoreactions must implicate a change in spin state, and second, they must contain bridging ligands.

3 OXALATE-BASED 2-D AND 3-D MAGNETS

In recent years, we have shown that it is possible to develop a strategy for the self-assembly of supramolecular systems based on transition metal oxalates, which typically behave as host/guest compounds with different lattice dimensionalities [3,79,80]. All of these classes of structure are formally composed of metal oxalate building blocks (Figure 16), and these compounds display a range of magnetic properties, since it is well known that the oxalate bridge is a good mediator of both antiferromagnetic and ferromagnetic interactions between paramagnetic metal ions.

3.1 Basic Principles of Specific 2-D and 3-D Network Configurations

The ambidentate coordinating ability of the oxalate ion (ox $= C_2O_4{}^{2-}$) has made it a useful and versatile choice as a bridging ligand for the design of two- and three-dimensional compounds. Indeed, the self-assembly of inorganic architectures

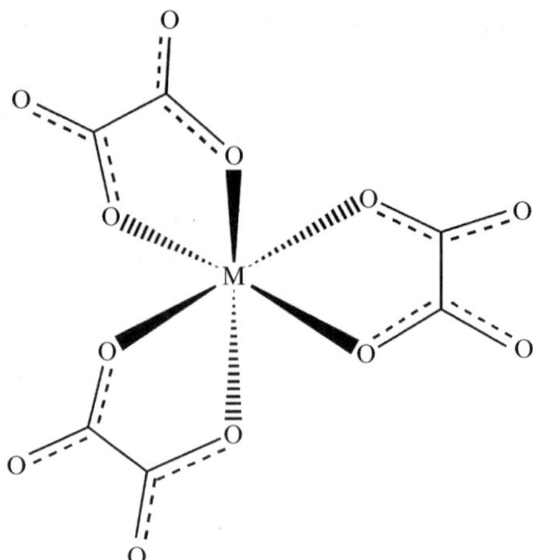

Figure 16 Schematic representation of the chiral (the Λ-isomer is shown) preorganised anionic coordination entity.

incorporating oxalate ions as the bridging species have to date yielded various types of frameworks which include one-dimensional chains [79,81], two-dimensional layers [82,83] and three-dimensional networks [84]. All of these classes of structure are formally composed of $[M^{z+}(ox)_3]^{(6-z)-}$ building blocks, where each of these units represents a three-connected point (Figure 1). As a consequence, it is the concept of connectivity which is important to address since it is the connectivity which defines the way in which a set of points are connected to construct a lattice which can be infinite in one- two- or three-dimensions. In the following discussion, we will focus on the fundamental ideas that are relevant to the formation of the two- and three-dimensional framework topologies.

Oxalate building blocks can in principle polymerise in two ways. The first alternative leads to a 2-D honeycomb layered structure, whereas the second possible arrangement results in the formation of an infinite 3-D framework. In the former case, building blocks of opposite chirality are alternately linked, which confines the bridged metal ions to lie within a plane and results in the formation of a two-dimensional layered motif (Figure 17a). In contrast, an assembly of building blocks of the same chiral configuration leads to the 3-D framework shown in Figure 17b.

After having considered the way in which the chirality of the subunits can influence the lattice dimensionality, it is then possible to apply a series of topological rules to define the number of building blocks that are needed to build closed circuits and, hence, extended framework motifs.

Generally, the formation of two-dimensionally linked assemblies affords subunits possessing either four-connected points or a combination of two three-connected points. Figure 18 illustrates the way in which two dimeric subunits may be combined to form a planar honeycomb network.

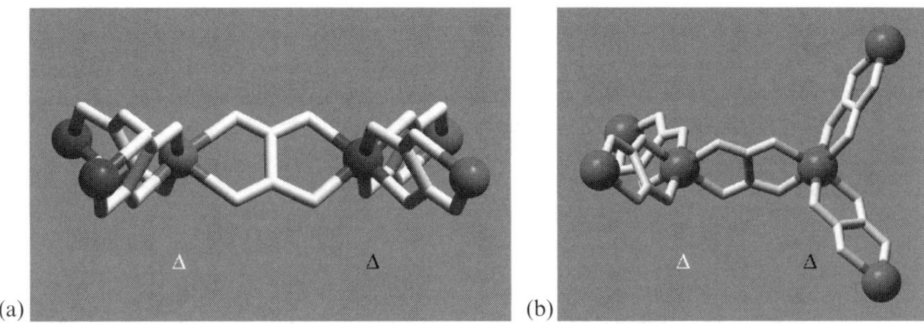

Figure 17 Chiral $[M^{z+}(ox)_3]^{(6-z)-}$ building blocks assembled with (a) alternating chiral configuration and (b) equal chiral configuration.

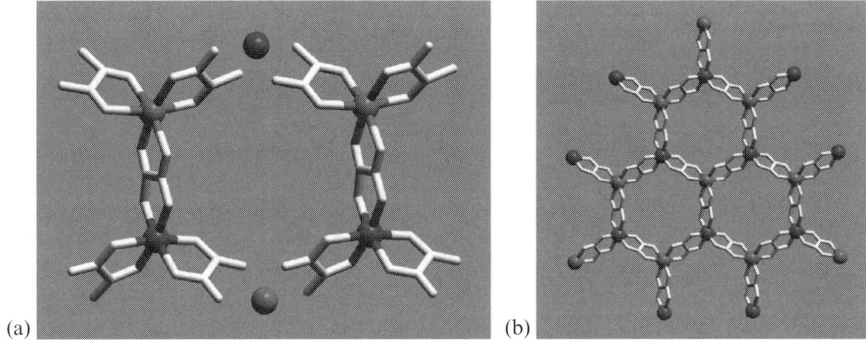

Figure 18 (a) Two dimeric units of the alternating chirality type are necessary to form a closed hexagon ring; (b) the resulting planar network motif.

In an analogous manner, it can easily be seen that two tetrameric subunits are needed to build closed circuits composed of 10 metal centres, which in sum define a three-dimensional decagon framework (Figure 19). Such tetrameric subunits are necessary, because only four three-connected points ($Z = 4$) combined together have the necessary number of six free links to build the 3-D net. Identically oriented links repeat at intervals of $(Z + 1)$ points, so that circuits of $2(Z + 1)$ points are formed. Hence, the structure represents a uniform net in the sense that the shortest path, starting from any point along any link and returning to that point along any other link, is a circuit of 10 points. In addition, the topological principle implies that for the 3-D case, only subunits of the same chirality are assembled. Consequently, the uniform anionic 3-D network type with stoichiometry $[M^{II}_2(ox)_3]_n{}^{2n-}$ or $[M^IM^{III}(ox)_3]_n{}^{2n-}$ ($M^{II/III}$ = transition metal ions and M^I = Li, Na) is chiral, since it is composed of $2n$ centres exhibiting the same kind of chirality. Naturally, this chiral topology is in line with the symmetry elements which are present in the crystalline state of these 3-D frameworks, which in sum

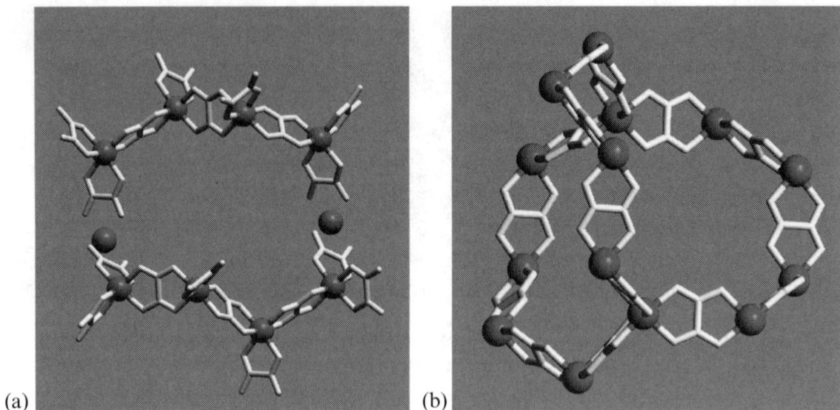

Figure 19 (a) Two tetrameric units of the same chirality type are necessary to form a closed decagon ring; (b) a fragment of the 3-D chiral framework.

constitute either one of the enantiomorphic cubic space groups $P4_332$ or $P4_132$ for the former and the cubic space group $P2_13$ for the latter stoichiometry. Thereby, the $2n$ metal ion centres occupy special sites with a threefold symmetry axis. Figure 20 shows a view of this network topology.

Extended helical geometries are also encountered through the three-dimensional repetitive assembling of subunits with helical chirality. Thus, the framework structure may alternatively be seen as composed of either right-handed (Δ-chirality) or left-handed (Λ-chirality) helices with a 4_1, 4_3or2_1 arrangement, running in three perpendicular directions while simultaneously being covalently bound to each other.

In contrast to the compounds that form chiral 3-D networks, the 2-D framework topology implies an assembly of coordination entities with alternating chirality between nearest neighbouring centres. Finally, the discrimination between the formation and crystallisation of either 2-D or 3-D frameworks with analogous network stoichiometries depends on the choice of the templating counter-ion. In particular, $[XR_4]^+$(X = N, P; R = phenyl, n-alkyl) cations have been found to initiate the growth of 2-D sheet structures containing $[M^{II}M^{III}(ox)_3]_n{}^{n-}$, network stoichiometries. Figure 21 shows a projection of a sector of a compound that crystallises in 2-D honeycomb layers.

This thinking applies, in particular, when planning the design of a chiral three-dimensional supramolecular host–guest system, since the mutual interaction of the two distinct complementary molecular units or coordination entities is necessary. Examples of this methodology include the above-described anionic, tris-chelated transition metal oxalato complexes $[M^{z+}(ox)_3]^{(6-z)-}$ which form the host system together with the cationic, tris-chelated transition metal diimino complexes, e.g. $[M(bpy)_3]^{2+/3+}$, bpy $= 2,2'$-bipyridine, which play the role of the guest compounds.

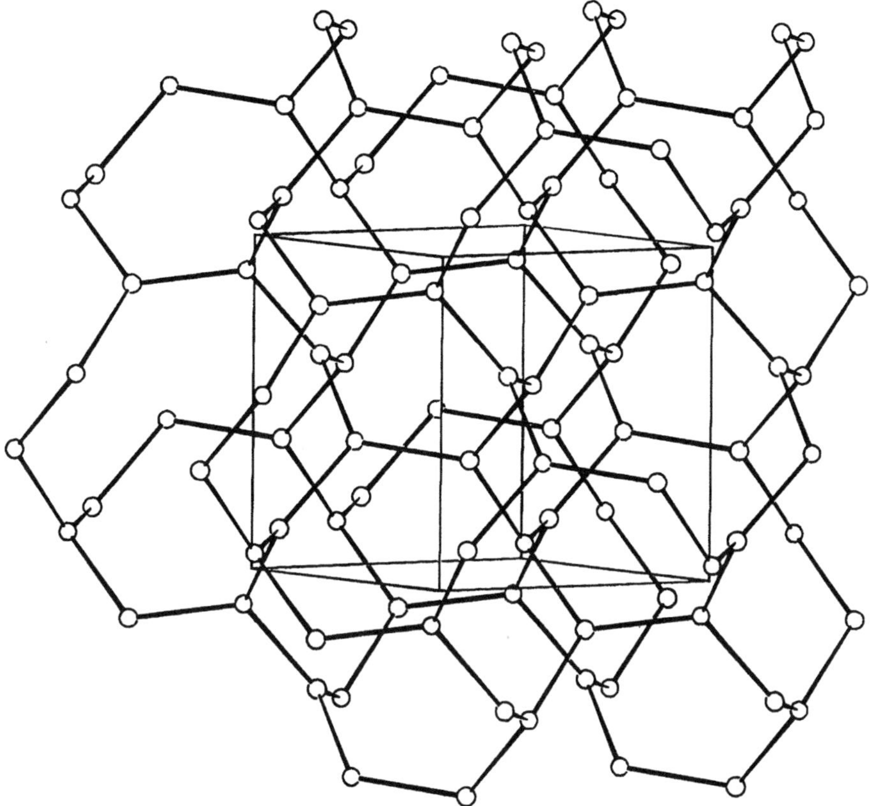

Figure 20 View of the three-connected 10-gon (10,3) network topology.

3.2 Structural Studies on 2-D Compounds

Several X-ray crystal structures of two-dimensional oxalato-bridged mixed-metal networks have been reported since 1993. The first structural information on 2-D oxalates was obtained by Atovmyan *et al.* [82], who succeeded in growing single-crystals of the compound $[NBu_4][Mn^{II}Cr^{III}(ox)_3]$ by slow diffusion of aqueous solutions of a mixture of NBu_4Br with $K_3[Cr(ox)_3] \cdot 3H_2O$ and $MnCl_2$ in an H-shaped tube. This structure revealed an anionic two-dimensional network consisting of μ-oxalato bridged Mn(II) and Cr(III) ions, with NBu_4^+ cations located between the layers. One of the butyl groups of the cations penetrates a void in the neighbouring anionic layer and the separation between two adjacent anionic layers is 8.95 Å.

Shortly thereafter, Decurtins *et al.* [83] obtained single crystals of the compound $[PPh_4][Mn^{II}Cr^{III}(ox)_3]$, the structure of which is comparable to that of Atovmyan *et al.* described above. It comprises of an alternating assembly of the

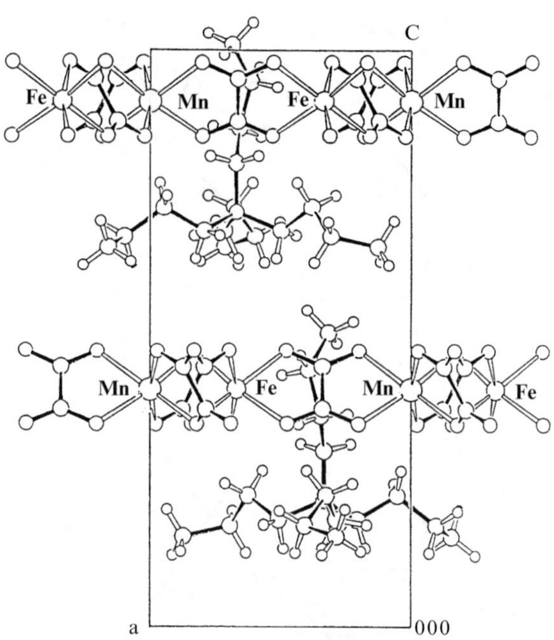

Figure 21 Sector from the $[N(Bu)_4][Mn^{II}Fe^{III}(ox)_3]]$ layer compound, [100] projection [80].

two types of chiral octahedron with Λ- and Δ-configurations adopting a two-dimensional honeycombed layered structure with the network stoichiometry $[Mn^{II}Cr^{III}(ox)_3]_n^{n-}$. Figure 22 shows an overall view of the zero layer of this crystal structure; there are six pairs of anionic and cationic layers per unit cell and the distance between two adjacent layers is reported to be 9.5 Å.

Those phenyl groups of the cations which point vertically in the direction of the anionic layers just fit into the slightly ellipsoidal shaped vacancies. In a subsequent communication, [80] the single-crystal X-ray crystal structures of two additional compounds with stoichiometries $[NPr_4][Mn^{II}Cr^{III}(ox)_3]$ and $[NBu_4][Mn^{II}Fe^{III}(ox)_3]$ were also reported. Interestingly, the compound with the latter stoichiometry crystallises in the hexagonal space group $P6_3$.

Day and co-workers [85] have carried out single-crystal X-ray crystallographic studies on the compound $[N\text{-}(n\text{-}C_5H_{11})_4][Mn^{II}Fe^{III}(ox)_3]$, which exhibits a two-dimensional network connectivity pattern in accordance with the above findings. In this case, the special disposition of the organic cations interleaving the honeycombed anionic layers (whereby the cavities are occupied by two CH_3 moieties approaching from opposite sides) leads to the molecule crystallising in the orthorhombic space group $C222_1$.

An alternative approach has been devised by Coronado and co-workers [86,87] which involves the combination of two magnetically active sub-lattices, namely

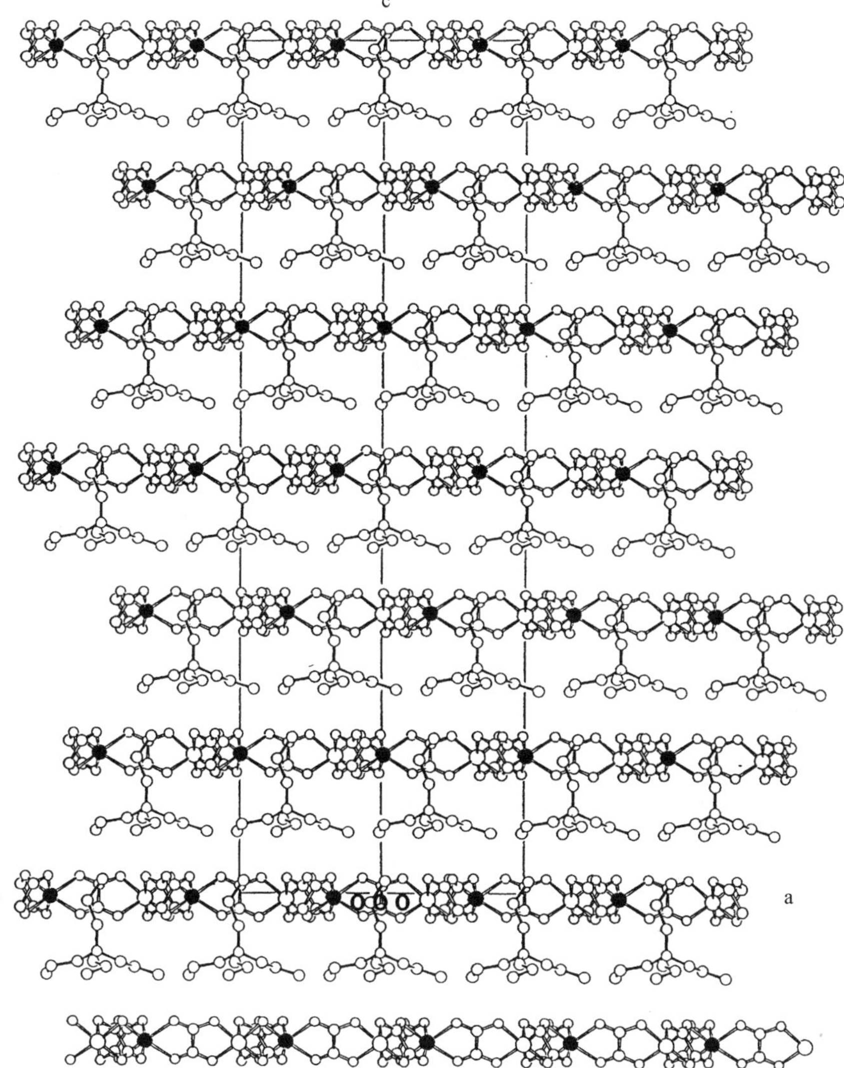

Figure 22 [110] projection of $[N(n\text{-}C_3H_7)_4][Mn^{II}Cr^{III}(ox)_3]$. Black atoms indicate Mn(II) positions [80].

the bimetallic oxalato-bridged honeycombed network $[M^{II}M^{III}(ox)_3]$(M^{II} = Mn, Fe, Co, Cr, Ni, Cu; M^{III} = Cr, Fe) and the organometallic cation decamethylferrocenium. The crystal structure was solved for the $Mn^{II}Fe^{III}$ derivative in the monoclinic space group *C2/m* and in this case the decamethylferrocenium cations intercalate the oxalato-bridged layers without penetrating into the vacancies. Day and co-workers [88,89,90] have reported detailed investigations of the synthesis, structural and magnetic characterisation of a series of mixed-valence compounds with stoichiometries of the type [A][$Cr^{II}Cr^{III}(ox)_3$] and [A][$Fe^{II}Fe^{III}(ox)_3$], together with mixed-metal compounds of stoichiometry [A][$Mn^{II}Fe^{III}(ox)_3$]. In their work, all of the phases prepared with a wide variety of organic template cations [A] were structurally characterised by X-ray powder diffraction, and the reflections indexed and unit cell constants refined on a hexagonal cell; all of the these structures have been proved to be consistent with a 2-D honeycomb network. The systematic variation of [A] was aimed at modulating the separation between the layers and hence varying the physical properties.

3.3 Magnetic Studies on 2-D Compounds

These 2-D polymeric materials all exhibit a wide range of magnetic behaviour and ferro-, ferri- or antiferromagnetic long-range ordering processes have all been observed. Generally, the magnetic interactions and magnetic ordering in the [A][$M^{II}M^{III}(ox)_3$] materials are strongly dependent on the character of the cations and the combination of $M^{II}M^{III}$ metal ions.

In 1992, Okawa and co-workers [91,92] reported a synthetic and magnetic study on ferri- and ferromagnetic mixed metal assemblies with the stoichiometry [NBu_4][$M^{II}M^{III}(ox)_3$] (M^{II} = Mn, Fe, Co, Ni, Cu, Zn; M^{III} = Cr, Fe). At that time, these compounds were falsely assumed to have a 3-D network structure, but in subsequent reports by Atovmyan *et al.* [82] and Decurtins *et al.* [83], the verification of the two-dimensional class of structure has since been proved. The series of Cr(III) derivatives show a ferromagnetic phase transition with $T_c < 14\,K$ and magnetic hysteresis loops with coercive fields $H_c < 320\,G$. The Fe(III) compounds with M^{II} = Fe, Ni have been shown to behave as ferrimagnets with magnetic phase transition temperatures of 43 and 28 K, respectively. [91]

In 1994, Day and co-workers [89] reported the results from a detailed investigation of the mixed-valence ferrimagnets [A][$Fe^{II}Fe^{III}(ox)_3$] (A = quaternary ammonium or phosphonium). In a later communication [93], these authors presented a more comprehensive study on the temperature-dependent magnetisation in these [A][$Fe^{II}Fe^{III}(ox)_3$] layer compounds, and discussed the manifestation of Néel N- and Q-type ferrimagnetism. The major aim of these studies was to map and identify the reasons to account for the presence of negative magnetisation at low temperature. According to Néel, the ground state of a ferrimagnet is determined by the saturation magnetisations of each magnetic sub-lattice and their relative ordering rates with respect to temperature. For the honeycomb lattice with alternating Fe(II) and Fe(III) ions, this situation corresponds to an initially steeper ordering on the Fe(II) sub-lattice.

The mixed-valence compound with stoichiometry [A][$Cr^{II}Cr^{III}(ox)_3$] (A = NBu_4, $P(Ph)_4$) has also been synthesised and fully characterised structurally and magnetically [88]. Again, although short-range antiferromagnetic correlations were observed below 100 K, no transitions to a long-range ordered state were observed above 2 K.

For the network of stoichiometry $[Mn^{II}Fe^{III}(ox)_3]$, one encounters the rare situation of a bimetallic lattice in which both metal ions have the same electronic ground state namely, $3d^5$, $S = \frac{5}{2}$, $^6A_1$. In this case, the compound has been found to mimic antiferromagnetic behaviour. Thus the temperature dependence of the magnetic susceptibility is that of a classical 2-D antiferromagnet, reaching a broad maximum at 55 K. Definitive evidence for this behaviour comes from a neutron powder diffraction study, which revealed a simple collinear antiferromagnetic alignment of Mn(II) and Fe(III) moments parallel to the *c*-axis or, in other words, perpendicular to the layers.

The first published work from neutron scattering studies on the magnetic structure of a layered bimetallic trisoxalate salt comes from Decurtins and co-workers [80] and refers to an $Mn^{II}Cr^{III}$ phase which orders ferromagnetically at $T_c = 6$ K. The best agreement between the observed and calculated magnetic neutron intensities was achieved with a collinear ferromagnetic arrangement of both of the Mn(II) and Cr(III) spins along the *c*-axis. This picture is consistent with the results of a single-crystal magnetisation experiment that showed the *c*-axis to be the easy axis of magnetisation. In fact, the same conclusions had already been drawn by Atovmyan and co-workers in two earlier publications [82, 94].

3.4 Structural Studies on 3-D Oxalato-bridged Compounds

The first structure of a 3-D transition metal network incorporating the oxalate ion was that of $[Ni(phen)_3][KCo^{III}(ox)_3] \cdot 2H_2O$, phen = 1,10-phenanthroline, reported by Butler and Snow in 1971. [95] The true dimensionality of this compound, however, went unrecognised during this period, and the potential of oxalate ions to form 3D networks was not fully realised until 1993, when Decurtins *et al.* published the crystal structure of the iron(II) oxalato complex with tris (2,2′-bipyridine)iron(II) cations. [84] This compound has an overall stoichiometry of $[Fe^{II}(bipy)_3]^{2+}_n[Fe^{II}_2(ox)_3]_n^{2n-}$ and forms a 3D anionic polymeric network that is best described with the three-connected 10-gon network topology. A view of this anionic network is given in Figure 20.

The possibilities for linking paramagnetic metal ions via oxalate bridges in 3-D lattices and their possible applications in the field of molecule-based magnets have since provided the momentum for subsequent investigations into the self-assembly of 3D oxalato-bridged metal complexes. In a subsequent communication, Decurtins *et al.* [96] reported the structural characterisation of three additional compounds, one homometallic with the same network stoichiometry, $[Mn^{II}_2(ox)_3]_2^{2n}$, as the Fe(II) compound above, and two bimetallic with network stoichiometries of $[NaFe^{III}(ox)_3]_n^{2n-}$ and $[LiCr^{III}(ox)_3]_n^{2n-}$. Both types of anionic networks form an analogous 3D pattern, but their differences in stoichiometry result in them crystallising in different space groups. The $[Mn^{II}_2(ox)_3]_2^{2n-}$ network is isostructural with the iron(II) compound and the structures of the two heterometallic compounds differ only in that the overall space group symmetry is lowered to $P2_13$. According to the lower symmetry, the asymmetric unit contains a complete oxalate ligand, both metals of the network and the complete bipyridine ligand. The planar oxalate ligands repeatedly bridge the metal ions in all three dimensions producing chiral left- or right-handed helical strands. Figure 23 shows the [100] projection of the $[LiCr^{III}(ox)_3]_n^{2n-}$ net. The *tris*-chelating $[M^{II}(bipy)_3]^{2+}$ cations occupy the vacancies within the network.

In 1996, a third related study came from Decurtins *et al.*, [97] the aim of which was to investigate the factors which contribute to the structural and chemical flexibility of this type of three-dimensional network. Since it had previously been established that the tris-chelated cations play an important role in initiating the formation and crystallisation of the 3D nets, the structural response towards varying the cations, i.e. using an $[M^{III}(bpy)_3]^{3+}(M^{III} = Cr,\ Co)$ template or the bulkier $[Ni^{II}(phen)_3]^{2+}$ cations, was investigated. In the former case, the system achieves this with an elaborate inclusion of an additional complex anion, while keeping the known three-dimensional topology (described above), crystal symmetry and lattice parameters fixed. It does so by encapsulating these anions in cubic-shaped empty

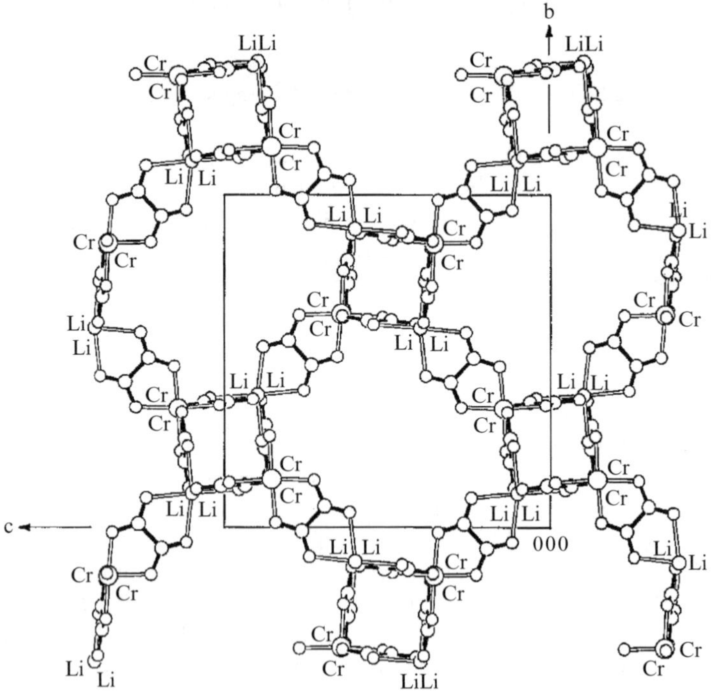

Figure 23 [100] projection of 3-D network of $[Fe^{II}(bipy)_3][LiCr^{III}(ox)_3]$ [96].

spaces formed by six of the planar bipyridine ligands from three adjacent cations. Figure 24 shows a view of the packing arrangement of three adjacent tris-chelated cations encapsulating a perchlorate anion. In the latter case, the framework demonstrates a marked flexibility in modifying both the shape and width of the cavities in order to accommodate the steric requirements of the bulky cations.

The possibility of using 1,2-dithiooxalate as an isosteric replacement to the oxalate bridging ligand has also been demonstrated, leading to the first reported example of a three-dimensional dithiooxalate-bridged compound. This compound with stoichiometry $[Ni^{II}(phen)_3][NaCo^{III}(dto)_3]$ has a topology which is once again consistent with the chiral three-dimensional (10,3) network. In contrast to the previously reported structures, this compound crystallises in the orthorhombic space group $P2_12_12_1$ and, as a consequence, there is no longer a threefold crystallographic axis imposed on the tris-chelated metal ions and the cations within the network. The 1,2-dithiooxalate ligands ($C_2O_2S_2{}^{2-}$) are approximately planar, but one important structural difference worthy of comment is the discrimination in the coordination behaviour of the bridging 1,2-dithiooxolate ligand, since bonding to cobalt(III) occurs through the sulfur atoms, whereas the sodium atoms bind to the oxygen ends of the bridging ligand. The interatomic Na–O distances with a mean value of 2.376(11) Å can be compared with the mean

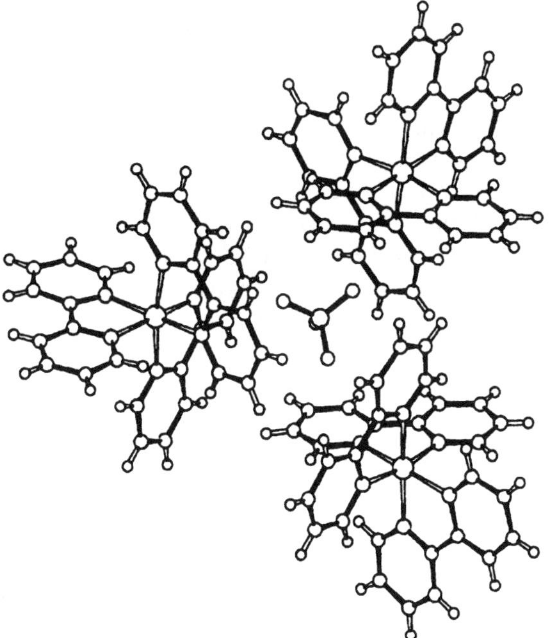

Figure 24 View of three adjacent $[Cr(bipy)_3]^{3+}$ complexes exhibiting the cavity formed by three pairs of parallel oriented bipyridine ligands. The centre of the cage is occupied by a $[ClO_4]^-$ anion [97].

value of 2.319(3) Å in the oxalate-bridged network $[NaFe^{III}(ox)_3]_n^{2n-}$ [96]. The cation packing arrangement is similar to that in the cubic oxalate-bridged compounds, namely three adjacent cations with their V-shaped pockets form a sufficiently large cavity to accommodate additional molecules. These series of compounds clearly illustrate that the 10-gon framework is indeed flexible and can readily distort to account for changes in the nature of the ligand or the cations.

In recent years, Decurtins and co-workers [98,99] have reported details regarding the crystal structures of several additional anionic 3D polymeric networks. The crystals of the compound $[Fe^{II}(bipy)_3]_n^{2+}[AgCr^{III}(ox)_3]_n^{2n-}$ have a regular tetrahedral morphology that is consistent with the cubic space group $P2_13$ and, indeed, this compound is isostructural with the previously described heterometallic $[M^IM^{III}(ox)_3]_n^{2n-}$ chiral 3-D polymeric networks. Furthermore, the above authors have repeatedly demonstrated that it is possible to apply a straightforward series of topological rules to realise the same 3D structural type for many different transition metal ions, the most recent being the compounds of stoichiometries $[Ru^{II}(bipy)_3]_n^{2+}[NaM^{III}(ox)_3]_n^{2n-}$ with $M^{III} = Ru^{III}$, Rh^{III}. The study of these compounds was of particular interest for their photophysical properties [100]. Applying this methodology, Julve and co-workers have recently synthesised and reported the crystal structure of the chiral 3-D oxalato-bridged compound $[Co^{III}(bpy)_3][Co^{II}_2(ox)_3]ClO_4$. [101] As expected, this compound is isostructural with the compounds of formula $[Cr^{III}(bpy)_3][Mn^{II}_2(ox)_3]ClO_4$ and $[Cr^{III}(bipy)_3][Mn^{II}_2(ox)_3]BF_4$ reported in 1996 by Decurtins *et al.* [97]

All of the above-described chiral 3-D structures were obtained from racemic starting materials. Recently however, Verdaguer and co-workers [102] have applied this methodology to exploit the feasibility of the self-assembly of chiral 2-D and 3-D oxalato-bridged polymeric networks starting from homochiral subunits, namely resolved $[Cr^{III}(ox)_3]^{3-}$ and $[M^{II}(bpy)_3]^{2+}$(M^{II} = Ni, Ru) building blocks. Naturally, the topology of the 3-D systems is consistent with the chiral 3-D three-connected 10-gon (10,3) network configuration achieved from the enantiomeric mixture of the precursors. Solid-state circular dichroism (CD) measurements demonstrate the enantiomeric character of the resulting polymers.

3.5 Magnetic Studies on 3-D Oxalato-bridged Compounds

Three-dimensional homo- and heterometallic oxalato-bridged frameworks are examples of supramolecular host–guest compounds which show long-range magnetic ordering behaviour and various kinds of photophysical properties [100,103]. The existence of a magnetically ordered phase was first deduced from magnetic susceptibility measurements on the three-dimensional compound $[Fe^{II}(bpy)_3][Mn^{II}_2(ox)_3]$ [96]. The experimental data revealed a rounded maximum at about 20 K in the χ_M versus T curve (thus $T_N < 20$ K) as well as a Weiss constant Θ of -33 K in the $1/_{\chi M}$ versus T plot. This long-range magnetic ordering originates from the exchange interactions between neighbouring Mn^{II} ions, mediated by the bridging oxalate ligands. This compound was the first three-dimensional oxalato network for which the magnetic structure could be solved [104]. Neutron diffraction experiments were performed on a polycrystalline sample in the temperature range 30–1.8 K. As expected, the appearance of magnetic neutron intensities due to the long-range antiferromagnetic ordering of the spins of the Mn(II) ions was detected.

Figure 25 depicts the pattern of the magnetic structure with the 3-D managanese(II) network. Despite the three-dimensional helical character of the framework structure incorporating the magnetic ions, a two-sub-lattice antiferromagnetic spin arrangement has been proved, hence no helimagnetic structure was apparent and this behaviour is in accordance with the typical isotropic character of the Heisenberg Mn(II) ion.

Magnetic studies on the isostructural three-dimensional networks with stoichiometries $[Co^{III}(bpy)_3][Co_2^{II}(ox)_3]ClO_4$ (**1**) and $[M^{II}(bpy)_3][Co_2^{II}(ox)_3]$ [M^{II} = Fe (**2**) and Ni (**3**)], recently carried out by Julve *et al.* [101], show the occurrence of a weak ferromagnetism at low temperatures for compounds **1** ($T_c = 9.0\,K$) and **2** ($T_c = 6.5\,K$), whereas no magnetic ordering was observed for compound **3**. This magnetic ordering at very low temperatures is attributed to the spin canting of the magnetic moments of the metal ions in the chiral three-dimensional oxalate-bridged cobalt(II) net. These results also suggest that the magnetic ordering of the three-dimensional $[Co^{II}{}_2(ox)_3]_n{}^{2n-}$ anionic framework is strongly dependent on the size and diamagnetic or paramagnetic character of the tris-chelated counter ions used to achieve the electroneutrality. In this report, the influence of the size of the $[M^{II/III}(bpy)_3]^{2+/3+}$ complex on the spin canted magnetic structure was also discussed, but further work is needed to investigate these ideas in greater detail.

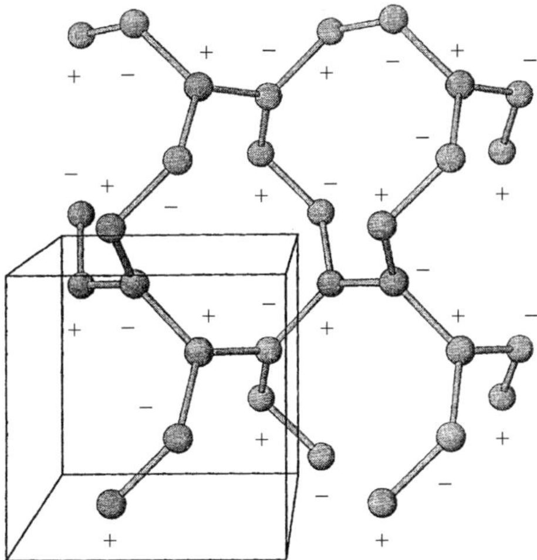

Figure 25 Scheme of antiferromagnetic collinear configuration of the magnetic moments originating from the Mn(II) ions, which constitute the chiral 3-D network compound [104].

4 THE SPIN CROSSOVER PHENOMENON

The spin crossover phenomenon provides us with one of the most impressive examples of molecular bistability. It was first observed in the early 1930s, but it took nearly 50 years before researchers realised that spin crossover compounds could be used as active elements in memory devices. A spin crossover transition arises in molecular species containing an octahedral coordinated transition metal ion with a $3d^n$ electronic configuration ($3 < n < 8$), which may 'cross over' between a low-spin (LS) and a high-spin (HS) state. Chemists have been able to induce this crossover by varying temperature or pressure, or by irradiating the molecules with light. The origin of the spin crossover phenomenon is molecular, but the temperature dependence of the HS molar fraction $\chi_{HS} = f(T)$ depends strongly on intermolecular interactions. The stronger these interactions are, the steeper are the $\chi_{HS} = f(T)$ curve around T_1. When the magnitude of these interactions overcomes a threshold value, the spin crossover may become cooperative; in other words, the crystal lattice as a whole is involved in the process. Thermally induced spin transitions between LS and HS states may not only be abrupt, but may also occur with hysteresis; the temperature of the LS $\longrightarrow$ HS transition in the warming mode, $T_c \uparrow$, is higher than the temperature of the HS $\longrightarrow$ LS transition in the cooling mode, $T_c \downarrow$. Between $T_c \uparrow$ and $T_c \downarrow$ the system is bistable, and its electronic state, LS or HS, depends on its history, and hence on the information that the system stores. This chemical flexibility can be used to design complexes the critical temperature of which is tunable up to or above room temperature. As the colour of the compound is determined by the electronic transitions implying d orbitals, the spin state change is frequently accompanied by a colour change. Bistable molecules of this type have introduced a 'new dimension' to molecular chemistry, namely memory. As a consequence, strategies that permit the rational design of molecule-based materials that can undergo a spin transition are very desirable.

4.1 Utilising the Spin Crossover Phenomenon – From Rational Design to Functional Molecular Materials

As already highlighted, three-dimensional oxalate networks of the type $[M^{II}(bpy)_3][M^I\text{–}M^{III}(ox)_3]$ possess a very intriguing three-dimensional structure that contains perfect cavities for tris-bipyridyl complex cations. Hauser and co-workers [105] have taken advantage of these observations to induce a thermal spin crossover transition in a $[Co(bipy)_3]$ cation by adjusting the size of this cavity. In the three-dimensional oxalate networks, this can be achieved by variation of the metal ions of the oxalate backbone. In $[Co(bipy)_3][NaCr(ox)_3]$, the $[Co(bpy)_3]^{2+}$ complex is in its usual ${}^4T_1(t_{2g}{}^5e_g{}^2)$ high-spin ground state. Substituting Na^+ by Li^+ reduces the size of the cavity. The resulting chemical pressure destabilises the high-spin state of $[Co(bpy)_3]^{2+}$ to such an extent that the ${}^2E(t_{2g}{}^6e_g{}^1)$ low-spin state becomes the actual ground state. As a result, $[Co(bpy)_3][LiCr(ox)_3]$ becomes a spin crossover system as shown by temperature-dependent magnetic susceptibility measurements, and also by single-crystal X-ray crystallographic data measured at 290 and 10 K.

4.2 Structural Changes

The three-dimensional network crystallises in the cubic space group $P2_13$, $Z = 4$. The general features of the crystal structure have been discussed earlier. As depicted in Figure 26, the structure is such that the three-dimensional oxalate backbone provides cavities which in terms of size and geometry provide a perfect fit for $[M^{II}(bpy)_3]^{2+}$ ions.

The site symmetry of both the tris-oxalate and the tris-bipyridyl compexes is C_3. Furthermore, the rather loose oxalate network is actively stabilized by the tris-bipyridyl complexes through electrostatic interactions from π-overlap between oxalate and bipyridine along the trigonal axis. Selected crystallographic data including relevant bond lengths and angles are presented in Table 1.

Comparing the bond lengths for the two temperature measurements, it is apparent that the M^I–O and Cr–O bond lengths are similar for the two temperatures. The M^{II}–N bond lengths, however, decrease from an average of 2.095 to 2.014 Å (that is, by 0.081 Å) between 293 and 10 K. This value is far too large to be due to thermal contraction alone and serves as a first indication that the $[Co(bpy)_3]^{2+}$ cation has become a spin crossover complex in this lattice.

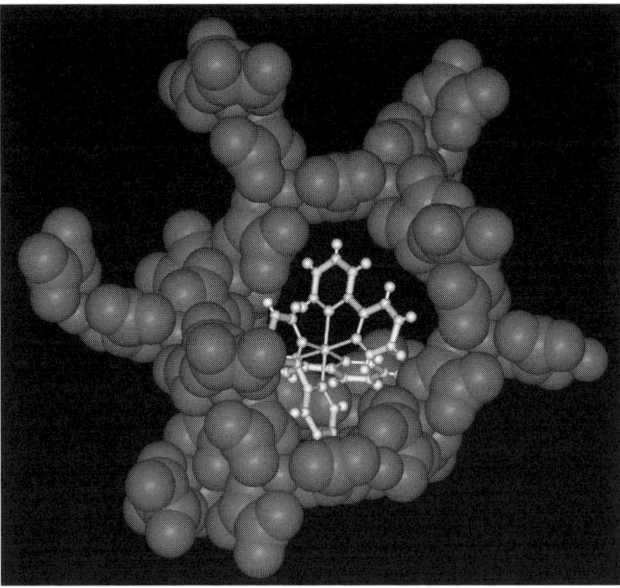

Figure 26 A chiral 3-D host–guest compound of stoichiometry $[M^{II}(bpy)_3][M^IM^{III}(C_2O_4)_3]_n$. Only one guest molecule is shown within the chiral framework.

Table 1 Selected crystallographic data for oxalate complexes [105]

Parameter	$[Co(bpy)_3][LiCr(ox)_3]$ at 293 K	$[Co(bpy)_3][LiCr(ox)_3]$ at 10 K
Unit cell parameters:		
A (Å)	15.3866(8)	15.2230(18)
V (Å^3)	3652.7	3527.8(7)
Bond lengths (Å):		
M^{II}–N1	2.096(2)	2.013(4)
M^{II}–N2	2.094(2)	2.015(3)
Cr–O1	1.978(2)	1.988(3)
Cr–O2	1.978(2)	1.981(3)
M^{I}–O3	2.143(6)	2.170(8)
M^{I}–O4	2.321(5)	2.222(7)

5 WORKING TOWARDS THE DESIGN OF DUAL-ACTION MATERIALS

One particularly attractive goal of chemists working in the area of molecule-based materials is the design of compounds combining two properties that would otherwise be difficult to achieve in a conventional inorganic solid with a continuous lattice. This approach is particularly attractive since the presence of two cooperative properties in the same crystal lattice might also result in new physical phenomena, which in turn could lead to novel applications. A promising strategy for creating this type of bifunctionality targets hybrid organic/inorganic crystals comprising two functional sub-lattices exhibiting distinct properties. Along these lines, two groups have recently explored the possibilities of using the 2-D type inorganic oxalate lattice to control the crystal packing of π-electron donor molecules [106,107]. In this way, bis(ethylenedithio)tetrathiafulvalene (BEDT-TTF) and its derivatives, which form the basis of most known molecular conductors and superconductors, have been combined with 2-D oxalate anionic frameworks to yield a range of materials which display magnetic in addition to conducting, superconducting and semiconducting properties.

Following this approach, Coronado *et al.* reported the first example of a molecule-based layered compound that is both magnetic and an electrical conductor [107]. Single crystals of composition [BEDT-TTF] $[MnCr(C_2O_4)_3]^-$ were grown via electrocrystallisation methods. The partial crystal structure of this compound reveals that the organic layers, namely the BEDT-TTF cations, are found to be alternating with the honeycomb layers of the bimetallic oxalato complex. The BEDT-TTF cations are tilted with respect to the inorganic layer by an angle of 45°. All the donor molecules are related by translations, and adopt a pure two-dimensional pseudohexagonal β-type packing motif. The anionic oxalate-based bimetallic layer is crystallographically disordered and as a consequence the solvent molecules could not be located. Studies have shown that this compound is metallic down to 4 K and is ferromagnetic below 5.5 K. There is, however, no evident interplay between the two sub-lattices (ferromagnetic and conducting), which means that from an electronic point of view, the two lattices are quasi-independent. Nevertheless, these advances indicate that the design of molecule-based materials with dual properties is bordering on the horizon. Consequently, one of the major challenges still to be realised in this field is the design of hybrid materials for which there is an interplay between at least two functional sub-lattices.

6 MAGNETISM – HISTORICAL OVERVIEW AND FUTURE OUTLOOK

For more than 2000 years, human fascination with magnets has been historically recorded. Around the eleventh century, the Chinese found that a steel needle stroked with 'lodestone' (Fe_3O_4) became magnetic and, furthermore, when this needle was freely suspended, it pointed north–south. Hence the magnetic compass was invented and its discovery soon spread to Europe. Indeed, Columbus made good use of this technology when he made his famous voyage across the Atlantic Ocean. At several points in history the field of magnetism has experienced a revival, the first of which came in the seventeenth century when William Gilbert, physician to Queen Elizabeth I of England, proposed that the Earth itself was a magnet, with its magnetic poles some distance away from its geographic poles. Gilbert's initial experiments and observations regarding the magnetic behaviour of iron resulted in the preparation of the first artificial magnet. Since these initial discoveries, the study and subsequent exploitation of magnets and magnetism have continued. To date, two broad uses of magnets exist, namely (i) permanent or 'hard' magnets, which produce an external magnetic field that in turn produces a force on either other magnets or on an electric current, and (ii) nonpermanent of 'soft' magnets, which guide or deflect magnetic fields and have large magnetic moments in response to small electric currents. With applications ranging from wristwatches to industrial motors, magnetic products impact all sectors of our economy. The global market for the production of magnetic materials is worth billions of dollars and was estimated in 2000 to be growing at an annual rate of 10% [1]. As a consequence, magnetic phenomena continue to be studied by many research groups worldwide.

Although discovered less than two decades ago, molecule-based magnets are a broad emerging class of magnetic materials that expand the materials properties typically associated with magnets to include low density, transparency, electrical insulation and low-temperature fabrication. Essentially, all of the common magnetic phenomena associated with conventional transition metal- and rare earth-based magnets can be found in molecule-based magnets. Although this field is still in its infancy on the magnetism time-scale, magnets with ordering temperatures exceeding room temperature, very high and very low coercivities and substantial remnant and saturation magnetizations have been achieved. In addition, new interesting phenomena such as spin crossover and photomagnetism have been discovered. The advent of molecule-based magnets offers new processing opportunities, and the discovery and development of new improved magnetic materials continues to be a major aim at the forefront of modern materials chemistry. In this chapter we have looked in detail at two families of molecule-based magnets, namely cyanometalates and transition metal oxalates. Our aim has been to demonstrate how suitable molecular building blocks combined with a rational synthetic approach can be successfully applied to engineer new classes of molecule-based magnets. Given the rapid growth in this field, it is clear that we can look forward to new advances in this multidisciplinary area of solid-state chemistry as magnetism continues its renaissance well into the twenty-first century.

REFERENCES

1. J. S. Miller, *Inorg. Chem.*, **2000**, *39*, 4392.
2. D. Braga, *J. Chem. Soc., Dalton Trans.*, **2000**, 3705.
3. J. -M. Lehn, *Supramolecular Chemistry: Concepts and Perspectives*, **1995**, VCH, Weinheim.
4. R. Robson, *Comprehensive Supramolecular Chemistry*, **1997**, Vol. 6, p.733. Pergamon Press, Oxford.
5. M. Eddaoudi, D. B. Moler, H. Chen, T. M. Reinke, M. O'Keeffe, O.M. Yaghi, *Acc. Chem. Res.*, **2001**, *34*, 319.
6. F. M. Tabellion, S. R. Siedel, A. M. Arif, P. Stang, *J. Am. Chem. Soc.*, **2001**, *123*, 11982.
7. M. Verdaguer, *Science*, **1996**, *272*, 698.
8. J. S. Miller, A. J. Epstein, *MRS Bull.*, **2000**, 21.
9. O. Kahn, *Molecular Magnetism*, **1993**, Wiley-VCH: Weinheim.
10. H. H. Wickman, A. M. Trozzolo, H. J. Williams, G. W. Hull, F. R. Merrett, *Phys. Rev.*, **1967**, *155*, 563.
11. H. H. Wickman, *J.Chem. Phys.*, **1972**, *56*, 976.
12. G. C. DeFohs, F. Palacio, C. J. O'Connor, S. N. Bhaatia, R. L. Carlin, *J. Am. Chem. Soc.*, **1977**, *99*, 8314.
13. N. Arai, M. Sorai, H. Suga, S. Seki, *J. Phys. Chem. Solids*, **1977**, *36*, 1231.
14. J. H. Helms, W. E. Hatfield, M. J. Kwiecien, W. M. Reiff, *J. Chem. Phys.*, **1986**, *84*, 3993.
15. R. Bumel, J. Casabo, J. Pons, D. E. Carnegie Jr, R. L. Carlin, *Physica*, **1985**, *132B*, 185.
16. O. Kahn, *Struct. Bonding (Berlin)*, **1987**, *68*, 89.
17. O. Kahn, *Angew. Chem., Int. Ed. Engl.*, **1985**, *24*, 834.
18. S. Decurtins, *Chimia*, **1988**, *10*, 539.
19. K. R. Dunbar, R. A. Heintz, *Prog. Inorg. Chem.*, **1997**, *45*, 283.
20. M. Verdaguer, A. Bleuzen, V. Marvaud, J. Vaissermann, M. Seuleiman, C. Desplanches, A. Scuiller, C. Train, R. Garde, G. Gelly, C. Lamenech, I. Rosenman, P. Veillet, C. Cartier, F. Villain, *Coord. Chem. Rev.*, **1999**, *190–192*, 1023.
21. M. Verdaguer, *Polyhedron*, **2001**, *20*, 1115.

22. M. Ohba, H. Okawa, *Coord. Chem. Rev.*, **2000**, *198*, 313.
23. *Misc. Berolinensia Incrementum Sci. Berlin*, **1710**, *1*, 337.
24. J. F. Keggin, J. F. Miles, *Nature (London)*, **1936**, *137*, 577.
25. A. Ludi, H-U. Güdel, *Struct. Bonding (Berlin)*, **1973**, *14*, 1.
26. J. H. Buser, G. Ron, A. Ludi, P. Engel, *J. Chem. Soc., Dalton Trans.*, **1974**, 2473.
27. O. Kahn, *Nature (London)*, **1995**, *373*, 667.
28. A. N. Holden, B. T. Matthias, P. W. Anderson, H. W. Lewis, *Phys. Rev.*, **1956**, *102*, 1463.
29. R. M. Bozworth, H. J. Williams, D. E. Walsh, *Phys. Rev.*, 1956, *103*, 572.
30. R. Klenze, B. Kanellopoulus, G. Trageser, H. H. Eysel, *J. Chem. Phys.*, **1980**, *72*, 5819.
31. W. D. Geiebler, D. Babel, *Z. Naturforsch., Teil B*, **1982**, *87*, 832.
32. D. Babel, *Comments Inorg. Chem.*, **1986**, *5*, 285.
33. S. Ferley, T. Mallah, R. Ouahès, P. Veillet, M. Verdaguer, *Nature (London)*, **1995**, *378*, 701.
34. O. Hatlevic, W. E. Buschmann, J. Zhang, J. L. Manson, J. S. Miller, *Adv. Mater.*, **1999**, *11*, 914.
35. S. Holmes, G. Girolami, *J. Am. Chem. Soc.*, **1999**, *121*, 5593.
36. P. Franz, H. Stoeckli-Evans, S. Decurtins, *Angew. Chem., Int. Ed. Engl.*, in preparation.
37. O. Sato, T. Iyoda, A. Fujishima, K. Hashimoto, *Science*, **1996**, *272*, 704.
38. M. Ohba, N. Maruono, H. Okawa, T. Enoki, J. M. Latour, *J. Am. Chem. Soc.*, **1997**, *119*, 1011.
39. T. Mallah, C. Auberger, M. Verdaguer, P. Veillet, *J. Chem. Soc., Chem. Commun.*, **1995**, 61.
40. M. Ohba, N. Maruono, H. Okawa, T. Enoki, J. M. Latour, *J. Am. Chem. Soc.*, **1994**, *116*, 11566.
41. R. J. Parker, D. C. R. Hockless, B. Moubaraki, K. S. Murray, L. Spicia, *J. Chem. Soc., Chem. Commun.*, **1996**, 2789.
42. H. Okawa, M. Ohba, in M. M. Turnbull, T. Sugimoto, L. K. Thompson (Eds), *Molecule-Based Magnetic Materials*, ACS Symposium Series No. 644. American Chemical Society: Washington, DC, **1966**, p.319.
43. M. Ohba, N. Uski, N. Fukita, H. Okawa, *Inorg. Chem.*, **1998**, *37*, 3349.
44. N. Re, E. Gallo, C. Floriani, H. Miyasaka, N. Matsumoto, *Inorg. Chem.*, **1996**, *35*, 6004.
45. E. Colacio, J. M. Dominguez-Vera, M. Ghazi, R. Kivekäs, M. Klinga, J. M. Moreno, *J. Chem. Soc., Chem. Commun.*, **1998**, 1071.
46. E. Colacio, M. Ghazi, H. Stoeckli-Evans, F. Lloret, J. M. Moreno, C. Pérez, *Inorg. Chem.*, **2001**, *40*, 4876.
47. S. Ferlay, T. Mallah, J. Vaissermann, F. Bartolome, P. Veillet, M. Verdaguer, *J. Chem. Soc., Chem. Commun.*, **1996**, 2482.
48. M. Ohba, H. Okawa, T. Ito, A. Ohto, *J. Chem. Soc., Chem. Commun.*, **1995**, 1545.
49. M. Ohba, H. Okawa, *Mol. Cryst. Liq. Cryst.*, **1996**, *286*, 101.
50. M. Ohba, N. Fukita, H. Okawa, Y. Hashimoto, *J. Am. Chem. Soc.*, **1997**, *119*, 1011.
51. H. Z. Kou, W. M. Bu, D. Z. Liao, Z. H. Jiang, S. P. Yan, Y. G. Fan, H. G. Wang, *J. Chem. Soc., Dalton. Trans.*, **1998**, 4161.
52. H. Miyasaka, N. Matsumoto, N. Re, E. Gallo, C. Floriani, *Angew. Chem., Int. Ed. Engl.*, **1995**, *34*, 1446.
53. H. Miyasaka, N. Matsumoto, H. Okawa, N. Re, E, Gallo, C. Floriani, *J. Am. Chem., Soc.*, **1996**, *118*, 981.
54. H. Miyasaka, H. Ieda, N. Matsumoto, N. Re, E. Gallo, C. Floriani, *Inorg. Chem.*, **1988**, *37*, 255.
55. H. Miyasaka, H. Okawa, A. Miyazaki, T. Enoki, *Inorg. Chem.*, **1988**, *37*, 4878.

56. N. Fukita, M. Ohba, H. Okawa, K. Matsuda, H. Iwnura, *Inorg. Chem.*, **1998**, *37*, 842.
57. M. S. El. Fallah, E. Rentschler, A. Caneschi, R. Sessoli, D. Gatteschi, *Angew. Chem. Int. Ed. Engl.*, **1996**, *35*, 1947.
58. M. Ohba, N. Usuki, N. Fukita, H. Okawa, *Angew. Chem., Int. Ed. Engl.*, **1995**, *34*, 1446.
59. N. Usuki, M. Yamada, M. Ohba, H. Okawa, *Polyhedron*, **2001**, *159*, 328.
60. J. A. Smith, J. R. G. Mascarós, R. Clérac, J.-S. Sun, X. Ouyang, K. R. Dunbar, *Polyhedron*, **2001**, *20*, 1727.
61. T. Mallah, C. Auberger, M. Verdaguer, P. Veillet, *J. Chem. Soc., Chem. Commun.*, **1995**, 61.
62. A. Scuiller,, T. Mallah, J. Vaissermann, F. Bartolomé, P. Veillet, M, Verdaguer, *J. Chem. Soc., Chem. Commun.*, **1996**, 2481.
63. J. Larionova, J. Sanchiz, S. Golhen, L. Ouahab, O. Kahn, *J. Chem. Soc., Chem. Commun.*, **1998**, 953.
64. J. Larionova, O. Kahn , S. Golhen, L. Ouahab, R. Clérac, *Inorg. Chem.*, **1999**, *38*, 3621.
65. J. Larionova, R. Clérac, J. Sanchiz, O. Kahn S. Golhen, L. Ouahab, *Inorg. Chem.*, **1999**, *38*, 3621.
66. O. Kahn, J. Larionova, L. Ouahab, *J. Chem. Soc., Chem. Commun.*, **1999**, 945.
67. A. K. Sra, M. Andruh, O. Kahn, S. Golhen, L. Ouahab, J. V. Yakhmi, *Angew. Chem., Int. Ed. Engl.*, **1999**, *38*, 2606.
68. M. Pilkington, S. Decurtins, *Chimia*, **2000**, *54*, 593.
69. G. Rombaut, S. Golhen, L. Ouahab, C. Mathionière, O. Kahn, *J. Chem. Soc., Dalton. Trans.*, **2000**, 3609.
70. J. Larionova, M. Gross, M. Pilkington, H.-P. Andres, H. Stoeckli-Evans, H. U. Güdel, S. Decurtins, *Angew. Chem., Int. Ed. Engl.*, **2000**, *39*, 1605.
71. Z. J. Zhong, H. Seino, Y. Mizobe, M. Hidai, A. Fujishima, S. Ohkoshi, K. Hashimoto, *J. Am. Chem. Soc.*, **2000**, *122*, 2952.
72. F. Bonadio, M. Gross, H. Stoeckli-Evans, S. Decurtins, *Inorg. Chem.*, **2002**, *22*, 5891.
73. M. Pilkington, M. Gross, L. Gilby, S. Decurtins, in preparation.
74. P. Franz, M. Pilkington, S. Decurtins, M. Verdaguer, K. Hashimoto, *Inorg. Chem.*, in preparation.
75. M. Pilkington, M. Biner, S. Decurtins, H. Stoeckli-Evans, unpublished results.
76. Z. J. Zhong, H. Seino, Y. Mizobe, M. Hidai, M. Verdaguer, S-I. Ohkoshi, K. Hashimoto, *Inorg. Chem.*, **2000**, *39*, 5095.
77. A. K. Sra, G. Rombaut, F. Lahitête, S. Golhen, L, Ouahab, C. Mathonière, J. V. Yakhmi, O. Kahn, *New J. Chem.*, **2000**, *24*, 871.
78. G. Rombaut, M. Verelst, S. Golhen, L. Ouahab, C. Mathonière, O. Kahn, *Inorg. Chem.*, **2001**, *40*, 1151.
79. J. J. Girerd, O. Kahn, M. Verdaguer, *Inorg. Chem.*, **1980**, *19*, 274.
80. R. Pellaux, H. W. Schmalle, R. Huber, P. Fischer, T. Hauss, B. Ouladdiaf, S. Decurtins, *Inorg. Chem.*, **1997**, *36*, 2301.
81. H. Oshio, U. Nagashima, *Inorg. Chem.*, **1992**, *31*, 3295.
82. L. O. Atovmyan, G. V. Shilov, R. N. Lyubovskaya, E. I. Zhilyaeva, N. S. Ovanesyan, S. I. Pirumova, I. G. Gusakovskaya, *JETP Lett.*, **1993**, *58*, 766.
83. S. Decurtins, H. W. Schmalle, H. R. Oswald, A. Linden, J. Ensling, P. Gütlich, A. Hauser, *Inorg. Chim. Acta*, **1994**, *216*, 65.
84. S. Decurtins, H. W. Schmalle, P. Schneuwly, H. R. Oswald, *Inorg. Chem.*, **1993**, *32*, 1888.
85. S. G. Carling, C. Mathonière, P. Day, K. M. A. Malik, S. J. Coles, M. B. Hursthouse, *J. Chem. Soc., Dalton Trans.*, **1996**, 1839.

86. M. Clemente-Léon, E. Coronado, J. Galan-Mascaros, C. Gomez-Garcia, *J. Chem. Soc., Chem. Commun.*, **1997**, 1727.
87. E. Coronado, J. Galan-Mascaros, C. Gomez-Garcia, J. Ensling, P. Gütlich, *Chem. Eur. J.*, **2000**, *6*, 552.
88. C. J. Nuttall, C. Bellitto, P. Day, *J. Chem. Soc., Chem. Commun.*, **1995**, 1513.
89. C. Mathonière, S. G. Carling, D. Yusheng, P. Day, *J. Chem. Soc., Chem. Commun.* **1994**, 1551.
90. C. Mathonière, C. J. Nuttall, S. G. Carling, P. Day, *Inorg. Chem.*, **1996**, *35*, 1201.
91. H. Tamaki, M. Mitsumi, K. Nakamura, N. Matsumoto, S. Kida, H. Okawa, S. Ijima, *Chem. Lett.*, **1992**, 1975.
92. H. Tamaki, Z. J. Zhong, N. Matsumoto, S. Kida, M. Koikawa, N. Achiwa, Y. Hashimoto, H. Okawa, *J. Am. Chem. Soc.*, **1992**, *114*, 6974.
93. C. J. Nuttall, P. Day, *Chem. Mater.*, **1998**, *10*, 3050.
94. L. O. Atovmyan, G. V. Shilov, N. S. Ovanesyan, A. A. Pyalling, R. N. Lyubovskaya, E. I. Zhilyaeva, Y. G. Morozov, *Synth. Met.*, **1995**, *71*, 1809.
95. K. R. Butler, M. R. Snow, *J. Chem. Soc. A*, **1971**, 565.
96. S. Decurtins, H. W. Schmalle, P. Schneuwly, J. Ensling, P. Gütlich, *J. Am. Chem. Soc.*, **1994**, *116*, 9521.
97. S. Decurtins, H. W. Schmalle, R. Pellaux, P. Schneuwly, A. Hauser, *Inorg. Chem.*, **1996**, *35*, 1451.
98. H. W. Schmalle, R. Pellaux, S. Decurtins, *Z. Kristallogr.*, **1996**, *211*, 533.
99. R. Pellaux, S. Decurtins, H. W. Schmalle, *Acta Crystallogr, Sect. C*, **1999**, *55*, 1057.
100. M. E. von Arx, L. van Pieterson, E. Burattini, A. Hauser, R. Pellaux, S. Decurtins, *J. Phys. Chem. A*, **2000**, *104–105*, 883.
101. M. Hernandez-Molina, F. Lloret, C. Ruiz, M. Julve, *Inorg. Chem.*, **1998**, *37*, 4031.
102. R. Andrés, M. Gruselle, B. Malézieux, M. Verdaguer, J. Vaissermann, *Inorg. Chem.*, **1999**, *38*, 4637.
103. M. E. von Arx, A. Hauser, H. Riesen, R. Pellaux, S. Decurtins, *Phys. Rev. B*, **1996**, *54*, 1500.
104. S. Decurtins, H. W. Schmalle, R. Pellaux, R. Huber, P. Fischer, B. Ouladdiaf, *Adv. Mater.*, **1996**, *8*, 647.
105. R. Sieber, S. Decurtins, H. Stoeckli-Evans, C. Wilson, D. Yufit, J. A. K. Howard, S. C. Capelli, A. Hauser, *Chem. Eur. J.*, **2000**, *2*, 361.
106. S. Rashid, S. S. Turner, P. Day, J. A. K. Howard, P. Guionneau, E. J. L. McInnes, F. E. Mabbs, R. J. H. Clark, S. Firth, T. Biggs, *J. Mater. Chem.*, **2001**, *11*, 2095.
107. E. Coronado, J. R. Galan-Mascarós, C. J. Gomez-Garcia, V. Laukhin, *Nature (London)*, **2000**, *408*, 447.

Chapter 8

Polymorphism, Crystal Transformations and Gas–Solid Reactions

DARIO BRAGA
University of Bologna, Bologna, Italy

FABRIZIA GREPIONI
University of Sassari, Sassari, Italy

... a new way for the atoms of water to stack and lock, to freeze....
It had a melting point of one hundred fourteen point four degrees Fahrenheit. [...]
that crystalline form of water, that blue–white gem, that seed of doom called ice-nine.

K. Vonnegut, *Cat's Cradle*, Penguin Books, London, 1965

1 INTRODUCTION

The controlled preparation and characterization of different crystal forms of the same substance has become one of the major issues of modern crystal engineering and solid-state chemistry. Even though the discovery of polymorphs of molecular crystals or of their diverse solvate forms (pseudo-polymorphs) is often serendipitous, crystal polymorphism can, to some extent, be controlled. The existence of more than one packing arrangement for the same molecular or ionic component(s) could be a major drawback for the purposed bottom-up construction of functional solids. Rather than attempting a thorough review of the subject, this

Crystal Design: Structure and Function, edited by G. R. Desiraju

chapter deals with polymorphism at large with a focus on interconverting and noninterconverting polymorphs, as well as on pseudo-polymorphs, with examples selected from the organic, organometallic and inorganic solid-state chemistry areas. A brief outline of major advantages and drawbacks of routine solid-state techniques for the investigation of molecular crystal polymorphism is also provided. Alternative ways to attain preparation of new polymorphs or pseudo-polymorphs are often afforded by nonsolution methods, which allow one to circumvent the often kinetic control of the crystallization process from solvent. Polymorphic and pseudo-polymorphic modifications of the same substance can be obtained by thermal and mechanical treatment and by solvation and desolvation. In these latter cases it is shown that there is a strict analogy (conceptual and practical) between the preparation of a pseudo-polymorph and the preparation of a crystalline product via gas uptake.

1.1 Crystal Design, Crystal Engineering and Polymorphism

Crystal engineering, the bottom-up construction of functional crystalline materials starting from molecules or ions, is one of the key issues of modern chemistry [1]. The idea that fuels the efforts of an ever increasing number of research groups is both intellectually challenging and utilitarian. The molecular crystal engineer aims to the design, synthesis, evaluation and exploitation of molecule-based materials with new or improved chemical and physical properties [2]. The paradigm of crystal design is, therefore, that of being able to predict the arrangement of molecules and ions on the basis of the knowledge of (i) the structure of the molecular or ionic building block and (ii) the type of intermolecular interactions controlling the recognition → aggregation → crystal growth process [3] (see Scheme 1). This approach implies a high degree of predictability and reproducibility to ensure that the desired result is obtained [4].

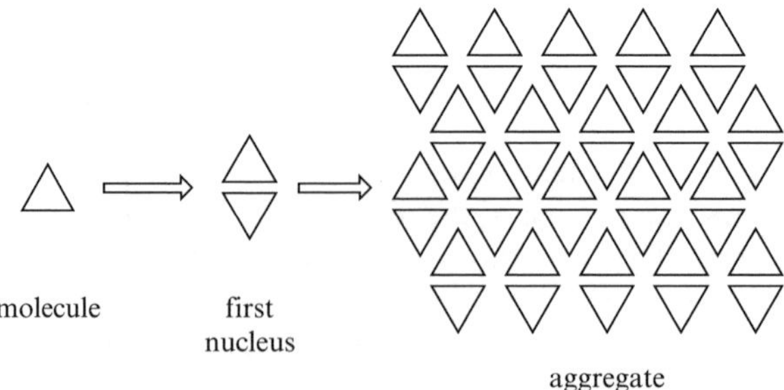

Scheme 1 The crystal as a result of the aggregation process, from the molecular unit via the recognition step.

Just as the civil engineer needs to know precisely how bricks and logs ought to be arranged and which mortar or cement will be necessary to keep them together in a stable structure, the crystal engineer needs to select proper supramolecular binding functionalities to glue together the building blocks in a predictable and stable manner [5]. Sometimes, though, the crystal engineers find that the same molecules or ions, with the same supramolecular bonding capacity, when crystallized (i.e. assembled) under what are believed to be exactly the same experimental conditions, yield two completely different crystal edifices, i.e. two polymorphic modifications of the same substance; an example is provided by Figure 1a, showing how the monoclinic and triclinic forms of crystalline $[Fe(C_5H_4COOH)_2]$ [6] differ in the relative orientation of the hydrogen-bonded molecular pairs.

Indeed, polymorphism [7] is one of the most fascinating phenomena of solid-state chemistry and represents a major challenge for crystal engineering (although, it is to be hoped, not its nemesis). It is a 'difficult' phenomenon, studied for many decades mainly, and separately, in the fields of organic and inorganic chemistry. In spite of the huge efforts of many researchers, our knowledge of the phenomenon is still embryonic, and the relationship between growth of thermodynamically stable (or metastable) crystalline phases and nucleation of the first crystallites is often mysterious. It is a fact that, in spite of the ambitions of the scientist, the construction of a crystalline material is not yet a carefully designed process strictly under human control. What crystal engineers do most of the times amounts to instructing bricks and logs how to recognize each other and self-assemble with the proper amount of cement to keep them together. *Ad absurdum*, we can imagine a building construction process where the desired function – say, the function of a bridge – is somewhat inscribed in the bricks and logs and imagine that these would self-aggregate to yield a bridge as if under the spell of Harry Potter's wand [8]. If the instructions are not very precise and/or if other (uncontrolled or less controlled) external factors affect the process, the result will be predictable only to some extent, as shown in Figure 1 b.

This is an admittedly naive description of the crystal engineering paradigms, but how far are we from reality? As long as scientists will come across a new and unexpected crystalline form of a given substance, it will signify that some knowledge is missing or that some critical stage of the process is not entirely under control. Many examples exist of appearing and disappearing polymorphs. The reader is addressed to the enticing paper by Dunitz and Bernstein for a compilation of examples [9].

A practical example of the possible consequences of the appearance of a new crystalline form is that of the HIV drug Ritonavir. The drug had to be taken off the market by Abbott because of the unexpected (and unwanted) appearance of a new, much less soluble crystal form, which caused a large portion of the drug to precipitate out of the commercialized, semisolid product. A concise history of the problem and how the problem was tackled and eventually solved has recently been published [10].

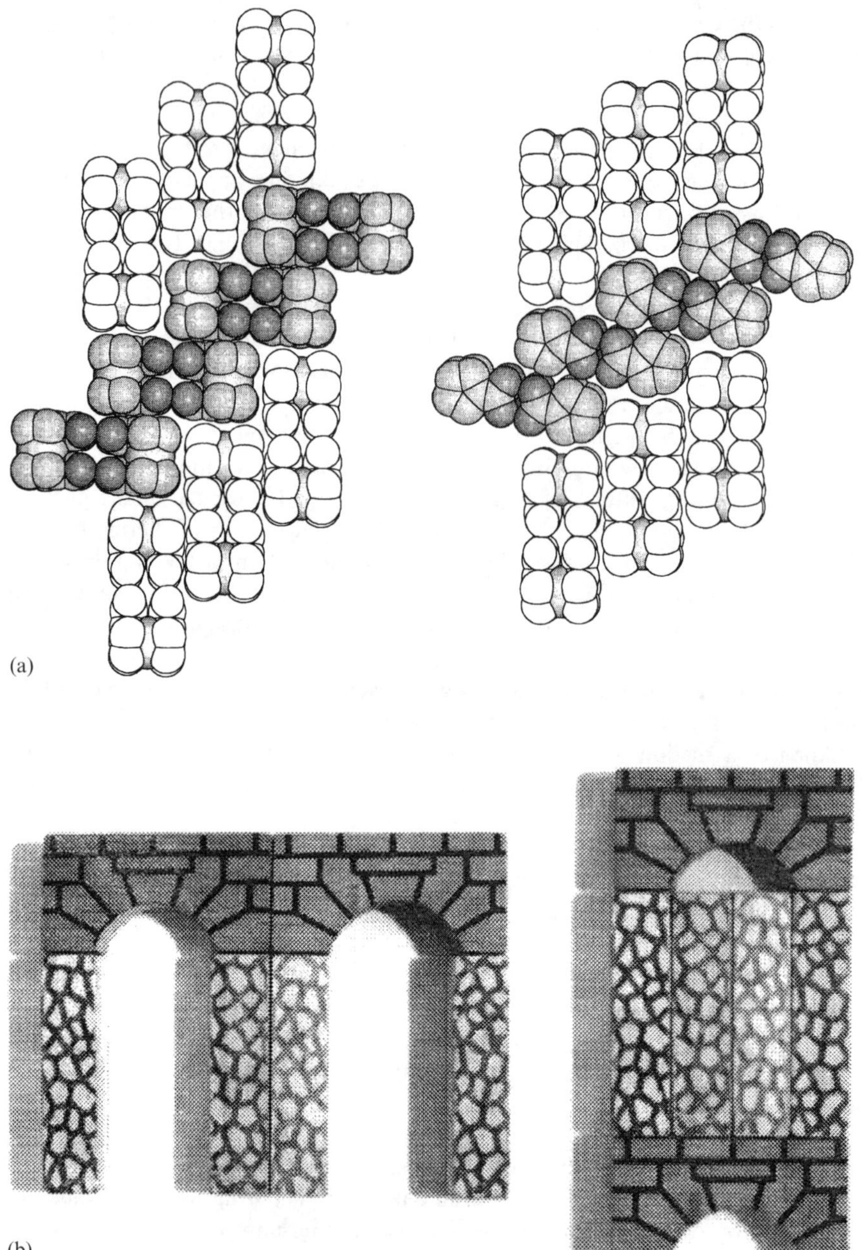

Figure 1 (a) Polymorphic modifications of the same building block, $[Fe(C_5H_4COOH)_2]$; (b) misinterpreted instructions may cause the 'function of a bridge' inscribed in the bricks and logs to result in the self-assembly of a bell tower.

Polymorphism is very often an unpredictable 'happening' of solid-state science and yet it is not a rare event. Actually, polymorphism should be a very likely event: the work of many theoretical solid-state scientists [11] has shown that there is usually a large number of possible arrangements of the same molecules or ions strictly within a few percent of the enthalpy of the observed crystal structure. However, while many compounds exist in several crystalline forms, for most substances only one crystalline form is known. Whether this is due to the simple fact that, in most cases, crystallographers and synthetic chemists have been satisfied with the determination of one molecular structure, thus making the occurrence of polymorphs an unrecognized, unsought or even undesired event (because a second molecular structure of the same substance is/was usually considered not worth publishing), or to some other more subtle structural feature, it is hard to say. It is a fact, however, that commercially relevant compounds (drugs, pigments, foods, materials) seem to show a propensity towards the formation of polymorphs.

It is consequential from these considerations that polymorphism occupies an extraordinarily relevant place in the solid-state sciences. Even the smallest step ahead in understanding the reasons for polymorphism, and of how to bend it to our wishes/needs, might represent a quantum leap in structural sciences with enormous conceptual, practical and economic consequences. One way to tackle this problem is the computational generation of theoretical crystal structures [11], which often goes under the epithet of crystal structure prediction. Even though this is a very important area of research, it will be touched upon only marginally in this chapter and mainly in connection with experimental results.

This chapter deals with polymorphism in broad terms. The focus will be mainly, but not exclusively, on crystals formed of neutral molecules or of molecular ions. Natural inorganic compounds, minerals, clays, etc., and extended covalent solids will not be discussed. Rather than attempting a thorough review of the subject, we will discuss polymorphism by means of a series of examples selected from the organic, organometallic and inorganic solid-state chemistry areas. We shall also look at the most important, routinely available, techniques for the characterization of polymorphs and of polymorph interconversion. We will then discuss alternative ways to generate new polymorphic or pseudo-polymorphic modifications of the same substance by 'nonsolution' methods, and the last section will draw an analogy between the formation of pseudo-polymorphs via gas uptake/release and solid–gas reactivity.

2 POLYMORPHISM, PSEUDO-POLYMORPHISM AND GAS–SOLID REACTIONS

Even though molecular crystal polymorphs contain exactly the same molecules or ions, they usually possess different chemical and physical properties such as density, diffraction pattern, solid-state spectra, melting point, stability and reactivity.

Many drugs can be obtained in different polymorphic modifications. These forms may possess fundamentally different properties, for instance different rates of assimilation because of differences in solubility [12]. The enormous economic implications of this aspect are witnessed, *interalia*, by the number of patent litigations involving drug companies. Polymorphs may also show different behaviour under mechanical stress, with relevant consequences on comminution and tableting, hence on their processing and marketing [13].

Indeed, pills and tablets are among the commonest forms in which pharmaceutical compounds are distributed. Hence problems associated with solid-state transformations and stability of solid drugs have always had a strong appeal, particularly for economic reasons. The 'Zantac story' has become almost a textbook example of the commercial implications of polymorphism. Zantac is the commercial name of ranitidine hydrochloride, an anti-ulcer drug produced by Glaxo Wellcome. The compound has been crystallized in two different forms, named polymorphs 1 and 2. Form 2 is Glaxo's commercial form and its patent ends in 2002, whereas form 1 had been patented by Glaxo earlier. The end of the patent covering form 1 in 1977 led many firms to attempt to enter the market with the generic product. However, attempts to make form 1 according to the procedure described in the Glaxo Wellcome patent invariably led to crystals of form 2. The issue led to a series of patent litigations between Glaxo and other companies, including Novopharm. The reader is directed to Bernstein's book [7] not only for a thorough account of a number of commercial issues related to drug polymorphism, but also for a complete state of the art overview of the field and for the historical background.

Definitions always imply that limitations are imposed on the subject matter. This is very much true of polymorphism, which spans all areas of chemistry and some of mineralogy. In the case of molecule-based polymorphs, the safest approach is probably to consider polymorphic crystal structures as those leading to identical liquid (whether melt or solution) or vapour states. The definition becomes less clear cut in the cases of solvate species and of amorphous materials. Table 1 provides a basic glossary of terms that will be used throughout this chapter.

The problem of whether an amorphous material may or may not be considered a 'form' of a solid substance has been discussed [18a]. Many compounds yield stable noncrystalline phases either as the exclusive product of a crystallization process, or in a mixture with crystals or as a consequence of the treatment of otherwise crystalline phases [18]. The fundamental drawback when dealing with amorphous phases is the dearth of techniques for the thorough characterization of an amorphous phase or for the discrimination between different amorphous phases. In general, diffraction techniques are of little help when dealing with amorphous materials and one has to rely mainly on spectroscopic means (see below). Clearly, if one takes the fulfilment of Bragg's law as the pre-condition for the existence of a crystalline phase, and hence for the existence of a molecular crystal polymorph, amorphous materials are not to be considered.

Table 1 Glossary of definitions on polymorphism.

	A (molecular) ***crystal polymorph*** is:
McCrone (1965) [14]	a solid crystalline phase of a given compound resulting from the possibility of at least two different arrangements of the molecules of that compound in the solid state
	Conformational polymorphs are:
McCrone (1965) [14] Bernstein and Hagler (1978) [15]	formed by molecules that can adopt different conformations in different crystal structures
Braga and Grepioni (2000) [16]	formed by coordination complexes where ligands bound in delocalized bonding modes adopt different relative orientations
	Concomitant polymorphs are:
Bernstein *et al.* (1999) [17]	polymorphic modifications of the same substance obtained from the same crystallisation process
	Pseudo-polymorphs are:
Threlfall (1995) [12b]	crystals formed by the same substance crystallized with different amounts or types of solvent molecules

Pseudo-polymorphs [12b,19], relevant for the following discussion, are characterized by the presence of solvent molecules that may differ in type and quantity and, of course, in the way of bonding within the crystal. Therefore, there is a difference in chemical composition between different solvates and between these and the unsolvated forms. Whether this would lead to a different behaviour in solution will depend on the nature of the encapsulated guest molecule in relation to that constituting the solvent. If the focus is on the chemically relevant species, the presence of solvent molecules will not be relevant; however, for the purposes of crystal engineering, the solvent molecules present in pseudo-polymorphs will not be innocent partners. The guest molecules participate in determining crystal properties, even if the structure of the chemically relevant molecule or ion is not influenced.

The occurrence of pseudo-polymorphism amongst the crystal structures deposited in the Cambridge Structural Database (CSD) [20] has been investigated independently by two groups of investigators. Görbitz and Hersleth investigated the abundance of hydrates and solvate forms in organic and metallo-organic crystals, separately [19a]. It turns out that, with reference to the structures deposited in the CSD until 1999, 8.1 % of the organic and 10.5 % of the metallo-organic crystals are hydrates, while the number of solvates is slightly higher, 7.0 % of the organic and 17.3 % of the metallo-organic structures contain uncoordinated solvent molecules other than water. In terms of relative abundance, in organic crystals the most abundant 'free' solvents are methanol (14.2 %), dichloromethane (8.52 %), benzene (8.27 %), ethanol (7.42 %), chloroform (7.36 %) and acetonitrile (7.32 %). In metallo-organic crystals dichloromethane becomes the most popular (23.47 %), followed by acetonitrile (9.46 %), benzene (9.12 %), toluene (8.71 %), tetrahydrofuran

(8.58 %) and methanol (7.62 %). Nangia and Desiraju have examined those solvates formed by solvent molecules (with the exclusion of water) capable of multi-point recognition with strong and weak hydrogen bonding units [19b]. It appears that multi-point hydrogen bonding facilitates the retention of organic solvents in crystals. When the solvent molecules are strongly attached to the solute, the entropy gain arising from extrusion of the solvent from the crystalline aggregate into the solution may no longer compensate for the enthalpic loss caused by the need to break solute–solvent hydrogen bonding interactions [19b].

The existence/formation of solvate crystals may actually be exploited in two different but closely related processes: (i) the preparation of new crystal forms of the same substance and (ii) the preparation of new compounds. Indeed, as the uptake/removal of solvent molecules via solid–vapour reactions is a viable route to new polymorphs or pseudo-polymorphs of a given substance [16], the reaction of a solid with a vapour is conceptually related to the supramolecular reaction of a crystalline material with a volatile substance to form a new crystalline solid. In a sense, the two processes, solid–gas reaction and solid–gas solvation, differ only in the energetic ranking of the interactions that are broken or formed through the processes. In solvent uptake/release we will be mainly dealing with noncovalent van der Waals or hydrogen bonding interactions whereas in chemical reactions we will be dealing with the formation of covalent bonds (see Scheme 2).

The processes described in Scheme 2 are also at the core of the investigation of solid-state sensors, reservoirs, filters and sieves for detecting or trapping small

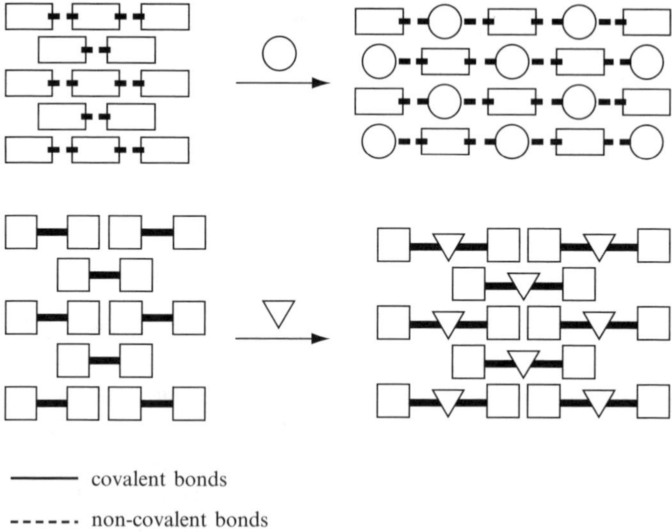

Scheme 2 Intermolecular bonds are broken and formed in a solvent uptake process (a), while covalent bonds are broken and formed in the reaction of a solid with a volatile substance (b).

molecules [21]. It has been pointed out that heterogeneous gas–solid reactions represent an alternative to nanoporosity (i.e. to zeolitic behaviour) for the controlled uptake and release of small molecules (see also Chapter 9 by Meijer *et al.* in this book) [22]. Solid–gas reactions between acids and bases are well known, although they have been extensively studied mainly in the field of organic solid-state chemistry [23]. These reactions often imply profound transformations in the chemical and physical nature of the solid material and rarely are of any practical utility for gas trapping unless fully reversible. A useful notion is that the techniques used to investigate the polymorph–pseudo-polymorph interconversion are essentially the same as those required in the study of gas–solid reactions. Those available routinely in solid-state chemistry laboratories will be described briefly in the context of this chapter.

2.1 Interconverting Isomers and Rate of Interconversion – The Time-scale Problem

Since polymorphs yield the same molecules upon transition to a less dense phase (liquid or vapour) and upon dissolution, the possibility of polymorph interconversion brings about the problem of structural nonrigidity in solution and in the solid state [24]. Isomerization processes in the solid state are well known and there is the intriguing possibility that less stable conformations are stabilized in the solid state by the packing.

From the viewpoint of the supramolecular chemist, molecular crystal [25a] polymorphism can be perceived as crystal isomerism. While the different distribution of chemical bonds for molecules of identical composition gives rise to molecular isomers (e.g. *cis*- and *trans*-isomers or, as is shown in Figure 2a, different intermediates of a merry-go-round process [26]), different distributions of intermolecular bonds give rise to isomers of the same molecular aggregate. Hence the change in crystal structure associated with an interconversion of polymorphs, i.e. a solid-to-solid phase transition, in which intermolecular bonds are broken and formed (as the intermolecular contacts among the S_8 molecules in the orthorhombic α-S_8 and the monoclinic γ-S_8 forms of solid sulfur [27]; see Figure 2b), can be regarded as the crystalline equivalent of an isomerization at the molecular level [25b]. In general, polymorph interconversion depends on whether the polymorphic modifications form an enantiotropic system (the polymorphs interconvert before melting, viz. the solid–solid transition between polymorphs is below the solid–liquid transition) or a monotropic system (the polymorphs melt before the solid–solid transition can occur). The chemical and physical properties of the crystalline material can change dramatically with the solid-state transformation.

The relationship between the rate of interconversion of isomers and the existence of polymorphic modifications has been discussed by Dunitz [25b]. Crystals of isomers that interconvert rapidly in solution would be classed as polymorphs whereas those of slowly interconverting isomers would be classed as different

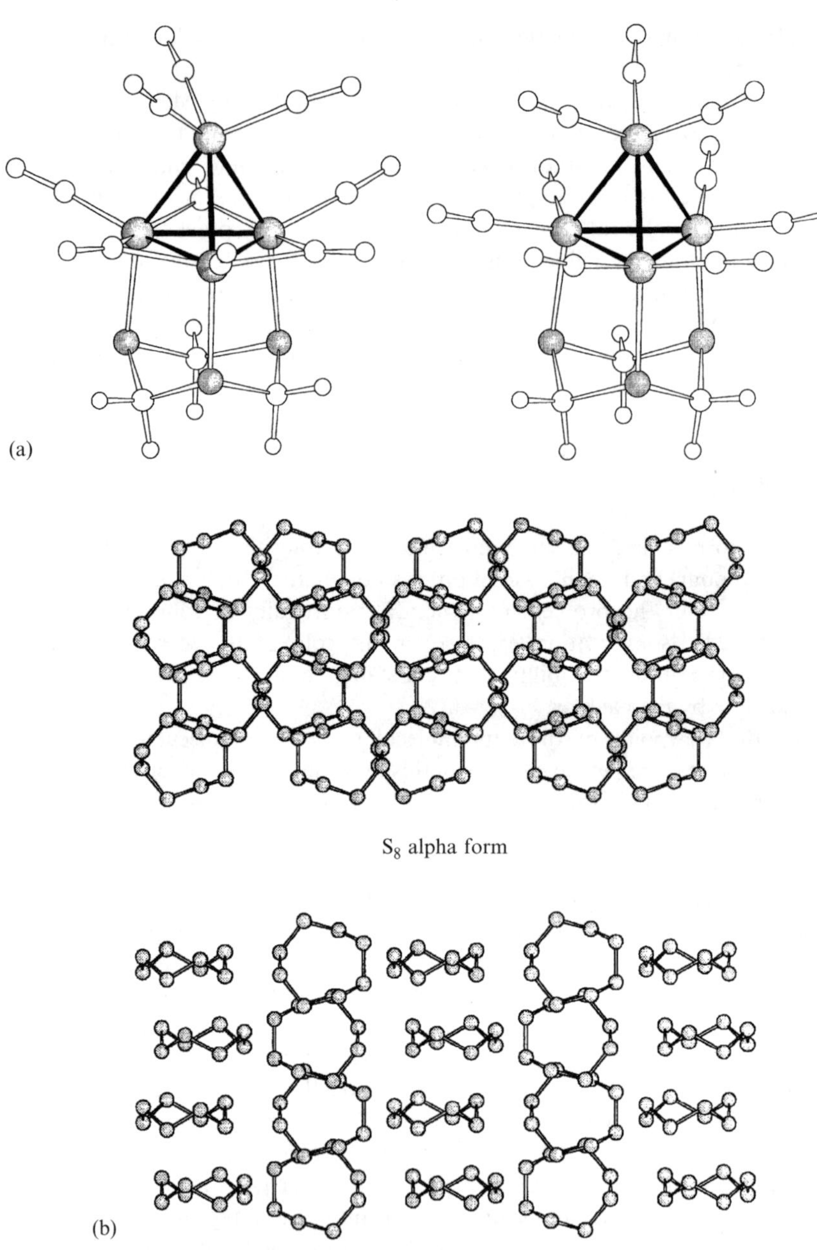

Figure 2 An example of molecular isomers (a), here the bridged and terminal forms of $[Ir_4(SCH_2)_3]$ [26] and a pair of supramolecular isomers (b), the orthorhombic α-S_8 and the monoclinic γ-S_8 polymorphic modifications of crystalline sulfur [27].

structural isomers. Separate crystallization may be attained if the rate at which two isomers interchange in solution is slow with respect to the time required for nucleation. Dunitz pointed out that the different behaviour can be understood on the basis of the Phase Rule: components are chemically distinct constituents when their concentration may be varied independently at a given temperature. The relative concentration of interconverting isomers depends on the temperature and cannot be varied independently. This concept is particularly relevant in the area of organometallic and coordination chemistry where, as will be apparent later, there are a number of dynamic processes that can take place, both in solution and in the solid state, that have no counterpart in the organic chemistry field. Processes such as reorientational motions of ligands, scrambling and fluxional processes, etc., all have implications for the type and nature of the crystalline materials that can be obtained and on their structural relationship.

3 METHODS OF INVESTIGATION

The methods and techniques available to the crystal engineer seeking new forms, or seeking how to prevent the formation of new forms, are those typical of solid-state research. We can group them in the classes of diffraction techniques [28], calorimetric and thermogravimetric techniques [29] and spectroscopic techniques [30]. Many good textbooks and compilations are available and there is no point in describing these methods in detail here. However, within the scope of this contribution, it is useful to point out aspects of the methods that are particularly relevant to the subject matter. Obviously, in addition to common techniques, there are nowadays available a wide range of sophisticated methods that cannot be discussed here owing to lack of space. These include scanning tunnelling microscopy (STM), atomic force microscopy (AFM), electron diffraction and electron microscopy [31]. Polymorphic modifications often differ markedly in some of their solid-state properties and/or in the response to the different techniques, both spectroscopic and diffractometric.

3.1 Diffraction Methods

Single-crystal diffraction methods (whether based on X-ray or neutron sources) are those that carry the most complete information on the intimate nature of the crystals and therefore provide the most valuable tools for identification, characterization and comparison of polymorphs and pseudo-polymorphs. It is difficult to deny that one of the reasons for the outpouring of new results in the field of crystal engineering is the quantum leap represented by the commercial availability of single-crystal diffractometers equipped with area detectors. These devices have not only reduced the time of data collection by an order of magnitude with respect to point detector diffractometers, but also allow the investigation of poorly diffracting or rapidly decaying materials.

However, diffraction methods have severe drawbacks. Disordered crystals are often difficult to tackle. If the disorder is of dynamic nature, e.g. arising from small- or large-amplitude motions in the crystal, the use of devices for variable-temperature measurements is compulsory and can also yield very useful information (see below for some examples) on the existence of enantiotropic systems related by phase transitions. In some, not frequent, cases the crystals are sufficiently robust to be used for direct phase transition measurements on the diffractometer. Figure 3 shows an example of multiple diffraction data sets collected on the same specimen

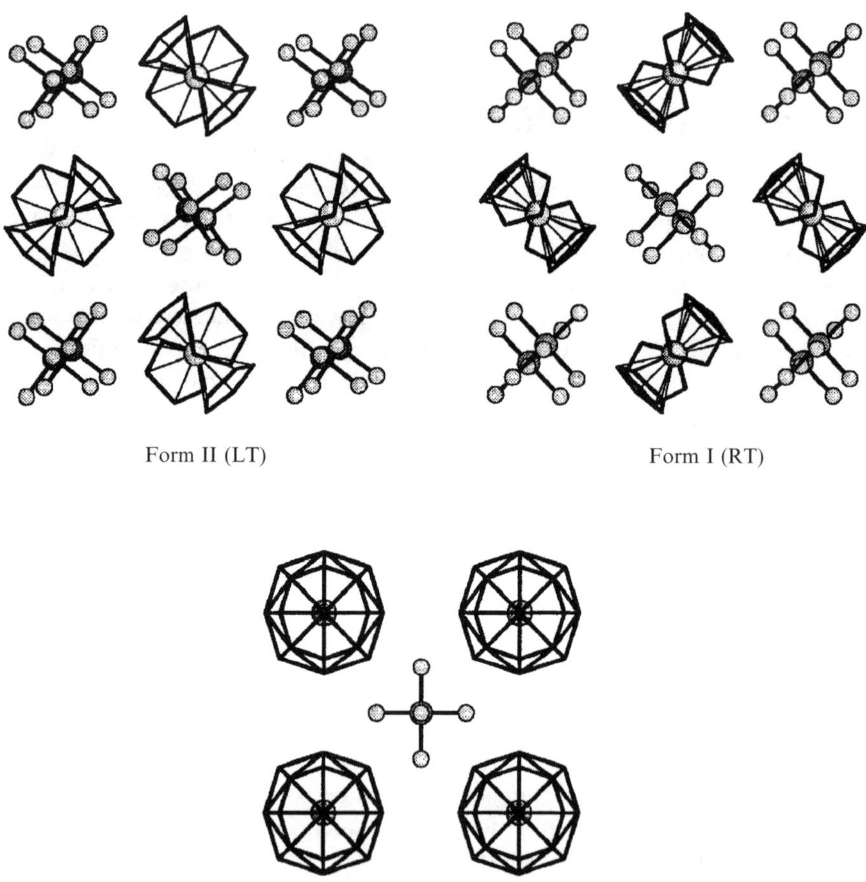

Figure 3 Schematic representation of the polymorph relationship in crystalline $[(\eta^5\text{-}C_5H_5)_2Co][PF_6]$. Diffraction data were collected on the same crystal specimen [32].

of crystalline $[(\eta^5\text{-}C_5H_5)_2Co][PF_6]$ through two solid to solid phase transitions [32]: the room temperature ordered monoclinic crystal (Form I) transforms, below 252 K, into another ordered monoclinic crystal (Form II) with different relative orientations of the two independent $[(\eta^5\text{-}C_5H_5)_2Co]^+$ cations, and into a semi-plastic cubic system (Form III) containing ordered PF_6^- anions and orientationally disordered $[(\eta^5\text{-}C_5H_5)_2Co]^+$ cations above 314 K. The crystal could be 'cycled' through the two phase transitions with only minimal loss of diffracting power.

However, it is often very difficult to grow single crystals of suitable size for a successful diffraction experiment and the researcher has to resort to powder diffraction experiments on a polycrystalline sample. Powder patterns are fingerprints of the solid materials and are therefore used to identify polymorphs. In the pharmaceutical industry, and in associated laboratories, it is becoming common practice to accompany the routine quality control analyses on the production line with the measurements of powder patterns.

Powder diffraction is certainly a straightforward technique, yet it is not without drawbacks. Problems with powder diffraction arise often from sample preparation. In fact, polycrystalline samples can be affected, at times severely, by preferential orientation effects, which alter the intensity sequence, as shown in Figure 4. As will be shown below, the common strategy of grinding the powder to destroy large crystals and increase the homogeneity of the sample, and hence to reduce the preferential orientation effect, often leads to more problems than it solves. It may induce a phase transition because of mechanical stress and/or thermal effects, or it may activate vapour uptake from the environment, thus leading to the formation of hydrated species. All these effects would generate powder diffractograms that differ from those of the original substance. Examples of each case will be provided in the following.

Another aspect of diffraction experiments that is often overlooked is the presence of amorphous materials, the amount of which cannot be easily quantified from a diffraction experiment. As pointed out above, there is no consensus on whether amorphous solids should be considered a 'form' of the substance under investigation. On the one hand, the amorphous component of a given substance is the same chemical as the crystalline material. When melted or dissolved in a solvent the amorphous phase of a molecular material will yield the same molecules as those in crystals, thus falling within McCrone's definition, but will behave as a different solid. On the other hand, an amorphous material may be several different systems. After all, the only difference between a microcrystalline powder and an amorphous powder is the failure to obey the restrictions of Bragg's law.

It is important to keep in mind that the precipitate from a solution is often a mixture of different materials: single crystals (that may or may not be a unique crystalline species), polycrystalline powder (which may or may not correspond to one of the single crystals) and amorphous material in varying amount. In the case of a unique single-crystal species and of a powder, it is often sufficient to compare the measured powder diffractogram with that calculated on the basis of the

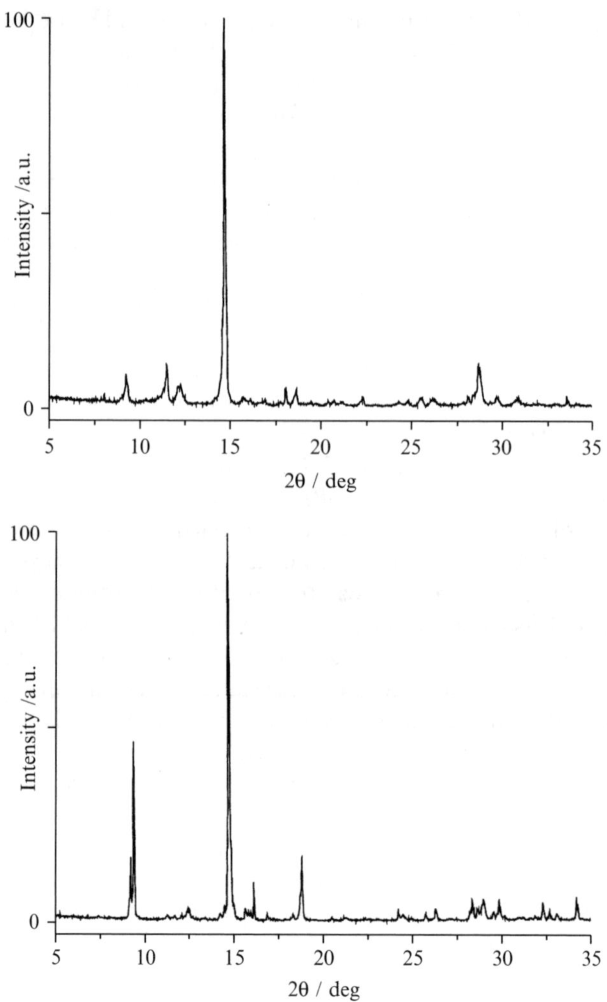

Figure 4 Preferential orientation effects shown by unground (top) and ground (bottom) $[(C_5H_5)_2Co]$ [fumarate] crystalline powder samples. Note how the positions of the peaks are not altered while the intensities change substantially as a consequence of the grinding process.

structure determined from the single-crystal data (Figure 5). If the two patterns correspond precisely, one knows that the product is unique and that, irrespective of the size of the crystallites, the solid residue possesses the same structure [33]. If the observed diffractograms contain extra peaks with respect to the calculated one, one may infer the presence, in the powder, of another phase, a by-product or an impurity. Another useful notion is that the calculated powder pattern provides

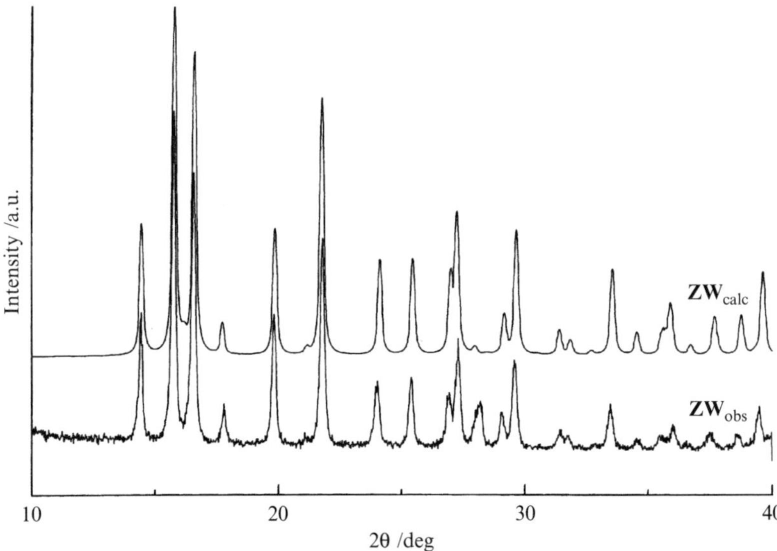

Figure 5 Comparison between the observed (bottom) powder diffractogram and that calculated (top) on the basis of the single-crystal X-ray structure of the organometallic zwitterion $[Co(\eta^5\text{-}C_5H_4COOH)(\eta^5\text{-}C_5H_4COO)]$ (ZW) [33].

a theoretical reference pattern for the powder diffraction experiment because the calculated pattern is not affected by preferential orientation or other sample preparation problems.

In order to address all these questions, the protocol shown in Chart 1 may be adopted.

Neutron diffraction, although much less accessible than single-crystal X-ray diffraction, has the advantage that hydrogen atoms can be detected easily even when heavy atoms are present. For this reason, the technique is useful in establishing polymorphism in hydrogen bonding patterns. However, neutron diffraction resources are not easily accessible and neutron diffraction experiments need large crystals, which are not always easy to obtain.

A special type of pseudo-polymorphism is that related to the proton transfer along an X–H···Y interaction. The motion may not be associated with a phase transition, but may well imply the transformation of a molecular crystal into a molecular salt. Wilson [34] has discussed, on the basis of an elegant neutron diffraction study, the migration of the proton along an O–H···O bond in a co-crystal urea–phosphoric acid (1:1) whereby the proton migrates towards the mid-point of the hydrogen bond as the temperature is increased, becoming essentially centred at $T > 300$ K. Wiechert and Mootz [35] isolated two crystalline materials composed of pyridine and formic acid of different composition. In the 1:1 co-crystal the formic acid molecule retains its proton and transfer to the basic N-atom on the

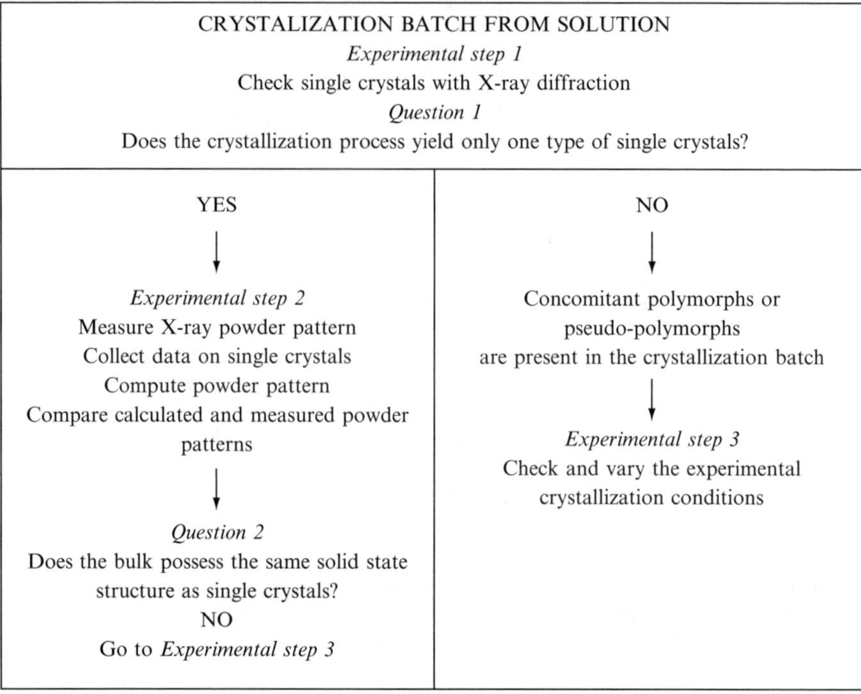

pyridine does not take place (hence molecules are linked by neutral O–H$\cdots$N interactions). In the 1:4 co-crystal, in contrast, one formic acid molecule releases its proton to the pyridine molecule establishing 'charge-assisted' N–H$^{(+)}\cdots$O$^{(-)}$ interactions. In a sense (see also below) this latter case could be regarded as an unusual case of a solvate crystal. Examples of this kind are rare, but serve to stress how the phenomenon of polymorphism can be, at times, full of ambiguity.

3.2 Thermal Analysis and Differential Scanning Calorimetry

Differential scanning calorimetry (DSC) is the most widely used thermal technique for the investigation of the relationship between polymorphic phases. The best instruments for this purpose are those that provide thermal control on both the heating and cooling cycles. The principle of DSC is very simple: any polymorph undergoing a phase transition will absorb heat when heated through the transition. Conversely, heat release will be observed on the cooling cycle. Heat absorptions/releases will show up as peaks in the DSC plot, the area of the peak being proportional to the ΔH of the transition. If a good baseline is determined, and if the sample is pure and crystalline and its weight known with precision, a good estimate of the ΔH involved in the process can be obtained. In the case of molecular crystals these values are always in the range of few kilojoules per mole [36,37].

Conditions have to be chosen with care in order to obtain reliable and reproducible results. In fact, DSC measurements may be affected by several experimental problems: baseline measurement, sample preparation, particle size, use of sealed or open sample holders, and heating and cooling rates, which also affect the thermal hysteresis between the endothermic and exothermic transitions. Baseline preparation may be critical: an imperfect baseline may be indicative of some imbalance between sample holder and reference but could also indicate some change in the heat capacity of the sample, which, in turn, may indicate decomposition or loss of solvent molecules. In the case of pseudo-polymorphs, the solvent loss will generally appear as an endothermic process. The desolvation can be very rapid, leading to a sharp peak, which, in turn, can be mistaken for a phase transition. If the experiment is carried out in an open pan, it will not be reversible and will not show the reverse endothermic peak on cooling. In a sealed pan, on the other hand, cooling may bring vapour back in the crystal. Chemical reactions can also take place during a DSC scan. A good example of an exothermic reaction is provided by crystalline bis-formylferrocene, $[(\eta^5\text{-}C_5H_4CHO)_2Fe]$ [38], which on heating shows a phase transition at 311K, followed by an exothermic peak at 453K, corresponding to a polymerization reaction (see Figure 6). This latter process is irreversible and the peak cancels out in the cooling cycle. If the reaction taking place during a DSC scan is rapid and endothermic, however, it may be confused with a solid-to-solid transition or with a melting process. The corresponding endothermic peaks, however, will disappear in the cooling cycles. This is why it is recommended to use DSC instruments that allow thermal control in the cooling cycle. The rate of heating/cooling is also very important. If the heating rate is fast an enantiotropic transition can be completely missed. It is not uncommon for a polymorph to melt at the correct temperature when heated rapidly while a transition to a higher melting form is observed when heated slowly. In such cases further heating my lead to a recrystallization, which appears as an exothermic peak, and then to a second melting. Decomposition under thermal treatment is also critical. Since decomposition is a chemical reaction, it may appear either as an endothermic or an exothermic process in the DSC plot [36b].

As mentioned above, polymorphs may also be related by order–disorder transitions, e.g. the onset of free rotation of a group of atoms, or local tumbling in semi-plastic or plastic phases. This may be due to random orientation of the molecules or ions, but is also diagnostic of the onset of a reorientational motion. Roughly spherical molecules and ions are likely to show order–disorder phase transitions to a plastic state. In the cases of co-crystals or of crystalline salts this process may affect only one of the components, leading to semi-plastic crystals (an example will be discussed below). Order–disorder phase transitions have often

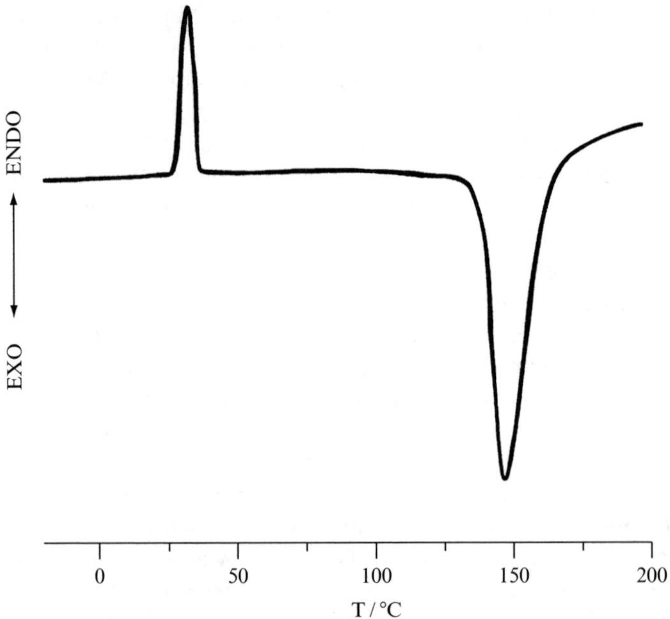

Figure 6 A DSC measurement on crystalline $[(\eta^5\text{-}C_5H_4CHO)_2Fe]$ shows both an endothermic peak at 311 K, corresponding to a phase transition, and an exothermic peak at 453 K, corresponding to a polymerization reaction [38].

been regarded as second-order transitions. On cooling, systems of this type often show a glass transition. This is also the behaviour shown by amorphous material.

Finally, it is worth mentioning that the DSC plot of a mixture of noninterconverting polymorphs undergoing independent phase transitions will show a superimposition of two independent plots.

3.3 Thermogravimetric Analysis

As will be discussed below, hydrates and solvates represent a relevant portion of the known polymorphic forms for many substances. While representing a complication in the study of polymorphism, their existence can be extremely advantageous. It is not uncommon to have several forms of the same molecule or salt that differ in the degree of solvation and the number and type of solvent molecules. The problem is complicated by the often nonstoichiometric, fractional or variable presence of solvent molecules.

The technique of choice for the study of solvates is thermogravimetric analysis (TGA), where the weight loss of a sample undergoing thermal treatment (in air or in a controlled atmosphere) allows the estimation, often with great precision, the amount of solvent that can be removed from the crystal and the temperature of desolvation. The hydrated zwitterion $[Co^{III}(\eta^5\text{-}C_5H_4COOH)(\eta^5\text{-}C_5H_4COO)]\cdot 3H_2O$ can be transformed quantitatively in the crystalline anhydrous form $[Co^{III}(\eta^5\text{-}C_5H_4COOH)(\eta^5\text{-}C_5H_4COO)]$. The TGA graph of the dehydration process shows how the first water molecule is lost at 378 K, while the other two water molecules are released at 506K [39] (see Figure 7). Furthermore, if coupled with other techniques, such as mass spectrometry, TGA may also provide essential analytical information, such as chemical characterization of the released solvent.

Experimental problems with TGA are usually connected with sample preparation: for instance, homogeneous or very disperse particle sizes may yield different results, while the presence of humidity adsorbed on the surface of the particles may mask or alter the response. Deliquescent or highly hygroscopic samples yield poorly reproducible results because it can be difficult to discriminate between removal of wetting solvent and removal of structural solvent. It is useful to accompany DSC experiments with TGA experiments. Heat absorption in a DSC plot may correspond to solvent loss and not to a phase transition (see above). Importantly, as shown below, a desolvation process may sometimes induce the formation of another polymorph or pseudo-polymorph not otherwise attainable.

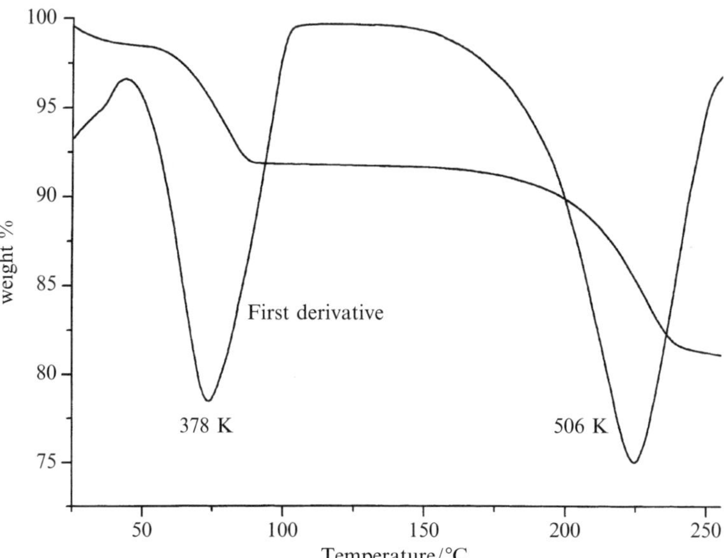

Figure 7 The hydrated zwitterion $[Co^{III}(\eta^5\text{-}C_5H_4COOH)(\eta^5\text{-}C_5H_4COO)]\cdot 3H_2O$ can be transformed quantitatively into the crystalline anhydrous form $[Co^{III}(\eta^5\text{-}C_5H_4COOH)(\eta^5\text{-}C_5H_4COO)]$. The TGA graph of the dehydration process shows how the first water molecule is lost at 378 K, while the other two water molecules are released at 506 K [39].

4 POLYMORPHISM IN METAL-CONTAINING COMPOUNDS VERSUS ORGANIC COMPOUNDS

Although the interest in molecular crystal polymorphism stems from organic solid-state chemistry, the field is now rapidly expanding to encompass metal-containing molecular crystals and molecular salts. The reason for this time lag is relatively simple. Organic solid-state chemistry reached maturity long ago and naturally the implications of polymorphism – as a phenomenon – have been perceived much sooner than in the somewhat younger areas of coordination and organometallic chemistry.

Many metal-containing coordination compounds share with organic compounds the 'periphery', i.e. the outer atoms – those that matter most in the formation and stability of the intermolecular interactions, and hence on the packing of molecules in the solid state [40]. However, coordination compounds are often charged species, sometimes made of large polynuclear ions. The presence of ions implies electrostatic interactions, which are usually much stronger (but less directional) than van der Waals and hydrogen bonding interactions, and very different physical properties of the solids (e.g. solubility and solvent type, melting points, etc.) [40]. The possibility that crystals of the same relevant anion or cation differ for the choice of counterion is very common in inorganic chemistry. Whether one may wish to regard these systems as special kinds of pseudo-polymorphs (differing in the counterion rather than in the co-crystallized solvent or guest molecules) is again a matter of opinion. Table 2 collects some of the situations that are commonly encountered in the study of crystal packings.

Table 2 Polymorphism and pseudo-polymorphism in crystalline coordination complexes.

Feature	Effects
Rigid molecules in different crystals	The molecular structure is essentially unaffected by the change in crystal structure
Conformationally flexible molecules in different crystals	Structurally nonrigid molecules can be deformed along soft deformational paths by the change in crystal structure
Structural isomers in different crystals	Less stable structural isomers of a fluxional process are isolated thanks to the compensatory effect of external interactions in the solid
Disordered versus ordered crystals	Ordered and disordered crystals of the same molecules may show different crystallographic symmetry, e.g. crystallize in different space groups
Same ion crystallized with different counterion	The same (chemically relevant) molecular ion is crystallized in different crystals with a different type/number of counterions
Polymorphism versus coordination network isomerism	The same network ligand yields different coordination networks with the same metal; the presence of a different number and type of molecules in the voids is also possible

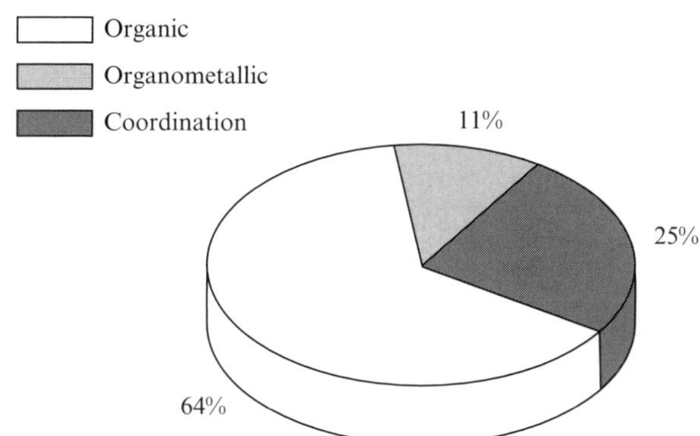

Scheme 3 Relative percentages of organic, organometallic and coordination polymorphic compounds present in the Cambridge Structural Database (CSD, version October 2001) [20a].

It is useful to compare the percentage of polymorphic forms in each class of compounds available in the Cambridge Structural Database [20a]. At the time of writing this chapter (April 2002, CSD version October 2001), organic compounds represent the 64% of all polymorphic forms deposited in the database, against 11% of organometallic compounds (containing at least one transition metal–carbon bond) and 25% of coordination compounds; see Scheme 3. Unfortunately, it is not possible to extract information on the abundance of polymorphic modifications among inorganic compounds because of the difficult interrogation of the Inorganic Crystal Structure Database (ICSD) [20b].

Since structural flexibility is a recurrent feature of coordination compounds, in view of the often delocalized nature of the metal–ligand bonding and/or the availability of almost isoenergetic, although geometrically different, bonding modes for the same ligand, the intriguing relationship between molecular nonrigidity and crystalline phase transitional behaviour needs to be taken into account when expanding from organic-only polymorphism to cases involving inorganic and coordination complexes. Many complexes exist in different isomeric forms that interconvert via low-energy processes (viz. reorientation, diffusion, scrambling and fluxionality) both in the gas phase and in the condensed state [41]. Since the structural isomers correspond to different energetic minima along the interconversion pathway, less thermodynamically stable isomers may be isolated in the solid state if the enthalpy difference is compensated by the packing energy [42]. It is consequential that 'organic-type' classifications of polymorphism tend to be somewhat restrictive when applied to nonorganic species [43].

A textbook example of conformational polymorphism is provided by ferrocene [44], for which one room temperature disordered and two low-temperature ordered crystalline forms are known. At the crystal level they differ in the relative orientation of the cyclopentadienyl rings and in small rotations of the molecules, so that the phase transition mechanism requires only low-energy reorientation of the rings and a limited motion of the molecules in the crystal structure.

Crystals of structural isomers related by low-energy interconversion pathways, such as transition metal cluster carbonyls with different distributions of bridging and terminal ligands, can be assimilated to conformational polymorphs. For instance, the two known forms of $Ru_6C(CO)_{17}$ contain a total of three isomers that differ in the rotameric conformation of the tricarbonyl units on the two apices and in the pattern of terminal, bridging and semi-bridging COs around the molecular equator (see Figure 8) [45]. Another interesting relationship between crystal and molecular structure of what may be regarded as a special sort of pseudo-polymorphs is provided by Zhao and Brammer [46a]. The study describes a case whereby structural isomers of transition metal compounds have been crystallized in two different coordination geometries. Pseudo-polymorphs have also been prepared by Soldatov and Ripmeester [46b] by using vinylpyridine ligands in the formation of a series of Ni and Co complexes capable of extensive inclusion properties. Out of 40 organic molecules tested as guests, 19 were shown to be encapsulated in the host, thus forming a plethora of pseudo-polymorphic modifications.

The ongoing intensive research on coordination network crystal engineering is opening up new avenues to investigations on polymorphism. The same network ligand may yield different coordination networks with the same metal and/or co-crystallize with a different number and type of molecules in the voids and channels. Solvent-dependent topological isomerism in coordination polymers has been reported by Champness and co-workers in the course of a study of three coordinated polymers [CuI_2 bis(4-pyridyl)disulfide] [47a]. Another example is provided by the square bis-chelate complex bis(1,1,1-trifluoro-5,5-dimethyl-5-methoxyacetylacetonato)copper(II), which was found to crystallize in two polymorphic modifications, namely the orthogonal form α and the trigonal form β. The two forms,

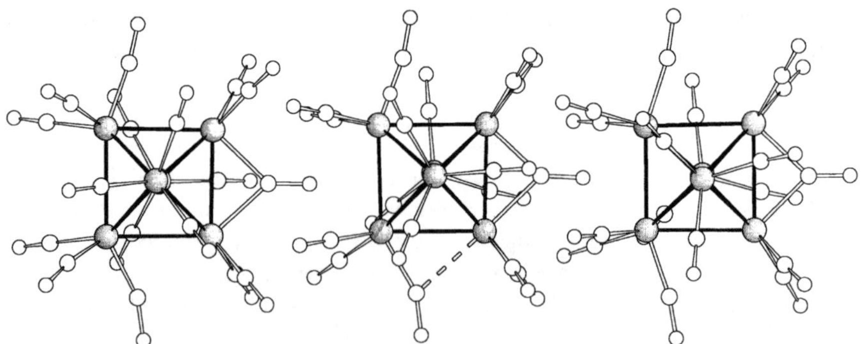

Figure 8 Conformational polymorphism in crystalline $Ru_6C(CO)_{17}$: the projections in the equatorial planes of the cluster evidence the difference in the rotameric conformation of the tricarbonyl units on the two apices, and in the pattern of terminal, bridging, and semi-bridging COs around the molecular equator [45].

which are characterized by the presence of extended networks, differ in the pattern of the additional coordination interactions: whereas the α-form is a dense solid, in the β-form the complex units assemble in a structure with open pores, which can accommodate a variety of guest molecules; the 'empty' porous β-form has also been obtained, via a pseudo-polymorph, i.e. the porous β-form with solvent in the cavities, and characterized [47b].

4.1 Noninterconverting Polymorphs

A textbook example of inorganic polymorphism is provided by the aragonite, calcite and vaterite forms of calcium carbonate. These well studied systems [48] differ, *inter alia*, in the coordination number of oxygen to calcium, as shown in Figure 9.

Polymorphism is also shown by hydrogen-bonded systems, of which there are several examples, e.g. oxalic acid that is known to form catemer-type motifs in the α-form [49] and chains of hydrogen-bonded rings in the β-form (see Figure 10) [50].

Two polymorphic modifications are also known for the complex $[(\eta^2\text{-fumaric acid})Fe(CO)_4]$ (called forms I and II) [51]. In form I, the fumaric acid ligands form ribbons of ligands joined by carboxylic rings. Interestingly, the same arrangement is observed in crystalline fumaric acid [52], which also possesses two polymorphic forms both based on molecular chains interlinked via hydrogen-bonded carboxylic rings. In form II the carboxylic rings form a catemer-type pattern [52]. The complex $HMn(CO)_5$, one of the first carbonyl hydrides to be structurally characterized by X-ray and neutron diffraction methods [53], is known in two polymorphic forms, both monoclinic, called α-$HMn(CO)_5$ and β-$HMn(CO)_5$. The complex $[Cr^0(\eta^6\text{-}C_6H_5COOH)_2]$ forms a dimer via a pair of typical hydrogen bonding rings involving the two –COOH groups of the acid [54].

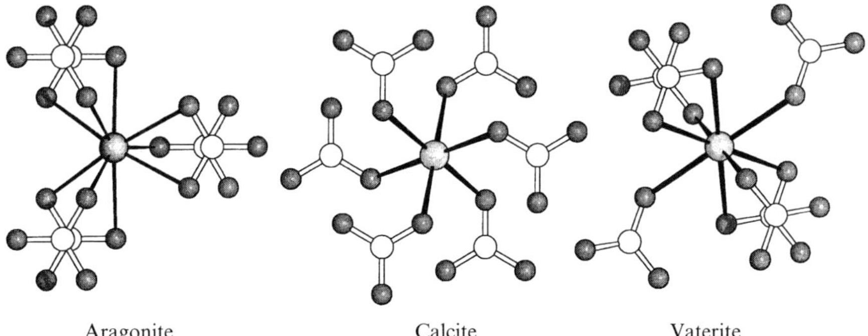

Figure 9 Coordination of nine, six and eight $O_{carbonate}$ atoms around the calcium ion in the aragonite, calcite and vaterite forms of calcium carbonate [48].

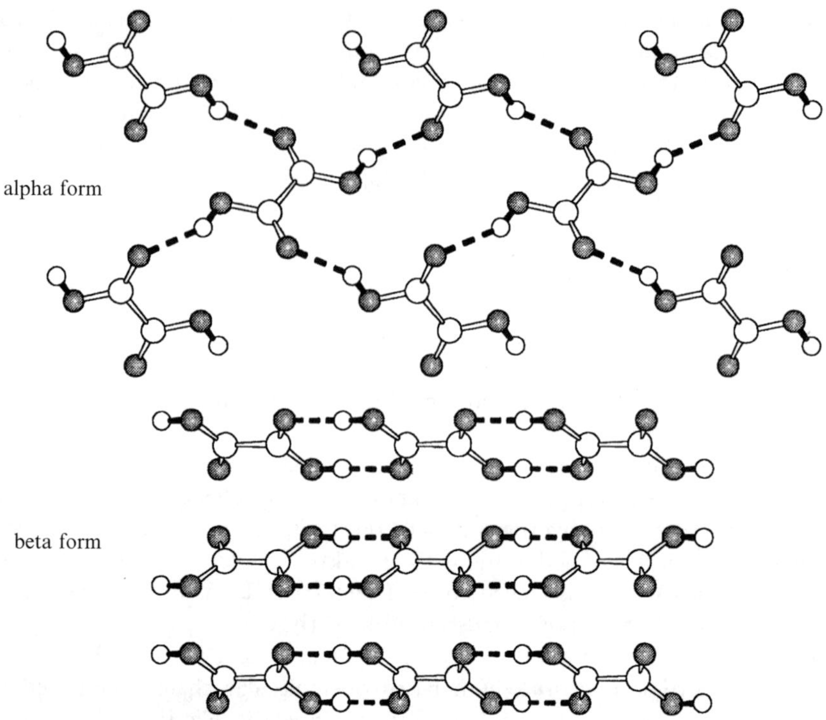

Figure 10 The oxalic acid forms catemer-type motifs in the α-form and chains of hydrogen bonded rings in the β-form [49].

Twin intermolecular hydrogen bonding of the type shown by $[Cr^0(\eta^6\text{-}C_6H_5COOH)_2]$ are not commonly observed with organic dicarboxylic acids, because dimer formation requires a degree of structural flexibility that cannot be afforded by molecules based on C–C σ-bonds. In fact, dicarboxylic organic acids tend to form chains joined by carboxylic rings in the solid state rather than self-assemble in dimeric units. A strictly comparable situation can be observed in crystalline ferrocene dicarboxylic acid, $[Fe^{II}(\eta^5\text{-}C_5H_4COOH)_2]$ [55]. This acid is known in two polymorphic modifications (a monoclinic and a triclinic form), which contain exactly the same type of dimeric unit, although in different arrangements in the solid state. The space-filling representations of the two forms of $[Fe^{II}(\eta^5\text{-}C_5H_4COOH)_2]$ (Figure 1a) and of $[Cr^0(\eta^6\text{-}C_6H_5COOH)_2]$ (Figure 11) show that the three crystals are formed by layers of dimers arranged in contiguous rows. Two polymorphic modifications of the oxidation product $[Cr^I(\eta^6\text{-}C_6H_5COOH)_2][PF_6]$ have been crystallized depending on the crystallization conditions [54]. Both forms contain chains of cations held together by hydrogen bonds between the carboxyl groups. The structure of the α-form can be ideally converted into that of the β-form by

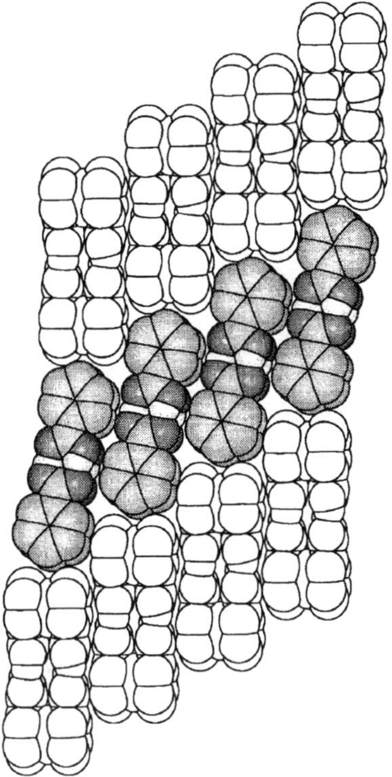

Figure 11 A comparison of the space-filling representation of the almost isostructural dimers $[Cr^0(\eta^6\text{-}C_6H_5COOH)_2]$ and $[Fe^{II}(\eta^5\text{-}C_5H_4COOH)_2]$.

sliding in opposite directions two out of every four layers formed by cationic chains and anions. While the β-form adopts a nearly cubic NaCl-like structure, the α-form shows the presence of twin rows of cations and anions (Figure 12).

A related structural situation is shown by $[MoH(\eta^5\text{-}C_5H_5)_2(CO)][Mo(\eta^5\text{-}C_5H_5)(CO)_3]$, which crystallizes in two different forms [56]. In the monoclinic form the intermolecular (Mo)H···O distances are short, suggesting intermolecular interactions, but in the triclinic form the different orientations of the cations lead to much longer (Mo)H···O distances and bring together neighbouring hydride ligands ($H\cdots H = 2.234$ Å). Another example of organometallic salt that can be obtained in different forms, depending on the crystallization conditions, is the complex $[Ru(\eta^6\text{-}C_6H_6)(\eta^6\text{-}C_6H_4(1\text{-}CH_3)(2\text{-}COOCH_3)][BF_4]_2$ [57].

Other examples are the two polymorphic malonato-bridged copper(II) complexes of formula $\{[Cu(bpy)(H_2O)][Cu(bpy)(mal)(H_2O)]\}(ClO_4)_2$, which have been prepared from different starting products; the two species, topologically similar in the solid state, have been studied for their ferromagnetic behaviour. The magnetic behaviour is due to exchange coupling through both the OCO and the OCCCO

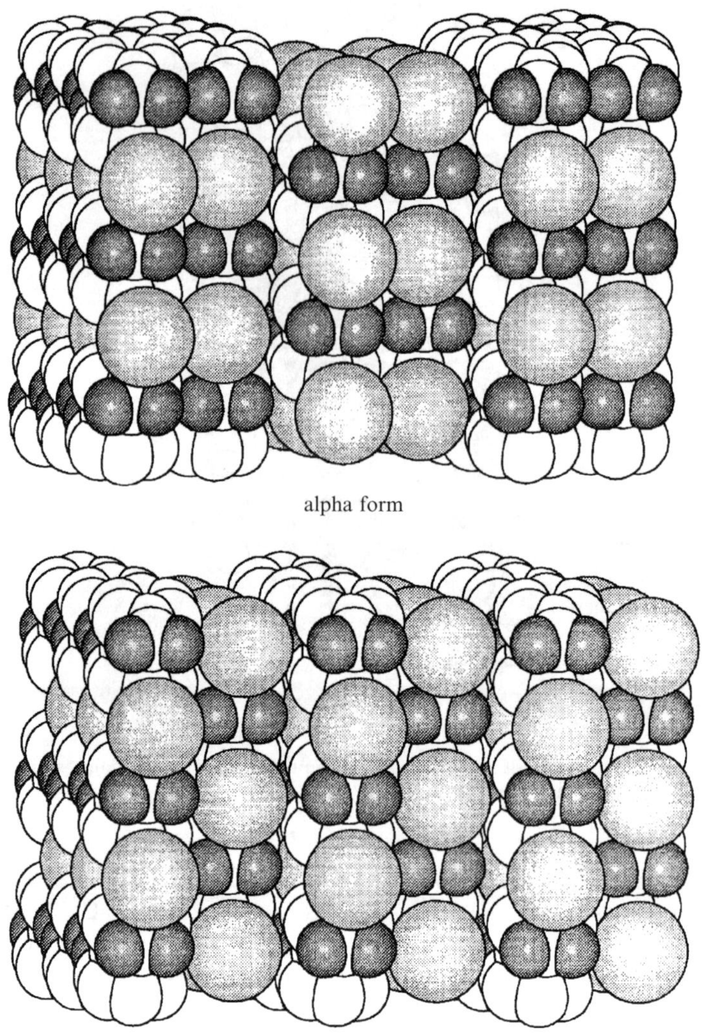

Figure 12 A comparison of the two forms of the salt $[Cr^{I}(\eta^6\text{-}C_6H_5COOH)_2][PF_6]$. Both forms contain chains of cations held together by hydrogen bonds between the carboxyl groups. The structure of the α-form can be ideally converted in that of the β-form by sliding in opposite directions two out of every four layers formed by cationic chains and anions. Furthermore, whereas the β-form adopts a nearly cubic NaCl-like structure, the α-form shows the presence of twin rows of cations and anions.

fragments of the bridging malonate ligands, and is similar in the two polymorphs [58]. Two distinct polymorphs of Cu(TCNQ) were prepared and characterized. The two solids are both constituted of extended networks, with the TCNQ ligands connecting the copper atoms, but the relative orientation of the ligands is markedly different in the two polymorphs. Phase I has been shown to be essentially diamagnetic and a good semiconductor at room temperature, whereas phase II displays Curie–Weiss behaviour down to very low temperature, and is a poor semiconductor. The two phases have also been obtained as thin films grown on copper, and conversion from phase I to the more thermodynamically stable phase II has been observed both in the crystals and in the thin films [59].

4.2 Polymorphism Arising from Static/Dynamic Disorder and from Order–Disorder Phase Transitions

As mentioned above, polymorphism may also arise for compounds formed by isomers whose structures are related by low-energy interconversion pathways, e.g. cluster carbonyls with different distributions of bridging and terminal ligands, or substituted metallo-arenes and metallo-cyclopentadienyl complexes [60]. In such cases, the structural isomers correspond to different energetic minima along the interconversion pathway and the cohesion of the respective crystals may stabilize the less thermodynamically stable isomers. Crystals of structural isomers may (or may not) interconvert via a phase transition.

It must be mentioned that dynamic processes may introduce disorder in the crystals [61] owing to the reorientational motions of whole molecules or groups of atoms within the crystal structure. In such cases, a phase transition often occurs when the temperature decreases and the dynamic process slows and finally stops. Phase transitions are generally classified as first or second order. In terms of Gibbs free energy, the transition is said to be first order if the first derivatives of $G(T,P)$ show an abrupt change at a given temperature, whereas second-order phase transitions show a discontinuity in the second derivatives of G (i.e. volume and entropy only show a change of slope at the transition point). The distinction between the two phenomena is often ambiguous. In general, beside melting, nucleation and growth of a new solid phase is first order, whereas transformations in which the molecules acquire orientational degrees of freedom, but with very limited translational motion, are second order. In such cases, the phase transition is associated with the onset of reorientational phenomena, in which the molecular (or molecular fragment) orientation is distributed between more than one potential energy configuration. As mentioned above, calorimetric measurements can be exploited to gain complementary information on the transformations detected by other methods, such as X-ray diffraction or NMR measurements. Both first- and second-order phenomena are accompanied by variations in the heat capacity curves, which can be studied by DSC as discussed above.

Ordered and disordered crystals of the same molecules may have different point symmetry arising from either static or dynamic disorder. It should be stressed that the presence of disorder does not *per se* indicate the existence of different patterns of intermolecular interactions, which is what identifies crystal isomers, viz. polymorphs. Since the appearance of disorder in crystals depends on the averaging process over time and space of the diffraction data, the same average image may result from superimposition of differently orientated crystallites with dimensions of a few nanometers (mosaic disorder), from microscopic or macroscopic twinning, or from overlap of a random distribution of unit cells containing molecules in different orientations (local disorder) or with orientations changing with time (dynamic disorder). Clearly, in the case of mosaic disorder the environment of each molecule is the same as in an ordered crystal. Therefore, if disordered and ordered forms of the same substance are available, they may not necessarily represent different polymorphic modifications even though the point symmetry may be different. In such cases only diffraction experiments carried out at variable temperatures, or auxiliary techniques sensitive to motion (such as SSNMR), can yield information on the nature of the disorder. Examples are available of organometallic molecules, globular in shape, undergoing phase transitions with formation of plastic or semi-plastic phases, characterized by short-range orientational disorder and long-range order.

A classical case study is that provided by the family of complexes $[M_3(CO)_{12}]$ ($M = Fe$, Ru, Os) [62]. The discussion of the molecular versus crystal structure features of these compounds has provoked heated debates through 30 years of cluster chemistry. Variable-temperature X-ray diffraction experiments have demonstrated that $[Fe_3(CO)_{12}]$, $[Fe_2Ru(CO)_{12}]$, $[Fe_2Os(CO)_{12}]$ and $[FeRu_2(CO)_{12}]$ show disorder of a dynamic nature, associated with metal core reorientation, accompanied by reversible order–disorder phase transitions. In the case of $[Fe_2Ru(CO)_{12}]$, for instance, the disordered room temperature structure becomes fully ordered at 220K and, on increasing the temperature to 313 K, the crystal becomes isomorphous with crystalline $[Fe_3(CO)_{12}]$ [62f,g]. These reversible changes are accompanied by an increase in the metal atom disorder, from the completely ordered structure at 223 K to a statistically disordered structure at 323 K (see Figure 13]. Crystals of $[FeRu_2(CO)_{12}]$ show a similar behaviour, passing from an ordered structure at 173 K to an extended disordered structure above the phase transition temperature (228 K). The parent cluster $[Fe_3(CO)_{12}]$ undergoes a phase transition at $\sim$210 K to another monoclinic phase with partial ordering of the metal atom triangles. The asymmetric unit is comprised of four complete molecules and one half-molecule of $[Fe_3(CO)_{12}]$, and one of these molecules is completely ordered.

4.3 Order-to-Order Phase Transitions Between Enantiotropic Systems

It has been argued above that molecular crystal polymorphism can be seen as a form of crystal isomerism, and that the change in crystal structure associated with polymorph interconversion, i.e. a solid-to-solid phase transition between ordered

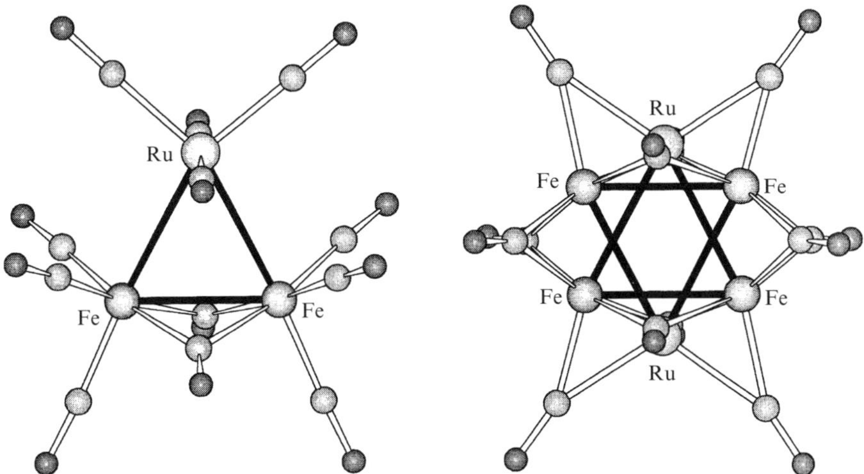

Figure 13 The dynamic disorder present in crystalline $[RuFe_2(CO)_{12}]$ and other isoelectronic species. The two structures represent the limiting fully ordered (223 K) and 50% disordered (323 K) arrangements of the metal atom triangles.

phases, in which intermolecular interactions are rearranged, can be regarded as the crystalline equivalent of an isomerization at the molecular level. As pointed out by Dunitz, however, phase transitions imply an extraordinary level of cooperativity, which is not relevant in solution. Moreover, when the phase transition takes place there is complete conversion of one crystal into the new one without equilibrium mixture of reactant and product.

The crystalline complexes $[(\eta^5\text{-}C_5H_5)_2M][PF_6](M = Co, Fe)$ afford a textbook example of interconverting polymorphs [63]. The two crystalline materials, which are isomorphous at room temperature, have been shown, by variable-temperature X-ray diffraction experiments and DSC, to undergo two reversible solid-to-solid phase changes towards a low temperature monoclinic phase and a high-temperature cubic phase. The only difference between the two crystalline systems is in the temperatures of these transitions: M = Fe, ca 213.1 and 347.1 K; M = Co, ca 251.8 and 313.9 K, measured in the heating cycles. Therefore, the range of thermal stability of the intermediate room temperature phase is ca 62 K in the case of Co and ca 134 K in the case of Fe. As mentioned above, in both cases, the phase transitions could be followed on a single-crystal X-ray diffractometer and diffraction data were collected on the same crystal specimen. While the room and low-temperature phases are ordered, the high-temperature phase of $[(\eta^5\text{-}C_5H_5)_2Co][PF_6]$ contains ordered $[PF_6]^-$ anions and orientationally disordered $[(\eta^5\text{-}C_5H_5)_2Co]^+$ cations (see Figure 3), whereas in the case of $[(\eta^5\text{-}C_5H_5)_2Fe][PF_6]$, both cations and anions are disordered.

Structurally similar systems, however, show very different phase transitional behaviours. The compound $[(\eta^5\text{-}C_5H_5)(\eta^6\text{-}C_6H_6)Ru][PF_6]$ [64] does not show a

Table 3 Phase transition behaviour and thermodynamic data for the family of complexes $[M(\eta^5\text{-}C_5H_5)][PF_6](M = Co, Fe)$, $[Cr(\eta^6\text{-}C_6H_6)_2][PF_6]$ and $[Ru(\eta^5\text{-}C_5H_5)(\eta^6\text{-}C_6H_6)][PF_6]$.

Species	Phase transition	ΔH(kJ mol^{-1})
$[(\eta^5\text{-}C_5H_5)_2Fe][PF_6]$	LT → RT at 213.1 K[a]	1.95
	RT → HT at 347.1 K[a]	4.50
$[(\eta^5\text{-}C_5H_5)_2Co][PF_6]$	LT → RT at 251.8 K[a]	1.27
	RT → HT 313.9 K[a]	3.06
$[(\eta^5\text{-}C_5H_5)(\eta^6\text{-}C_6H_6)Ru][PF_6]$	No RT → LT transition[b]	—
	RT → HT at 332.5 K[a]	4.16
$[(\eta^6\text{-}C_6H_6)_2Cr][PF_6]$	No RT → LT transition[b]	—

[a]Differential scanning calorimetry data, heating cycle.
[b]Differential scanning calorimetry data, cooling cycle down to 223 K.

low-temperature phase transition on decreasing the temperature to 223 K on the DSC instrument and to 100 K on the diffractometer, but undergoes an order–disorder phase transition on increasing the temperature to 332.5 K. The bis(benzene)-chromium analogue, $[Cr(\eta^6\text{-}C_6H_6)_2][PF_6]$, on the other hand, even though it crystallizes in a manner that is strictly related to that of the low-temperature phases of Co and Fe, does not appear to undergo phase changes either on cooling or on heating. Table 3 collects information on the phase transitional behaviour of several metallocene and metallobenzene hexafluorophosphate salts.

Hexafluoroarseniate salts show similar behaviour. $[(\eta^6\text{-}C_6H_6)(\eta^5\text{-}C_5H_5)Fe][AsF_6]$, for instance, undergoes phase transition between three different crystal forms [65]. Variable-temperature solid-state NMR measurements have shown that rotation of the entire cation takes place in a cubic phase above 310 K, whereas in the intermediate β-phase the rotational motion is restricted to a 90° in-plane reorientation. Below 270 K the crystal is in a low-symmetry phase, in which whole-body rotation does not take place although the rings execute a jumping motion that persists down to 200 K. Transition from a rotational jumping state to a whole-body reorientation has also been detected from the Mössbauer spectra of the $[PF_6]^-$ salt of the same complex [65b]. Activation energies lower than 20 kJ mol^{-1} were estimated for the reorientations around the three rotational axes. If mesitylene is substituted for benzene, the rotational process is stopped [65b]. The effect of substituting fluorobenzene for benzene has also been studied [65c]. The temperature dependence of the Mössbauer spectra of the $[PF_6]^-$ salt has been explained in terms of isotropic reorientation of the cation. The X-ray diffraction results indicate a primitive cubic cell with totally disordered cations at room temperature, while the DSC measurements are consistent with the occurrence of a first-order phase transition at 255 K. Analogous behaviour is shown by the $[AsF_6]^-$ and $[SbF_6]^-$ salts, whereas no phase transition is shown by the $[BF_4]^-$ salt [65c].

Recently, it has been shown by Hollingsworth *et al.* that urea inclusion compounds of formula $X(CH_2)_6CN$/urea (X = Cl, Br) undergo reversible crystal-to-crystal phase transitions associated with large translational motion of the guest molecules [66].

As mentioned above, the metastable room temperature phase of crystalline bis-formylferrocene, $[Fe(\eta^5\text{-}C_5H_4CHO)_2]$, on heating undergoes a first transition at ca 311 K. On cooling, however, the reverse process leads to a new phase. Subsequent cycles of heating and cooling show that the new phase reversibly switches between the room temperature and high-temperature phases, without reverting to the initial phase. This behaviour suggests that the first room temperature phase is a kinetic product of the crystallization process: on heating, the sample undergoes an order $\rightarrow$ disorder phase transition to the plastic phase, which transforms on cooling to a thermodynamically more stable form. Once this latter phase is formed, the initial phase can no longer be obtained, unless the compound is redissolved and recrystallized. The ΔH of transition (14.0 kJ mol^{-1}) is comparable to the value (12.1 kJ mol^{-1}) found for the mono-formyl derivative $[Fe(\eta^5\text{-}C_5H_5)(\eta^5\text{-}C_5H_4CHO)]$ [67], which shows a mesophase between 316 K and the melting point (396 K). On further heating crystalline $[Fe(\eta^5\text{-}C_5H_4CHO)_2]$ above the phase transition temperature, an exothermic peak is observed on the DSC plot at 453 K (see Figure 6); this peak was taken as indicative of the occurrence of a polymerization reaction in the solid state.

5 INDUCING POLYMORPHS VIA NONSOLUTION METHODS

Thus far we have discussed polymorphism and interconversion between polymorphs. While these examples should suffice to show the many facets and practical implications of polymorphism, they have not provided insight into the problem of making new crystal forms of the same substance. As mentioned in the Introduction, any step forward in the understanding of how to make polymorphs on purpose may represent a substantial advancement in a field.

In synthetic laboratories, the most common method to obtain crystals is crystallization from solution. In a way this is also the most difficult process to control. The solvent evaporation rate can be affected by several factors such as ambient humidity, temperature and ventilation, thermal excursion during the 24 h in the case of 'open air' room temperature crystallizations, and vibrations of the building (footsteps, traffic, lifts, etc.) and of the refrigerators. There is no unique and transferable recipe for the crystallization of a substance: Philip Ball concluded his *Nature* Editorial in 1996 by saying, 'the precipitation of good single crystals remains a black magic' [68].

This section will focus on some examples of induced polymorphism. In particular, we will discuss how the formation of new crystalline phases may be achieved by mechanical grinding and/or by thermal dehydration in TGA experiments and how crystallization of these elusive phases may be obtained via seeding.

5.1 Seeding

A common laboratory practice to control the outcome of a crystallization process is via seeding. The idea is that of providing nucleation 'seeds' that may act as templating units during crystal growth. Seeding is also a common industrial practise in the large-scale production of desired polymorphic forms of substances such as drugs and pigments. Of course, involuntary seeding may also alter the crystallization process in an undesired matter.

We have shown, for example, that single crystals of the anhydrous neutral zwitterion $[Co^{III}(\eta^5\text{-}C_5H_4COOH)(\eta^5\text{-}C_5H_4COO)]$ [39] can be obtained by seeding the water solution with seeds prepared by stepwise dehydration of the hydrated species $[Co^{III}(\eta^5\text{-}C_5H_4COOH)(\eta^5\text{-}C_5H_4COO)] \cdot 3H_2O$. Both anhydrous and hydrated species have been structurally characterized by single crystal and powder diffraction, and the conversion of the initial hydrated product into the anhydrous zwitterion has been monitored by DSC and TGA. TGA shows that the hydrated form reversibly releases one water molecule at 378 K, while the loss of the two remaining water molecules occurs at ca 506 K (see Figure 7) and is immediately followed by a phase transition. Powder diffraction shows that the first dehydration leaves the crystal structure almost unchanged. Crystallization from a water solution of the powder obtained from TGA at 506 K in the presence of seeds (a small portion of the same powder) leads to the growth of single crystals of the anhydrous species. The whole process is shown in Scheme 4. The structure of

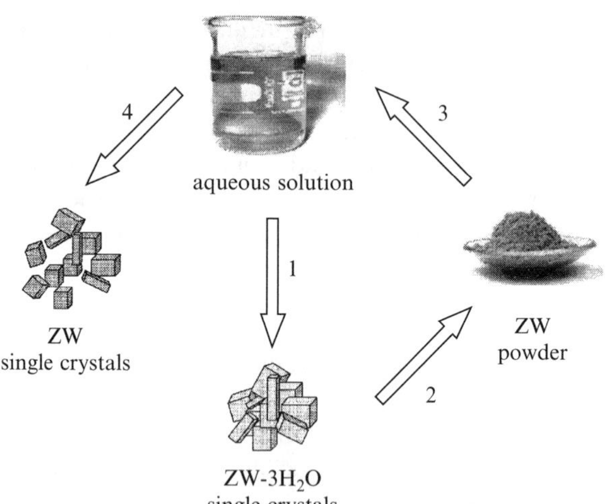

Scheme 4 1: Crystallization of the hydrated zwitterion $[Co^{III}(\eta^5\text{-}C_5H_4COOH)(\eta^5\text{-}C_5H_4COO)] \cdot 3H_2O(ZW \cdot 3H_2O)$ from its aqueous solution. 2: The dehydration process yields crystalline powder of the anhydrous form $[Co^{III}(\eta^5\text{-}C_5H_4COOH)(\eta^5\text{-}C_5H_4COO)]$ (ZW). 3: Seeding of the original solution with powder of ZW. 4: The seeding process yields single crystals of ZW [39].

$[Co^{III}(\eta^5\text{-}C_5H_4COOH)(\eta^5\text{-}C_5H_4COO)]$ is based on a one-dimensional network of O–H···O bonded zwitterion molecules. Comparison of the calculated and measured powder diffractograms of the anhydrous phase confirms that the powder obtained at 506 K and the single crystals precipitated at room temperature after seeding possess the same structure (see Figure 5). Importantly, whereas crystallization, in the presence of seeds of the anhydrous form, leads to the isolation of single crystals of the anhydrous material, in the absence of seeds the hydrated species is obtained. This latter form can, therefore, be seen as a pseudo-polymorphic modification of $[Co^{III}(\eta^5\text{-}C_5H_4COOH)(\eta^5\text{-}C_5H_4COO)]$.

5.2 Heteromolecular Seeding

Deliberate seeding can also be exploited by using seeds of isostructural or quasi-isostructural species that crystallize well to achieve the crystallization of an unyielding material. The same procedure can also be exploited to freeze a given conformation of flexible molecules. A textbook example of this latter approach is the nucleation of orthorhombic ferrocene, which possesses eclipsed cyclopentadienyl rings, obtained by seeding a ferrocene solution at low temperature with ruthenocene crystals [69]. Seeding has been used in experiments of shape mimicry [70a], whereas chiral co-crystals of tryptamine and hydrocinnamic acid were prepared by crystallization in the presence of seeds of different chiral crystals [70b].

Heteromolecular seeding has been instrumental in the separation of two concomitant polymorphs [71]. Precipitation of $[(\eta^5\text{-}C_5H_5)_2Fe]^+$ as its $[AsF_6]^-$ salt generates two types of crystals: a trigonal phase (Fe-T) and a monoclinic phase (Fe-M), which proved difficult to separate. From another experiment, it was known that crystallization of the congener $[(\eta^5\text{-}C_5H_5)_2Co]^+[AsF_6]^-$ leads only to a trigonal form (Co-T) isomorphous with Fe-T. In addition, the monoclinic phase Fe-M is isomorphous with the room temperature monoclinic phase of the pair $[(\eta^5\text{-}C_5H_5)_2Fe][PF_6]$ and $[(\eta^5\text{-}C_5H_5)_2Co][PF_6]$. The relationship between crystalline Fe-T and Fe-M is shown in Figure 14. It is worth noting that, besides the differences in packing arrangements of the forms, the cyclopentadienyl ligands are eclipsed in Fe-T, whereas they are all staggered in Fe-M. In order to drive the crystallization process towards the formation of the separate polymorphs, heteromolecular seeding was used. Crystals of trigonal $[(\eta^5\text{-}C_5H_5)_2Co][AsF_6]$ were used to grow the trigonal form of $[(\eta^5\text{-}C_5H_5)_2Fe][AsF_6]$, while crystals of monoclinic $[(\eta^5\text{-}C_5H_5)_2Fe][PF_6]$ were used to obtain the monoclinic form of $[(\eta^5\text{-}C_5H_5)_2Fe][AsF_6]$. The seeding was successful, even though a small amount of the alternative phase could always be detected in the powder diffractograms and in the DSC scans. What is more, the two seeding processes yielded good-quality single crystals of Fe-T and Fe-M, which proved to be sufficiently robust to undergo a full cycle of four phase transitions directly on the diffractometer, (Fe-T → Fe-M → Fe-C

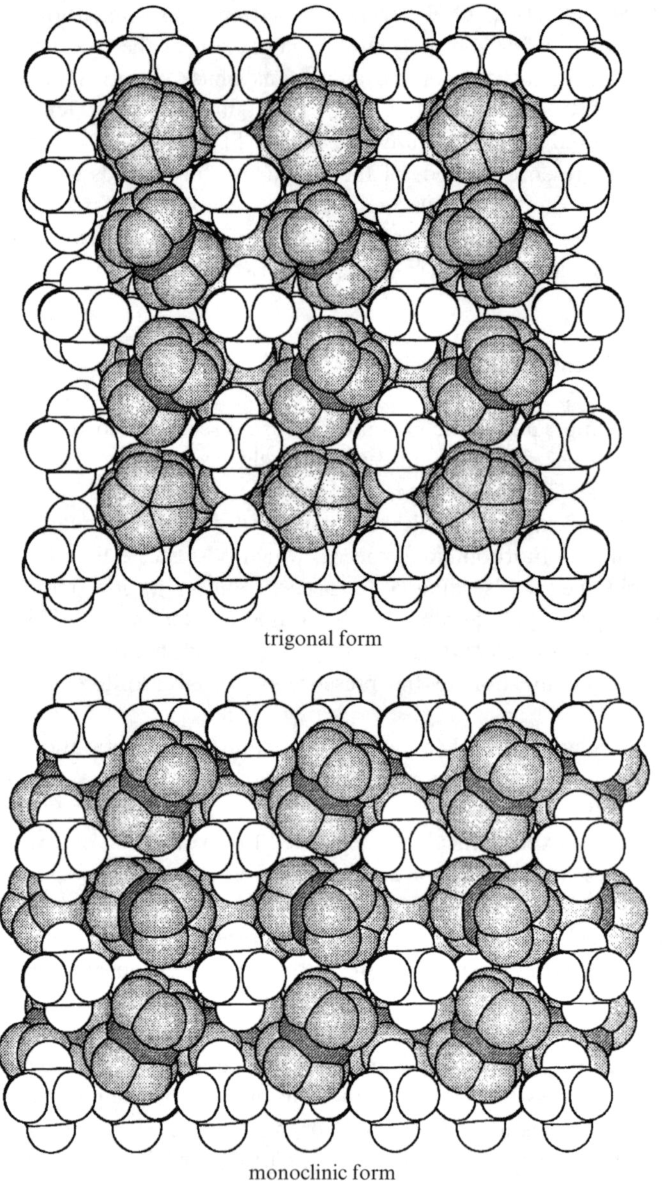

Figure 14 Precipitation of $[(\eta^5\text{-}C_5H_5)_2Fe]^+$ as its $[AsF_6]^-$ salt generates two concomitant crystals: a trigonal phase (Fe-T) and a monoclinic phase (Fe-M), which can be separated out by heteromolecular seeding with isomorphous crystals of trigonal $[(\eta^5\text{-}C_5H_5)_2Co][AsF_6]$ and of monoclinic $[(\eta^5C_5H_5)_2Fe][PF_6]$. The crystals are sufficiently robust to undergo a full cycle of four phase transitions directly on the diffractometer, Fe-T → Fe-M → Fe-C → Fe-M → Fe-T.

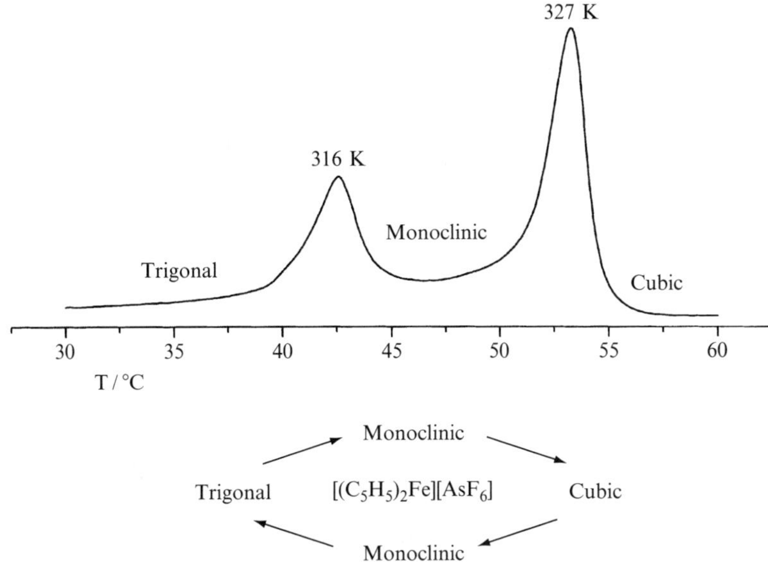

Figure 14 (*continued*)

→ Fe-M → Fe-T), an uncommon situation that permitted a whole rationalization of the phase transitional behaviour shown by the DSC.

5.3 Desolvation

Many molecular and ionic systems crystallize out of solutions as solvates. This is often a totally unavoidable event because solvent molecules fill the voids of otherwise less dense crystal packings, or the solid solvate precipitates out for kinetic reasons, as solvate nuclei are likely to be formed first with respect to unsolvated ones. The presence of solvent molecules trapped more or less tightly within the crystal structure may be turned to advantage if the solvent can be removed by low-pressure or high-temperature treatments or by other means. An example of polymorphs obtained via desolvation is provided by the hexagonal form of a C_{60} polymorph that can be obtained from desolvation of cubic 1:1 solvate C_{60} grown from dichloromethane [72].

Analogously, the sandwich cation $[(\eta^6\text{-}C_6H_6)_2Ru][BF_4]_2$ can be crystallized from nitromethane as the solvated form $[(\eta^6\text{-}C_6H_6)_2Ru][BF_4]_2 \cdot MeNO_2$. The solvate crystals, if exposed to air, rapidly convert to a different crystalline material, as ascertained by powder diffraction [73].

Recrystallization from water of this powder afforded crystals suitable for single-crystal X-ray analysis, which were characterized as the unsolvated form

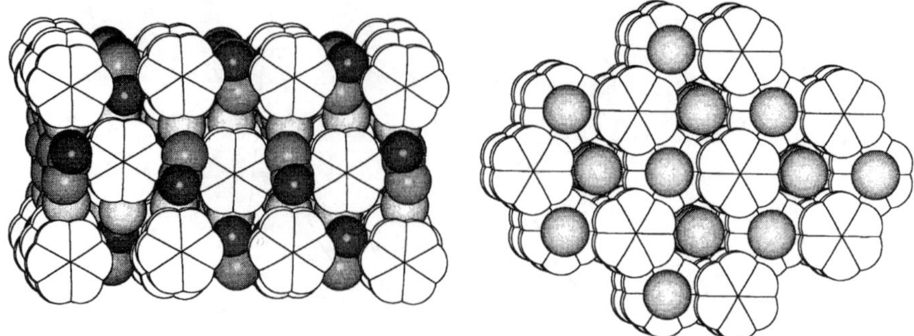

Figure 15 The relationship between the packings in $[(\eta^6\text{-}C_6H_6)_2Ru][BF_4]_2 \cdot MeNO_2$ and $[(\eta^6\text{-}C_6H_6)_2Ru][BF_4]_2$. Upon removal of the nitromethane molecule, the solvated form converts into the unsolvated form.

$[(\eta^6\text{-}C_6H_6)_2Ru][BF_4]_2$. Comparison of the calculated X-ray powder diffractogram of the unsolvated form with that measured for the crystalline powder obtained from the solvated form confirms that the solvated form converts into the unsolvated form. The pseudo-polymorph $[(\eta^6\text{-}C_6H_6)_2Ru][BF_4]_2 \cdot MeNO_2$ appears to be kinetically favoured when crystallization occurs from nitromethane, whereas the unsolvated form is more stable thermodynamically. Figure 15 shows a comparison between the two packings. Even though mechanistic hypotheses on interconversion between crystalline phases have to be considered with great caution, it seems likely that loss of nitromethane, loosely bound to the sandwich cations via C–H ··· O interactions, occurs via the channels in between the anions, which can then 'close around' the cations without extensive rearrangement.

5.4 Mechanical Stress

The notion that mechanical stress may cause solid-to-solid phase transitions or other physicochemical transformations is well known in materials sciences. However, examples of molecular crystals undergoing mechanically induced phase transitions or polymorph → pseudo-polymorph transformations between crystalline phases are not very abundant; some recent entries in the literature can be found in Ref. 74.

In the context of this discussion, it is useful to recall that solvent uptake may be attained by mechanical treatment of unsolvated crystals. Even gentle grinding of a powder product to prepare a sample for powder diffraction may lead to the formation of a hydrated product.

The case of the hydrated crystalline material $[(\eta^5\text{-}C_5H_5)_2Co]^+[(\eta^5\text{-}C_5H_4COOH)(\eta^5\text{-}C_5H_4COO)Fe]^- \cdot H_2O$ will be described in some detail [75]. The hydrated form is obtained by simply grinding the crystalline powder that precipitates from the solvent (THF) on reacting $[(\eta^5\text{-}C_5H_5)_2Co]$ with $[(\eta^5\text{-}C_5H_4COOH)_2Fe]$. The same result is obtained by grinding single crystals of the anhydrous salt $[(\eta^5\text{-}C_5H_5)_2Co]^+[(\eta^5\text{-}C_5H_4COOH)(\eta^5\text{-}C_5H_4COO)Fe]^-$ obtained by recrystallization of the same powder from nitromethane. Once $[(\eta^5\text{-}C_5H_5)_2Co]^+[(\eta^5\text{-}C_5H_4COOH)(\eta^5\text{-}C_5H_4COO)Fe]^- \cdot H_2O$ has been obtained by grinding, its single crystals can be grown from water or nitromethane, while crystals of the anhydrous form are no longer observed. However, on heating, the hydrated form loses water at 373 K and reverts to the starting material. The process has been investigated by TGA and by powder and single-crystal X-ray diffraction. Although the effect of grinding samples is well known, what appears to be noteworthy in the case of the anhydrous → hydrated transformations is the fact that water molecules can be inserted in stoichiometric amount into a complex and highly organized crystal edifice without loss of crystallinity or disruption of the anionic organization. The relationship between polymorph and pseudo-polymorph is shown in Figure 16.

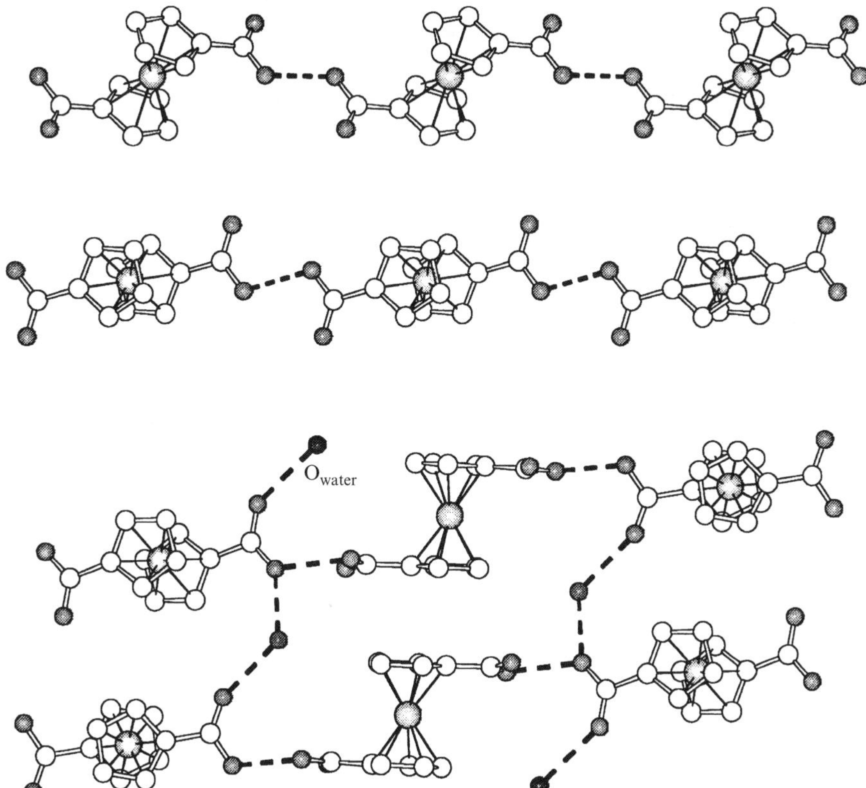

Figure 16 Quantitative 1:1 hydration of $[(\eta^5\text{-}C_5H_5)_2Co]$ with $[(\eta^5\text{-}C_5H_4COOH)_2Fe]$ to generate $[(\eta^5\text{-}C_5H_5)_2Co]^+[(\eta^5\text{-}C_5H_4COOH)(\eta^5\text{-}C_5H_4COO)Fe]^- \cdot H_2O$ obtained by grinding the crystalline powder that precipitates from THF.

6 SOLID-STATE REACTIVITY – GAS–SOLID REACTIONS AND SOLVATION PROCESSES

The examples provided in the previous sections show that one of the means to obtain new pseudo-polymorphs is via solvent uptake. Since the co-crystallizing solvent may be present in the vapour phase (say water), one may regard the process that leads from an unsolvated to a solvated species (and reverse) as a supramolecular reaction whereby a given set of noncovalent bonds (those between molecules in the nonsolvated form, for example) are broken and a new set of noncovalent bonds (those between host and guest molecules) are formed as shown at the beginning of this chapter in Scheme 2.

On this conceptual premise, uptake and release of solvent can be paralleled to a solid–gas reaction, whereby the reactants are the molecules in the crystalline solid and the molecules in the gas phase and the product is the solvated crystal. Clearly, the same reasoning applies to the reverse process, i.e. generation of a new crystalline form by means of gas release. In gas–solid reactions, gases are reacted directly with crystals or amorphous phases to give complete conversion and usually quantitative yields. What would then be the difference between a solvation reaction and a reaction leading to new molecular/ionic species if not the energetic scale of the processes and the fact that in solvation processes molecules retain their chemical identity?

Heterogeneous gas–solid reactions are well known in chemistry. It has been known for a long time that graphite reacts with metal vapours to form intercalation compounds. The rusting of iron by atmospheric oxygen is a trivial example of a heterogeneous solid–gas reaction, whereas the generation of a large quantity of nitrogen in the reaction of solid NaN_3 and KNO_3 is exploited for the functioning of air-bags in automobiles. It is also well known that thermal treatment of calcium carbonate to remove CO_2, as well as the reverse reaction during the drying process of hydrated $Ca(OH)_2$, is at the basis of the use of mortar in construction.

Furthermore, in the reactions between gases and solids the costs of removing and reprocessing solvents are eliminated; this goes along with the great pressure on developing solvent-free, i.e. environmentally more friendly reactions [76]. Beside gas–solid reactions, topochemical solid-state reactions of the type explored by Schmidt [77] in the early days of crystal engineering are now experiencing a wave of renewed interest [76,78]. A discussion of these aspects is beyond the scope of this contribution. It is worth noting, however, that in spite of the wide range of applications in both fundamental studies and in materials chemistry, heterogeneous reactions are still regarded as an oddity by chemists used to thinking of the solution as the appropriate medium for reactivity.

In the following we shall focus on heterogeneous acid–base reactions, which were among the first processes to be studied. The Italian scientist Pellizzari studied the reaction of dry ammonia vapour with dried and pulverized phenols and carboxylic acids [79]. This subject was taken up and extended by Paul, Curtin and co-workers in a series of elegant studies [80–82]. One of the best known case studies is the reaction of crystalline benzoic acid with ammonia. This process has been studied on both powder and single crystals and also on a number of related compounds. The reactions lead to quantitative formation of 1:1 ammonium salts. They were able to demonstrate that certain crystal faces are attacked preferentially by the ammonia vapour, and the resulting reaction front travels more rapidly through the crystals along directions corresponding to specific molecular arrangements [81,82]. Crystalline *p*-chlorobenzoic anhydride reacts with gaseous ammonia to give the corresponding amide and ammonium salt [83]; similar reaction has been investigated in the case of optically active cyclopropanecarboxylic acid crystals [84]. The possibility of reacting solid phenols with ammonia has been known since the beginning of the last century. Even those phenols which are too weakly acidic to react in solution form ammonium salts in heterogeneous processes [31]. Interestingly, the reverse reactions, namely removal of gases from the crystalline salts, has not been examined in detail.

A series of solid-state reactions have been explored by Kaupp *et al.* [85], in which gaseous amines are reacted with aldehydes to give imines. Analogous reactions with solid anhydrides, imides, lactones or carbonates and isothiocyanates have been used to give, respectively, diamides or amidic carboxylic salts or imides, diamides, carbamic acids and thioureas. In a number of cases investigated the yields were found to be quantitative and no workup (except for washing in few cases) is required in the 100 % yield reactions. Upscaling to the kilogram scale was also explored with good results. Ammonia and other gaseous amines, in particular methylamine, have also been shown to aminolyse thermoplastic polycarbonates. Such degradation processes have a high impact on the stability of data storage compact disks [86].

Kaupp *et al.* have also exploited heterogeneous reactions with ClCN and BrCN in the quantitative synthesis of cyanamides, cyanates, thiocyanates and derivatives [87]. Various types of cyanamides can be prepared by the interaction of gaseous cyanogen chloride or cyanogen bromide with crystalline primary or secondary amines and with imide salts. All reactions were found to perform best in the presence of gaseous trimethylamine, which binds the acid formed and makes possible a quantitative use of the reacting base. The trimethylamine hydrohalide that forms can be washed out with water, while the reaction product remains solid. If the same reaction is carried in the liquid phase the yield is reduced to 45 %, showing that the solid–gas reaction is superior. It has been argued that the possibility of obtaining quantitative reactions depends on the crystal packing, which has to permit the occurrence of three distinct reaction stages: (1) phase rebuilding with long-range molecular movements, (2) phase transformation into the product lattice and (3) crystal disintegration to form fresh surfaces. In many instances, these mechanisms could be verified by AFM measurements and correlated with the known crystal structures.

Gaseous acids have been shown to form salts with strong and weak solid nitrogen bases. Solid hydrohalides are formed quantitatively by reaction with vapours of HCl, HBr and HI; the same applies to dibases such as *o*-phenylenediamines. The products are much more easily handled than when they are formed in solution. The solid products can in turn be used to react with gaseous acetone to form the corresponding dihydrohalides of 1,5-benzodiazepines [88].

In other applications, gaseous acids have been used for direct solid–gas reactions. For instance, conversion of secondary and tertiary alcohols into the corresponding chlorides has been shown to proceed efficiently when the powdered solid alcohol is exposed to HCl in a desiccator [89,90].

6.1 Amphoteric Gas Traps

A recent application of the principles outlined above has been possible in the organometallic chemistry area [91]. The zwitterion $[Co^{III}(\eta^5\text{-}C_5H_4COOH)(\eta^5\text{-}C_5H_4COO)]$ can be quantitatively prepared from the corresponding dicarboxylic cationic acid $[Co^{III}(\eta^5\text{-}C_5H_4COOH)_2]^+$. The amphoteric behaviour of the zwitterion depends on the presence of one–COOH group, which can react with bases, and one–COO^- group, which can react with acids. The organometallic zwitterion $[Co^{III}(\eta^5\text{-}C_5H_4COOH)(\eta^5\text{-}C_5H_4COO)]$ undergoes fully reversible heterogeneous reactions with the hydrated vapours of a variety of acids (e.g. HCl, CF_3COOH, HBF_4) and bases (e.g. NH_3, NMe_3, NH_2Me), with formation of the corresponding salts, as shown in Table 4 [91]. For instance, complete conversion of the neutral crystalline zwitterion into the crystalline chloride salt is attained in 5 min of exposure to vapour of 36% aqueous HCl. Formation of the salt in the heterogeneous reaction is easily assessed by comparison of the observed X-ray powder diffraction pattern with that calculated on the basis of the single-crystal structure determined previously. Crystalline $[Co^{III}(\eta^5\text{-}C_5H_4COOH)_2]Cl \cdot H_2O$ can be converted back to neutral $[Co^{III}(\eta^5\text{-}C_5H_4COOH)(\eta^5\text{-}C_5H_4COO)]$ by heating the sample for 1 h at 440 K in an oil-bath under low pressure (10^{-2} mbar). TGA was used to demonstrate that the solid releases one water molecule and one HCl molecule per molecular unit at 394 and 498 K, respectively (see Figure 17). The powder diffractogram of the final product corresponds precisely to that of the anhydrous $[Co^{III}(\eta^5\text{-}C_5H_4COOH)(\eta^5\text{-}C_5H_4COO)]$.

The behaviour of the zwitterion towards NH_3 is similar: the neutral system quantitatively transforms into the hydrated ammonium salt $[Co^{III}(\eta^5\text{-}C_5H_4COO)_2][NH_4] \cdot 3H_2O$ upon 5 min of exposure to vapour of 30% aqueous ammonia. Single crystals of the ammonium salt for X-ray structure determination can be obtained if the reaction of the zwitterions with ammonia is carried out in aqueous solution. As in the case of the chloride salt, formation of $[Co^{III}(\eta^5\text{-}C_5H_4COO)_2][NH4] \cdot 3H_2O$ in the heterogeneous reaction is assessed by comparison of the observed and calculated X-ray powder patterns. Absorption of ammonia is also fully reversible: upon

Table 4 Reaction of solid $[Co^{III}(\eta^5\text{-}C_5H_4COOH)(\eta^5\text{-}C_5H_4COO)]$ with hydrated acid and base vapours.

Acid[a]	Exposure time[b]	$\nu_{CO}(cm^{-1})$[c]	Base[a]	Exposure time[b]	$\nu_{CO}(cm^{-1})$[c]
HCl (37%)	5 min	1733(s) 1707(s)	NH_3 (30%)	5 min	1610(s)
CF_3COOH (99%)	1 h	1723(s) 1706(s)	NH_2Me (30%)	1 h	1610(s)
HBF_4 (54%)	16 h	1736(s) 1708(s)	NMe_3 (30%)	1 h	1612(s)

[a]Concentration of the solutions in parentheses.
[b]Uptake time for a complete conversion of 10 mg of $[Co^{III}(\eta^5\text{-}C_5H_4COOH)(\eta^5\text{-}C_5H_4COO)]$; all processes are fully reversible.
[c]Diagnostic CO stretching frequencies (KBr) for reaction products. Neutral $[Co^{III}(\eta^5\text{-}C_5H_4COOH)(\eta^5\text{-}C_5H_4COO)]$ is characterized by two bands at 1717 and 1654 cm^{-1}(Nujol).

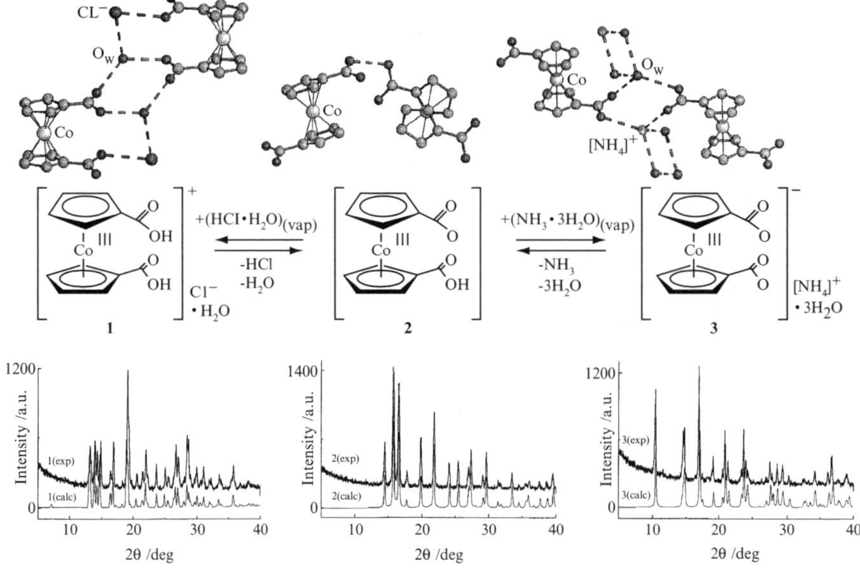

Figure 17 Schematic representation of the uptake and release of hydrated vapours of HCl and NH_3 by crystalline $[Co^{III}(\eta^5\text{-}C_5H_4COOH)(\eta^5\text{-}C_5H_4COO)]$. Top: the different relative arrangements of the organometallic moieties in the solid state are shown. Bottom: comparison of calculated and observed powder diffraction patterns.

thermal treatment (1 h at 373 K, ambient pressure) the salts converts quantitatively into the neutral zwitterion. The material can be cycled through several absorption and release processes of HCl or ammonia, as well as other gaseous acids and bases (see Table 4) without decomposition or detectable formation of amorphous material. Infrared spectroscopy can also be utilized to detect quickly the formation of the organometallic cation or anion upon reaction with volatile acids or bases (see Table 4). In summary, the organometallic zwitterionic system $[Co^{III}(\eta^5\text{-}C_5H_4COOH)\ (\eta^5\text{-}C_5H_4COO)]$ behaves as a fully reversible 'amphoteric trap' towards a variety of hydrated vapours of acids and bases.

7 CONCLUSION

The objective of this chapter has not been to provide a review on polymorphism, but rather to attempt to bridge different aspects of solid-state chemistry and molecular crystal engineering that have in common the existence of different crystal forms of the same principal species. The question of whether a crystalline material contains only molecules A in different arrangements (e.g. polymorphs), or molecules A *and* molecules B, and whether this latter association ought to be considered a co-crystal or a pseudo-polymorph, is somewhat semantic. It is unquestionable, however, that crystals of different composition not only have different crystal structures but also may possess different physical properties, such as solubility, thermal behaviour, resistance to mechanical stress, gas-absorption/release capacity, colour, melting point, etc. These differences may be relevant, and carry economic and practical implications, when considering, for instance, the bioavailability of a drug or the thermal stability of a pigment.

In spite of the great interest in the phenomenon of polymorphism and of the increased research activity beyond the boundaries of organic solid-state chemistry, it is a fact that only a few molecular compounds possess several crystalline forms, whereas for many other tens of thousands of molecular compounds only one crystalline form is known. In other words, why are there so few molecular crystals polymorphs? The often quoted association between number of known forms and 'the time and energy spent in searching for them' put forward by McCrone probably contains the answer to this question. It is probable that if thorough (combinatorial?) crystallization experiments were carried out on any given molecular species or molecular salt, alternative crystalline forms would be found. It is probable but not certain.

Reproducibility and predictability are paradigms in the exact sciences. This is why, apart from all the utilitarian reasons associated with the marketing of solid-state materials, we would like to learn how to make polymorphs, or, which is the same, how effectively to prevent polymorph formation. One could argue that polymorphism, with its high degree of serendipity, could be the nemesis of crystal engineering because polymorphism is in logical contrast with a discipline that aims to control and reproduce univocally a given crystal structure. Paul and Curtin concluded their paper in *Science* in 1975 [80] by saying, 'The recent ready availability of crystal structure data for molecular crystals should lead to a more rapid development of the principles governing the packing of such crystals and their reactivity. The resulting control of such reactions should not only provide the means of causing them when they are desired but also a means to prevent them when they are to be avoided...'. Here is where we still stand almost 30 years later. Studies of crystal engineering are expected to shed light on the principles governing packing hence on the way to make different crystals, and obtain different properties, with the same molecules or ions.

In this chapter, we have described some common laboratory techniques and practices for tackling problems of polymorphism and pseudo-polymorphism. In the case of solids, routine analytical and spectroscopic laboratory tools are much less useful than in solution chemistry. The utilization of powder diffraction is, sometimes, the only way to ascertain whether the whole solid material has the same structure as that characterized by single-crystal diffraction, i.e if they correspond to the same crystal form. The notion that powder grinding, a common method of sample preparation, may lead to solid-state transformations and to the formation of new polymorphic modifications could be crucial for the success of many crystal engineering processes and for many materials chemistry uses. While scientists in the pharmaceutical industries are aware of this phenomenon and of the possible consequences of tableting and other mechanical processing on solid drugs, this awareness is not diffuse in the field of crystal engineering, which still has an essentially academic basis and originates from synthetic chemistry and crystallography rather than from applied research in materials chemistry. Furthermore, since the crystallization product may not always be the most (thermodynamically) stable system, it appears that any new solid material should be subjected routinely to a DSC run in order to ascertain the possibility of phase transitions. This is also a way to check if the crystalline product is a stable thermodynamic form or only the one quickest to grow. One may envisage a general approach to the relationship between polymorphs and pseudo-polymorphs. If seeds of a polymorphic modification can be obtained from nonsolution methods (e.g. mechanical, thermodynamic, perhaps solid-state reactions), these can be used in the seeding process, which may allow the growth of less kinetically favoured (although often more thermodynamically stable) crystal forms. Seeding may be valuable not only to obtain the desired crystalline form, but also to prevent crystallization of undesired forms (e.g. *ice-nine* in *Cat's Cradle*; see the quotation at the beginning of the chapter).

We have argued that the reaction of a solid with a vapour is conceptually related to the supramolecular reaction of a crystalline material with a volatile solvent to form a new crystalline solid. The two processes, solid–gas reaction and solid–gas solvation, differ only in the energetic ranking of the interactions that are broken or formed through the processes. In solvation–desolvation processes one is dealing mainly with noncovalent van der Waals or hydrogen bonding interactions, whereas in chemical reactions covalent bonds are broken or formed. The possibility of switching between interactions between charged species and molecules as a consequence of proton transfers or redox processes should also be considered.

This awareness is useful to devise crystal engineering studies whereby gas uptake is exploited not only to produce new crystalline forms of a given substance but also as a means to produce new materials in crystalline form. Clearly the (conceptual) borderline between the two types of processes is very thin. One may then purposefully plan to assemble molecules that are capable of absorbing molecules from the gas phase and, possibly, to react with them. Reaction implies sensing and could be exploited to detect molecules if there is a measurable response from the solid state. If the reaction is quantitative and reversible, the same processes (whether based on weak noncovalent bonding or on some type of covalent/ionic, high-enthalpy, process) can be used to trap gases and deliver them where appropriate. Clearly, the control of solid-state reactions, that can be used to trap environmentally dangerous or poisonous molecules, is an attractive goal for solid-state chemistry and crystal engineering.

The conclusion of this contribution is rather obvious. All in all, *Natura non facit saltus*: solid-state reactivity and molecular materials engineering obey the same basic rules as those governing polymorphism and solid-to-solid transformations.

I closed my eyes.
There was a sound like that of the gentle closing of a portal as big as the sky, the great door of heaven being closed softly.
It was a grand AH-WHOOM.
I opened my eyes – and all the sea was ice-nine.
The moist green earth was a blue-white pearl.

K. Vonnegut, *Cat's Cradle*, Penguin Books, London, 1965, pp. 34, 37, 163.

ACKNOWLEDGMENTS

We gratefully acknowledge the support of MURST (projects *Supramolecular Devices* 1999–2000 and *Solid Supermolecules* 2000–2001) and of the Universities of Bologna (project *Innovative Materials*) and Sassari. We thank Professor Joel Bernstein for continuing enthusiasm and for many useful discussions and Dr Vincenzo Cannata (Zambon Group) for having brought many relevant problems of polymorphism to our attention.

REFERENCES

1. D. Braga, F. Grepioni and A. G. Orpen (Eds), *Crystal Engineering: from Molecules and Crystals to Materials*, Kluwer, Dordrecht, 1999.
2. See, for example (a) D. W. Bruce and D. O'Hare (Eds), *Inorganic Materials*, Wiley, Chichester, 1992; (b) S. R. Marder, *Inorg. Mater.*, **115** (1992); (c) N. J. Long, *Angew.*

Chem., Int. Ed. Engl., **34**, 21 (1995); (d) T. J. Marks and M.A. Ratner, *Angew. Chem., Int. Ed. Engl.*, **34** (1995) 155; (e) D. R. Kanis, M. A. Ratner and T. J. Marks, *Chem. Rev.*, **94**, 195 (1994); (f) O. Kahn, *Molecular Magnetism*, VCH, New York, 1993; (g) D. Gatteschi, *Adv. Mater.*, **6**, 635 (1994); (f) J. S. Miller and A. J. Epstein, *Chem. Eng. News*, **73**, 30 (1995); (g) Proceedings of the Dalton Discussion 'Inorganic Crystal Engineering', *J. Chem. Soc., Dalton Trans.*, 3705 (2000), and references therein.

3. J.-M. Lehn (Ed.), *Supramolecular Chemistry: Concepts and Perspectives*, VCH, Weinheim, 1995.
4. (a) D. Braga and F. Grepioni, *Coord. Chem. Rev.*, **183**, 19 (1999); (b) D. Braga, G. Coiazzi, L. Maini, M. Polito, L. Scaccianoce and F. Grepioni, *Coord. Chem. Rev.*, **216**, 225 (2001); (c) G. R. Desiraju, *Angew. Chem., Int. Ed. Engl.*, **34**, 2311 (1995); (d) C. B. Aakeroy and K. R. Seddon, *Chem. Soc. Rev.*, 397 (1993); (e) A. N. Khlobystov, A. J. Blake, N. R. Champness, D. A. Lemenovskii, A. G. Majouga, N. V. Zyk and M. Schröder, *Coord. Chem. Rev.*, **222**, 155 (2001); (f) M. Eddaoudi, D. B. Moler, H. L. Li, B. L. Chen, T. M. Reineke, M. O'Keeffe and O. M. Yaghi, *Acc. Chem. Res.*, **34**, 319 (2001).
5. D. Braga, and F. Grepioni *Acc. Chem. Res.*, **33**, 601 (2000).
6. F. Takusagawa and T. F. Koetzle, *Acta Crystallogr., Sect. B*, **35**, 2888 (1979).
7. J. Bernstein, *Polymorphism in Molecular Crystals*, Oxford University Press, Oxford, 2002.
8. J. K. Rowling, *Harry Potter and the Philosopher's Stone*, Bloomsbury Publishing, London, 1997.
9. J. Dunitz, and J. Bernstein, *Acc. Chem. Res.*, **28**, 193 (1995); see also J.-O. Henck, J. Bernstein, A. Ellern and R. Boese, *J. Am. Chem. Soc.*, **123**, 1834 (2001).
10. S. R. Chemburkar, J. Bauer, K. Deming, H. Spiwek, K. Patel, J. Morris, R. Henry, S. Spanton, W. Dziki, W. Porter, J. Quick, P. Bauer, J. Donaubauer, B. A. Narayanan, M. Soldani, D. Riley and K. McFarland, *Org. Process Res. Dev.*, **4**, 413 (2000); J. Bauer, S. Spanton, R. Henry, J. Quick, W. Dziki, W. Porter and J. Morris, *Pharm. Res.*, **18**, 859 (2001).
11. (a) A. Gavezzotti, *Acc. Chem. Res.*, **27**, 309 (1994); (b) J. P. M. Lommerse, W. D. S. Motherwell, H . L. Ammon, J. D. Dunitz, A. Gavezzotti, D. W. M. Hofmann, F. J. J. Leusen, W. T. M. Mooij, S. L. Price, B. Schweizer, M. U. Schmidt, B. P. van Eijck, P. Verwer and D. E. Williams, *Acta Crystallogr., Sect. B*, **56**, 697 (2000); (c) T. Beyer, T. Lewis, and S. L. Price, *CrystEngComm*, **44**, 1 (2001); (d) A. Gavezzotti, *J. Am. Chem. Soc.*, **122**, 10724 (2000).
12. (a) A. Burger, in *Topics in Pharmaceutical Sciences*, Eds D. D. Breimer and P. Speiser, Elsevier, Amsterdam, 1983, p. 347; (b) T. L.Threlfall, *Analyst*, **120**, 2435 (1995).
13. S. R. Byrn, *Solid State Chemistry of Drugs*, Academic Press, New York, 1982, p. 79.
14. W. C. McCrone, in *Polymorphism in Physics and Chemistry of the Organic Solid State*; Eds D. Fox, M. M. Labes and A. Weissberger, Interscience, New York, 1965, vol. II, p. 726.
15. (a) J. Bernstein, in *Organic Solid State Chemistry*, Ed. G. R. Desiraju, Elsevier, Amsterdam, 1987, p. 471, and references therein; (b) J. Bernstein and A. T. Hagler, *J. Am. Chem. Soc.*, **100**, 673 (1978); (c) J. Bernstein, in *Structure and Properties of Molecular Crystals, Ed. M.* Pierrot, Elsevier, Amsterdam, 1990.
16. (a) D. Braga and F. Grepioni, *Chem. Soc. Rev.*, **4**, 229 (2000); (b) see also D. Braga and F. Grepioni, in *Static and Dynamic Structures of Organometallic Molecules and Crystals*, Eds J. M. Brown and P. Hofmann, *Topics in Organometallic Chemistry*, vol. 4, Springer Berlin, 1999, p. 48; (c) D. Braga, G. Cojazzi, A. Abati, L. Maini, M. Polito, L. Scaccianoce and F. Grepioni, *J. Chem. Soc., Dalton Trans.*, 3969 (2000).
17. (a) J. Bernstein, R. J. Davey and J.-O. Henck, *Angew. Chem., Int. Ed. Engl.*, **38**, 3440 (1999); (b) see also N. Bladgen and R. J. Davey, *Chem. Br.*, **35**, 44 (1999).

18. (a) P. H. Poole, T. Grande, F. Sciortino, H. E. Stanley and C. A. Angell, *Comput. Mater. Sci.*, **4**, 373 (1995); (b) B. C. Hancock and G. Zograf, *J. Pharm. Sci.*, **86**, 1 (1997); (c) M. Yoshioka, B. C. Hancock and G. Zograf, *J. Pharm. Sci.*, **83**, 1700 (1994); (d) M. Brinkmann, G. Gadret, M. Muccini, C. Taliani, N. Masciocchi and A. Sironi, *J. Am. Chem. Soc.*, **122**, 5147 (2002).
19. (a) C. H. Görbitz and H.-P. Hersleth, *Acta Crystalloger., Sect. B*, **56**, 526 (2000); (b) A. Nangia and G. R. Desiraju, *Chem. Commun.*, 605 (1999); (c) B. Bechtloff, S. Nordhoff and J. Ulrich, *Cryst. Res. Technol.*, **36**, 1315 (2001).
20. (a) F. H. Allen and O. Kennard, *Chem. Des. Autom. News*, **8**, 31 (1993); (b) Inorganic Crystal Structure Database (ICSD), Fachinformationszentrum (FIZ), Karlsruhe, and Gmelin Institut, Release 2000/2.
21. (a) V. A. Russell, C. C. Evans, W. Li and M. D. Ward, *Science*, **276**, 575 (1997); (b) K. T. Holman, A. M. Pivovar, J. A. Swift, and M. D. Ward, *Acc. Chem. Res.*, **34**, 107 (2001); (c) O. M. Yaghi, G. Li and H. Li, *Nature (London)*, **378**, 703 (1995); (d) S.-I. Noro, S. Kitagawa, M. Kondo and K. Seki, *Angew. Chem. Int. Ed. Engl.*, **39**, 2082 (2000); (e) B. F. Abrahams, B. F. Hoskins, D. M. Michail and R. Robson, *Nature (London)*, **369**, 727 (1994); (f) C. N. R. Rao, N. Srinivasan, A. Choudhury, S. Neeraj and A. A. Ayi, *Acc. Chem. Res.*, **34**, 80 (2001); (g) M. Fujita, *Acc. Chem. Res.*, **32**, 53 (1999).
22. (a) M. Albrecht, M. Lutz, A. L. Speck and G. van Koten, *Nature (London)*, **406**, 970 (2000); (b) M. Albrecht, R. A. Gossage, M. Lutz, A. L. Speck and G. van Koten, *Chemistry*, **6**, 1431 (2000); (c) M. Albrecht, A. M. M. Schreurs, M. Lutz, A. L. Speck and G. van Koten, *J. Chem. Soc., Dalton Trans.*, 3797 (2000).
23. (a) I. C. Paul and D. Y. Curtin, *Acc. Chem. Res.*, **6**, 217 (1973); (b) R. S. Miller, D. Y. Curtin and I. C. Paul, *J. Am. Chem. Soc.*, **96**, 6340 (1974).
24. D. Braga *Chem. Rev.*, **92**, 369 (1992).
25. (a) G. R. Desiraju (Ed.), *The Crystal as a Supramolecular Entity. Perspectives in Supramolecular Chemistry*, vol. 2, Wiley, Chichester, 1996, p. 107; (b) J. D. Dunitz, *Acta Crystallogr., Sect. B*, **51**, 619 (1995).
26. D. Braga and F. Grepioni, *J. Chem. Soc., Dalton Trans.*, 1223 (1993).
27. (a) P. Coppens, Y. W. Yang, R. H. Blessing and W. F. Cooper, *J. Am. Chem. Soc.*, **99**, 760 (1977); (b) A. C. Gallacher and A. A. Pinkerton, *Acta Crystallogr., Sect.C*, **49**, 125 (1993).
28. (a) J. D. Dunitz, *X-ray Analysis and the Structure of Organic Molecules*, VHCA, Basel, 1995; (b) D. L. Bish and J. E. Post (Eds), 'Modern Powder Diffraction', *Rev. Miner.*, **20**, 19 (1989); (c) K. D. M. Harris, and M. Tremayne, *Chem. Mater.*, **8**, 2554 (1996); (d) K. D. M. Harris, M. Tremayne and B. M. Kariuki, *Angew. Chem., Int. Ed. Engl.*, **40**, 1626 (2001).
29. (a) A. K. Cheetham (Ed.), *Solid State Chemistry Techniques*, Oxford University Press, Oxford, 1987; (b) F. H. Herbstein, *J. Mol. Struct.*, **374**, 111 (1996); (c) S. Materazzi, *Appl. Spectrose. Rev.* **33**, 189 (1998).
30. (a) D. E. Bugay, *Adv. Drug Deliv. Rev.*, **48**, 43, (2001); (b) C. A. Fyfe, *Solid State NMR for Chemists*, ACFC Press, Ontario, 1983.
31. G. Kaupp, in *Comprehensive Supramolecular Chemistry*, Ed. J. E. D. Davies, Elsevier, Oxford, vol. 8, 1996, p. 381.
32. D. Braga, L. Scaccianoce, F. Grepioni and S. M. Draper, *Organometallics*, **15**, 4675 (1996).
33. D. Braga, L. Maini, M. Polito, M. Rossini and F. Grepioni, *Chem. Eur. J.*, **6**, 4227 (2000).
34. C. C. Wilson, *Acta Crystallogr., Sect. B*, **57**, 435 (2001).
35. D. Wiechert and D. Mootz, *Angew. Chem., Int. Ed. Engl.*, **38**, 1974 (1999).

36. (a) S. L. Simon, *Thermochim. Acta*, **374**, 55 (2001); (b) D. Dollimore, and P. Phang, *Anal. Chem.*, **72**, 27R (2000).
37. S. D. Clas, C. R. Dalton and B. C. Hancock, *Pharm. Sci. Technol. Today*, **2**, 311 (1999).
38. D. Braga, F. Paganelli, E. Tagliavini, S. Casolari, G. Cojazzi and F. Grepioni, *Organometallics*, **18**, 4191 (1999).
39. D. Braga, G. Cojazzi, L. Maini, M. Polito and F. Grepioni, *Chem. Commun.*, 1949 (1999).
40. D. Braga and F. Grepioni, *Acc. Chem. Res.*, **27**, 51 (1994).
41. See, for instance, L. J. Farrugia, *J. Chem. Soc., Dalton Trans.*, 1783 (1997), and references therein.
42. M. J. Calhorda, D. Braga and F. Grepioni, in *Metal Clusters in Chemistry*, Eds P. Braunstein, L. Oro and P. R. Raithby, Wiley-VCH, Weinheim, 1999.
43. D. Braga and F. Grepioni, in *Organometallic Bonding and Reactivity – Fundamental Studies*, Eds J. M. Brown and P. Hofmann, Springer, Berlin, 1999, p. 47.
44. (a) P. Seiler and J. D. Dunitz, *Acta Crystallogr., Sect. B*, **35**, 2020 (1979); (b) F. Takusagawa, and T. F. Koetzle, *Acta Crystallogr., Sect. B*, **35**, 1074 (1979); (c) P. Seiler and J. D. Dunitz, *Acta Crystallogr., Sect. B*, **35**, 1068 (1979); (d) D. Braga, and F. Grepioni, *Organometallics*, **11**, 711 (1992); (e) J. D. Dunitz, *Acta Crystallogr., Sect B*, **51**, 619 (1995); (f) J. D. Dunitz, in *Organic Chemistry: Its Language and its State of the Art*, Ed. M. V. Kisakürek, Verlag Chemie, Basel, 1993, p. 9.
45. D. Braga, F. Grepioni, B. F. G. Johnson, P. Dyson, P. Frediani, M. Bianchi, F. Piacenti and J. Lewis, *J. Chem. Soc., Dalton Trans.*, 2565 (1992).
46. (a) D. Zhao and L. Brammer, *Inorg. Chem.*, **33**, 5897 (1994); (b) D. V. Soldatov and J. A. Ripmeester, *Chem. Eur. J.*, **7**, 2979 (2001).
47. (a) A. J. Blake, N. R. Brooks, N. R. Champness, M. Crew, A. Deveson, D. Fenske, D. H. Gregory, L. R. Hanton, P. Hubberstey and M. Schroder, *Chem. Commun.*, **16**, 1432, (2001); (b) D. V. Soldatov, J. A. Ripmeester, S. I. Svergina, I. E. Sokolov, A. S. Zanina, S. A. Gromilov and Y. A. Dyadin, *J. Am. Chem. Soc.*, **121**, 4179 (1999).
48. (a) H. Effenberger, K. Mereiter and J. Zemann, *Z. Kristallogr.*, **156**, 233 (1981); (b) D. Jarosch and G. Heger, *Tschermaks Mineral. Petrogr. Mitt.*, **35**, 127 (1986); (c) H. J. Meyer, *Z. Kristallogr.*, **128**, 183 (1969).
49. V. R. Thalladi, M. Nusse and R. Boese, *J. Am. Chem. Soc.*, **122**, 9227 (2000).
50. J. L. Derissen and P. H. Smit, *Acta Crystallogr., Sect. B*, **30**, 2240 (1974).
51. (a) C. Pedone and A. Sirigu, *Inorg.Chem.*, **7**, 2614 (1968); (b) Y. Hsiou,Y. Wang and L.-K. Liu *Acta Crystallogr., Sect. C*, **45**, 721 (1989).
52. (a) C. J. Brown, *Acta Crystallogr.*, **21**, 1 (1966); (b) A. L. Bednowitz, and B. Post, *Acta Crystallogr.*, **21**, 566 (1966).
53. S. J. La Placa, W. C. Hamilton, J. A. Ibers and A. Davison, *Inorg. Chem.*, **8**, 1928 (1969).
54. D. Braga, L. Maini, F. Grepioni, C. Elschenbroich, F. Paganelli and O. Schiemann, *Organometallics*, **20**, 1875 (2001).
55. F. Takusagawa and T. F. Koetzle, *Acta Crystallogr., Sect. B*, **35**, 2888 (1979).
56. (a) A. S. Antsyshkina, L. M. Dikareva, M. A. Porai- Koshits, V. N. Ostrikova, Yu. V. Skripkin, O. G. Volkov, A. A. Pasynskii and V. T. Kalinnikov, *Koord. Khim.*, **11**, 82 (1985); (b) J. A. Marsella, J. C. Huffman, K. G. Caulton, B. Longato and J. R. Norton, *J. Am. Chem. Soc.*, **104**, 6360 (1982).
57. D. Braga, A. Abati, L. Scaccianoce, B. F. G. Johnson and F. Grepioni *Solid State Sci.*, **3**, 783 (2001).
58. C. R-Pérez, M. Hernández-Molina, P. Lorenzo-Luis, F. Lloret, J. Cano and M. Julve, *Inorg. Chem.*, **39**, 3845 (2000).

59. R. A. Heintz, H. Zhao, X. Ouyang, G. Grandinetti, J. Cowen and K. Dunbar, *Inorg. Chem.*, **38**, 144 (1999).
60. D. Braga, P. J. Dyson, F. Grepioni, and B. F. G. Johnson *Chem. Rev.*, **94**, 1585 (1994).
61. N. G. Parsonage and L. A. K. Staveley, *Disorder in Crystals*, Clarendon Press, Oxford, 1978.
62. (a) D. Braga, F. Grepioni, B. F. G. Johnson and J. Farrugia, *J. Chem. Soc., Dalton Trans.*, 2911 (1994); (b) H. Dorn, B. E. Hanson, and E. Motell, *Inorg. Chim. Acta*, **54**, L71 (1981); (c) B. E. Hanson, E. C. Lisic, J. T. Petty and G. A. Iannacone, *Inorg. Chem.*, **25**, 4062 (1986); (d) B. E. Mann, *J. Chem. Soc., Dalton Trans.*, 1457 (1997); (e) B. F. G. Johnson, *J. Chem. Soc., Dalton Trans.*, 1473 (1997); (f) D. Braga, J. Farrugia, F. Grepioni and A. Senior, *J. Chem. Soc., Chem. Commun.*, 1219 (1995); (g) D. Braga, J. Farrugia, A. L. Gillon, F. Grepioni and E. Tedesco, *Organometallics*, **15**, 4684 (1996).
63. (a) D. Braga, L. Scaccianoce, F. Grepioni and S. M. Draper, *Organometallics*, **15**, 4675 (1996); (b) F. Grepioni, G. Cojazzi, S. M. Draper, N. Scully and D. Braga, *Organometallics*, **17**, 296 (1998).
64. F. Grepioni, G. Cojazzi, D. Braga, E. Marseglia, L. Scaccianoce and B. F. G. Johnson *J. Chem. Soc., Dalton Trans.*, 553 (1999).
65. (a) B. W. Fitzsimmons and A. R. Hume, *J. Chem. Soc., Dalton Trans.*, 180 (1980); (b) B. W. Fitzsimmons and W. G. Marshall, *J. Chem. Soc., Dalton Trans.*, 73 (1992); (c) B. W. Fitzsimmons and I. Sayer, *J. Chem. Soc., Dalton Trans.*, 2907 (1991).
66. M. D. Hollinsgworth, M. L. Peterson, K. L. Pate, B. D. Dinkelmeyer and M. E. Brown, *J. Am. Chem. Soc.*, **124**, 2094 (2002).
67. M. F. Daniel, A. J. Leadbetter, R. E. Meads and W. G. Parker, *J. Chem. Soc., Faraday Trans.*, **74**, 456 (1978).
68. P. Ball, *Nature (London)*, **381**, 20 (1996).
69. P. Seiler and J. D. Dunitz, *Acta Crystallogr., Sect. B*, **38**, 1741 (1982).
70. (a) R. J. Davey, N. Blagden, G. D. Potts and R. Docherty, *J. Am. Chem. Soc.*, **119**, 1767 (1997); (b) H. Koshima and M. Miyauchi, *Cryst. Growth Des.*, **1**, 355 (2001).
71. D. Braga, G. Cojazzi, D. Paolucci and F. Grepioni, *CrystEngComm*, **38**, 1 (2001).
72. R. Céolin, J. Ll. Tamarit, M. Barrio, D. O. López, S. Toscani, H. Allouchi, V. Agafonov and H. Szwarc, *Chem. Mater.*, **13**, 1349 (2001).
73. D. Braga, G. Cojazzi, A. Abati, L. Maini, M. Polito, L. Scaccianoce and F. Grepioni, *J. Chem. Soc., Dalton Trans.*, 3969 (2000).
74. (a) M. R. Caira, *Top. Curr. Chem.*, **198**, 163 (1998); (b) M. R. Caira, L. R. Nassimbeni and M. Timme, *J. Pharm. Sci.*, **4**, 884 (1995); (c) W. H. Ojala and M. C. Etter, *J. Am. Chem. Soc.*, **114**, 10288 (1992); (d) M. M. Devilliers, J. G. Vanderwatt and A. P. Lotter, *Drug Dev. Ind. Pharm.*, **17**, 1295 (1991); (e) R. J. Webb, T. Y. Dong, C. G. Pierpont, S. R. Boone, R. K. Chadha and D. N. Hendrickson, *J. Am. Chem. Soc.*, **113**, 4806 (1991).
75. D. Braga, L. Maini and F. Grepioni, *Chem. Commun.*, 937 (1999).
76. K. Tanaka and F. Toda, *Chem. Rev.*, **100**, 1025 (2000), and references therein.
77. G. M. J. Schmidt, *Pure Appl. Chem.*, **27**, 647 (1971).
78. L. R. MacGillivray, *CrystEngComm*, **7**, 1 (2000), and references therein.
79. G. Pellizzari, *Gazz. Chim. Ital.*, **14**, 362 (1884).
80. I. C. Paul and D. Y. Curtin, *Science*, **187**, 19 (1975).
81. (a) R. S. Miller, D. Y. Curtin and I. C. Paul, *J. Am. Chem. Soc.*, **93**, 2784 (1971); (b) R. S. Miller, D. Y. Curtin and I. C. Paul, *J. Am. Chem. Soc.*, **94**, 5117 (1972); (c) R. S. Miller, D. Y. Curtin and I. C. Paul, *J. Am. Chem. Soc.*, **96**, 6329 (1974); (d) C. C. Chiang, C.-T. Lin, A. H.-J. Wang, D. Y. Curtin and I. C. Paul, *J. Am. Chem. Soc.*, **99**, 6303 (1977).

82. R. S. Miller, I. C. Paul and D. Y. Curtin, *J. Am. Chem. Soc.*, **96**, 6334 (1974).
83. C. T. Lin, I. C. Paul and D. Y. Curtin, *J. Am. Chem. Soc.*, **96**, 3699 (1974).
84. R. S. Miller, D. Y. Curtin and I. C Paul, *J. Am. Chem. Soc.*, **96**, 6340 (1974).
85. G. Kaupp, J. Schmeyers and J. Boy, *Tetrahedron*, **56**, 6899 (2000).
86. (a) G. Kaupp and A. Kuse, *Mol. Cryst. Liq. Cryst.*, **313**, 36 (1998); (b) G. Kaupp, *Mol. Cryst. Liq. Cryst.*, **242**, 153, (1994); (c) G. Kaupp, J. Schmeyers, M. Haak, T. Marquardt and A. Herrmann, *Mol. Cryst. Liq. Cryst.*, **276**, 315 (1996).
87. G. Kaupp, J. Schmeyers and J. Boy, *Chem. Eur. J.*, **4**, 2467 (1998).
88. G. Kaupp, U. Pogodda, and J. Schmeyers, *Chem. Ber.*, **127**, 2249 (1994).
89. F. Toda, H. Takumi, and M. Akehi, *J. Chem. Soc., Chem. Commun.*, 1270 (1990).
90. (a) F. Toda and K. Okuda, *J. Chem. Soc., Chem. Commun.*, 1212 (1991): (b) K. Tanaka, D. Fujimoto, T. Oeser, H. Irngartinger and F. Toda, *Chem. Commun.*, 413 (2000).
91. (a) D. Braga, G. Cojazzi, D. Emiliani, L. Maini and F. Grepioni, *Chem. Commun.*, 2272 (2001); (b) D. Braga, G. Cojazzi, D. Emiliani, L. Maini and F. Grepioni, *Organometallics*, **21**, 1315 (2002).

Chapter 9

Solid–Gas Interactions Between Small Gaseous Molecules and Transition Metals in the Solid State. Toward Sensor Applications

MICHEL D. MEIJER, ROBERTUS J. M. KLEIN GEBBINK, AND GERARD VAN KOTEN

Debye Institute, Utrecht University, Utrecht, The Netherlands

1 INTRODUCTION

The application of (new) materials in the macroscopic world, viz., as useful devices, is one of the ultimate goals a materials scientist may achieve. Research towards interactions between molecules and external stimuli, such as electrons, photons and chemicals, ultimately leads to the development of new devices, such as molecular motors, switches, energy-transducing materials and chemical sensors [1]. In particular, molecular recognition of guest molecules by materials can be used in the design of sensor materials. Chemical sensors [2] require a number of features for their successful functioning. A selective interaction between the target species and sensor material has to take place. In addition, this interaction should be fast and reversible, thereby regenerating the starting sensor material. More importantly, it should cause a change in the physical properties of the sensor material, which is preferentially detectable by optical spectroscopy (UV–Vis, IR) or by electronic measurements (conductivity).

Crystal Design: Structure and Function, edited by G. R. Desiraju

Organometallic complexes may fulfill these requirements, as chelated transition metal centers may display a very selective and fast coordination to gases such as CO, O_2, H_2, SO_2, olefins, etc. Moreover, these reactions may be accompanied by a detectable change in the physical properties of the metal center or the complete organometallic complex. Sensor materials are generally applied as solid-state material, such as monolayers [3,4], and therefore detection of the sensor response usually proceeds in the solid state. For example, Langmuir–Blodgett films of metallophthalocyanins [5] and metalloporphyrins [6] have been tested for their gas-sensing capability of NO_2 and SO_2 gas, respectively. Furthermore, interactions between gases and crystalline organometallic complexes have been reported and, in a few cases, applied for sensory reasons. In this overview, the design and potential application of these crystalline organometallic complexes as chemical sensor material will be discussed. In particular, the sensor requirements will be points of discussion, such as stability of the (crystalline) material, reversibility of the sensor 'mode' and the ways to detect the formation of the desired host–guest complexes.

2 ACCESS OF SMALL MOLECULES INTO SOLIDS

In solid–gas interactions, the bulk of a crystalline material has to be easily accessible for gaseous molecules. For sensor applications, it is important that each individual metal center (sensing unit) can undergo the desired interaction with a guest molecule, in order to obtain a sensor response as high as possible. Therefore, the transport of guest molecules through the solid material plays an important role. A possible way to improve this transport concerns the inclusion of micropores in the material. The success of zeolites as microporous materials has been one of the stimulating factors in the design of their organometallic counterparts, with potential applications in gas storage, gas separation and heterogeneous catalysis [7]. A prerequisite and challenge in the design of these materials is not only the presence of the nanoporous channels, which should be available to guest molecules, but also the structural stability of the material upon absorption and desorption of guests. Only a small number of transition metal-based porous materials fulfilling these prerequisites have been prepared and tested for gas–solid interactions and absorption characteristics. In general, they consist of networks built up from metal centers or clusters, linked together with organic ligands or 'struts' [8]. Examples of such crystalline metal–organic frameworks are the complexes $Zn_4O(BDC)_3 \cdot (DMF)_8(C_6H_5Cl)$ (**1**) [9] and $[Cu(4,4'\text{-bpy})_2](SiF_6)$ (**2**) [10] (Figure 1, BDC = 1,4-benzenedicarboxylate, 4,4′-bpy = 4,4′-bipyridine). These 3-D networks are built up from a square grid coordination polymer based on metal-consisting corners (Zn_4O and Cu for **1** and **2**, respectively) connected by organic ligands. In the third dimension, these square grids are then cross-linked with BDC ligands for **1** and SiF_6^{2-} counterions for **2**.

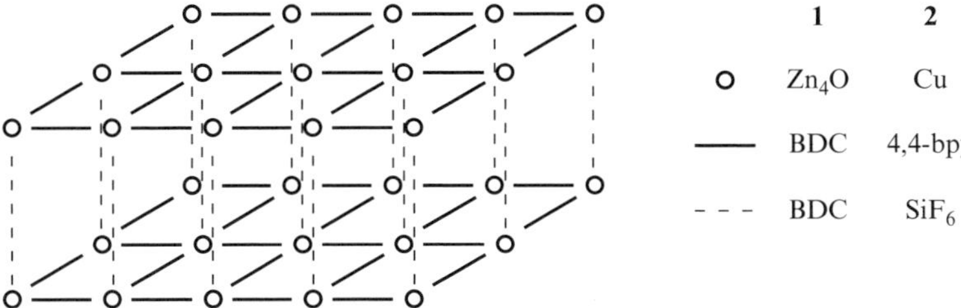

Figure 1 Schematic structure of metal–organic frameworks **1** and **2**.

During the synthesis of **1** and **2**, guest molecules, such as solvent or co-reagent molecules, are included, which can be removed from the crystalline material at elevated temperatures (up to 300 °C for **1**, for example). In this process, the crystals do not collapse, resulting in the desired microporous materials with more than 50% of the internal space available for guest molecules and with high internal surface areas. Moreover, various guest molecules, such as N_2, Ar, CH_2Cl_2, $CHCl_3$, CCl_4, C_6H_6 and C_6H_{12}, can reversibly be absorbed in these materials.

Although the uptake and release of gas molecules in this type of metal–organic frameworks is well regulated, i.e. the network stays intact during repetitive absorption cycles, no chemical interaction between the material and the potential guest molecule is reported, which is one prerequisite for sensor application. An interaction between the included metal centers and absorbed guests would even weaken the crystalline structure of the material and could eventually lead to its complete destruction. This type of microporous material is thus more suitable in gas separation or gas storage applications, and not as a chemical sensor. Additionally, it has been found that the presence of extensive hydrophobic regions in crystals and the exposure of solids to gases at elevated temperatures also enhance the diffusion of small organic molecules into the crystalline bulk, thereby substituting the need for channels in the material [11].

3 GAS–SOLID REACTIONS

As mentioned before, the direct interaction of gas molecules with transition metal centers, inducing changes in the physical properties of the host compounds, is one of the prerequisites in sensor design. Bianchini's group studied solid–gas reactions between a range of gases and transition metal complexes in detail. In particular, a range of metal complexes of tripodal polyphosphine ligands, such as $P(CH_2CH_2PPh_2)_3$ [12], $MeC(CH_2CH_2PPh_2)_3$ [13] and $N(CH_2CH_2PPh_2)_3$ [14] (abbreviated as PP_3, triphos and NP_3, respectively) were tested (Figure 2). Off-white coloured crystals of the (η^2-ethene)dihydride complex $[(triphos)Ir(H)_2(C_2H_4)]BPh_4$

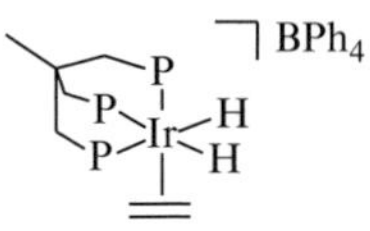

3

Irreversible reaction with C_2H_4, H_2 and CO

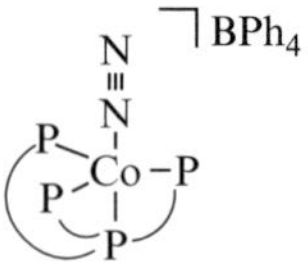

4

Irreversible reaction with C_2H_4, H_2CO and MeCHO

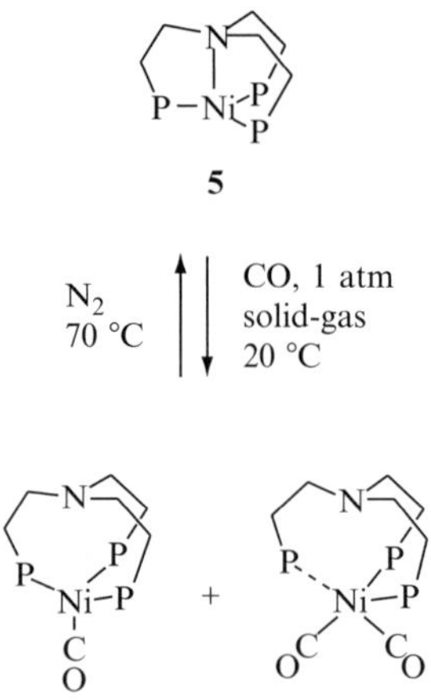

5

N_2 70 °C

CO, 1 atm solid-gas 20 °C

6 + 7

Figure 2 Solid–gas reactions using metal complexes of tripodal phosphine ligands.

(**3**) reacted with various gases, such as CO, ethene, H_2 and ethylene [13]. Analogously, when ethyne or gaseous formaldehyde was passed over purple–red crystals of the dinitrogen cobalt(II) complex $[(PP_3)Co(N_2)]BF_4$ (**4**) (90 °C), irreversible adduct formation occurred [12]. Solid–gas reactions with **3** and **4** were accompanied by a colour change of the starting materials. Solid–gas reaction between red crystals of $(NP_3)Ni$ (**5**) and CO (1 atm) at room temperature afforded a yellow product, consisting of **6** and a new complex, **7** (see Figure 2) [^{31}P cross-polarization magic angle spinning (CP MAS) NMR and in situ infrared measurements] [14]. The latter was suggested to be a metastable species in which a free phosphine arm from the NP_3 ligand is located close to the nickel center as a result of the constraining environment of the crystal lattice.

During these solid–gas reactions, no substantial fragmentation of the exposed crystals was observed, although the products appear slightly opaque as observed by cross-polarization microscopy. The absence of fragmentation indicates that the reaction does not proceed via reaction at the surface, followed by subsequent breakdown of the crystal to facilitate further reaction at the newly formed surface. Instead, the reactions take place in the bulk of the crystal, after diffusion of the gas molecules from the surface to the bulk, which is enhanced by the presence of the hydrophobic regions in the crystals provided by the PPh_2 groups of the ligands. Furthermore, applying higher (reaction) temperatures (up to 100 °C) further accelerated gas diffusion into the crystal. Complexes **3** and **4** are not suitable as transition metal sensors for the gases mentioned in Figure 2 because of the irreversible nature of the reported solid–gas reactions. Some of these complexes were successfully applied as solid transition metal complex catalysts [13]. However, the interaction between **4** and H_2 takes place in a reversible fashion, as will be discussed in the next paragraph. A colour change from red to yellow was observed for the nickel complex 5 during uptake of CO. CO could also be thermally released from the solid mixture of **6** and **7** to regenerate the starting complex **5**. Although this complex seems to be a candidate for the development of a solid-state CO sensor, no further studies have so far been reported concerning this subject.

3.1 Reversible Gas–Solid Reactions Using Cobalt Complexes

A large number of reports have considered gas interactions using cobalt complexes. More importantly, various cobalt complexes display reversible uptake and release of gases, such as O_2, H_2 or NO, and may therefore be suitable as gas-sensing materials. One of the first examples described the reversible uptake of molecular oxygen by cobalt salcomine and cobalt porphyrin [15,16] complexes in order to mimic biological O_2 carriers, such as iron hemoglobin. Cobalt(II) cyanide complexes, such as $Co(CN)_5^{3-}$, were bound inside zeolite Y, and studied for their O_2 absorption capacity for potential application in gas separation or gas storage devices [17,18].

Borovik's group studied the concept of cobalt complexes in a polymeric matrix as sensor in more detail. Four-coordinated Co(II) metal centers were incorporated into a porous methacrylate network by copolymerization of a styrene-substituted cobalt(II)(salen) complex with ethylene glycol dimethacrylate (see Figure 3) [19]. These complexes were specifically studied for their sensor capacity for NO [20].

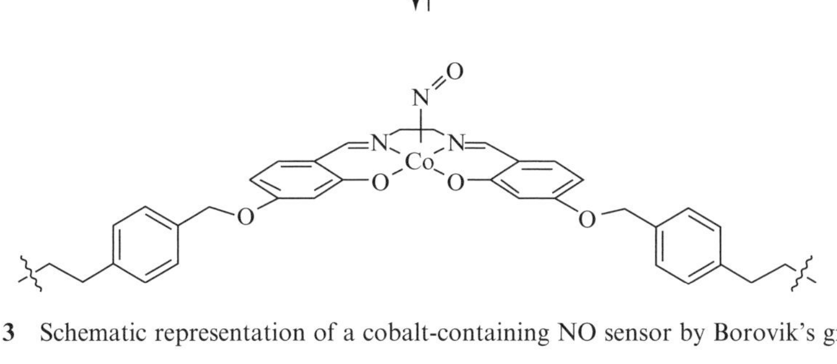

Figure 3 Schematic representation of a cobalt-containing NO sensor by Borovik's group.

The polymer matrix isolates the metal sites, preventing undesirable side-reactions, such as the formation of bimolecular metal-based adducts. UV–Vis and EPR measurements showed that the cobalt centers in the polymer matrix form 1:1 NO–cobalt adducts on exposure to NO gas, but are relatively inert towards O_2, CO and CO_2. The uptake of NO by this sensing polymer caused an immediate colour change of the initial orange material to the brown–green cobalt–NO adducts. The formation of this sensor-'on' function occurred at room temperature. However, the release of NO was found to be slow at room temperature in a NO free environment (80% release in 30 days). Under vacuum at 120 °C, the regeneration of the starting material was achieved within 1 h.

Lithium pentacyanocobaltate polymers, $Li_3[Co(CN)_5] \cdot 4DMF$ (**10**) and $Li_3[Co(CN)_5] \cdot 2DMF$ (**11**), were studied for their dioxygen gas uptake [21]. In particular, crystals of **11** reacted reversibly with O_2, affording 1:1 cobalt superoxo complexes. By X-ray powder measurements, it was observed that the crystallinity of the material remained intact. Since the diffractograms of the starting material and the O_2 adduct did not differ significantly, it was suggested that in the crystal of **10** and **11** enough space is present to host an O_2 molecule, without breaking down the crystal lattice. Furthermore, the crystalline material had decreased its O_2-uptake efficiency by only 15% after 600 uptake and release cycles, probably owing to decomposition of the crystal by moisture. Also, a colour change from green to red was observed during the uptake of O_2 by **10**, but this was not further related to sensor application. Therefore, although these materials are very promising in the development of oxygen-carrier materials, they could also be excellent candidates as crystalline O_2 sensors.

Successful reversible hydrogen and ethene uptake using the nitrogen cobalt(I) complex $[(PP_3)Co(N_2)]BPh_4$ (**4**) was reported by Bianchini and co-workers. Upon treatment of purple–red crystals of **4** with H_2 or ethene (90 °C, 1 h), the off-white dihydride complex $[(PP_3)Co(H)_2]BPh_4$ (**12**) or the yellow–orange π-ethylene complex $[(PP_3)Co(C_2H_4)]BPh_4$ (**13**) was formed quantitatively (Figure 4). Treatment with H_2 of the analogous PF_6^- complex (**14**) resulted in the reversible formation of the red η^2-H_2 complex **15**.

The gaseous reaction of **4** did not occur below 65 °C. Above that temperature, reactivity was indicated by colour changes of the crystalline material from red–purple to white. The reaction could also be monitored by IR spectroscopy, probing the ν(N–N) absorption at 2125 cm^{-1}, which was replaced by two ν(Co–H) absorptions at 1967 and 1833 cm^{-1}. Treatment of the dihydrogen complexes with N_2 afforded the starting material **4**.

3.2 PtCl(NCN) Complexes as SO_2 Sensors

In 1986, it was reported that organoplatinum complexes based on the monoanionic terdentate ligand $[C_6H_3(CH_2NMe_2)_2\text{-}2,6]^-$, abbreviated as NCN, reversibly

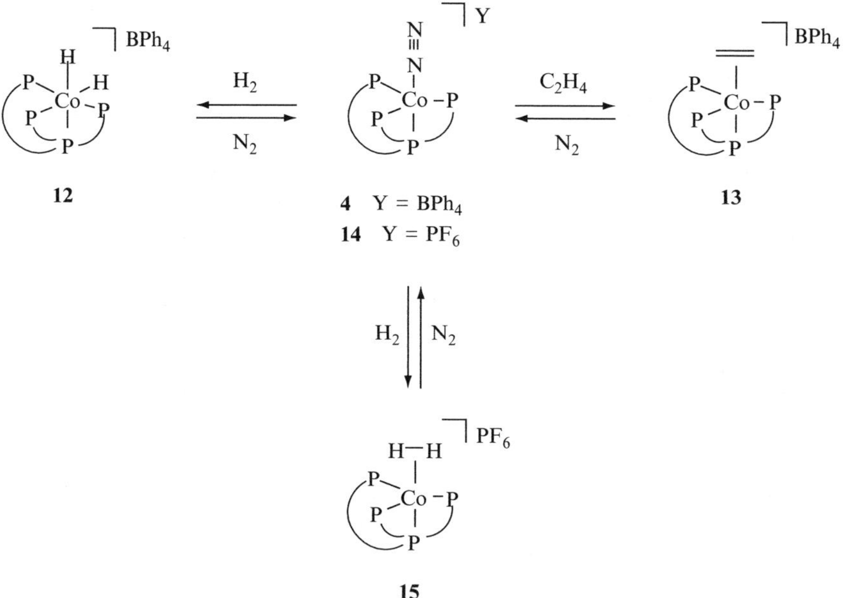

Figure 4 Reversible reaction of H_2 or C_2H_4 with tripodal polyphosphine cobalt complex **4**.

bind SO_2 gas in solution (Figure 5) [22]. In this reaction, a stable pentacoordinated platinum center is formed, in which the SO_2 molecule is bound via the S-atom to the platinum center in an η^1-fashion. This reaction also proceeds in the solid state and is accompanied by an immediate colour change of the colourless starting material to the bright orange SO_2 adduct. The detailed mechanism of the reversible binding process was studied using density functional theory calculations and low-temperature experiments [23], which revealed that exchange of SO_2 proceeds via a dissociative mechanism (first order in NCN–platinum and zeroth order in SO_2).

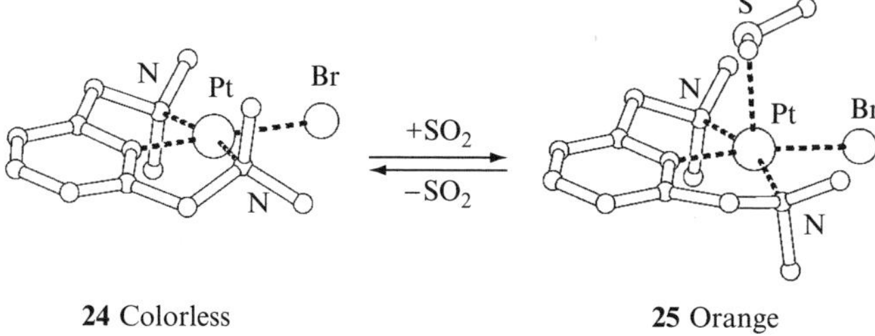

Figure 5 Bonding of SO_2 to NCN–PtBr.

Desorption of SO_2 can be achieved by storage of the SO_2 adducts in an SO_2-free environment. This is accelerated by exposure to air or SO_2-free solvents and under reduced pressure. These processes regenerate the starting four-coordinate NCN–Pt complexes quantitatively. Various NCN–Pt complexes were studied in which the organic groups on the nitrogen donor atoms, the platinum-bound halide atoms or the *para*-substituents on the aryl ring were varied (Figure 6) [24]. In all cases, the complexes displayed sensing properties for SO_2, thus forming the orange SO_2 adducts after exposure to SO_2 gas, both in solution and, more interestingly, in the solid state. Since the response of the platinum complex was hardly influenced by modifications of the NCN ligand backbone, the position *para* to the platinum center (E) was used to attach the sensing unit to macromolecular supports, such as dendrimers [25].

In the solid state, the same diagnostic colour change (from colourless to bright orange) is observed when crystals of platinum–NCN complexes are exposed to SO_2. Interestingly, the dimensions of the aryl platinum crystals change dramatically during the uptake and release processes, as was deduced from X-ray structure analysis. A special situation occurs in the solid-state structure of the organoplatinum complex [PtCl(NCN–OH)] (E = OH, R = R′ = Me, X = Cl, **26**; see Figure 6) and its SO_2 adduct [26]. The platinum metal centers in these crystals are also involved in second-order interactions between monomeric units in the crystal, and additionally perform the sensing task. As can be seen in the single crystal structure of **26**, α-type chains are formed via intermolecular Pt–Cl···H–O hydrogen bonds (Figure 7). In the square-pyramidal, orange SO_2 adduct [PtCl(NCN–OH)(SO_2)], **27**, additional orthogonal Pt–$S(O)_2$···Cl interactions between the α-type chains are present, thus interconnecting these giving a β-type network (Figure 7).

The applicability of this system in the development of a crystalline SO_2 sensor was studied in greater detail. While binding of gaseous substrates by nonporous crystalline materials may lead to the destruction of the long crystalline order, this did not happen in the case of **26**. The crystallinity of the sensor material is maintained during the uptake and release of SO_2, as was shown by time-resolved powder diffraction experiments (Figure 8).

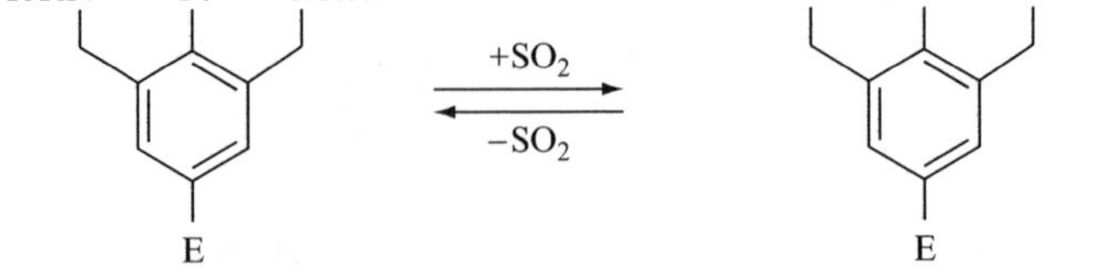

Figure 6 Various tested NCN–Pt complexes. E = H, OH, OSiMe$_2t$Bu, or Fréchet dendron; X = Cl, Br, I or C_6H_6Me-4;R = R′ = Me or Et.

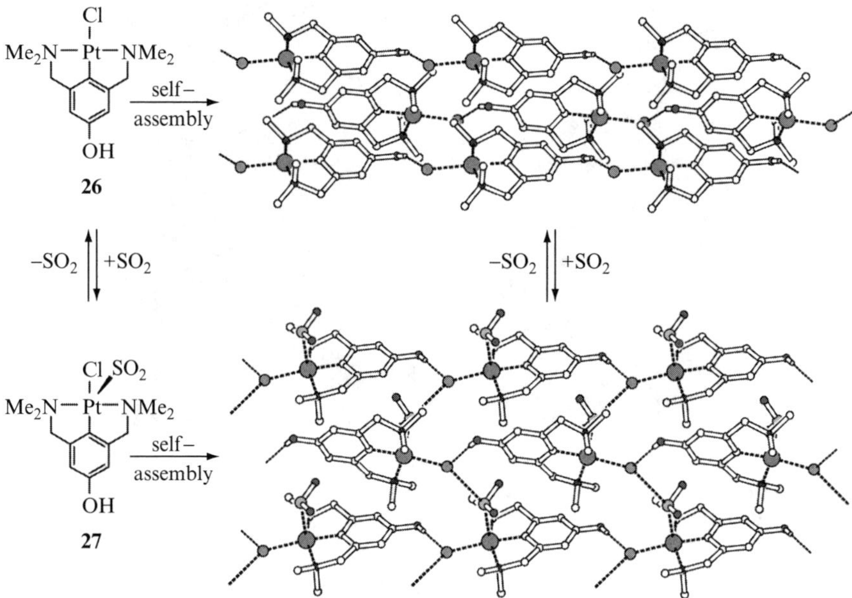

Figure 7 Crystal structures of organoplatinum complexes **26** and **27**.

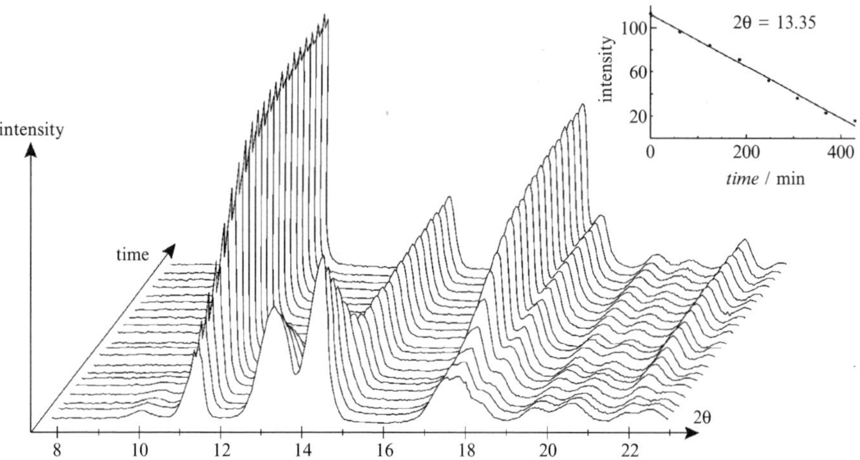

Figure 8 Time-resolved powder refraction measurements during the release of SO_2 from crystalline **27**.

In this experiment, crystalline **27** releases SO_2, yielding crystalline **26** (when a single crystal of **27** was exposed to air, **26** was obtained as a microcrystalline product; however, this was no longer suitable for a single-crystal X-ray analysis).

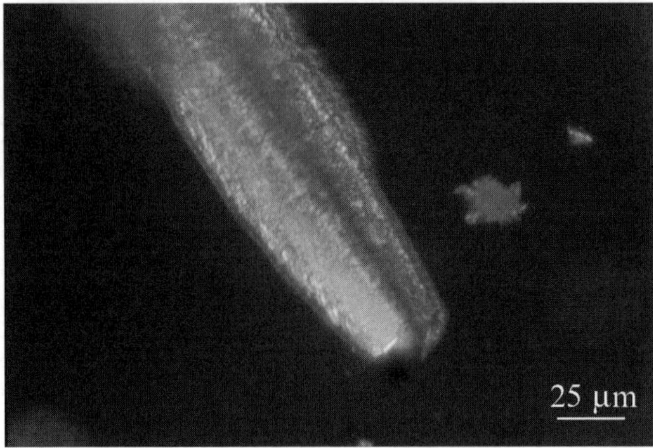

Figure 9 Snapshot of a crystal of **27** during the release of SO_2, forming **26**. Reproduced by permission of *Nature* from M. Albrecht, M. Lutz, A. L. Spek and G. van Kotem, *Nature*, **2000**, *406*, 970 (see also **Plate 27**).

Therefore, local dissolution and recrystallization seem to play an important role in the gas uptake mechanism in these type of sensor materials. The coordination of SO_2 to the platinum center (and the reverse reaction) is therefore likely to take place in temporarily and very locally formed solutes in the crystalline material, whereas the overall material remains crystalline. The full reversibility of the solid-state reaction was, furthermore, demonstrated with time-resolved solid-state infrared spectroscopy (observation at the metal-bound SO_2 vibration, $\nu_s = 1072\,cm^{-1}$), even after several repeated cycles. Exposure of crystalline samples of **26** alternatingly to an atmosphere of SO_2 and air did show no loss in signal intensities, e.g. due to the formation of amorphous powder. The release of SO_2 from a crystal of **27** was also observed using optical cross-polarization microscopy. A colourless zone (indicative of **26**) is growing from the periphery of the crystal whereas the orange colour (indicative for **27**) in the core of the crystal diminishes (see Figure 9).

4 CONCLUSIONS

The application of solid-state organometallic complexes as sensing units in chemical sensors seems to be a promising field of organometallic chemistry and materials science. However, the prerequisites for successful application imply that a difficult set of properties have to be combined in one sensor system. Reversible and repetitive interaction of guest molecules with a solid-state (crystalline) material is desired in combination with a structurally stable and sensing organometallic complex, i.e. a complex that yields an optical or electronic response upon this interaction. So far in most successful cases, the finding of good sensory properties of an organometallic crystal is based more on serendipity than on (empirical) prediction. The design of (selective) organometallic sensors, therefore, poses an interesting, multidisciplinary challenge. To derive selection rules for the designing process, expertises from organometallic chemistry, crystal engineering and solid-state chemistry have to join forces. From an organometallic point of view, the molecular reactivity of concrete complexes has to be translated to solid-state reactivity. In addition, special features have to be incorporated into metal complexes to ensure the build-up of an open and stable crystalline framework. From the crystal engineer/solid state point of view, strategies have to be developed for the construction of crystalline (coordination) networks that contain reactive (organo)metallic sites. Therefore, the research performed in the field of (organometallic) coordination polymers that still contain coordinately unsaturated metal sites may in the end lead to suitable organometallic sensors.

ACKNOWLEDGEMENTS

The authors thank Drs Martin Albrecht and Martin Lutz and Professor Ton Spek for the fruitful collaboration on crystalline, organometallic gas sensors.

REFERENCES

1. See, for example, *Acc. Chem. Res.* **2001**, *34*(6), special issue on molecular machines.
2. See, for example, *Acc. Chem. Res.* **1998**, *31*(5), special issue on chemical sensors.
3. Crooks, R. M.; Ricco, A. J. *Acc. Chem. Res.* **1998**, *31*, 219.
4. Nalwa, H. S.; Kakuta, A. *Appl. Organomet. Chem.* **1992**, *6*, 645.
5. See, for example, (a) Wright, J. D. *Inst. Phys. Conf. Ser* **1990**, *111*, 333; (b) Sadaoka, Y.; Jones, T. A.; Revell, G. S.; Göpel, W. *J. Mater. Sci.* **1990**, *25*, 5257; (c) Göpel, W. *Synth. Met.* **1991**, *41–43*, 1087; (d) Guillaud, G.; Simon, J.; Germain, J. P. *Coord. Chem. Rev.* **1998**, *178–180*, 1433.
6. (a) Papkovsky, D. B.; Desyaterik, I. V.; Ponomarev, G. V.; Kurochkin, I. N.; Korpela, T. *Anal. Lett.* **1995**, *28*, 2027; (b) Natale, C. D.; Macagnano, A.; Repole, G.; Saggio, G.; D'Amico, A.; Paolesse, R.; Boschi, T. *Mater. Sci. Eng.* **1998**, *C5*, 209.
7. For a review on transition metal porous solids, see Yaghi, O. M.; Li, H.; Davis, C.; Richardson, D.; Groy, T. L. *Acc. Chem. Res.* **1998**, *31*, 474.
8. Zaworotko, M. J. *Angew. Chem., Int. Ed. Engl.* **2000**, *39*, 3052.
9. (a) Yaghi, O. M.; Li, G.; Li, H. *Nature (London)* **1995**, *378*, 703; (b) Eddaoudi, M.; Li, H.; Yaghi, O. M. *J. Am. Chem. Soc.* **2000**, *122*, 1391; (c) Li, H.; Eddaoudi, M.; O'Keeffe, M.; Taghi, O. M. *Nature (London)* **1999**, *402*, 276.
10. Noro, S.-I.; Kitagawa, S.; Kondo, M.; Seki, K. *Angew. Chem., Int. Ed. Engl.* **2000**, *39*, 2082.
11. (a) Siedle, A. R.; Newmark, R. A. *Organometallics* **1989**, *8*, 1442; (b) Siedle, A. R.; Newmark, R. A.; Sahyun, M. R. V.; Lyon, P. A.; Hunt, S. L.; Skarjune, R. P. *J. Am. Chem. Soc.* **1989**, *111*, 8346.
12. Bianchini, C.; Peruzzini, M.; Zanobini, F. *Organometallics* **1991**, *10*, 3415.

13. (a) Bianchini, C.; Frediani, P.; Graziani, M.; Kaspar, J.; Meli, A.; Peruzzini, M.; Vizza, F. *Organometallics* **1993**, *12*, 2886; (b) Bianchini, C.; Farnetti, E.; Graziani, M.; Kaspar, J.; Vizza, F. *J. Am. Chem. Soc.* **1993**, *115*, 1753; (c) Bianchini, C.; Graziani, M.; Kaspar, J.; Meli, A.; Vizza, F. *Organometallics* **1994**, *13*, 1165; (d) Bianchini, C.; Meli, A.; Peruzinni, M.; Vizza, F.; Frediani, P.; Herrera, V.; Sanchez-Delgado, R. A. *J. Am. Chem. Soc.* **1993**, *115*, 7505.
14. Bianchini, C.; Zanobini, F.; Aime, S.; Gobetto, R.; Psaro, R.; Sordelli, L. *Organometallics* **1993**, *12*, 4757.
15. Collman, J. P.; Gagne, R. R.; Kouba, J.; Ljusberg-Wahren, H. *J. Am. Chem. Soc.* **1974**, *96*, 6800.
16. Collman, J. P.; Brauman, J. I.; Doxsee, K. M.; Halbert, T. R.; Hayes, S. E.; Suslick, K. S. *J. Am. Chem. Soc.* **1978**, *100*, 2761.
17. Taylor, R. J.; Drago, R. S.; George, J. E. *J. Am. Chem. Soc.* **1989**, *111*, 6610.
18. Taylor, R. J.; Drago, R. S.; Hage, J. P. *Inorg. Chem.* **1992**, *31*, 253.
19. Krebs, J. F.; Borovik, A. S. *Chem. Commun.* **1998**, 553.
20. Padden, K. M.; Krebs, J. F.; MacBeth, C. E.; Scarrow, R. C.; Borovik, A. S. *J. Am. Chem. Soc.* **2001**, *123*, 1072.
21. Ramprasad, D.; Pez, G. P.; Toby, B. H.; Markley, T. J.; Pearlstein, R. M. *J. Am. Chem. Soc.* **1995**, *117*, 10694.
22. Terheijden, J.; van Koten, G.; Mul, W. P.; Stufkens, D. J.; Muller, F.; Stam, C. H. *Organometallics* **1986**, *5*, 519.
23. Albrecht, M.; Gossage, R. A.; Frey, U.; Ehlers, A. W.; Baerends, E. J.; Merbach, A. E.; van Koten, G. *Inorg. Chem.* **2001**, *40*, 850.
24. Albrecht, M.; Gossage, R. A.; Lutz, M.; Spek, A. L.; van Koten, G. *Chem. Eur. J.* **2000**, *6*, 1431.
25. (a) Albrecht, M.; Gossage, R. A.; Spek, A. L.; van Koten, G. *Chem. Commun.* **1998**, 1003; (b) Albrecht, M.; van Koten, G. *Adv. Mater.* **1999**, *11*, 171; (c) Albrecht, M.; Hovestad, N. J.; Boersma, J.; van Koten, G. *Chem. Eur. J.* **2001**, *7*, 1289.
26. (a) Albrecht, M.; Lutz, M.; Schreurs, A. M. M., Lutz, E. T. H.; Spek, A. L.; van Koten, G. *J. Chem. Soc.*, Dalton Trans. **2000**, 3797; (b) Albrecht, M.; Lutz, M.; Spek, A. L.; van Koten, G. *Nature (London)* **2000**, *406*, 970.

Cumulative Author Index

This index comprises the names of contributors to Volumes 1–7 of **Perspectives in Supramolecular Chemistry**.

Cumulative Title Index

This index comprises the titles and authors of all chapters appearing in Volumes 1–7 of **Perspectives in Supramolecular Chemistry**

Index